State of the Apes

Disease, Health and Ape Conservation

As the Anthropocene unfolds, the impact of humanity on all ecosystems on the planet is becoming more visible and better understood. Deforestation, encroachment into natural habitat and other human activities are driving an increase in the frequency of interactions between people and wildlife, including viruses, parasites and bacteria. One consequence is a heightened risk of disease transmission, with serious implications for biodiversity protection and human health. Indeed, infectious disease is often listed among the principal threats to ape conservation, along with habitat loss and hunting, which can also expose apes to health risks. In captive settings such as sanctuaries and zoos, apes face similar health risks from increased human contact, as well as geriatric and psychological disorders. Spillover of wildlife pathogens into sanctuaries can also occur.

This volume of *State of the Apes* brings together original research and analysis with topical case studies and emerging best practice to further the ape conservation agenda around disease and health. It provides an overview of relevant disease and health issues and explores factors such as the ethics of intervening in and managing ape health; the impact of research and tourism on apes; the One Health approach; and disaster management and the protection of apes. It shows how the welfare of apes is interrelated with that of the people who share their habitats, while also demonstrating the benefits of integrating ape conservation in health, socioeconomic activities (such as in the extractive industries, industrial agriculture and infrastructure development), and regulatory policy and practice at all levels, from the local to the international.

This volume is available as an open access eBook via Cambridge Core and at www.stateoftheapes.com.

State of the Apes

Series editors

Alison White   Arcus Foundation

Steve Unwin   University of Birmingham then
Wildlife Health Australia

Annette Lanjouw   Arcus Foundation

Katy Scholfield   Arcus Foundation

Helga Rainer   Arcus Foundation at time of writing

The world's primates are among the most endangered of all tropical species. All great ape species – gorilla, chimpanzee, bonobo and orangutan – are classified as either Endangered or Critically Endangered. Furthermore, nearly all gibbon species are threatened with extinction. Whilst linkages between ape conservation and economic development, ethics and wider environmental processes have been acknowledged, more needs to be done to integrate biodiversity conservation within broader economic, social and environmental communities if those connections are to be fully realized and addressed.

Intended for a broad range of policy-makers, industry experts and decision-makers, academics, researchers and NGOs, the *State of the Apes* series looks at the threats to these animals and their habitats within the broader context of economic and community development. Each publication presents a different theme, providing an overview of how these factors interrelate and affect the current and future status of apes, with robust statistics, welfare indicators, official and various other reports providing an objective and rigorous analysis of relevant issues.

State of the Apes

Disease, Health and Ape Conservation

Shaftesbury Road, Cambridge CB2 8EA, United Kingdom

One Liberty Plaza, 20th Floor, New York, NY 10006, USA

477 Williamstown Road, Port Melbourne, VIC 3207, Australia

314–321, 3rd Floor, Plot 3, Splendor Forum, Jasola District Centre, New Delhi – 110025, India

103 Penang Road, #05–06/07, Visioncrest Commercial, Singapore 238467

Cambridge University Press is part of Cambridge University Press & Assessment, a department of the University of Cambridge.

We share the University's mission to contribute to society through the pursuit of education, learning and research at the highest international levels of excellence.

www.cambridge.org
Information on this title: www.cambridge.org/9781009069984

DOI: 10.1017/9781009071727

First published 2023

A catalogue record for this publication is available from the British Library

ISBN 978-1-316-51307-1 Hardback
ISBN 978-1-009-06998-4 Paperback

Cambridge University Press & Assessment has no responsibility for the persistence or accuracy of URLs for external or third-party internet websites referred to in this publication and does not guarantee that any content on such websites is, or will remain, accurate or appropriate.

Credits

Editors
Alison White, Steve Unwin, Annette Lanjouw, Katy Scholfield and Helga Rainer

Production Coordinator
Alison White

Designer
Rick Jones, StudioExile

Cartographer
Jillian Luff, MAP*grafix*

Illustrators
Daly Design

Editorial Consultant and Copy-editor
Tania Inowlocki

Fact-checkers
Magdalena Svensson and Nadine A. Sugianto

Reference Editor
Eva Fairnell

Proofreaders
Sarah Binns and Elaine Ross

Indexer
Caroline Jones, Osprey Indexing

Foreword

Over ten years ago, the ape conservation team at the Arcus Foundation came to me with the innovative idea to develop a series of reports regarding the *State of the Apes*. This was inspired by their work with our many grantees and partners around the world who had invaluable insights into the huge spectrum of complex issues impacting ape populations, but whose knowledge was widely dispersed and difficult to access. In addition, many industry leaders, investors, government agencies and other stakeholders were making decisions that had serious, deleterious impacts on apes and their habitats. This was due to a lack of awareness, poor access to information and analysis that would enable them to make more informed, sustainable choices, and a lack of signposting to help them source that information and identify relevant experts to advise them.

I had always envisioned that Arcus could do more than simply provide financial support to our grantees on the frontlines. This project seemed to me a useful opportunity to invest in collective action to help turbocharge the generation of knowledge in the field and thus truly advance great and small ape conservation. The Arcus Foundation had never undertaken a project like this, and there was certainly an element of risk in the proposition; but it seemed a risk well worth taking. It has been our hope that the series would become an important and well used compendium of knowledge for the field, but it was also envisioned for the series to be a resource that would help inform key global-level financial, governmental and industry leaders who often make extremely influential decisions that impact ape populations around the world.

Just months after those early conversations, we forged a partnership with Cambridge University Press, and here we are: 10 years and five volumes later. I couldn't be more thrilled to have been a part of this project, particularly this final volume, the timeliness of which is breathtaking. It often seems that many among the world's population who live in large cities of highly developed countries feel little connection with non-human animals beyond their domesticated pets. For them, at least, before the beginning of the COVID-19 pandemic, the threat that disease posed to non-human animals had hardly been top-of-mind or even relevant to the lives they were living. The principal editors and their collaborators began developing this latest book in the series in 2020, just four years after the Ebola pandemic of 2014–2016 and just as the global debate raged regarding whether the novel SARS-CoV-2 virus had originated in a Chinese market, a laboratory, or through other means. Our species was getting a refresher course on the connections between human and non-human health on a greater scale than at any other moment, certainly in my lifetime. We humans were confronted with the fact that our health and that of other species are intertwined in an extremely high-stakes fashion.

Of course, conservationists, scientists and ecologists have long been interested in understanding and addressing the threat that disease represents to the health and survival of millions of species who define and sustain critical ecosystems that support all life on earth and define the world as we know it. Each day, we lose between 150 and 200 species to extinction, and climate change is unquestionably accelerating that pace. Their loss is colossal in terms of both the intrinsic value of the species and of the role they play in ensuring the biological diversity and integrity of the planet. The magnificent and charismatic nature of endangered apes, who share precarious landscapes with

countless other endangered species and marginalized human communities, and who are particularly vulnerable to extinction due to their slow reproductive rate, make them compelling subjects for a series like this one. The importance of this volume and those that preceded it—to ape conservationists like its authors, to me and to anyone committed to building a sustainable future —cannot be exaggerated.

The contributing authors have done an amazing job of presenting the complexity of threats to ape health and wellbeing, as well as the trade-offs presented by the matrix of approaches to managing these threats. The authors present important information about the similarities between human and non-human ape susceptibility to disease, the differences between disease and disease management scenarios in captive apes and wild apes, the role of non-infectious threats to ape health, and the infectious risks that a strategy like tourism or research can introduce, even as it serves to mitigate other negative forces affecting ape health.

Throughout each chapter, the authors share actionable insights, but they also iden-

tify knowledge gaps and critical questions worthy of further research and inquiry and present frameworks and constructs that will serve future work. For example, Chapter 2 on the One Health movement explains in compelling detail how the health of humans, non-human animals, and every element of the natural environment from plants to oceans is just that—one health. Like a kinetic sculpture, any disturbance in its design has the potential to wreak havoc. Ensuring its function is a delicate imperative that will require the kind of disciplined, fact-based approaches reflected in the case studies of this volume.

Of course, attending to One Health is not just about science. It requires the engagement and collective action of an inestimable number of people and institutions, and the authors rightly emphasize the importance of forging and leveraging alignment with the 2030 Sustainable Development Goals adopted by all member states of the United Nations in 2015. The beauty of this publication is that it bundles science, strategy and hope into a single package comprising baseline data and analysis to support work

toward progress that no single individual or entity can achieve alone.

As a committed ape conservationist and philanthropist, I have been proud and excited to support this series through the foundation I founded more than 20 years ago. I am confident that *State of the Apes* will be an important resource for years to come, and I offer my congratulations and thanks to all who brought their extraordinary brains and hearts to this project.

Jon Stryker

President and Founder,
Arcus Foundation

IUCN Patron of Nature (International Union for the Conservation of Nature)

Contents

Section 1
Disease, Health and Ape Conservation

Section 2
The Status and Welfare of Great Apes and Gibbons

Other Titles in this Series

Arcus Foundation. 2020. *State of the Apes: Killing, Capture, Trade and Conservation.* Cambridge: Cambridge University Press.

Arcus Foundation. 2018. *State of the Apes: Infrastructure Development and Ape Conservation.* Cambridge: Cambridge University Press.

Arcus Foundation. 2015. *State of the Apes: Industrial Agriculture and Ape Conservation.* Cambridge: Cambridge University Press.

Arcus Foundation. 2014. *State of the Apes: Extractive Industries and Ape Conservation.* Cambridge: Cambridge University Press.

Other Language Editions

All available from www.stateoftheapes.com

Bahasa Indonesia

Arcus Foundation. 2020. *Negara Kera: Pembunuhan, Penangkapan, Perdagangan, dan Konservasi.*

Arcus Foundation. 2018. *Negara Kera: Pembangunan Infrastruktur dan Konservasi Kera.*

Arcus Foundation. 2015. *Negara Kera: Pertanian Industri dan Konservasi Kera.*

Arcus Foundation. 2014. *Negara Kera: Industri Ekstraktif dan Konservasi Kera.*

Chinese

类人猿现状：捕杀、捕捉、贸易和类人猿保护

类人猿现状：基础设施开发与类人猿保护

类人猿现状：采掘业与类人猿保护

French

Arcus Foundation. 2020. *La planète des grands singes : La destruction, la capture, le trafic et la conservation.*

Arcus Foundation. 2018. *La planète des grands singes : Le développement des infrastructures et la conservation des grands singes.*

Arcus Foundation. 2015. *La planète des grands singes : L'agriculture industrielle et la conservation des grands singes.*

Arcus Foundation. 2014. *La planète des grands singes : Les industries extractives et la conservation des grands singes.*

The Arcus Foundation

The Arcus Foundation is a private grant-making foundation that advances social justice and conservation goals. The Foundation works globally and has offices in New York City, USA and Cambridge, UK. For more information visit:

- arcusfoundation.org.

Or connect with Arcus at:

- twitter.com/ArcusGreatApes and
- facebook.com/ArcusGreatApes.

Great Apes and Gibbons Program

The long-term survival of humans and the great apes is dependent on how we respect and care for other animals and our shared natural resources. The Arcus Foundation seeks to increase respect for and recognition of the rights and value of the great apes and gibbons, and to strengthen protection from threats to their habitats. The Arcus Great Apes Program supports conservation and policy advocacy efforts that promote the survival of great apes and gibbons in the wild and in sanctuaries that offer high-quality care, safety and freedom from invasive research and exploitation.

Contact details

New York office:

445 5th Avenue, 7th Floor
New York, New York 10016, United States

+1 212 488 3000 / phone

**Cambridge office
(Great Apes and Gibbons Program):**

95 Regent Street
Cambridge, CB2 1AW
United Kingdom

+44 (0)1223 653040 / phone

Notes to Readers

Acronyms and Abbreviations

A list of acronyms and abbreviations can be found at the back of the book, starting on p. 314.

Annexes

All annexes can be found at the back of the book, starting on p. 274, except for the Abundance Annex, which is available from the *State of the Apes* website:

- www.stateoftheapes.com.

Glossary

There is a glossary of scientific terms and keywords at the back of the book, starting on p. 317.

Chapter Cross-referencing

Chapter cross-references appear throughout the book, either as direct references in the body text or in brackets.

Human and Non-human Apes

To simplify and enhance the flow of the text, this volume uses the term "apes" to refer exclusively to non-human apes, unless noted otherwise.

Pathogens and Disease

While this volume identifies pathogens, it places more emphasis on the diseases they cause. For example, Chapter 1 mentions that severe acute respiratory syndrome coronavirus 2 (SARS-CoV-2) causes COVID-19, but the book focuses more generally on the health impacts of the disease and options for mitigating transmission risks.

Since the content of this volume was drafted, following a directive from the World Health Organization owing to issues around the use of racist and stigmatizing language, the nomenclature for monkeypox has changed. The term 'monkeypox' is being phased out and the preferred term is now 'mpox'.

Ape Abundance Estimates

Definitive, up-to-date abundance estimates are not available for all ape species. The most recent Abundance Annex—available at www.stateoftheapes.org—presents estimates at the site level. For great apes, it uses abundance classes to indicate population ranges; for gibbons, it employs a mix of abundance classes and density estimates. In this volume, the Apes Overview and some chapters feature abundance information based on other geographic scales, drawn from a variety of sources, including forthcoming Red List assessments. Consequently, some figures may not align exactly.

Ape Range Maps

The ape range maps throughout this volume show the extent of occurrence (EOO) of each species. An EOO includes all known populations of a species contained within the shortest possible continuous imaginary boundary. Some areas within these boundaries may be unsuitable for apes or unoccupied.

The Arcus Foundation commissioned the ape distribution maps in the Apes Overview, Figures AO1 and AO2, to provide the most accurate and up-to-date illustration of range data. These maps were created by the IUCN Primate Specialist Group, which manages the A.P.E.S. portal and database. This volume also features maps created by contributors who used ape range data from other sources. As a consequence, the maps may not all align exactly.

Acknowledgments

Pulling together the contents of a volume of *State of the Apes* is an extensive undertaking. As we embarked on this volume, the 5th in the series, the COVID-19 pandemic complicated the task further, adding significant delays. We extend special thanks to all involved for their understanding, flexibility and support in bringing this volume to completion.

Our aim is not only to encourage the critical engagement of all stakeholders—including conservation organizations, medical institutions, wider civil society, industry, donor and financial institutions, and government—but also to increase support for great apes and gibbons.

Jon Stryker and the Arcus Foundation Board of Directors have been instrumental in enabling the production of the *State of the Apes* series. We are grateful to them for their invaluable support over the past decade, particularly in the face of the unforeseen circumstances that affected this volume's publication schedule. We also thank Megan Keirnan, Jenny van der Meijden and the team at Cambridge University Press for their commitment to the series.

As this is the final volume in the series, we would like to take this opportunity to express our deep gratitude to our colleagues in the Arcus Great Apes & Gibbons Program—Matthew Lewis, Adam Phillipson and Marie Stevenson—for their input, help and patience. We are also extremely grateful to the production consultants who have worked with us on the series from the start: Sarah Binns, proofreader; Eva Fairnell, reference editor; Tania Inowlocki, editorial consultant and copy-editor; Caroline Jones at Osprey Indexing, indexer; Rick Jones at StudioExile, graphic designer; and Jillian Luff at MAP*grafix*, cartographer. You have all been amazing.

In addition to the thematic content, each publication provides an overview of the status of apes, both in their natural habitats and in captivity. Our sincere appreciation goes to all the great ape and gibbon scientists who contribute their valuable data to build the A.P.E.S. database and to Tenekwetche Sop who manages it. Such collaborative efforts are key to effective conservation action. Thanks also go out to the rescue and rehabilitation centers, sanctuaries, zoos and other captive-ape organizations that provided detailed information and data.

This volume would not exist without the essential input of the *State of the Apes* Publication Committee and those who participated in the preparatory stakeholder meeting. In addition to refining the contents and structure of the book, these experts helped us to identify potential authors and contributors. They are Peter Apell, Caroline Asiimwe, Dirck Byler, Anne-Lise Chaber, Susan Cheyne, Elizabeth (Annie) Cook, Tom Gillespie, Gladys Kalema-Zikusoka, Fabian Leendertz, Elizabeth Lonsdorf, Linda May, Sen Nathan, Joachim Nieuwland, George Omondi Paul, Joshua Rukundo, Benard Ssebide and Chris Walzer.

Authors and contributors and those who provided essential data are named at the end of each chapter, and we thank them again here. We could not have produced this book without them.

We also wish to acknowledge the critical role of those who reviewed the content; they provided constructive feedback and helped ensure quality control. They are Liv Baker, Anna Behm Masozera, Marc Bekoff, Raquel Costa, Gemma Davies, Rosa Garriga, Kirsten Gilardi, Tom Gillespie, Harold Goodwin, Melvin Gumal, Sonya Hill, Tatyana Humle, Jenny Jaffe, Sonya Kahlenberg, Aili Kang, Sascha Knauf, Fabian Leendertz, Nancy Lung, Marie McIntyre, Matthew McLennan, Michael Muehlenbein, Lawrence Mugisha, Alexandra Palmer, Clare Palmer, Andrew Plumptre, Ben Rawson, Fabrice Renaud, Cindy Rizzo, Steve Ross, Jessica Rothman, Anne Russon, Sohel Saikat, Karmele Llano Sánchez, Lesa Thompson, Chris Whittier, Elizabeth A. Williamson and Steve Wise.

Most of the photographs in this volume were generously contributed by their creators; the rest were provided at a significant discount. In addition, a number of organizations kindly gave us permission to reproduce figures and extracts from previously published books, journals and reports, as well as from internal documents. Photographers and copyright holders are credited alongside images throughout this book.

To ensure that the *State of the Apes* series is accessible to as many stakeholders as possible, it is published under an open access agreement with Cambridge University Press. Each volume is also available in multiple languages. For the translated editions of this volume—in Bahasa Indonesia, Chinese (Mandarin) and French—we wish to recognize our team of translators, graphic designers and proofreaders: Alboum Translation Services, Nelly Aubaud Davies, Xuezhu (Bamboo) Huff, MAP*grafix*, Anton Nurcahyo, OWLingua, Hélène Piantone Beaudin, StudioExile, Erica Taube, Beth Varley and Rumanti Wasturini. Much gratitude goes to the Arcus Communications team—Heather Antonissen, Angela Cave, Jovahn Huertas, Sebastian Naidoo and Bryan Simmons—for managing the website and for coordinating launch activities for each volume in the series.

Many others contributed in various ways, including by providing introductions, anonymous input, strategic advice and essential, if sometimes tedious, administrative tasks. We also thank all those who provided much-appreciated moral support.

The Arcus Foundation is pleased to pass the baton on to the International Union for the Conservation of Nature (IUCN) Section on Great Apes and Section on Small Apes. *State of the Apes* will continue in a mix of formats under their expert coordination and guidance. We look forward to seeing how it develops.

**Alison White, Steve Unwin,
Annette Lanjouw, Katy Scholfield
and Helga Rainer**
Editors

Apes Overview

Apes Index

All information is drawn from the *Handbook of the Mammals of the World, Volume 3: Primates* (Mittermeier, Rylands and Wilson, 2013), unless otherwise cited.

Bonobo (*Pan paniscus*)

Distribution and Numbers in the Wild

The bonobo is present only in the Democratic Republic of Congo (DRC), bio-geographically separated from chimpanzees and gorillas by the Congo River (see Figure AO1). The population size is unknown, as only 30% of the species' historical range has been surveyed; however, estimates from the four geographically distinct bonobo strongholds suggest a minimum population of 15,000–20,000 individuals, with numbers decreasing (Fruth *et al.*, 2016).

The bonobo is included in Appendix I of the Convention on International Trade in Endangered Species of Wild Fauna and Flora (CITES) and is categorized as endangered on the International Union for Conservation of Nature (IUCN) Red List (Fruth *et al.*, 2016; see Box AO1). The causes of population decline include poaching; habitat loss and degradation; disease; and people's lack of awareness that hunting and eating bonobos is unlawful. Poaching, which is mainly carried out as part of the commercial wild meat trade and for some medicinal purposes, has been exacerbated by the ongoing effects of armed conflict, such as military-sanctioned hunting and the accessibility of modern weaponry and ammunition (Fruth *et al.*, 2016).

Physiology

Male adult bonobos reach a height of 73–83 cm and weigh 36–43 kg, while females are slightly smaller, weighing 26–36 kg. Bonobos are moderately sexually dimorphic and similar in size and appearance to chimpanzees, although with a smaller head and lither appearance. Their life expectancy in the wild is estimated at about 40 years, while the oldest known captive bonobo was 60 years of age (Hohmann, Robbins and Boesch, 2006; Lowenstine, McManamon and Terio, 2016; Robson and Wood, 2008).

The bonobo diet is mainly frugivorous (more than 50% fruit), supplemented with leaves, stems, shoots, pith, seeds, bark, flowers, honey and fungi. Only a very small part of the diet consists of animal matter—such as insects, small reptiles, birds and medium-sized mammals, including other primates.

Social Organization

Bonobos live in fission–fusion communities of up to 120 individuals, consisting of multiple males and females. When foraging, they split into smaller mixed-sex subgroups, or parties, averaging 5–23 individuals. Male bonobos cooperate with and tolerate one another; however, lasting bonds between adult males are rare, in contrast to the bonds between adult females, which are strong and potentially last for years. A distinguishing feature of female bonobos is that they are co-dominant with males and form alliances against certain males within the community. Among bonobos, the bond between mother and son is the strongest, lasts into the son's adulthood and proves highly important for his social status. During intergroup encounters, bonobos, especially females, tend to exhibit relaxed interactions, such as feeding in the same fruiting tree (Idani, 1990). Female bonobos are the dispersal sex and usually immigrate to neighboring communities (Ishizuka, Toda and Furuichi, 2020). Together with chimpanzees, bonobos are the closest living relatives of humans, with whom they share 98.8% of their DNA (Smithsonian Institute, 2022; The Chimpanzee Sequencing and Analysis Consortium, 2005).

Chimpanzee (*Pan troglodytes*)

Distribution and Numbers in the Wild

Chimpanzees are widely distributed across equatorial Africa, with discontinuous populations from southern Senegal to western Uganda and Tanzania. They can be found in forest-dominated landscapes as well as ones dominated by savannah and agricultural mosaics (Humle *et al.*, 2016b; see Figure AO1).

Chimpanzees are listed in CITES Appendix I, and all four subspecies are categorized as either endangered or critically endangered on the IUCN Red List. There are about 114,200–317,000 central chimpanzees (*Pan troglodytes troglodytes*); 17,600–96,600 western chimpanzees (*Pan t. verus*); 170,000–

250,000 eastern chimpanzees (*Pan t. schweinfurthii*); and probably fewer than 9,000 Nigeria–Cameroon chimpanzees (*Pan t. ellioti*).[1]

All populations are believed to be declining, but the rate has not yet been quantified for all (Humle *et al.*, 2016b). The western chimpanzee population decreased at a rate of 6% per year from 1990 to 2014, which corresponds to a population decline of 80.2% over the study period (Kühl *et al.*, 2017). Decreases in chimpanzee numbers are mainly attributed to increased poaching for the commercial wild meat trade, habitat loss and degradation, and disease (Humle *et al.*, 2016b).

Physiology

Male chimpanzees are 77–96 cm tall and weigh 28–70 kg, while females measure 70–91 cm and weigh 20–50 kg. They share many facial expressions with humans, although forehead musculature is less pronounced and they have more flexible lips. Chimpanzees live for up to 60 years in the wild.

Chimpanzees are mainly frugivorous. Some communities include 200 species of food items in a diet of fruit supplemented by bark, flowers, fungi, honey, leaves, pith, seeds, shoots, stems and animal prey, such as ants and termites, but also small mammals, including other primates. Chimpanzees are the most carnivorous of all the apes.

Social Organization

Chimpanzees show fission–fusion, multi-male–multi-female grouping patterns. A large community includes all individuals who regularly associate with one another; such communities comprise an average of 35 individuals, with the largest-known group exceeding 150, although this size is rare. The community separates into smaller, temporary subgroups, or parties. The parties can be highly fluid, with members moving in and out quickly or a small number of individuals staying together for a few days before rejoining other members of the community.

Home ranges are typically defended by highly territorial males, who may attack or even kill neighboring chimpanzees. Male chimpanzees are dominant over female chimpanzees and are generally the more social sex, sharing food and grooming each other more frequently. Female chimpanzees are the dispersal sex and usually immigrate to a neighboring community during adolescence or in early adulthood. Chimpanzees are noted for their ability to use tools, especially to access embedded foods such as termites or the kernel of a nut, and for their sophisticated forms of cooperation, such as in hunting and territorial defense. The level of cooperation in social hunting activities and their tool-use repertoire vary across communities, however.

Gorilla (*Gorilla* species (spp.))

Distribution and Numbers in the Wild

The western gorilla (*Gorilla gorilla*) is distributed throughout western equatorial Africa and has two subspecies: the western lowland gorilla (*Gorilla g. gorilla*) and the Cross River gorilla (*Gorilla g. diehli*). The eastern gorilla (*Gorilla beringei*) is found in the DRC and across the border in Uganda and Rwanda. There are two subspecies of the eastern gorilla: the mountain gorilla (*Gorilla b. beringei*) and Grauer's gorilla (*Gorilla b. graueri*) (see Figure AO1).

Three of the four gorilla taxa are listed as critically endangered on the IUCN Red List.[2] The first range-wide population estimate for the western lowland gorilla put the total population at nearly 362,000 in 2013; since then, the figure has decreased considerably. In 2018 scientists estimated that by 2020 the total population would be just 300,000 (Strindberg *et al.*, 2018; Williamson, Strindberg and Maisels, 2018). As few as 250–300 Cross River gorillas remain in the wild (Bergl *et al.*, 2016; Dunn *et al.*, 2014). The most recent population estimate for Grauer's gorilla is 6,800, which is up from the 2016 estimate due to improvements in the accuracy of assessments, but still indicates an overall loss of 60% and in the Kahuzi-Biega National Park, a loss estimated at 80%, since the mid-1990s (Maisels, Plumptre and Strindberg, 2021; Plumptre *et al.*, 2021). An estimated 1,000 mountain gorillas remain in the wild (Granjon *et al.*, 2020b; Hickey *et al.*, 2019a).

The main threats to both the eastern and western gorilla are poaching for the commercial wild meat trade, habitat destruction and degradation, and disease (for the western gorilla, the Ebola virus in particular) (Maisels, Bergl and Williamson, 2018; Plumptre, Robbins and Williamson, 2019). Grauer's gorilla is also threatened by civil unrest (Plumptre, Robbins and Williamson, 2019). Another threat is the impact of climate change on the gorilla's forest habitats (Maisels, Bergl and Williamson, 2018; Plumptre, Robbins and Williamson, 2019).

Physiology

The adult male of the eastern gorilla is slightly larger (159–196 cm, 120–209 kg) than the western gorilla (138–180 cm, 145–191 kg). Both species are highly sexually dimorphic and females are about half the size of males. Their life span ranges from 30 to 40 years in the wild. Mature males are known as "silverbacks" due to the development of a gray saddle on their back when they attain maturity.

The gorillas' diet consists predominantly of ripe fruit, tree leaves and terrestrial, herbaceous vegetation. More herbaceous vegetation is ingested while fruit is scarce, in line with seasonality. Gorillas do not eat meat but occasionally consume ants and termites. Mountain gorillas have less fruit in their environment than lowland gorillas, so they feed mainly on leaves, pith, stems, bark and, occasionally, ants.

Social Organization

Western gorillas live in stable groups with multiple females and one adult male (silverback). In contrast, eastern gorillas are poly-gynous and can be polygynandrous, with groups that comprise one or more silverbacks, multiple females, their offspring and immature relatives. The average group consists of ten individuals, but eastern gorillas can live in groups of up to 65 individuals, whereas the maximum group size for the western gorilla is 22. Gorillas are not territorial and home ranges overlap extensively. Groups that live in the same areas normally adopt a strategy of mutual avoidance. When neighboring silverbacks come into contact, they generally use chest beats and vocalizations, although intergroup encounters can escalate into physical fights.

Orangutan (*Pongo* spp.)

Distribution and Numbers in the Wild

Although orangutans were once present throughout much of southern Asia, their range is now limited to the two islands of Sumatra and Borneo (Delgado and van Schaik, 2000; Wang *et al.*, 2014; see Figure AO2).

In Sumatra in 2015, fewer than 14,000 Sumatran orangutans (*Pongo abelii*) were ranging over 17,000 km² (1.7 million ha), primarily in the Leuser Eco-system (Singleton *et al.*, 2017). Fewer than 1,000 individuals of the newly described species, the Tapanuli orangutan (*Pongo tapanuliensis*), are found in about 1,000 km² (100,000 ha) of fragmented forests in Batang Toru (Nowak *et al.*, 2017; Wich *et al.*, 2019). In Borneo, between 80,000 and 100,000 Bornean orangutans (*Pongo pygmaeus* spp.) remain in the wild over more than 100,000 km² (10 million ha) (Ancrenaz *et al.*, 2016; Voigt *et al.*, 2018). As a result of continuing habitat loss and hunting, both the Sumatran orangutan and the Bornean orangutan are classified as critically endangered (Ancrenaz *et al.*, 2016; Nowak *et al.*, 2017; Singleton *et al.*, 2017). All three species are listed in Appendix I of CITES.

The main threats to all orangutan species are habitat loss and fragmentation, as well as killings in the context of human–ape conflict, hunting and the interna-tional live animal trade.[3] For the Bornean orangutan, additional threats include forest fires and people's lack of awareness that they are protected by law (Sherman *et al.*, 2020). The main threat to the Sumatran orangutan is the cur-rent land use plan issued by the government of Aceh, which does not recognize the Leuser Ecosystem's National Strategic Area status (Singleton *et al.*, 2017). As a result, illegal logging and habitat loss represent high risks to the species' long-term survival (Wich *et al.*, 2016). The small size of the only remaining population of Tapanuli orangutans is a serious concern as it may lead to inbreeding and genetic depression (Nater *et al.*, 2017). The proposed hydroelectric dam and its associated infrastruc-ture would further fragment and isolate the population by splitting the Sibual-Buali Nature Reserve (Laurance *et al.*, 2020; Nasution, Perwitasari-Farajallah and Utami-Atmoko, 2018, 2020; Wich *et al.*, 2019).

Physiology

Orangutans are highly sexually dimorphic. Adult males can reach a height of 94–99 cm and weigh 60–85 kg (flanged) or 30–65 kg (unflanged). Females are far smaller than males, reaching about 64–84 cm and weighing 30–45 kg. In the wild in Sumatra, the life expectancy is 58 years for males and 53 years for females. Bornean orangutans may live as long, although no accurate data are yet available for this species.

Fully mature males develop a short beard and protruding cheek pads, termed "flanges." Some male orangutans experience "developmental arrest," maintaining a female-like size and appearance for many years past sexual maturity; they are known as "unflanged" males. Orangutans are the only mammal species to exhibit male bimaturism.

The orangutan diet consists mainly of fruit, although it also features leaves, shoots, seeds, bark, pith, flowers, eggs, soil and invertebrates, such as termites and ants. Carnivorous behavior has also been observed, especially in Sumatra, but at a very low frequency (preying on species such as slow lorises).

Social Organization

The mother–offspring unit is the only permanent social unit among orangutans. Social interaction between independent individuals does occur, with varying frequency across populations and taxa; they are more common in the two Sumatran species than the Bornean species (Fröhlich *et al*., 2020; Roth *et al*., 2020). While females are usually relatively tolerant of each other, flanged males are intolerant of other flanged and unflanged males (Utami-Atmoko *et al*., 2009). Orangutans on Sumatra are generally more social than those on Borneo and live in overlapping home ranges, with flanged males emitting "long calls" to alert others to their location (Delgado and Van Schaik, 2000; Spillmann *et al*., 2017). Orangutans are characterized by an extremely slow life history, with the longest interbirth interval of any primate species, an average of 7.6 years (van Noordwijk *et al*., 2018).

Gibbons (*Hoolock* spp.; *Hylobates* spp.; *Nomascus* spp.; *Symphalangus* spp.)

The four genera of gibbon share ecological and behavioral attributes, such as social monogamy in territorial groups; vocalization through elaborate song (including complex duets); and frugivory and brachiation (moving through the canopy using only the arms). Gibbons primarily consume fruit but have a varied diet that includes insects, flowers, leaves and seeds. Female gibbons have a single offspring every 2.5–3 years. Gibbons are diurnal and sing at sunrise and sunset; they dedicate a significant part of the day to finding fruit trees within their territories.

Hoolock genus

Distribution and Numbers in the Wild

Three species comprise the *Hoolock* genus: the western hoolock (*Hoolock hoolock*), the eastern hoolock (*Hoolock leuconedys*) and the Gaoligong or Skywalker hoolock (*Hoolock tianxing*), first described in 2017 (Fan *et al*., 2017; Fan, Turvey and Bryant, 2020). The Mishmi Hills hoolock (*Hoolock h. mishmiensis*), the most recently discovered subspecies of western hoolock, was officially named in 2013 (Choudhury, 2013).

The western hoolock's distribution spans Bangladesh, India and Myanmar. The eastern hoolock lives in China and Myanmar (see Figure AO2). To date, the Gaoligong hoolock has only been seen in eastern Myanmar and southwestern China (Fan *et al*., 2017). The Gaoligong hoolock comprises an estimated nine subpopulations and about 200 individuals in China. No recent population estimates exist for Myanmar (P.-F. Fan, personal communication, 2019). Current estimates suggest that the population in Myanmar may be several thousands; additional surveys are needed to gather detailed data. Density estimates range from 0.76–1.0 groups/km² (Aung *et al*., 2023).

With an estimated maximum contiguous population of 15,000 individuals, the western hoolock is listed as endangered on the IUCN Red List (Brockelman, Molur and Geissmann, 2019). The eastern hoolock has a population of 10,000–50,000 and is listed as vulnerable on the IUCN Red List (Brockelman and Geissmann, 2019). Both species are listed in CITES Appendix I, with the main threats identified as habitat loss and fragmentation, as well as hunting for food, pets, tourism and medicinal purposes. The Gaoligong hoolock is categorized as endangered on the IUCN Red List (Fan, Turvey and Bryant, 2020).

Physiology

An individual hoolock can have a head and body length of 45–81 cm and weigh 6–9 kg; males are slightly heavier than females. Like most gibbons, the *Hoolock* genus is sexually dichromatic, with the pelage (coat) of females and males differing in terms of patterning and color. Pelage also differs across species: unlike the western hoolock, the eastern one features a complete separation between the white brow markings and a white preputial tuft.

The diet of the western hoolock is primarily frugivorous, supplemented with vegetative matter such as leaves, shoots, seeds, moss and flowers. While little is known about the diet of the eastern hoolock, it probably resembles that of the western hoolock.

Social Organization

Hoolocks live in family groups of 2–6 individuals, consisting of a mated adult pair and their offspring. They are thought to be territorial, although this assumption has yet to be corroborated by data. Hoolock pairs vocalize a "double solo" rather than the more common "duet" of various gibbons.

Hylobates genus

Distribution and Numbers in the Wild

Nine species are currently included in the *Hylobates* genus, although there remains some dispute about whether Abbott's gray gibbon (*Hylobates abbottii*), the Bornean gray gibbon (*Hylobates funereus*) and Müller's gibbon (*Hylobates muelleri*) represent full species (see Table AO2).

This genus of gibbon was and is found in discontinuous tropical and subtropical forests, from southwestern China—where the genus has been extirpated—through Indochina, the Malay Peninsula and Thailand to the islands of Borneo, Java and Sumatra (Fan, 2017; Wilson and Reeder, 2005; see Figure AO2). The least abundant species is the moloch gibbon (*Hylobates moloch*); the most abundant, collectively, are the "gray gibbons" (Abbott's, the Bornean and Müller's gibbons), although no accurate population numbers are available for Abbott's gray gibbon.

All *Hylobates* species are listed as endangered on the IUCN Red List and are in CITES Appendix I. Three hybrid zones occur naturally and the hybrids continue to coexist with the unhybridized species in the wild. The main collective threats facing the genus are deforestation, hunting and the illegal pet trade.

Physiology

Average height for both sexes of all species is approximately 46 cm and their weight ranges between 5 kg and 7 kg. With the exception of the pileated gibbon (*Hylobates pileatus*), species in the genus are not sexually dichromatic, although the lar gibbon (*Hylobates lar*) has two color phases, which are not related to sex or age.

Gibbons are mainly frugivorous. Figs are an especially important part of their diet and are supplemented by leaves, buds, flowers, shoots, vines and insects, while small animals and bird eggs form the protein input.

Social Organization

Hylobates gibbons are largely socially monogamous, forming family units of two adults and their offspring; however, polyandrous and polygynous units have been observed, especially in hybrid zones. Territorial disputes are predominantly led by males, who become aggressive towards other males, whereas females tend to lead daily movements and ward off other females.

Nomascus genus

Distribution and Numbers in the Wild

Seven species make up the *Nomascus* genus (see Table AO2). Somewhat less widely distributed than the *Hylobates* genus, the *Nomascus* genus is present in Cambodia, the Lao People's Democratic Republic, Viet Nam and southern China, including Hainan Island (see Figure AO2).

Population estimates exist for some taxa: there are approximately 5,000 western black crested gibbons (*Nomascus concolor*), about 200 Cao Vit gibbons (*Nomascus nasutus*) and 37 Hainan gibbons (*Nomascus hainanus*) (Fan, 2017).[4] Population estimates for the northern and southern white-cheeked crested gibbons (*Nomascus leucogenys* and *Nomascus siki*, respectively) are available for some sites, and overall numbers are known to be severely depleted. The northern and southern yellow-cheeked crested gibbons (*Nomascus annamensis* and *Nomascus gabriellae*, respectively) have the largest populations among the *Nomascus* gibbons.

All species are listed in CITES Appendix I. On the IUCN Red List, five are categorized as critically endangered: the Cao Vit and Hainan gibbons, the northern and southern white-cheeked crested gibbons, and the western black crested gibbon. Two are listed as endangered: the northern and southern yellow-cheeked crested gibbons (IUCN, 2022). Major threats to these populations include hunting for food, pets and medicinal purposes, as well as habitat loss and fragmentation.

xix

Physiology

Average head and body length across all species of this genus, for both sexes, is approximately 47 cm; individuals weigh around 7 kg. All *Nomascus* species have sexually dimorphic pelage; adult males are predominantly black while females are a buffy yellow. Their diet is much the same as that of the *Hylobates* genus: mainly frugivorous, supplemented with leaves and flowers.

Social Organization

Gibbons of the *Nomascus* genus are mainly socially monogamous, although most species have also been observed in polyandrous and polygynous groups. More northerly species appear to engage in polygyny to a greater degree than southern taxa. Copulations outside monogamous pairs have been recorded, albeit infrequently.

Symphalangus genus

Distribution and Numbers in the Wild

Siamang (*Symphalangus syndactylus*) are found in several forest blocks across Indonesia, Malaysia and Thailand (see Figure AO2). The species faces severe threats to its habitat across its range. No accurate estimates exist for the total population size. The species is listed in CITES Appendix I and is classified as endangered on the IUCN Red List (IUCN, 2022; Nijman *et al.*, 2020).

Physiology

The siamang's head and body length is 75–90 cm, and adult males weigh 10.5–12.7 kg, while adult females weigh 9.1–11.5 kg. The siamang is minimally sexually dimorphic, and the pelage is the same across the sexes: black. The species has a large inflatable throat sac.

Siamang rely heavily on figs and somewhat less on leaves—a diet that allows them to be sympatric with *Hylobates* gibbons in some locations, since the latter focus more on fleshy fruits. The siamang diet also includes flowers and insects.

Social Organization

Males and females call territorially, using their large throat sacs, and males give chase to neighboring males. One group's calls inhibit nearby groups and, consequently, they take turns to vocalize. The groups are usually based on monogamous pairings, although polyandrous groups have been observed. Males may also adopt the role of caregiver for infants.

Photo credits

Bonobo: © Takeshi Furuichi, Wamba Committee for Bonobo Research

Chimpanzee: © Arcus Foundation and Jabruson, 2014.
All rights reserved. www.jabruson.photoshelter.com

Gorilla: © Annette Lanjouw

Orangutan: © Perry van Duijnhoven 2013

Gibbons:
Hoolock: © Dr. Axel Gebauer/naturepl.com;
Hylobates: © International Primate Protection League (IPPL);
Nomascus: © IPPL;
Symphalangus: © Pete Oxford/naturepl.com

Ape Socioecology[5]

This section presents an overview of the socioecology of apes: bonobos; chimpanzees; eastern and western gorillas; Bornean, Sumatran and Tapanuli orangutans; and gibbons (including siamangs).

Chimpanzees are the most wide-ranging ape species in Africa, occurring across 21 countries, while bonobos are restricted to the Democratic Republic of Congo (DRC) (Fruth *et al.*, 2016; Humle *et al.*, 2016b). Gorillas live in ten African countries (Maisels, Bergl and Williamson, 2018; Plumptre, Robbins and Williamson, 2019). Orangutans are found in Asia—in both Indonesia and Malaysia—and are the only apes to have two distinct male types (Ancrenaz *et al.*, 2016;

Nowak *et al.*, 2017; Singleton *et al.*, 2017). Gibbons are the most geographically widespread group of apes. Currently, 20 species of gibbon in four genera are recognized across Asia: 9 *Hylobates* species, 7 *Nomascus* species, 3 *Hoolock* species and the single *Symphalangus* species (Fan *et al.*, 2017; IUCN, 2019b; Thinh *et al.*, 2010).

Social Organization

Apes vary considerably in their social organization. While orangutans lead semi-solitary lives, some gibbons form family groups with monogamous pairs, and African great apes—bonobos, chimpanzees and gorillas—live in larger social groupings.

BOX AO1

IUCN Red List Categories and Criteria, and CITES Appendices

The International Union for Conservation of Nature (IUCN) Species Survival Commission assesses the conservation status of each species and subspecies using IUCN Red List categories and criteria. As all great apes and gibbons are categorized as vulnerable, endangered or critically endangered, this box presents details on a selection of the criteria for these three categories (see Table AO1). A summary of the five criteria is provided in Annex I. Full details of the IUCN Red List categories and criteria (in English, French, Japanese and Spanish) can be viewed and downloaded :

https://www.iucnredlist.org/resources/categories-and-criteria.

Detailed guidelines on their use are available at:

https://www.iucnredlist.org/resources/redlistguidelines.

Appendices I, II and III to the Convention on International Trade in Endangered Species of Wild Fauna and Flora (CITES) are lists of species afforded different levels or types of protection from overexploitation. All non-human apes are in Appendix I, which comprises species that are the most endangered among CITES-listed animals and plants.

CITES prohibits international trade in species that are threatened with extinction, except under specified circumstances, including for certain types of scientific research, so long as the transfers do not contravene national legislation. Such exceptional trade requires both an import permit and an export permit, or a re-export certificate—which authorities grant only if they determine that the transfers are not likely to have a

Table AO1

Principal Criteria for the Red List Categories Vulnerable, Endangered and Critically Endangered

IUCN Red List category	Risk of extinction in the wild	Number of mature individuals in the wild	Rate of population decline over the past 10 years or 3 generations (whichever is longer)
Vulnerable	High	<10,000	≥30%
Endangered	Very high	<2,500	≥50%
Critically endangered	Extremely high	<250	≥80%

negative impact on the survival of the species in the wild, that the specimens to be transferred have been acquired legally and that the trade is not for primarily commercial purposes— so long as the transfers do not contravene national legislation (see Chapters 6 and 8). Article VII of the Convention provides for a number of exemptions to this general prohibition. For more information, see https://www.cites.org/eng/disc/text.php#VII.

Table AO2

Great Apes and Gibbons

IUCN Red List category: ● Vulnerable ● Endangered ● Critically endangered

Great ape genus	Species	Scientific name	Range countries
Pan	Bonobo	*Pan paniscus*	■ Democratic Republic of Congo (DRC)
	Central chimpanzee	*Pan troglodytes troglodytes*	■ Angola ■ Cameroon ■ Central African Republic ■ DRC ■ Equatorial Guinea ■ Gabon ■ Republic of Congo
	Eastern chimpanzee	*Pan troglodytes schweinfurthii*	■ Burundi ■ Central African Republic ■ DRC ■ Rwanda ■ South Sudan ■ Tanzania ■ Uganda
	Nigeria–Cameroon chimpanzee	*Pan troglodytes ellioti*	■ Cameroon ■ Nigeria
	Western chimpanzee	*Pan troglodytes verus*	■ Ghana ■ Guinea ■ Guinea-Bissau ■ Ivory Coast ■ Liberia ■ Mali ■ Senegal ■ Sierra Leone
Gorilla	Cross River gorilla	*Gorilla gorilla diehli*	■ Cameroon ■ Nigeria
	Grauer's gorilla	*Gorilla beringei graueri*	■ DRC
	Mountain gorilla	*Gorilla beringei beringei*	■ DRC ■ Rwanda ■ Uganda
	Western lowland gorilla	*Gorilla gorilla gorilla*	■ Angola ■ Cameroon ■ Central African Republic ■ Equatorial Guinea ■ Gabon ■ Republic of Congo
Pongo	Northeast Bornean orangutan	*Pongo pygmaeus morio*	■ Indonesia ■ Malaysia
	Northwest Bornean orangutan	*Pongo pygmaeus pygmaeus*	■ Indonesia ■ Malaysia
	Southwest Bornean orangutan	*Pongo pygmaeus wurmbii*	■ Indonesia
	Sumatran orangutan	*Pongo abelii*	■ Indonesia
	Tapanuli orangutan	*Pongo tapanuliensis*	■ Indonesia

▶

Gibbon genus	Species (excluding subspecies)	Scientific name	Range countries
Hoolock	Eastern hoolock	*Hoolock leuconedys*	■ China ■ Myanmar
	Gaoligong hoolock (a.k.a. Skywalker hoolock)	*Hoolock tianxing*	■ China ■ Myanmar
	Western hoolock	*Hoolock hoolock*	■ Bangladesh ■ India ■ Myanmar
Hylobates	Abbott's gray gibbon	*Hylobates abbotti*	■ Indonesia ■ Malaysia
	Agile gibbon (a.k.a. dark-handed gibbon)	*Hylobates agilis*	■ Indonesia ■ Malaysia
	Bornean gray gibbon (a.k.a. northern gray gibbon)	*Hylobates funereus*	■ Brunei ■ Indonesia ■ Malaysia
	Bornean white-bearded gibbon (a.k.a. Bornean agile gibbon)	*Hylobates albibarbis*	■ Indonesia
	Kloss's gibbon (a.k.a. Mentawai gibbon)	*Hylobates klossii*	■ Indonesia
	Lar gibbon (a.k.a. white-handed gibbon)	*Hylobates lar*	■ Indonesia ■ Lao People's Democratic Republic (PDR) ■ Malaysia ■ Myanmar ■ Thailand
	Moloch gibbon (a.k.a. Javan gibbon, silvery gibbon)	*Hylobates moloch*	■ Indonesia
	Müller's gibbon (a.k.a. Müller's gray gibbon, southern gray gibbon)	*Hylobates muelleri*	■ Indonesia

Bonobos and chimpanzees form multi-male and multi-female dynamic communities or groups that can fission into smaller groups (known as parties) or fuse to form larger ones. These parties can vary in size throughout the day, depending on food availability and the presence of reproductively active females (Wrangham, 1986). Parties, especially in chimpanzees, tend to be smaller during periods of fruit scarcity (Furuichi, 2009). Adult female chimpanzees often spend time alone with their offspring or in a party with other females, while adult female bonobos tend to associate more extensively with their adult sons. Chimpanzee communities average 35 members, with some even exceeding 150 members (Mitani, 2009; Mittermeier, Rylands and Wilson, 2013). Bonobo communities usually comprise 10–120 individuals (Fruth, Williamson and Richardson, 2013). In both species, females are typically the dispersing sex, emigrating from their native community to a neighboring one upon sexual maturity, around 8 years of age in bonobos, and between 8 and 14 in chimpanzees (Hashimoto, 1997; Walker *et al.*, 2018).

Gorillas live in stable, cohesive social units, or groups, with a median size of ten. Most groups consist of one or more "silverback" males with several females and their offspring. Mountain gorillas differ in that

	Pileated gibbon (a.k.a. capped gibbon, crowned gibbon)	*Hylobates pileatus*	■ Cambodia ■ Lao PDR ■ Thailand
Nomascus	Cao Vit gibbon (a.k.a. eastern black crested gibbon)	*Nomascus nasutus*	■ China ■ Viet Nam
	Hainan gibbon (a.k.a. Hainan black crested gibbon, Hainan black gibbon, Hainan crested gibbon)	*Nomascus hainanus*	■ China (Hainan Island)
	Northern white-cheeked crested gibbon (a.k.a. northern white-cheeked gibbon, white-cheeked gibbon)	*Nomascus leucogenys*	■ Lao PDR ■ Viet Nam
	Northern yellow-cheeked crested gibbon (a.k.a. northern buffed-cheeked gibbon)	*Nomascus annamensis*	■ Cambodia ■ Lao PDR ■ Viet Nam
	Southern white-cheeked crested gibbon (a.k.a. southern white-cheeked gibbon)	*Nomascus siki*	■ Lao PDR ■ Viet Nam
	Southern yellow-cheeked crested gibbon (a.k.a. red-cheeked gibbon, buff-cheeked gibbon, buffy-cheeked gibbon)	*Nomascus gabriellae*	■ Cambodia ■ Viet Nam
	Western black crested gibbon (a.k.a. black crested gibbon, black gibbon, concolor gibbon, Indochinese gibbon)	*Nomascus concolor*	■ China ■ Lao PDR ■ Viet Nam
Symphalangus	Siamang	*Symphalangus syndactylus*	■ Indonesia ■ Malaysia

Sources: IUCN (2023); Mittermeier, Rylands and Wilson (2013); author knowledge and experience; S. Wich, personal communication, 2021 and E.A. Williamson, personal communication, 2022

they frequently contain more than 20 individuals and have a multi-male structure (Robbins and Robbins, 2018). Their largely vegetation-based diet enables mountain gorillas to live in areas with limited amounts of fruit. Western gorillas typically form one-male groups with one silverback, although multi-male and all-male groups (non-reproductive groups that contain no females) occur occasionally. Multi-male groups contain more than one silverback, but only rarely contain more than two.

Gorillas are among the few primate species in which both males and females disperse from their natal groups. Males emigrate to become solitary when they are blackbacks or young silverbacks (about 13–15 years of age). Males may be solitary for several years before forming a group. Male western gorillas tend to acquire groups around age 18, a few years later than mountain gorillas, who typically become dominant around 15 years of age. Western gorilla males almost exclusively follow the path of becoming solitary and forming new groups after females join them. Mature males never join established groups, so multi-male groups are extremely rare among western gorillas. When the silverback of a one-male group dies, the group disintegrates, and the adult females and immature offspring join a solitary male or another group. In contrast to western gorillas, about 40% of mountain gorilla groups are multi-male. Mountain gorilla males follow one of two strategies to become the leader of a group: either they remain in the group and attempt a takeover from within, or they emigrate to become a solitary male and eventually form new groups (Robbins and Robbins, 2018).

The three orangutan species are semi-solitary, do not exhibit territorial behavior

Figure AO1

Ape Distribution in Africa[6]

State of the Apes Disease, Health and Ape Conservation

and have loosely defined communities. They are not considered "social" animals, but they exhibit an individual-based fission–fusion social organization (Roth *et al.*, 2020; van Schaik, 1999). The basic social

unit is a single individual, although adult females are usually found with one baby or one baby and an adolescent. Flanged adult males, characterized by fatty cheek pads and large size, lead a semi-solitary existence and are rather intolerant of other flanged males and, to a lesser degree, unflanged ones (Emery Thompson, Zhou and Knott, 2012; Spillmann *et al.*, 2017; Utami-Atmoko *et al.*, 2009). Smaller, unflanged adult males are more tolerant of other orangutans. Adult females are the most social individuals and sometimes travel together for a few hours to several days, especially in Sumatra, where orangutans occasionally congregate when food is abundant (van Schaik, 1999; Wich *et al.*, 2006). Females associate more with their maternal kin; unrelated individuals show low social tolerance and sometimes exhibit aggression, which can lead to death (Knott *et al.*, 2008; Marzec *et al.*, 2016). Male orangutans are the dispersing sex: upon reaching sexual maturity, they leave the area where they were born to establish their own range (Arora *et al.*, 2012). Female orangutans show a strong natal philopatry and, upon adulthood, part of their range overlaps with their natal range (Ashbury *et al.*, 2020; van Noordwijk *et al.*, 2012). Orangutans have a polygynandrous mating system, within which males pursue a roving strategy. Although females tend to prefer flanged males for mating, unflanged males also produce a significant number of offspring in a population (Goossens *et al.*, 2006; Utami-Atmoko *et al.*, 2009).

Gibbons are highly territorial and live in semi-permanent family groups, defending a territory to the exclusion of other gibbons. Both male and female gibbons disperse from their natal groups and establish their own territories (Leighton, 1987). Gibbons have been typified as forming socially monogamous family groups. Other studies, however, have revealed they are not necessarily sexually monogamous (Palombit, 1994).

Figure AO2

Ape Distribution in Asia[7]

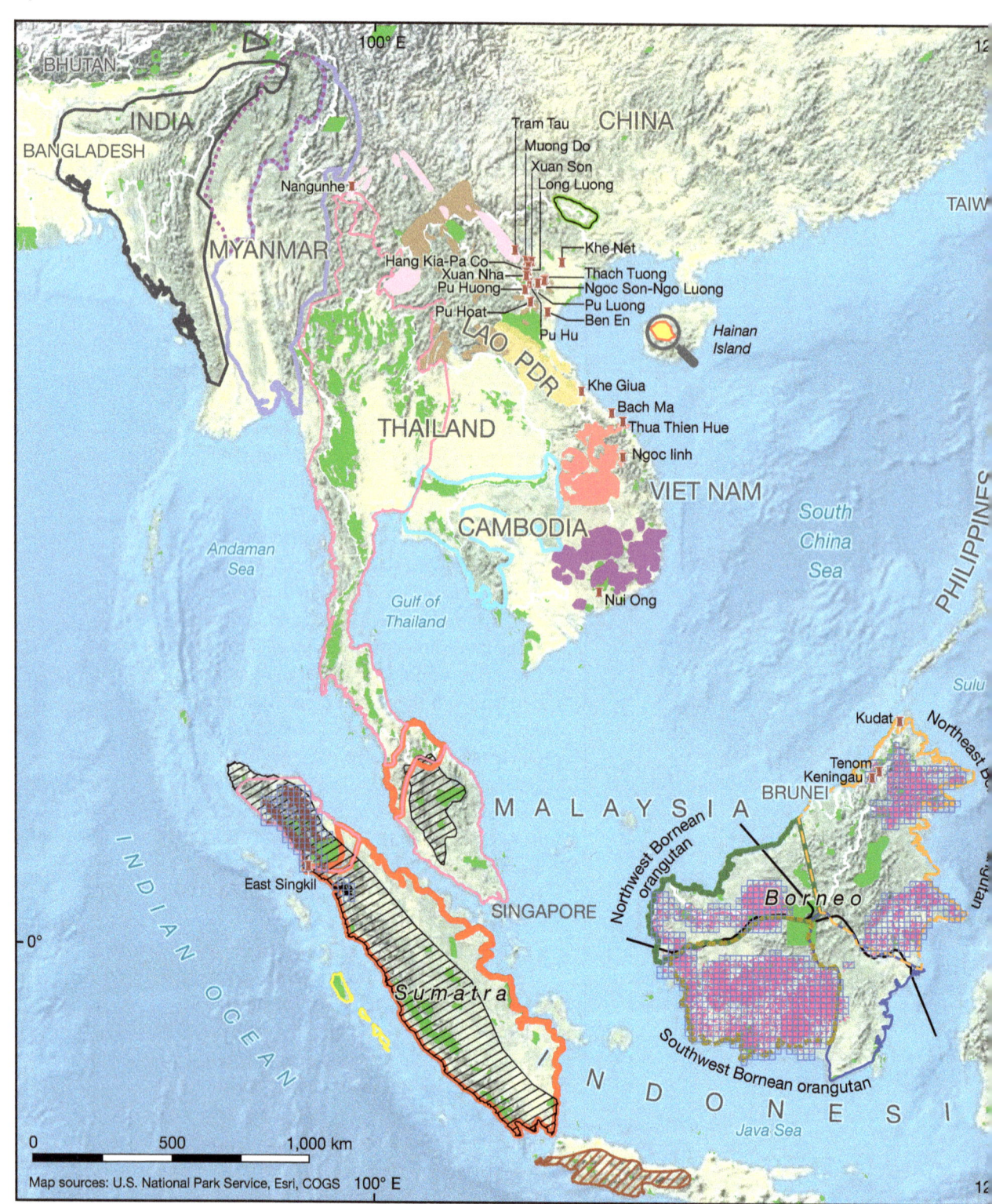

Map sources: U.S. National Park Service, Esri, COGS

Notable exceptions include extra-pair copulations (mating outside of the pair bond), departure from the home territory to take up residence with neighboring individuals and male care of infants (Lappan, 2008; Palombit, 1994; Reichard, 1995). Research also indicates that the more northerly Cao Vit, Hainan and western black crested gibbons commonly form polygynous groups (Fan and Jiang, 2010; Fan *et al.*, 2010; Zhou *et al.*, 2008). There is no consensus regarding the underlying reasons for these variable social and mating structures; they may be natural or a by-product of small population sizes, compression scenarios or sub-optimal habitats. Group demography only changes in the event of a death of one of the adults; there is no regular immigration into or emigration from these social groups. Gibbons in habitat fragments are isolated from other groups and thus their dispersal is compromised, which can threaten the long-term sustainability of these populations. There is insufficient information about dispersal distances for subadult gibbons to determine maximum distances over which gibbons can disperse (perhaps with the assistance of canopy bridges).

Habitat Type and Status

Most apes live in various types of closed-canopy, moist, mixed tropical forest, including lowland, swamp, seasonally inundated, gallery, coastal, submontane, montane and secondary regrowth forests. Some bonobo populations and eastern and western chimpanzees also live in forest–savannah mosaic landscapes. The largest populations of great apes are found below 500 m elevation, in the vast swamp forests of Asia and Africa (Williamson *et al.*, 2013). Bonobos have a discontinuous distribution at 300–700 m above sea level (asl) across undulating terrain in the DRC, south of the Congo River (Fruth *et al.*, 2016; Fruth, Williamson and

Legend

N

- Apes locally extirpated
- IUCN A.P.E.S. Database survey coverage
- Protected areas (IUCN categories I to IV)

Hoolock gibbons [3]
- Eastern hoolock (*Hoolock leuconedys*)
- Western hoolock (*Hoolock hoolock*)
- Gaoligong hoolock (*Hoolock tianxing*)

Hylobates gibbons
- Abbott's gray gibbon (*Hylobates abbotti*)
- Agile gibbon (*Hylobates agilis*)
- Bornean gray gibbon (*Hylobates funereus*)
- Bornean white-bearded gibbon (*Hylobates albibarbis*)
- Kloss's gibbon (*Hylobates klossii*)
- Lar gibbon (*Hylobates lar*)
- Moloch gibbon (*Hylobates moloch*)
- Müller's gibbon (*Hylobates muelleri*)
- Pileated gibbon (*Hylobates pileatus*)

Nomascus gibbons
- Cao Vit gibbon (*Nomascus nasutus*)
- Hainan gibbon (*Nomascus hainanus*)
- Northern white-cheeked crested gibbon (*Nomascus leucogenys*)
- Northern yellow-cheeked crested gibbon (*Nomascus annamensis*)
- Southern white-cheeked crested gibbon (*Nomascus siki*)
- Southern yellow-cheeked crested gibbon (*Nomascus gabriellae*)
- Western black crested gibbon (*Nomascus concolor*)

Symphalangus genus
- Siamang (*Symphalangus syndactylus*)

Pongo genus
- Bornean orangutan subspecies boundaries
- Bornean orangutan (*Pongo pygmaeus*)
- Sumatran orangutan (*Pongo abelii*)
- Tapanuli orangutan (*Pongo tapanuliensis*)

0°

Note: Active collection of population data is ongoing for apes in various locations across their entire range. Updated information is available on the A.P.E.S. Portal (IUCN SSC PSG SGA, n.d.-b).

Richardson, 2013). Eastern chimpanzees and eastern gorillas can range above 2,000 m altitude. On average, Bornean orangutans prefer low-ranging areas between sea level and 500 m asl, although they can sometimes be found in forests above 500 m asl (Nowak *et al.*, 2017; Payne, 1988). In Sumatra, great apes generally live at higher elevations: the Sumatran orangutan is found from sea level to more than 1,500 m asl, and the Tapanuli orangutan is typically found above 500 m asl, but this is due to deforestation in their historic range, which included much lowland forest (Meijaard *et al.*, 2021; Wich *et al.*, 2016).

Most chimpanzees and bonobos inhabit evergreen forests, but some populations also live in deciduous woodland and drier savannah-dominated habitats interspersed with gallery forest. Although many populations are found in protected areas, a great number of chimpanzee communities occur outside. Indeed, the majority of chimpanzees in West Africa—in countries such as Guinea, Liberia and Sierra Leone—are present outside protected areas, and approximately 80% of central chimpanzees and western gorillas live outside of protected areas in Central Africa (Brncic, Amarasekaran and McKenna, 2010; Kormos *et al.*, 2003; Strindberg *et al.*, 2018; Tweh *et al.*, 2015). In Indonesian Borneo, half of the wild orangutan population is surviving outside of protected forests, in areas that are prone to human development and transformation. In Sabah and Sarawak—the Malaysian parts of Borneo—orangutans are mostly found in protected primary and degraded forests (Ancrenaz *et al.*, 2016). The majority of Sumatran orangutans occur in primary protected forests (Nowak *et al.*, 2017; Singleton *et al.*, 2017). Several species of great apes are increasingly found in human-made mosaic landscapes dominated by agriculture and other types of land uses (Spehar *et al.*, 2018). Gibbons range from montane to lowland peat swamp habitats from sea level to elevation of up to 2,700 m (Hu *et al.*, 2018). Many gibbons exist outside protected areas, but as they are wholly arboreal, they are unable to survive in anthropogenic mosaic landscapes (Cheyne *et al.*, 2016; Geissmann *et al.*, 2013; Sarma, Krishna and Kumar, 2015).

> Great apes are adapted to a plant diet, but all taxa consume insects, and some kill and eat small mammals.

Diet

Great apes are adapted to a plant diet, but all taxa consume insects, and some kill and eat small mammals. On the whole, apes may also target cultivars—that is, crops in fields or fruit and trees in orchards and plantations—especially when wild foods are scarce, but also because these may be preferred, since they are highly nutritious and easy to access (Campbell-Smith *et al.*, 2011b; Seiler and Robbins, 2016). Succulent fruits are the main source of nutrition for bonobos, chimpanzees and orangutans, as well as gorillas, except at altitudes where few fleshy fruits are available. Although bonobos are mainly fruit eaters, they consume more terrestrial herbaceous vegetation, as well as aquatic plants, than chimpanzees (Fruth *et al.*, 2016).

Gorillas across their range rely more heavily than any other ape species on herbaceous vegetation, such as the leaves, stems and pith of understory vegetation, as well as leaves from shrubs and trees.[8] Early research suggested that gorillas ate very little fruit, a finding that can be attributed to the fact that initial studies of their dietary patterns were conducted in the Virunga Volcanoes, the only habitat in which gorillas eat almost no fruit as it is virtually unavailable. These conclusions were adjusted once detailed studies were conducted on gorillas living in lower-altitude habitats (Doran-Sheehy *et al.*, 2009; Masi, Cipolletta and Robbins, 2009; Williamson *et al.*, 1990). While gorillas incorporate a notable amount of fruit into their diets when it is available, they are less frugivorous than chimpanzees, consuming vegetative matter even at times of high fruit availability (Head *et al.*, 2011; Morgan and Sanz, 2006; Yamagiwa and Basabose, 2009). Mountain gorillas are primarily terrestrial. Although western gorillas are more arboreal, they still primarily travel on the ground and not through the tree canopy. Wherever gorillas and chimpanzees are sympatric, dietary divisions between the species limit direct competition for food (Head *et al.*, 2011).

Similarly, in Asia, orangutans feed primarily on fruits, but they consume more bark and young leaves when fruit sources become scarce, adapting their diet to what is available in the forest. Sumatran orangutans are more frugivorous than their Bornean relatives. In Borneo, they are known to feed on more than 1,500 plant species from 453 genera and 131 families (Russon *et al.*, 2009). The list continues to grow as more data are collected. For example, a number of tree species that had never been recorded as part as the Sumatran orangutan diet are regularly consumed by the newly described Tapanuli orangutans (Wich *et al.*, 2014b). The resilience of the species and its ability to cope with drastic habitat changes is illustrated by records of species' presence in acacia plantations in East Kalimantan; mosaics of mixed agriculture in Sumatra; agricultural landscapes dominated by oil palm plantations in Borneo; and forests exploited for timber (Ancrenaz *et al.*, 2010, 2015; Campbell-Smith *et al.*, 2011a; Meijaard *et al.*, 2010; Wich *et al.*, 2016). In these disturbed landscapes, Bornean orangutans rely more on young shoots and leaves than they do in primary forest.

Gibbons are reliant on forest ecosystems for food. Their diets are characterized by high levels of fruit intake, dominated by figs and supplemented with young and mature leaves, as well as flowers, although siamangs are more folivorous (Bartlett, 2011; Cheyne, 2008b; Elder, 2009; Palombit, 1997). Reliance on other protein sources, such as insects, bird eggs and small vertebrates, is probably underrepresented in the literature. The diet composition changes with the seasons and habitat type; flowers and young leaves dominate during the dry season in peat swamp forests, while figs dominate in dipterocarp forests (Cheyne, 2010; Fan and Jiang, 2008; Lappan, 2009; Marshall and Leighton, 2006). Gibbons have not been observed to forage on crops (either on plantations or small-scale farms), yet it is possible that they exploit disturbed areas if necessary.

Home and Day Range

Foraging in complex forest environments requires spatial memory and mental mapping. Daily searches for food are generally restricted to a particular location, an area of forest that an individual ape or group knows well. Chimpanzees are capable of memorizing the individual locations of thousands of trees over many years (Normand and Boesch, 2009). The other ape species are likely to possess similar mental capacities. The area used habitually by an individual, group or community of a species is referred to as a home range. Establishing a home range helps apes to secure access to resources within it (Delgado, 2010; Mittermeier, Rylands and Wilson, 2013).

Chimpanzee home ranges vary dramatically in size, ranging from around 10 km² to 90 km² (1,000–9,000 ha), depending on the habitat and resource distribution. Populations in drier and more open habitats have larger home ranges (Herbinger, Boesch and Rothe, 2001; Pruetz and Herzog, 2017). Male chimpanzees are typically highly territorial and patrol the boundaries of their ranges. Parties of males may attack members of neighboring communities and some populations are known for their aggression (Williams *et al.*,

> **"** Foraging in complex forest environments requires spatial memory and mental mapping. **"**

2008). Victors benefit by gaining females or increasing the size of their range. Chimpanzees are generally highly intolerant of neighboring groups and intergroup encounters can result in lethal attacks among males in particular (Mitani, Watts and Amsler, 2010; Watts et al., 2006; Wilson et al., 2014b). The frequency of such encounters can be exacerbated by shifts in home ranges linked to habitat loss, changes in habitat quality and disruptions in the chimpanzees' environment (such as road construction or logging).

Home ranges of bonobos also vary significantly, from 20 km² to 60 km² (2,000–6,000 ha), typically with extensive overlap between the ranges of different communities (Fruth, Williamson and Richardson, 2013). Bonobos do not engage in territorial defense or cooperative patrolling; encounters between members of different communities are more often characterized by tolerance than conflict (Lucchesi et al., 2020).

Eastern gorillas range over areas measuring 6–34 km² (600–3,400 ha), and western gorilla home ranges average 10–20 km² (1,000–2,000 ha)—and potentially up to 50 km² (5,000 ha).[9] Gorillas are not territorial; they have overlapping home ranges that they do not actively defend. There is evidence, however, that they have distinct, exclusive core areas (the parts used the most by a group), suggesting that groups do partition their habitat (Seiler et al., 2017). Gorillas preferentially use areas of their home range that have higher food availability, which means their movement patterns may vary seasonally according to fruit availability (Seiler et al., 2018; Seiler and Robbins, 2020). As the density of gorillas increases, the degree of home range overlap can increase dramatically, as can the frequency of intergroup encounters, which may lead to increased fighting, injuries and mortality (Caillaud et al., 2014).

A male orangutan's range encompasses several female ranges. As high-status flanged males are able to monopolize both food and females to a degree, they may temporarily reside in a relatively small area—4–8 km² (400–800 ha) for Bornean males—even though the actual size of their home range could be much larger than 10 km² (1,000 ha). Orangutan home range overlap is usually extensive, but flanged male orangutans establish personal space by emitting long calls (Spillmann et al., 2017). Unflanged adult males have no strictly defined home range and move over large distances (Utami-Atmoko et al., 2009). Bornean flanged adult males and adult females move an average of 200 m each day; unflanged adult males usually cover twice that distance. Sumatran orangutans move farther, but still less than 1 km each day on average (Singleton et al., 2009). Female orangutans have longer daily travel distances and larger home rages when they are young and nulliparous—during their "exploration phase"—than during adulthood (Ashbury et al., 2020). As long as distance between males is maintained, physical conflicts are rare; however, close encounters between adult males can trigger aggressive displays that sometimes lead to fights and possibly death (Knott, 1998). Females show a life-long site fidelity and live in overlapping and rather stable home ranges, but they may defend the core area of their range against female intruders, especially unrelated individuals (Ashbury et al., 2020; Knott et al., 2008).

Territorial apes whose habitats are destroyed encounter great difficulties establishing a new territory nearby, where other animals are already established. Indeed, in these situations the animals whose territory has been destroyed usually slowly die off.

African apes are semi-terrestrial and often rest on the ground during the daytime. In contrast, orangutans are almost exclusively arboreal, although in Borneo, they can walk on the ground for considerable distances in all types of natural and human-

> Territorial apes whose habitats are destroyed encounter great difficulties establishing a new territory nearby, where other animals are already established.

made habitats (Ancrenaz *et al.*, 2014; Loken, Boer and Kasyanto, 2015; Loken, Spehar and Rayadin, 2013). Consequently, they are able to cross open, artificial infrastructure to a certain extent. In Sabah, for example, orangutans have been seen crossing sealed and dust roads as long as the traffic is not too heavy (Ancrenaz *et al.*, 2021). High terrestriality in orangutans increases sanitary concerns and the risk of contracting diseases to which they are not usually exposed in the tree canopy. At this stage, there is a dearth of information about such sanitary and health risks.

The semi-terrestrial African apes range considerably longer distances and the most frugivorous roam several kilometers each day: mountain gorillas travel about 500 m–1 km per day; bonobos and western lowland gorillas average 2 km but sometimes reach 5–6 km; and chimpanzees travel 2–3 km, although they occasionally venture out on 10-km excursions. Savannah-dwelling chimpanzees generally range farther daily than their forest-dwelling counterparts. The distance travelled by western gorillas declines with increasing availability of understory vegetation, varying between approximately 500 m and 3 km per day (Seiler and Robbins, 2020). As a result of their dietary patterns, both eastern and western gorillas are restricted to moist forest habitats (at altitudes ranging from sea level to more than 3,000 m) and are not found in forest–savannah mosaics or gallery forests inhabited by chimpanzees and bonobos (Robbins, 2011).

Hylobates gibbon territories average 0.42 km² (42 ha) in size, but there is considerable variation. The more northerly *Nomascus* taxa can maintain larger territories—from about 0.13 to 1.3 km² (13–130 ha)—possibly in line with lower resource abundance at certain times of year in these more seasonal forests (Fan *et al.*, 2010). Few seasonal forests have increased resource abundance, yet gibbon density and territory size

may not be directly correlated with these factors (Bryant *et al.*, 2015; Hamard, Cheyne and Nijman, 2010; Zhang *et al.*, 2014).

Nesting

Most apes do not only feed in trees, but also rest, socialize and sleep in them, although gorillas are largely terrestrial. Being large-brained, highly intelligent mammals, they need long periods of sleep. All weaned great apes build nests or beds in which they spend the night. Orangutans nest only in trees; bonobos and chimpanzees may also build daytime nests in trees or on the ground to rest; and gorillas nest primarily on the ground (Prasetyo *et al.*, 2009).

Tree nests are usually constructed 10–20 m above ground. Variation in nesting height is influenced by environmental variables such as rainfall, temperature, habitat structure, availability of material, predator presence and demographic parameters such as the sex or the age of the individual, as well as social factors such as transferred habits (Fruth, Tagg and Stewart, 2018). Great apes reuse nests on rare occasions; the frequency of reuse depends largely on the availability of sleeping site locations and material for construction (Ancrenaz, Calaque and Lackman-Ancrenaz, 2004; Fruth, Tagg and Stewart, 2018). Bonobos prefer to nest in areas with abundant food, while sleeping site association with fruiting trees is more variable in chimpanzees (Fruth, Tagg and Stewart, 2018; Serckx *et al.*, 2014). Both chimpanzees and bonobos show tree species preferences when it comes to nesting (Fruth and Hohmann, 1996).

Reproduction

Male great apes reach sexual maturity between 8 and 18 years of age, while chimpanzees attain adulthood at 8–15 years, bonobos at 10, eastern gorillas around 12–16

> Most apes do not only feed in trees, but also rest, socialize and sleep in them, although gorillas are largely terrestrial.

and western gorillas at 18 (Williamson *et al.*, 2013). Orangutan males mature between the ages of 8 and 16 years, but they may not develop flanges for another 20 years (Utami-Atmoko *et al.*, 2009). Female great apes become reproductively active between the ages of 6 and 12 years: gorillas at 6–7 years, chimpanzees at 7–8, bonobos at 9–12 and orangutans at 10–11. They tend to give birth to their first offspring between the ages of 8 and 16: gorillas at 10 (with an average range of 8–14 years), chimpanzees at 13.5 years (with a mean of 9.5–15.4 years at different sites), bonobos at 13–15 years and orangutans at 15–16 years (with a median around 14.5 years) (van Noordwijk *et al.*, 2018; Williamson *et al.*, 2013).

Pregnancy length in gorillas and orangutans is about the same as for humans; it is slightly shorter in chimpanzees and bonobos, at 7.5–8 months (Peacock and Rogers, 1959; Stevens, 2020; van Noordwijk *et al.*, 2018). Apes usually give birth to one infant at a time, although twin births do occur (Goossens *et al.*, 2011). Births are not seasonal; however, conception requires females to be in good health. Chimpanzees and bonobos are more likely to ovulate when fruit is abundant, so in some populations there are seasonal peaks in the number of conceiving females, with contingent peaks in birth rate during particular months (Anderson, Nordheim and Boesch, 2006; Emery Thompson and Wrangham, 2008). Bornean orangutans living in highly seasonal dipterocarp forests are most likely to conceive during mast fruiting events, when fatty seeds are plentiful (Knott, 2005). Sumatran orangutans do not face such severe constraints (Marshall *et al.*, 2009). Meanwhile, gorillas are less dependent on seasonal foods and show no seasonality in their reproduction.

Gibbon females have their first offspring at around 9 years of age. Data from captivity suggest that gibbons become sexually mature as early as 5.5 years of age (Geissmann, 1991). Interbirth intervals are in the range of 2–4 years, and gestation lasts about seven months (Bartlett, 2011). Captive individuals have lived upwards of 40 years; gibbon longevity in the wild is unknown but thought to be considerably shorter. Since gibbons mature relatively late and have long interbirth intervals, their reproductive lifetime may be only 10–20 years (Palombit, 1992). Population replacement in gibbons is therefore relatively slow.

All apes have slow reproductive rates; mothers invest considerable time in a single offspring and infants are slow to develop and mature. Infants sleep with their mothers until they are weaned (4–5 years in African apes; 5–6 years in Bornean orangutans; 7 years in Sumatran orangutans) or a sibling is born. Weaning marks the end of infancy for African apes around the age of 3–6 years, but orangutan infants remain dependent on their mothers until they reach 7–9 years of age (Knott, 2001; van Noordwijk *et al.*, 2009; Williamson *et al.*, 2013). Generally, females cannot become pregnant while an infant is nursing because suckling inhibits the reproductive cycle (Stewart, 1988; van Noordwijk *et al.*, 2013). Consequently, births are widely spaced, occurring on average every 4–7 years in African apes and every 7–8 years in orangutans (Emery Thompson *et al.*, 2007; Robbins *et al.*, 2009; Stoinski *et al.*, 2013; van Noordwijk *et al.*, 2018). Interbirth intervals are longer for orangutans than for African apes and gibbons, indicating higher reproductive investment from the mother and conservative growth, development and reproduction related to the hypometabolism detected in this species (Pontzer *et al.*, 2016; van Noordwijk *et al.*, 2018).

Interbirth intervals can be shortened if a member of the same species—typically an unrelated adult male—kills unweaned

> ❝ All apes have slow reproductive rates; mothers invest considerable time in a single offspring and infants are slow to develop and mature. ❞

offspring (Harcourt and Greenberg, 2001; Hrdy, 1979). While infanticide has not been observed in bonobos and orangutans, it can occur if a female great ape with an infant transfers to a different group. A male in the new group may kill the offspring, triggering an early resumption of her reproductive cycle (Knott *et al.*, 2019; Watts, 1989).

Long-term research on mountain gorillas and chimpanzees has allowed female lifetime reproductive success to be evaluated. The mean birth rate is 0.2–0.3 births per adult female per year, or one birth for every adult female every 3.3–5.0 years. The birth rate is lower and infant mortality is higher for western gorillas than mountain gorillas (Robbins *et al.*, 2022). Mountain gorilla females produce an average of 3.6 offspring during their lifetimes; similarly, chimpanzees produce 1.0–4.3 offspring who survive into adulthood (Emery Thompson and Wrangham, 2013; Robbins *et al.*, 2011a).

Key points to be noted are that: 1) documenting the biology of long-lived species takes decades of research due to their slow rates of reproduction, and 2) ape populations that have declined in numbers are likely to take several generations to recover (generation time among apes is 15–25 years) (IUCN, 2022). These factors make apes far more vulnerable than smaller, faster-breeding species. Orangutans have the slowest life history of any mammal, with later age at first reproduction, longer interbirth intervals and longer generation times than African apes; as a result, they are the most susceptible to loss (van Noordwijk *et al.*, 2018; Wich *et al.*, 2009a, 2009b).

Acknowledgments

Principal authors: Annette Lanjouw,[10] Helga Rainer[11] and Alison White[12]

Socioecology section: Marc Ancrenaz,[13] Susan M. Cheyne,[14] Tatyana Humle,[15] Benjamin M. Rawson,[16] Martha M. Robbins[17] and Elizabeth A. Williamson[18]

Endnotes

1   Heinicke *et al.* (2019); Humle *et al.* (2016a); Maisels *et al.* (2016); Oates *et al.* (2016); Plumptre *et al.* (2010, 2016a); Strindberg *et al.* (2018).

2   Bergl *et al.* (2016); Hickey *et al.* (2020); IUCN (2022); Maisels, Bergl and Williamson (2018); Plumptre *et al.* (2016b).

3   Davis *et al.* (2013); Gaveau *et al.* (2014); Sherman *et al.* (2020); Singleton *et al.* (2017); Wich *et al.* (2012a, 2019).

4   For the Cao Vit and Hainan gibbons, the updated population numbers are taken from unpublished data seen by the authors.

5   The information provided in this section is largely drawn from Emery Thompson and Wrangham (2013), Mittermeier, Rylands and Wilson (2013), Reinartz, Ingmanson and Vervaecke (2013), Robbins (2011), Robbins and Robbins (2018), Wich *et al.* (2009a), Williamson and Butynski (2013a, 2013b) and Williamson *et al.* (2013).

6   The Arcus Foundation commissioned the ape distribution maps (Figures AO1 and AO2) for *State of the Apes*, so as to provide accurate and up-to-date illustrations of range data. This volume also features maps created by contributors who used ape range data from different sources. As a consequence, the maps may not all align exactly.

7   See Endnote 4.

8   Doran-Sheehy *et al.* (2009); Ganas *et al.* (2004); Masi, Cipolletta and Robbins (2009); Robbins, Ortmann and Seiler (2022); Wright *et al.* (2015); Yamagiwa and Basabose (2009).

9   Caillaud *et al.* (2014); Head *et al.* (2013); Robbins (2011); Seiler *et al.* (2018); Williamson and Butynski (2013a, 2013b).

10   Arcus Foundation (www.arcusfoundation.org).

11   At the time of writing: Arcus Foundation (www.arcusfoundation.org).

12   Arcus Foundation (www.arcusfoundation.org).

13   HUTAN–Kinabatangan Orang-utan Conservation Programme (www.hutan.org.my).

14   Borneo Nature Foundation (www.borneonaturefoundation.org).

15   At the time of writing: University of Kent (www.kent.ac.uk/sac). Now: Re:wild (www.rewild.org/).

16   World Wide Fund for Nature in Asia Pacific (asiapacific.panda.org).

17   Max Planck Institute for Evolutionary Anthropology (www.eva.mpg.de).

18   University of Stirling (www.stir.ac.uk/about/faculties/natural-sciences).

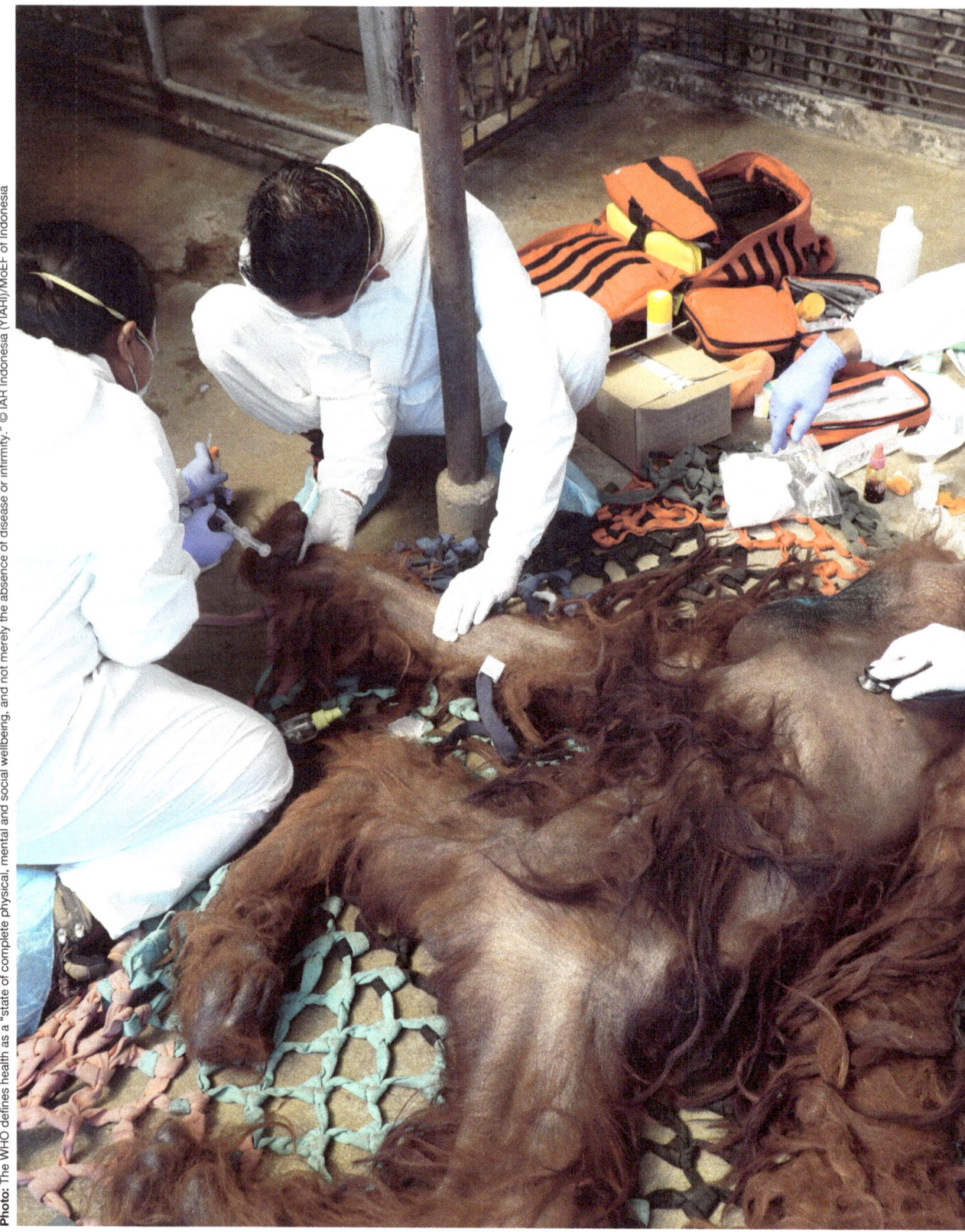

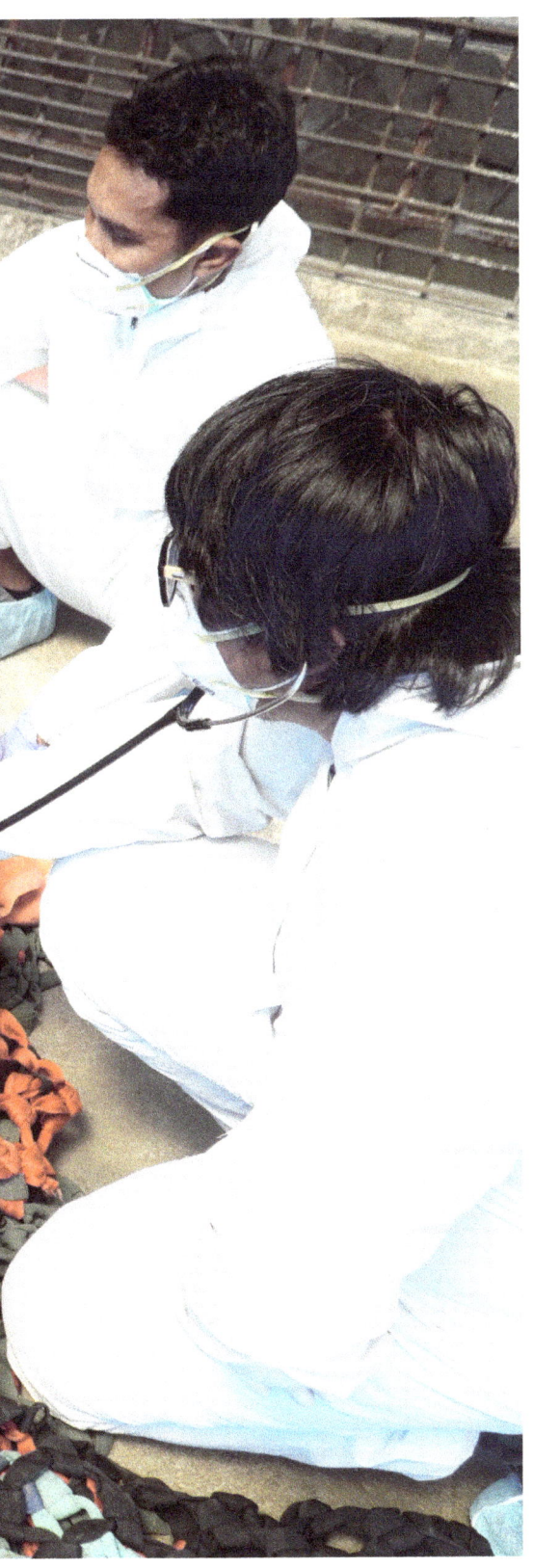

INTRODUCTION

Section 1:
Disease, Health and
Ape Conservation

The *State of the Apes* Series[1]

Commissioned by the Arcus Foundation, the *State of the Apes* series aims to raise awareness about the status of apes around the world and the impacts of human activities on apes and ape habitat. Apes are closely related to humans and they – and their habitats – are vulnerable to many anthropogenic threats. To build an understanding of both the severity and extent of these threats, as well as potential means for preventing and mitigating them, the volumes in this series feature original research and analysis, case studies and best practice from leading scholars and practitioners active in various sectors, including conservation, industry and academia.

This Arcus Foundation initiative is designed to inform debate, practice and policy in ways that reconcile ape conservation and welfare with socioeconomic development. Robust statistics on the status and welfare of apes are derived from the Ape Populations, Environments and Surveys (A.P.E.S.) database, available at iucngreat apes.org/apes-database.

This volume—the final one in the *State of the Apes* book series—examines ape conservation and welfare through the lens of disease and health. It explores factors such as the ethics of intervening in and managing ape health; the impact of research and tourism activities on apes; the One Health approach; and disaster management and the protection of apes. It shows how the welfare of apes is interrelated with that of the people who share their habitats, while also demonstrating the benefits of integrating ape conservation in health, social and economic development (including sectors such as extractive industries, industrial agriculture, infrastructure development) and regulatory policy and practice at all levels, from the local to the international. Moreover, the analysis and findings presented throughout the chapters are intended to help conservation specialists enhance their own practice.

The volume's focus is on all non-human ape species: bonobos, chimpanzees, gorillas, orangutans and gibbons. The analysis concentrates on apes in range countries, which encompass much of the tropical belt of Africa and Southeast Asia, although it also reviews the welfare of captive apes in facilities worldwide. To ensure a holistic overview of current thinking and practice regarding health and disease in relation to ape survival and welfare, the Arcus Foundation commissioned contributions from a range of experts working in academia, conservation, the private sector, sanctuaries, health and veterinary medicine.

Disease, Health and Ape Conservation

The World Health Organization defines health as a "state of complete physical, mental and social well-being, and not merely the absence of disease or infirmity" (WHO, 2020a). While the determinants of health are complex and interconnected for every individual species, comparisons of health needs are possible across taxa that are anatomically and physiologically similar, such as humans and apes. Species with close genetic relationships are susceptible to many of the same infectious and non-infectious diseases, and disease-causing organisms (pathogens) can move between them.

Managing the spread of disease within or between individuals, groups and species requires an understanding of the nature

of the disease, its ability to spread and its potential impacts. The COVID-19 pandemic has demonstrated the challenges of managing disease spread and impact at the global scale, including at the political, social and structural levels.

As the Anthropocene unfolds, the impact of humanity on all ecosystems on the planet is becoming more visible and better understood. Climate and ecological systems are collapsing and one million species are threatened with extinction (IPBES, 2019;

IPCC, 2023). Ecosystems are struggling to cope with continued and cumulative stresses. All the while, deforestation, encroachment into natural habitat and other human activities are driving an increase in the frequency of interactions between people and various forms of wildlife, including viruses, parasites, bacteria and fungi (Nellemann and Newton, 2002). The consequence is a heightened risk of disease transmission, with serious implications for conservation, biodiversity protection and human health (Balasubramaniam

Photo: Historically, wild apes have shared their natural environment with various wildlife species and their associated pathogens, but had limited contact with humans until the current age. Now, many wild apes live in habitats that are subject to different degrees of anthropogenic encroachment. Chimpanzees crossing a road in Bulindi, Uganda. © Jacqueline Rohen

et al., 2022; Conover and Conover, 2022; Marrana, 2022; Muehlenbein, 2013).

Like humans, many apes are social beings who live in groups and interact closely with each other. The health of any group is closely tied to the health of individuals within that group, as well as to the size of the group. The smaller a population, the more vulnerable it is to disease (Prado-Martinez *et al.*, 2013). Moreover, species-specific behaviors and social systems can influence the risk and the impact of diseases. The settings in which apes live—be they natural forested habitat, sanctuaries or rehabilitation centers in their range countries, or zoos and sanctuaries elsewhere in the world—also come with unique sets of health-related risks and challenges (see Annex II for a summary table of ape health and disease issues).

Viewed through a conservation lens, disease represents a major threat to the survival of apes and other species that are threatened with extinction (Gilardi *et al.*, 2015). Indeed, infectious disease is often listed among the principal threats to ape conservation, along with habitat loss and hunting, which can also expose apes to health threats. The burning of habitats, for example, can pose non-infectious health risks, while the degradation of landscapes can change species composition and behavior, potentially altering primates' exposure to infectious agents (Erb *et al.*, 2018; Herrera and Nunn, 2019). Similarly, encroachment of hunters into wildlife habitats heightens not only the threat that apes may be injured or killed, but also the likelihood of exposure to human pathogens, to which apes can be highly susceptible (Köndgen *et al.*, 2008; Laurance *et al.*, 2006; see Figure 1.2). Diseases that can be transmitted from animals to humans and from humans to animals are called "zoonoses" (Hubálek, 2003).

Historically, wild apes have shared their natural environment with various wildlife species and their associated pathogens, yet

they probably had limited contact with humans until the current age. Nowadays, many wild apes live in habitats that are subject to different degrees of anthropogenic encroachment; in fact, more than 70% of wild ape populations are found outside protected areas (see Chapter 7). Within national parks or similarly protected areas, contact with humans can be relatively controlled, such as when people are involved in research, tourism or park protection and hunting. Outside of national parks, apes sometimes inhabit forests that are heavily used by humans, be it for hunting, fishing or the harvesting of other forest resources. Apes also enter human habitats, for example to forage crops. Under these circumstances, the likelihood of pathogen transmission rises, and apes are more likely to sustain human-caused injuries, including from snares.

In zoos, apes live in artificial conditions and in close contact with people, which can significantly increase the likelihood of infection with human pathogens (see Figure 1.2). Strategies to mitigate this risk include the use of masks, gloves and other personal protective equipment, as well as disinfectant footbaths (Kalter, 1989). While the spillover of pathogens between zoo animals may be limited, captive apes can come into contact with rodents and pathogens from other enclosures or if they are kept in mixed-species enclosures (Hardgrove et al., 2021). Confinement in enclosures and associated stress can increase the burden of certain ape-specific pathogens, such as gastrointestinal parasites (Toft, 1986). Many of the captive apes in sanctuaries, particularly in the United States, have come from laboratories where they were deliberately infected with various pathogens for research (Knight, 2008; Morimura, Idani and Matsuzawa, 2011). Non-infectious diseases are also common in captivity. Some are the direct result of stress, diet or other factors linked to captivity, while ape geriatric

diseases are linked to relatively long life-spans, which are rarely observed in the wild (Lowenstine, McManamon and Terio, 2016).

Apes in range country sanctuaries and rehabilitation centers occupy an intermediate position between captive and wild apes, although this volume generally groups them together with captive populations. Most of these "semi-captive" animals were born in the wild, and some are eventually released back into the wild, which comes with the risk of pathogen carryover to wild individuals and groups (Köndgen et al., 2017; Sherman et al., 2021). Sanctuaries and zoos face similar health threats, such as human diseases and geriatric disorders; however, spillover of wildlife pathogens can and does occur, as evidenced by the transmission of the monkeypox virus to sanctuaries (see Figure 1.2 and Chapters 1, 7 and 8).

Regardless of the setting, apes encounter myriad (micro)organisms that can impact their health in various ways. These include bacteria, viruses, viruses of bacteria, fungi, protozoa and a variety of macroparasites (Gogarten et al., 2018, 2021; Nishida and Ochman, 2019). Those that cause visible signs of disease were historically the most studied, but they are vastly outnumbered by commensal (micro)organisms. Scientists have characterized only a fraction of the latter and have examined an even smaller proportion with respect to their effects on ape hosts (Bueno de Mesquita et al., 2021; Gogarten et al., 2021).

With the advent of next-generation sequencing technologies, the human gut microbiome received intense research interest. Studies suggest that this gut ecosystem has various impacts on health, including in terms of digestion, immunity and psychological wellbeing (Hooper, Littman and Macpherson, 2012; Tremaroli and Bäckhed, 2012; Winter et al., 2018). Yet, even for humans, the understanding of how the microbiome interacts with host health is

> **Many wild apes live in habitats that are subject to different degrees of anthropogenic encroachment; in fact, more than 70% of wild ape populations are found outside protected areas.**

rudimentary and often relies on extrapolation from experimentation with model organisms (iHMP Research Network Consortium, 2019). While it seems likely that the gut microbiome of apes has a comparable importance for their health and wellbeing, associated research is still in its infancy (Björk *et al.*, 2019). Similarly, researchers have documented clinical signs of gastrointestinal parasites in captive apes, yet their impact on wild ape health remains largely unknown (Gogarten *et al.*, 2020).

This volume of *State of the Apes* examines what steps can be taken to mitigate disease risks to apes, improve the resilience of their populations and restore their role in ecosystem function. While highlighting the continuum between welfare and conservation, it reviews how human-caused changes to ape habitats, food availability, sociality and behavior, along with proximity to people, affect ape health and wellbeing.

The volume identifies pathogens, yet it places more emphasis on the diseases they cause. For example, Chapter 1 mentions that severe acute respiratory syndrome coronavirus 2 (SARS-CoV-2) causes COVID-19, but the book focuses more generally on the health impacts of the disease and options for mitigating transmission risks. COVID-19 is referenced frequently throughout the volume, not only in view of the far-reaching consequences of the pandemic, but also because associated concepts and guidance apply to all infectious diseases. The COVID-19 pandemic, caused by SARS-CoV-2, has lent a sense of urgency to much of this research and debate.

Thematic Chapters

The thematic section of this edition of *State of the Apes* comprises six chapters, which are briefly described below. The introduction to Section 2 presents Chapters 7 and 8.

Chapter 1 provides an overview of the factors that impact the health of wild and captive apes, providing evidence of known pathogens and highlighting the areas that require greater research and investigation.

Chapter 2 contextualizes the health of an individual within that of the population and the larger ecosystem, linking human, animal and environmental health. It presents the One Health approach as a way of considering the interconnections between humans, domestic animals and wildlife and the social and ecological environments that they share. The chapter provides a series of case studies that demonstrate how disease can be prevented through system-level approaches.

Chapter 3 examines two areas of human activity that can have a particular impact on ape health: tourism and research. Both of these activities bring humans in close and repeated contact with apes. Tourists and researchers who come from different geographical regions are especially likely to increase apes' risk of naive contacts with pathogens. These activities can significantly increase the likelihood and potential severity of health consequences for apes, whether in captivity or in free-ranging settings. The chapter examines the potential for disease spillover in both directions—from apes to humans and vice versa.

Chapter 4 considers the main strategies for managing ape health and the linkages with public health. The authors review decision-making processes designed to help veterinarians assess potential limitations and benefits of proceeding with health interventions.

Chapter 5 explores some of the ethical considerations that are relevant to ape health care and protection, including in regions of the world where medical services are often severely limited for human populations. In examining challenges inherent in balancing the value of the individual against that of the population, the chapter points to

> 66 Non-infectious diseases are common in captivity. Some are the direct result of stress, diet or other factors linked to captivity. 99

compassionate conservation as a method for resolving such tensions. These concepts are complex and context-specific and have value in raising awareness about the need for explicit consideration of the ethics of ape protection and care.

Chapter 6 analyzes disease outbreaks and other health-related crises from the perspective of disaster management. It covers risk mapping, risk mitigation and capacity in disaster management, prevention, preparedness, response and recovery.

Acknowledgments

Principal authors: Annette Lanjouw,[2] Katy Scholfield[3] and Alison White[4]

Contributors: Fabian H. Leendertz,[5] Ariane Düx,[6] Jan F. Gogarten,[7] Livia V. Patrono,[8] Kamilla Pleh[9] and Joost Philippa[10]

Endnotes

1 Adapted from the Introduction to the first volume, *State of the Apes: Extractive Industries and Ape Conservation* (Arcus Foundation, 2014, pp. 2–5).

2 Arcus Foundation (www.arcusfoundation.org).

3 Arcus Foundation (www.arcusfoundation.org).

4 Arcus Foundation (www.arcusfoundation.org).

5 Helmholtz Institute for One Health, Helmholtz-Centre for Infectious Research (www.helmholtz-hzi.de/en) and Robert Koch Institute (www.rki.de).

6 Helmholtz Institute for One Health, Helmholtz-Centre for Infectious Research (www.helmholtz-hzi.de/en) and Robert Koch Institute (www.rki.de).

7 Helmholtz Institute for One Health, Helmholtz-Centre for Infectious Research (www.helmholtz-hzi.de/en), Robert Koch Institute (www.rki.de) and University of Greifswald (zoologie.uni-greifswald.de/en/organization/departments/applied-zoology-and-nature-conservation).

8 Helmholtz Institute for One Health, Helmholtz-Centre for Infectious Research (www.helmholtz-hzi.de/en) and Robert Koch Institute (www.rki.de).

9 Helmholtz Institute for One Health, Helmholtz-Centre for Infectious Research (www.helmholtz-hzi.de/en) and Robert Koch Institute (www.rki.de).

10 At the time of writing: International Animal Rescue (www.internationalanimalrescue.org).

CHAPTER 1

Review of Ape Disease and Health

Introduction

Ape health is understudied when compared to human health, but it is by no means less complex (see Annex III). Different wild ape species live in a variety of social systems, ranging from semi-solitary orangutans and pair-bonded gibbons to the more gregarious African great apes (see the Apes Overview). The formation of social groups can provide a number of health benefits, from mental and social health to physical health, including via group coordination, increased protection from predators, and enhanced access to grooming partners for the removal of ecto-parasites (Akinyi *et al.*, 2013; Janson and Goldsmith, 1995; Samuni *et al.*, 2018; Wittig *et al.*, 2016). Sociality can also create costs, especially by heightening the risk of exposure to communicable diseases (see Box 1.1).

Simulation studies comparing disease spread among chimpanzee and orangutan social networks suggest that chimpanzees might generally be more susceptible to the spread of a variety of infectious diseases than orangutans. Accumulated evidence from wild populations, coupled with modeling results, broadly supports the idea that solitary orangutans are less susceptible to communicable diseases, such as ebolaviruses and respiratory diseases, although no systematic comparison of pathogen richness across ape species has yet been undertaken (Carne *et al.*, 2014). Anecdotal reports of mortality associated with disease transmission to orangutans in sanctuaries and zoos are difficult to substantiate and disseminate in the research and ape health practitioner communities. By publishing and documenting cases, practitioners who work in orangutan health can help to fill the knowledge gap concerning the relationship between ape sociality and health.

While sociality can affect disease spread in ape populations, species-specific behaviors can influence exposure to disease. For example, chimpanzees and bonobos (*Pan paniscus*) are known to hunt other mammals, including primates, which can expose them to their prey's pathogens (Leendertz *et al.*, 2011; Samuni, Wegdell and Surbeck, 2020; see Apes Overview). Similarly, apes' settings and degrees of habituation can influence their exposure to pathogens from humans (Grützmacher *et al.*, 2016; Köndgen *et al.*, 2008).

Diseases that can be spread from animals to humans and vice versa are called "zoonoses" (Hubálek, 2003). The past decades have witnessed a rise in emerging zoonotic diseases, the majority of which originate in wildlife (Jones *et al.*, 2008). Disease transmission between species is known as a "spillover event" (Ellwanger and Chies, 2021; see Figure 1.1 and Annex III). As humans' closest living relatives, apes share many of the same genetic, anatomical and physiological features; accordingly, humans and apes tend to be susceptible to similar diseases (Calvignac-Spencer *et al.*, 2021). From a public health perspective, apes are therefore seen as sources or sentinels for human diseases (Calvignac-Spencer *et al.*, 2012). Major human pathogens that originated in apes include the malaria-causing *Plasmodium falciparum* from gorillas and the pandemic HIV-1 group M from chimpanzees (Liu *et al.*, 2010; Sharp and Hahn, 2011). Meanwhile, chimpanzees and gorillas can suffer from diseases that also affect humans, such as yaws and leprosy (Hockings *et al.*, 2021; Mubemba *et al.*, 2020). They have also served as amplifying hosts for ebolaviruses, which has led to several outbreaks in humans (Leroy *et al.*, 2004).

A comprehensive review of the diverse organisms associated with apes would easily fill an entire book and go beyond the scope of this chapter. Rather, the focus here is on health-related issues with available data and a known or likely impact on the conservation or welfare of wild and captive apes. While all apes share certain health issues, those of semi-captive and captive apes largely overlap; the chapter discusses them together to highlight similar challenges as well as management and treatment options. A separate section examines the health of wild apes.

Stakeholders in the fields of public health and conservation usually consider health at the population scale. In contrast, veterinary clinicians, who work mainly in zoos and sanctuaries, focus on individual apes and the groups in their care, with the goal of maximizing animal welfare. This chapter examines aspects of these perspectives for apes in captive and wild settings, based on available data.

Annex III features evidence of confirmed transfers of human pathogens to apes in the wild, excluding anecdotal information. The shortness of this table highlights the critical

FIGURE 1.1

Examples of Pathogen Spillovers between Wildlife and Humans

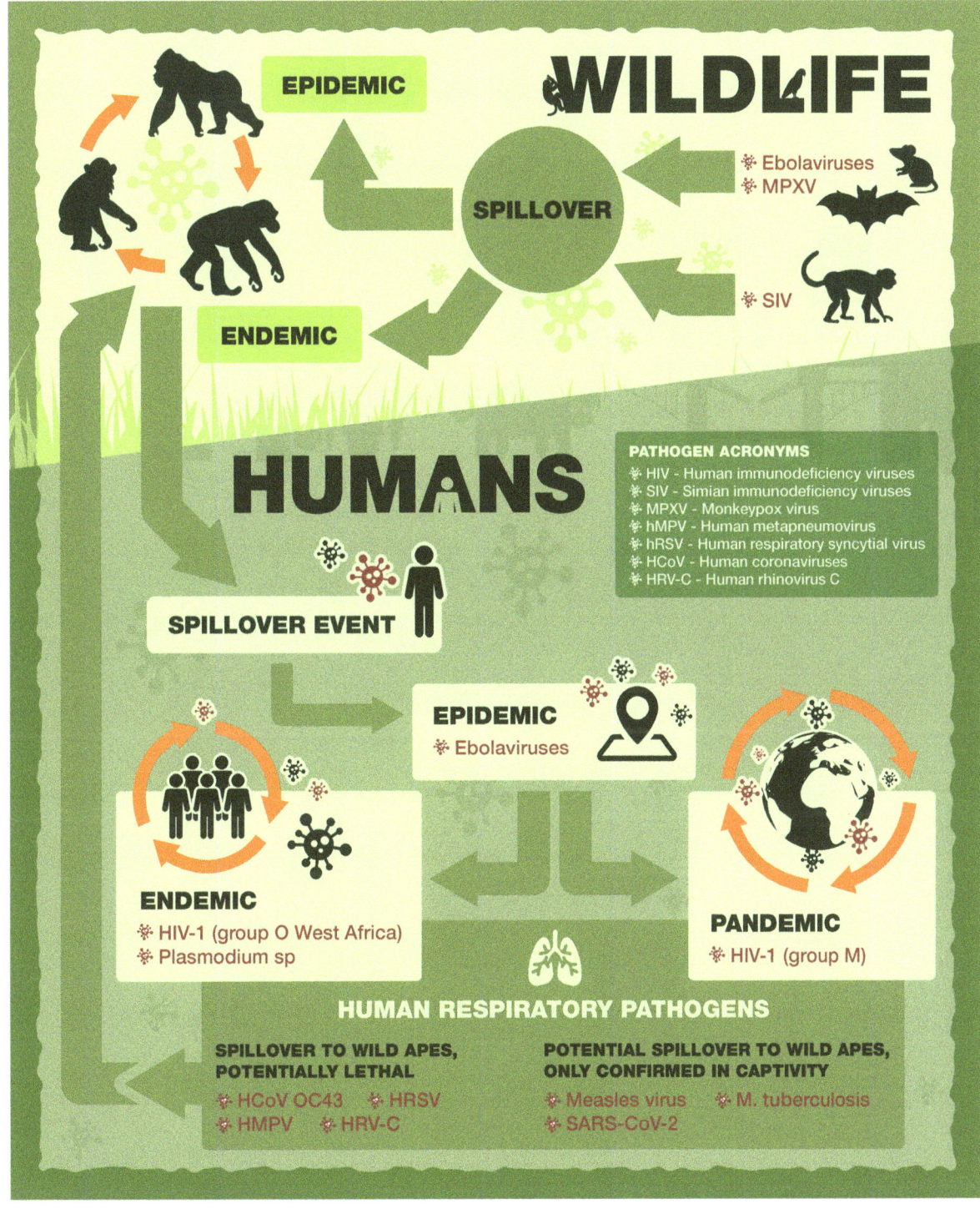

Notes: Arrowheads show directionality of spillover. Potentially lethal pathogens appear in red. Suspected pathogens, which have only been confirmed in captivity, appear in italics. Details and references are available in the main text.

FIGURE 1.2

Properties of Different Settings and Implications for Disease Transmission

Note: Details and sources provided in the main text.

need for those working with apes in situ and ex situ to fill the many remaining data gaps on ape diseases. Box 1.2 explores measures, protocols and procedures for the prevention of infectious diseases, as does Chapter 4. Box 1.3 provides an overview of methods used for sample collection from wild and captive apes, which can be used to study different aspects of their health. More information can be found in the relevant literature and through consultation with experts, who can guide the design of prevention strategies, health-monitoring protocols and related systems.

Key findings of the chapter include:

- Apes and humans are susceptible to similar diseases, which allows for spillovers in both directions (see Figure 1.1).

- Health-related risks, challenges and management options vary across wild and captive apes (see Figure 1.2).

- The transmission of respiratory pathogens from humans to apes in both wild and captive settings is common and can cause high morbidity and mortality.

- Infectious diseases are a major conservation threat for wild apes, especially in gregarious species.

- Non-infectious diseases play an important role in the health of captive apes.

- Further research is required to fill knowledge gaps, including with respect to specific diseases, such as myocardial fibrosis, and regarding the relationship between sociality and health.

Wild Apes

Infectious Diseases with a Plausible Effect on Fitness

This section considers several pathogens that have been shown to impact wild ape health or cause mortality. The extreme scarcity of relevant data and observations, however, precludes certainty regarding the extent to which these pathogens affect the survival of ape populations and the frequency with which they cause mortality. More data on the prevalence of these pathogens across ape populations are needed to build a better understanding of their conservation implications.

Monkeypox

Monkeypox is a viral disease that causes smallpox-like symptoms in humans (Bunge et al., 2022). The causative agent of this zoonotic disease was first discovered in a Danish primate laboratory and was therefore named monkeypox virus (MPXV) (von Magnus et al., 1959). Just like great apes and humans, however, monkeys are accidental hosts of the virus, which is thought to have a rodent reservoir (Di Giulio and Eckburg, 2004). MPXV is endemic in West and Central African tropical forested regions, but a recent surge in monkeypox cases across the globe is of grave concern (Zumla et al., 2022).

Today, MPXV is the most relevant *Orthopoxvirus* (family *Poxviridae*) since the eradication of smallpox in 1980 (Di Giulio and Eckburg, 2004; Shchelkunov et al., 2001). The recent marked increase in human monkeypox cases has been attributed to several concurring factors: human encroachment into wildlife habitats, better disease surveillance and declining global smallpox immunity, which previously had a cross-protective effect against MPXV infections.

Clinically, monkeypox and smallpox are hardly distinguishable, although the former has a lower mortality (10%) and human-to-human transmission rate. In humans, the disease starts with fever, malaise and respiratory symptoms, followed by the appearance of a maculo-papular rash; in certain cases, the eyes are affected and severe respiratory distress can occur (Di Giulio and Eckburg, 2004; Sklenovská and Van Ranst,

2018). The ongoing surge in global cases appears to be driven primarily by sexual contact, but in the past MPXV was thought to be transmitted predominantly via direct contact and respiratory droplets (Zumla *et al.*, 2022). While smallpox vaccines provide a partial protection against infection with MPXV, no licensed treatments are currently available (Brown and Leggat, 2016).

In wild primates, MPXV was first detected in 2012 in a sooty mangabey (*Cercocebus atys*), found dead in Taï National Park, Ivory Coast (Radonić *et al.*, 2014). Between 2017 and 2018, three outbreaks occurred in wild chimpanzees inhabiting the same ecosystem. The affected chimpanzees were habituated and followed on a daily basis by the staff of the Taï Chimpanzee Project, which allowed for close observation and sample collection (Patrono *et al.*, 2020). In total, 14 chimpanzees from three neighboring communities developed clinical signs. Four infants had severe illness and exhibited a typical maculo-papular rash, with one fatal case. The other ten chimpanzees showed mild to severe respiratory signs, with no or only a few visible skin lesions. In addition, 11 chimpanzees shed the virus without any clinical signs (Patrono *et al.*, 2020). These findings show that MPXV infections can have diverse clinical manifestations and that they may merit consideration in the differential diagnosis of respiratory infections in African great apes.

Sarcoptic Mange

Sarcoptic mange (also known as scabies) is a highly contagious skin disease caused by the mite *Sarcoptes scabiei*. *Sarcoptes* mites show a certain level of host-specificity but can infect other species under favorable conditions. The human parasite is ubiquitous in tropical Africa and can survive several days in the environment, providing opportunities for indirect transmission, such as through clothing (Arlian, Vyszenski-Moher and Pole, 1989; Browne *et al.*, 2021; Graczyk *et al.*, 2001). *S. scabiei* burrows tunnels into the outer skin layers of its host, causing intense itching and producing red papules that can develop into severe skin alterations, including crusts, hair loss, thickening and inflammation of the skin, as well as secondary infections. Left untreated, scabies can be fatal and have devastating effects on endangered wildlife populations (Pence and Ueckermann, 2002). Pathologies are due to the severe immune response triggered by *S. scabiei*, and disease progression depends on the host's individual health status (Bhat *et al.*, 2017).

While a clear scabies diagnosis requires invasive sampling (skin scrapings), which is problematic with respect to wild apes, the disease can often be recognized based on clinical signs (Engelman *et al.*, 2020). Treatment is relatively simple, usually involving a single dose of Ivermectin administered via darting, complemented with antibiotics in case of secondary bacterial infections (Rowe, Whiteley and Carver, 2019). Given the social nature of apes and the transmissibility of *S. scabiei*, treatment of all cohabiting individuals is advised (Graczyk *et al.*, 2001).

The first *S. scabiei* outbreak among apes was observed in 1996 and involved four habituated mountain gorillas (*Gorilla beringei beringei*) in the Bwindi Impenetrable National Park, Uganda. Three were successfully treated by Ivermectin dart; however, the most affected infant succumbed to the disease and died (Kalema-Zikusoka, Kock and Macfie, 2002). During the second outbreak, five juveniles from two groups were affected and successfully darted (Graczyk *et al.*, 2001). The only observed scabies outbreak among wild chimpanzees took place in 1997, in the Gombe National Park, Tanzania, and resulted in the death of three suckling infants (Dunay *et al.*, 2018; Wallis and Lee, 1999). In view of the high prevalence in surrounding human populations, the contagiousness of the pathogen and the curious

nature of great apes, human-to-ape transmission is the probable source of infection (Kalema-Zikusoka, Kock and Macfie, 2002).

Yaws

Treponema pallidum—the bacterium responsible for venereal syphilis (*Treponema p. pallidum*), bejel (*Treponema p. endemicum*) and yaws (*Treponema p. pertenue*, TPE) in humans—causes a yaws-like disease in primates (Čejková *et al.*, 2012; Centurión-Lara *et al.*, 2006; Marks, Solomon and Mabey, 2014). Since the first mention of the disease in Guinea baboons (*Papio papio*) in the 1960s, the bacterial subspecies TPE has been found to cause necrotizing dermatitis of the face, extremities and anogenital region in various primates across sub-Saharan Africa (Chuma *et al.*, 2019; Fribourg-Blanc and Mollaret, 1969; Fribourg-Blanc, Mollaret and Niel, 1966; Knauf *et al.*, 2018).

TPE infections of wild gorillas and chimpanzees have long been suspected based on clinical signs; they could not be confirmed until recently because of ethical considerations regarding invasive sampling (Harper and Knauf, 2013). The first evidence of TPE infection in great apes was based on the detection of TPE DNA in chimpanzee bones from Ivory Coast and gorilla feces from the Republic of Congo (Chuma *et al.*, 2019; Gogarten *et al.*, 2016). As the samples came from unknown individuals, however, there was no direct link between diagnostics and clinical signs. The link was finally established in 2020, when a wild chimpanzee with yaws-like facial lesions from the Sangaredi area in Guinea was gravely injured by hunters and had to be euthanized by a veterinarian, who was able to perform a necropsy (F. Leendertz, personal observation, 2021). A diagnosis based on these samples provided conclusive evidence that TPE infections had caused yaws-like disease in chimpanzees (Mubemba *et al.*, 2020).

Successful treatment of yaws in sooty mangabeys with long-acting antibiotics has been reported but requires darting (F. Leendertz, personal observation, 2021). Many questions remain regarding TPE transmission in great ape and other primate populations, but direct contact with an unknown animal reservoir or environmental source seems likely (Baylet *et al.*, 1971; Chuma *et al.*, 2018). TPE may also spread within groups via direct contact, sexual transmission or possibly via flies (Gogarten *et al.*, 2019a; Kumm and Turner, 1936; Satchell and Harrison, 1953). As there is no clear distinction between TPE strains that infect primates and humans, zoonotic transmission could potentially occur, hampering the ongoing World Health Organization campaign to eradicate human yaws (Knauf, Liu and Harper, 2013).

Leprosy

Leprosy is one of the oldest diseases known to humankind, and most people are familiar with its horrifying images of disfigured faces and crippled limbs (Schuenemann *et al.*, 2018). These symptoms are the ultimate consequence of nerve damage caused by the bacterium *Mycobacterium leprae* and occasionally by the more recently discovered *M. lepromatosis* (Han *et al.*, 2008, 2009). Leprosy was long considered a purely human disease, but this notion had to be revised after its detection in nine-banded armadil-

los (*Dasypus novemcinctus*), red squirrels (*Tamiasciurus hudsonicus*) and captive primates, including a chimpanzee (Avanzi *et al.*, 2016; Gormus *et al.*, 1991; Meyers *et al.*, 1985; Suzuki *et al.*, 2011; Truman, 2005; Walker, Withington and Lockwood, 2014). While the infections in armadillos and squirrels are thought to be the results of human-to-animal spillovers, it is unclear whether the captive primates were infected by humans or a different source.

Recent findings of leprosy in wild chimpanzees suggest that a non-human source of *M. leprae* exists (Hockings *et al.*, 2021). Western chimpanzees (*Pan troglodytes verus*) from two wild populations in Cantanhez National Park in Guinea-Bissau and Taï National Park in Ivory Coast presented with leprosy-like lesions, including nodules on the face, hair loss and skin depigmentation, as well as abnormal nail growth and hand deformity. These clinical signs showed a progression over time comparable to advanced leprosy in humans. The *M. leprae* strains detected in fecal and necropsy samples of chimpanzees at the two sites differ from one another, but both are rare and have not been observed in humans from either country. In humans, leprosy is transmitted through direct and prolonged contact, which is extremely unlikely between wild chimpanzees and humans at either site (Hockings *et al.*, 2021). The wild chimpanzees may therefore have been infected with *M. leprae* by an unidentified animal or environmental source.

It remains unknown whether chimpanzee-to-chimpanzee transmission occurs, whether the pathogen is present in other great ape habitats and what impact the disease may have on great ape populations. While treatment with antibiotics is possible in humans and potentially in primates in captive settings, it is not feasible for wild apes because it requires repeated drug administration over a period of several months (CDC, 2017).

Infectious Diseases with a Measured Effect on Fitness

Research has conclusively demonstrated that certain pathogens have a measurable effect on wild great ape fitness and survival, as well as the potential long-term persistence of impacted populations. This section discusses the effects of four infectious diseases on wild ape populations, highlighting key studies and potential options for prevention and treatment. The last part of this section focuses on respiratory diseases that have had a devastating impact on wild apes. All documented cases involve viruses that are endemic in humans.

Anthrax

Classical anthrax caused by *Bacillus anthracis* is a severe bacterial disease of domestic and wild herbivorous ungulates that sporadically infects humans. Depending on its entry route, the bacterium causes the milder cutaneous form (which has a 20% case fatality rate if untreated) or the often-fatal inhalation or gastrointestinal form (CDC, 2020b). In contrast, sylvatic anthrax is caused by the bacterium *Bacillus cereus* biovar *anthracis*, abbreviated here as Bcbva (Klee *et al.*, 2010). Bcbva was first discovered in Taï National Park in Ivory Coast, where it caused clusters of chimpanzee mortality in 2001 and 2002 (Leendertz *et al.*, 2004). Chimpanzees who appeared healthy hours before were found dead, and post-mortem examination showed internal bleeding. One of the chimpanzees reportedly experienced a sudden onset of unspecific signs, including weakness and vomiting; the individual died within two hours (Leendertz *et al.*, 2004).

Since then, Bcbva was also detected in several chimpanzee carcasses and a gorilla carcass in Cameroon, as well as in the Central African Republic (Antonation *et al.*, 2016; Leendertz *et al.*, 2006a). While the exact geographic and host ranges of Bcbva are

unknown, the pathogen was also isolated from a carrion fly in Liberia and a goat in the Democratic Republic of Congo (DRC); moreover, it was found to infect a broad range of other species, including various monkeys, duiker antelopes, mongooses, porcupines and forest elephants (*Loxodonta cyclotis*) (Antonation *et al.*, 2016; Hoffmann *et al.*, 2017). While no Bcbva infections have been reported for bonobos, the infected goat in DRC highlights that the pathogen may be present within the bonobo range (Antonation *et al.*, 2016). It is probably present throughout rainforests in West and Central Africa, but nowhere else is it known to have had as devastating an impact as in Taï National Park (Romero-Alvarez *et al.*, 2020).

In the hyperendemic area of Taï National Park, Bcbva was shown to be the number one mammal killer of infectious origin. It was detected in 40% (81/204) of all wildlife carcasses found between 1996 and 2015 (Hoffmann *et al.*, 2017). Since the start of the veterinary monitoring program in 2001, 38 anthrax-infected chimpanzee carcasses have been detected in the research area of the Taï Chimpanzee Project, which covers the home ranges of four chimpanzee groups (Hoffmann *et al.*, 2017; A. Düx, personal observation, 2022). Given that many chimpanzees disappeared from the habituated groups and their carcasses were never found, the real toll of anthrax on the Taï chimpanzee population is probably higher. Modeling of the long-term survival of Taï National Park chimpanzees has shown that without intervention, Bcbva is likely to lead to their extirpation (Hoffmann *et al.*, 2017).

It is unclear how apes get infected with Bcbva. Classical anthrax is generally considered a point-source infection that occurs in endemic regions (Turner *et al.*, 2014). The bacterium usually does not spread from animal to animal but forms infectious spores that survive for long periods in the environment (Beyer and Turnbull, 2009). Less is

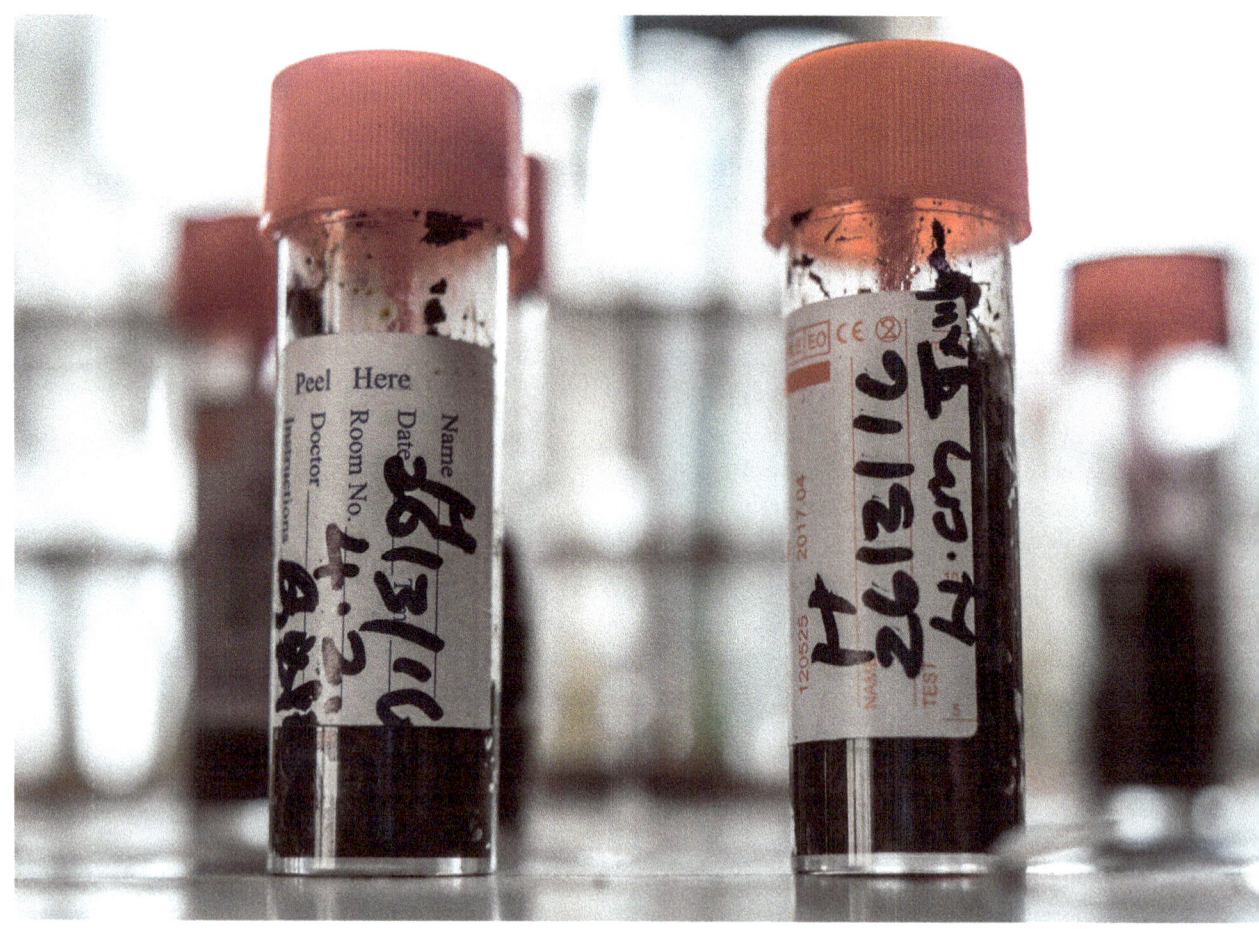

known about Bcbva, but spore-contaminated soil on fruit or plants are considered a likely source of infection (Zimmermann *et al.*, 2017; F. Leendertz, personal observation, 2021). Carrion flies may contribute to the spread by feeding on carcasses and subsequently regurgitating Bcbva-containing material on surrounding vegetation. While viable bacteria could be cultured from flies, it is not clear whether they contain sufficient infectious material to cause disease (Gogarten *et al.*, 2019a; Hoffmann *et al.*, 2017). Observed clusters of anthrax cases support the hypothesis that chimpanzees are infected when feeding from the same contaminated source (Hoffmann *et al.*, 2017; Leendertz *et al.*, 2004; F. Leendertz, personal observation, 2021). For omnivorous

chimpanzees, hunting of infected animals could play a role (Leendertz *et al.*, 2004). The risk of within-group transmission is generally low but may be elevated when great apes touch, groom or bite carcasses of their conspecifics (Beyer and Turnbull, 2009; Gonçalves and Carvalho, 2019).

In humans, anthrax can successfully be treated with antibiotics, but due to the rapid progression of the disease in chimpanzees, a timely treatment following the observation of anthrax signs is not feasible (CDC, 2020a). Preventive treatment of individuals who were in close contact with sick and deceased animals might be feasible in some cases and vaccination of animals in hyperendemic regions may become an option in the future.

Ebola

Ebola virus disease (EVD) outbreaks in humans, which occur sporadically in West and Central Africa, have become larger and more frequent over the past decade (CDC, 2022). The disease is often fatal, and symptoms can include fever, vomiting, diarrhea, internal bleeding and multiorgan failure (Jacob *et al.*, 2020). Apes are also susceptible to ebolaviruses, but observations of infected wild apes are rare; recorded signs include lethargy, abnormal behavior and abdominal pain, while post-mortems have shown internal bleeding (Formenty *et al.*, 1999; Georges *et al.*, 1999). Ebolaviruses differ from anthrax, which threatens apes in endemic hotspots, causing isolated cases or mortality clusters infected from the same point source (Hoffmann *et al.*, 2017; Leendertz *et al.*, 2004). In contrast, ebolaviruses are present in unknown animal reservoirs (presumably bats) throughout African range states and can cause large outbreaks among great apes.[1]

Due to the obscure nature of Ebola reservoirs, it can only be speculated how spillover to great apes occurs. If bats are indeed Ebola reservoirs, they could contaminate fruit and leaves with saliva and excretions when roosting in or feeding on trees that great apes frequent (Formenty *et al.*, 1999; Leendertz *et al.*, 2016). Since certain monkeys who hunt and eat bats are themselves prey for chimpanzees, their consumption could be a route of infection (Tapanes, Detwiler and Cords, 2016). Anecdotal reports of great apes catching and playing with bats indicate that direct contact with infected bats could also play a role (M.H. Surbeck, personal communication, 2019).

Irrespective of the route of initial spillover, once a great ape contracts EVD, the disease can spread within and probably between groups (and potentially even species), causing large epidemics (Bermejo *et al.*, 2006; Caillaud *et al.*, 2006). Theoretically, great ape populations' different social structures influence their ability to sustain a large outbreak. At the same time, the effects of an outbreak on different social structures may vary across great ape species (see Box 1.1).

In 1994, *Tai Forest ebolavirus* caused an EVD outbreak among chimpanzees in Taï National Park, killing 25% of the affected social group (Formenty *et al.*, 1999). In Central Africa, *Zaire ebolavirus* caused massive die-offs among chimpanzees and gorillas (Bermejo *et al.*, 2006; Leroy *et al.*, 2004; Walsh *et al.*, 2003). Between 1994 and 2003, the border region between Gabon and the Republic of Congo was hit by several EVD outbreaks in humans, most of which may have been linked to contact with sick or deceased wildlife, in particular chimpanzees and gorillas (Georges *et al.*, 1999; Georges-Courbot *et al.*, 1997; Leroy *et al.*, 2004). During this period almost 200 great ape carcasses were detected in the region and the chimpanzee and gorilla populations shrank considerably (Lahm *et al.*, 2007; Leroy *et al.*, 2004; Rouquet *et al.*, 2005). During just four months in 2002–3, for example, 32 great ape carcasses were detected in the Lossi Gorilla Sanctuary, in the Republic of Congo. Samples from 12 carcasses were analyzed and nine tested positive for *Zaire ebolavirus*. At the same time, 130 of 143 habituated gorillas in the Lossi Sanctuary disappeared (Bermejo *et al.*, 2006).

The total impact EVD has on great apes can only be guessed, as population densities in some remote regions are unknown and veterinary surveillance exists only in a few areas. While no data exist for EVD in bonobos, they are almost certainly at risk as they live in regions of the DRC where EVD outbreaks have occurred. Bonobos are probably susceptible, considering that all other hominins and many other primates can be infected (Inogwabini and Leader-Williams, 2012).

For orangutans, the situation is less clear. Asia is not known to harbor any human-pathogenic ebolaviruses; however, *Reston ebolavirus*, which circulates in bats in the Philippines, can cause disease in primates (Demetria *et al.*, 2018; Jayme *et al.*, 2015). Other, more distantly related viruses from the same family as ebolaviruses (*Filoviridae*) have been detected in Chinese bats (He *et al.*, 2015; Yang *et al.*, 2017). No filovirus infections in wild or captive orangutans (or gibbons) have been documented. The one publication that suggests there is serologic evidence of exposure is disputed for a few reasons, including the origin of

BOX 1.1

Ebola and the Social Structure of Gorilla Populations

The diversity of great ape social structures suggests that a pathogen may not spread in the same way among bonobos, chimpanzees, gorillas and orangutans (Carne *et al.*, 2014). Similarly, a pathogen's effects on social structures may vary across great ape species. The rarity of observations of naturally occurring infectious agents with a proven pathogenic potential currently precludes a thorough comparison based on real-world data.

Outbreaks of ebola virus disease (EVD) have allowed researchers to study the impact of a lethal disease on social structures in great apes—and, conversely, the influence of social structures on disease risk. Between 2001 and 2005 in the Republic of Congo, multiple EVD outbreaks severely affected populations of western lowland gorilla (*Gorilla gorilla gorilla*) (and, probably to a lesser extent, central chimpanzees (*Pan troglodytes troglodytes*)) (Bermejo *et al.*, 2006; Walsh *et al.*, 2003). The composition and size of gorilla groups can vary markedly, and male gorillas may be solitary. A group may comprise a single male and multiple females, or multiple males and multiple females, or only males. Observations made at the Lokoué site in the Odzala-Kokoua National Park before, during and after an EVD outbreak in 2004 show that individuals living in groups suffered from a higher death rate (97%) than solitary individuals (77%), pinpointing a clear cost of group living (Caillaud *et al.*, 2006). Accordingly, at the population scale, the proportion of gorillas with a solitary lifestyle was markedly higher after the outbreak. Importantly, this risk imbalance resulted in a reversal of the overall sex ratio, as adult females (all living in groups) were more affected than adult males, 8% of whom were solitary (Caillaud *et al.*, 2006).

These changes were not permanent, however. Ten years after the outbreak, both the proportion of solitary gorillas and the overall sex ratio were back to their pre-outbreak values, reflecting transiently altered social dynamics (Genton *et al.*, 2015, 2017). While they may represent an extreme example, EVD outbreaks clearly show the potentially complex interactions of great ape social systems and the pathogens that affect them.

samples and methods; an official "expression of concern" now accompanies this publication (Nidom *et al.*, 2012). The interpretation of serological results requires caution as unspecific reactivity and cross-reactivity are common for the serological ebolavirus assays used (Allela *et al.*, 2005; Natesan *et al.*, 2016). Since orangutans are less gregarious than African great apes, the risk of disease outbreaks of epidemic proportions is probably comparatively low (Carne *et al.*, 2014). The other human-pathogenic African ebolaviruses (*Bundibugyo ebolavirus* and *Sudan ebolavirus*) can also infect primates, but no cases have been observed in the wild (Leendertz *et al.*, 2017).

Since it is not possible to predict where the next Ebola outbreak will occur, management of EVD in great apes is particularly challenging. While no treatment option is available for great apes, different vaccination strategies have been discussed. Even with a safe and effective vaccine, however, the broadscale vaccination of wild great apes in remote areas would be difficult to accomplish.

Simian Immunodeficiency Virus in Chimpanzees

The simian immunodeficiency virus in chimpanzees (SIVcpz) is a retrovirus that causes an illness similar to the progression of human immunodeficiency virus (HIV) infections in humans (Sharp and Hahn, 2011). The latter stages of infection develop into simian acquired immunodeficiency syndrome, similar to when HIV develops into acquired immune deficiency syndrome, or AIDS (Keele *et al.*, 2009).

Central chimpanzees (*Pan troglodytes troglodytes)* have been identified as the reservoir from which two lineages of HIV-1 independently emerged, the pandemic group M and very rare group N. SIVcpz, the closest relative of HIV-1 group M, is found in multiple sites across the chimpan-

zee range (Sharp and Hahn, 2011). The most recent common ancestor of HIV-1 group M viruses dates back to the late 19th or early 20th century, suggesting that HIV-1 group M passed from chimpanzees into the human population in the region during the colonial period (Gryseels *et al.*, 2019; Keele *et al.*, 2006; Van Heuverswyn *et al.*, 2007). Central chimpanzees transmitted SIVcpz to western lowland gorillas (*Gorilla gorilla gorilla*), giving rise to simian immunodeficieny virus in gorillas (SIVgor). Gorillas later became the proximal source of two additional HIV-1 lineages, the epidemic group O (mostly restricted to Cameroon) and extremely rare group P (D'arc *et al.*, 2015; Plantier *et al.*, 2009). Chimpanzee and gorilla hunting is the most likely route of SIVcpz and SIVgor transmission to humans (Pepin, 2021).

The evolutionary history of SIVcpz is one of cross-species transmission, which is thought to result from the predatory behavior of chimpanzees, who often prey on monkeys. Male chimpanzees from Taï National Park consume as much as 45 kg of monkey meat per year, yet this behavior has not led to the transmission of the simian immunodeficiency virus that infects the western red colobus (*Piliocolobus badius*) to the chimpanzee population (Gogarten *et al.*, 2014; Leendertz *et al.*, 2011). Chimpanzees may thus be resistant to infection with this SIV strain.

SIVcpz was long thought to be non-pathogenic in its natural hosts, central and eastern chimpanzees (*Pan troglodytes schweinfurthii*). Using a longitudinal study extending over almost a decade, however, Keele *et al.* (2009) show that eastern chimpanzees belonging to two habituated communities in Gombe National Park (Tanzania) incurred an increased likelihood of death and lowered fertility when infected by SIVcpz. The virus also led to clinical manifestations suggestive of AIDS. A subsequent investigation of the impact of SIVcpz on chimpanzee population dynamics in the same communities and in an additional, non-habituated community suggested that SIVcpz probably played a role in the marked decline of the non-habituated community. While simulations showed that even low SIVcpz prevalence significantly increased the risk of community extinction, female intercommunity migration was found to reduce this risk considerably. These findings indicate that the survival of an infected community can strongly depend on connectivity with other social units (Rudicell *et al.*, 2010).

Respiratory Disease

Respiratory pathogens are recognized as a major cause of mild to severe disease in wild great apes. In the past two decades, continuous veterinary monitoring within conservation programs and the progressive improvement of diagnostic tools applicable to non-invasive samples have allowed for the gathering of solid evidence on the risk of pathogen transmission from humans. Over the same period, common human endemic viruses have been identified across great ape species and habitats. Among the first to be identified in wild human-habituated great apes suffering from severe respiratory disease were viruses of the family *Pneumoviridae*, such as the human metapneumovirus (HMPV) and types A and B of the human orthopneumovirus, previously known as the human respiratory syncytial virus, or HRSV (Köndgen *et al.*, 2008; Rima *et al.*, 2017). Both viruses have been repeatedly detected since. HMPV has been transmitted to western chimpanzees in Ivory Coast, eastern chimpanzees in Tanzania and Uganda, and mountain gorillas (*Gorilla beringei beringei*) in Rwanda (Kaur *et al.*, 2008; Köndgen *et al.*, 2008; Negrey *et al.*, 2019; Palacios *et al.*, 2011). HRSV has been found in western chimpanzees in Ivory Coast, in

western lowland gorillas in the Central African Republic and in bonobos in the DRC (Grützmacher *et al.*, 2016, 2018b; Köndgen *et al.*, 2008, 2017).

More recently, infections with members of other viral families were also reported in wild ape populations. These included the human rhinovirus C (family *Picornaviridae*), the human respirovirus 3 (family *Paramyxoviridae*) in chimpanzees in Uganda and the human coronavirus OC43 (family *Coronaviridae*) in chimpanzees in Ivory Coast (Negrey *et al.*, 2019; Patrono *et al.*, 2018; Scully *et al.*, 2018).

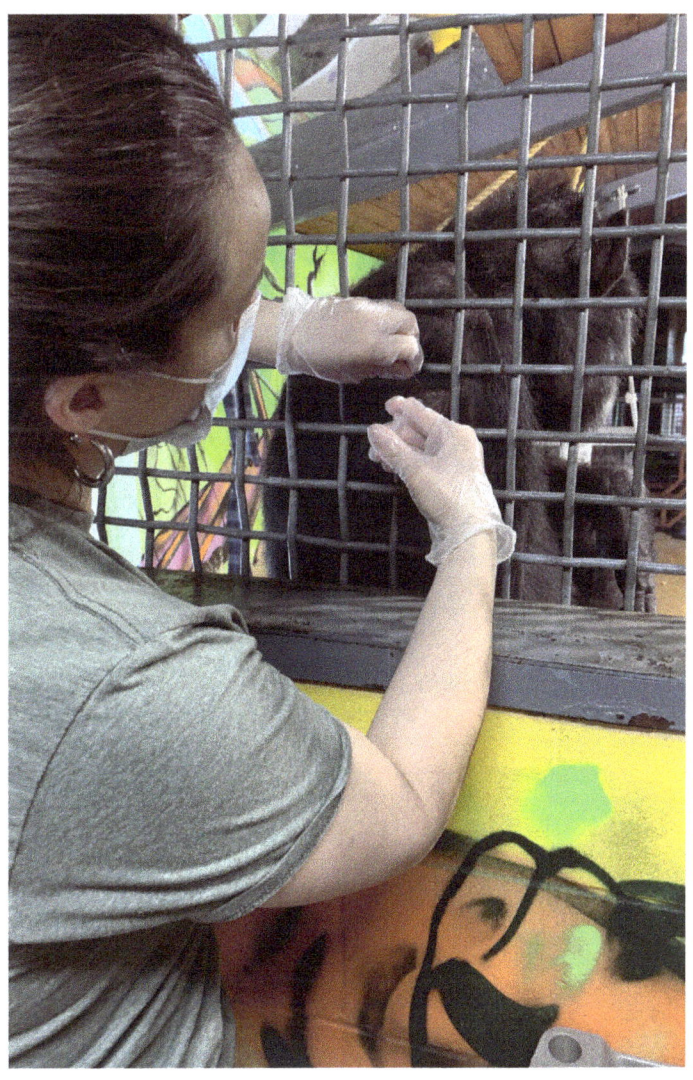

Photo: In assessing how best to manage the health of apes in captive settings, practitioners may opt to vaccinate them, especially against pathogens with a high regional prevalence. Guidelines differ widely but are often based on procedures followed in the country where the apes are kept in captivity. Chimpanzee having an injection. © Justin Taus/ Fauna Foundation

Phylogenetic analyses on the partial or complete viral genome sequences detected in these different outbreaks have consistently confirmed that the strains found in

great apes fall within the diversity of human lineages, clearly indicating human-to-ape spillover. Due to a lack of data on the circulation of these pathogens in local human populations, however, it has not been possible to establish more precise links to the geographical origins of the transmitted strains (Patrono *et al.*, 2022).

The aforementioned outbreaks were associated with mortality events, which contributed significantly to raising awareness on the risks posed by habitat overlap with humans and the need for establishing hygiene rules and surveillance systems within great ape research and tourism projects (Macfie and Williamson, 2010; see Box 1.2). Morbidity varied greatly across outbreaks but was generally high, reaching up to 100% during an HMPV outbreak in western chimpanzees (Köndgen *et al.*, 2010). In contrast, no mortality was associated with the human coronavirus OC43, which caused only mild clinical signs, whereas at least one death occurred in all other cases (Patrono *et al.*, 2018). The highest mortality rates were recorded during outbreaks caused by pneumoviruses, with up to 18% of the population succumbing to infection (Köndgen *et al.*, 2010). The true figure may be even higher given the difficulties in finding carcasses in the rainforest and their rapid decomposition due to environmental conditions, which strongly influences sampling possibilities (Köndgen *et al.*, 2017).

Viral infections often paved the way for secondary bacterial ones, to which mortality was ultimately attributed. Among bacteria, *Streptococcus pneumoniae* (or pneumococcus) has been found in several lethal outbreaks (Chi *et al.*, 2007; Grützmacher *et al.*, 2018b; Köndgen *et al.*, 2017). This opportunistic bacterium is part of the commensal nasal flora and can occasionally become pathogenic, following primary damage of the airway epithelium, which leads to pneumonia (Morris, Cleary and Clarke, 2017). Genomic analyses on some of the pneumococcal strains found in lungs of deceased

BOX 1.2

Prevention of Infectious Diseases

The prevention of infectious diseases encompasses a large range of measures, protocols and procedures to minimize the risk of natural and unintentional infections of humans and animals. Preventive measures work only when there is broad compliance, which requires repeated educational efforts for all involved with ape populations. To be wholly effective, work on preventing infectious diseases requires consultation with appropriate professionals; this chapter is in no way meant to replace such collaboration with experts.

Broadly, disease risk assessments carried out by professionals may help evaluate potential dangers associated with particular situations. When an animal enters a captive setting, a quarantine period allows for the monitoring of behavior and the potential insurgence of clinical signs. During this period, an assessment of the individual's health status is critical to minimizing the risk of novel pathogens entering and spreading in a facility (Gilardi *et al.*, 2015; see Chapter 4). Priorities of enclosure design thus include ensuring a physical separation between new animals and the resident population, as well as separate waste disposal and the disinfection of food or enrichment items brought in from outside. Although there is no standard duration, quarantines are typically imposed for 60 to 90 days, depending on diagnostic capacities as well as the ecology and prevalence of pathogens of most concern, as defined in the relevant disease risk assessment. Involving trained professionals in the design and implementation of these procedures can help to safeguard the psychological wellbeing of apes during the period of isolation, as well as during the preceding move between captive settings or from the wild into captivity.

To reduce the risk of infection, a facility can ensure that staff members who attend to captive apes are healthy and vaccinated, limit the number of staff during the quarantine period, and incorporate staffing decisions into its disease risk assessment and disease mitigation strategy. Similarly, it can integrate the use of personal protective equipment into its mitigation strategy. In all captive situations where there is long-term close contact between the (rehabilitant) apes and their caregivers, the use of masks and gloves is advisable during as well as after the quarantine period, especially during periods of high risk, such as flu season (Stevens, 2020; see Chapter 2).

In assessing how best to manage the health of apes in captive settings, practitioners may opt to vaccinate them, especially against pathogens with a high regional prevalence. Guidelines differ widely but are often based on procedures followed in the country where the apes are kept in captivity; detailed guidance may thus need to be requested from the relevant national ministry of health. For all those working with wild ape populations, a key resource on preventive measures overall is the International Union for Conservation of Nature (IUCN) publication *Best Practice Guidelines for Health Monitoring and Disease Control in Great Ape Populations* (Gilardi *et al.*, 2015).

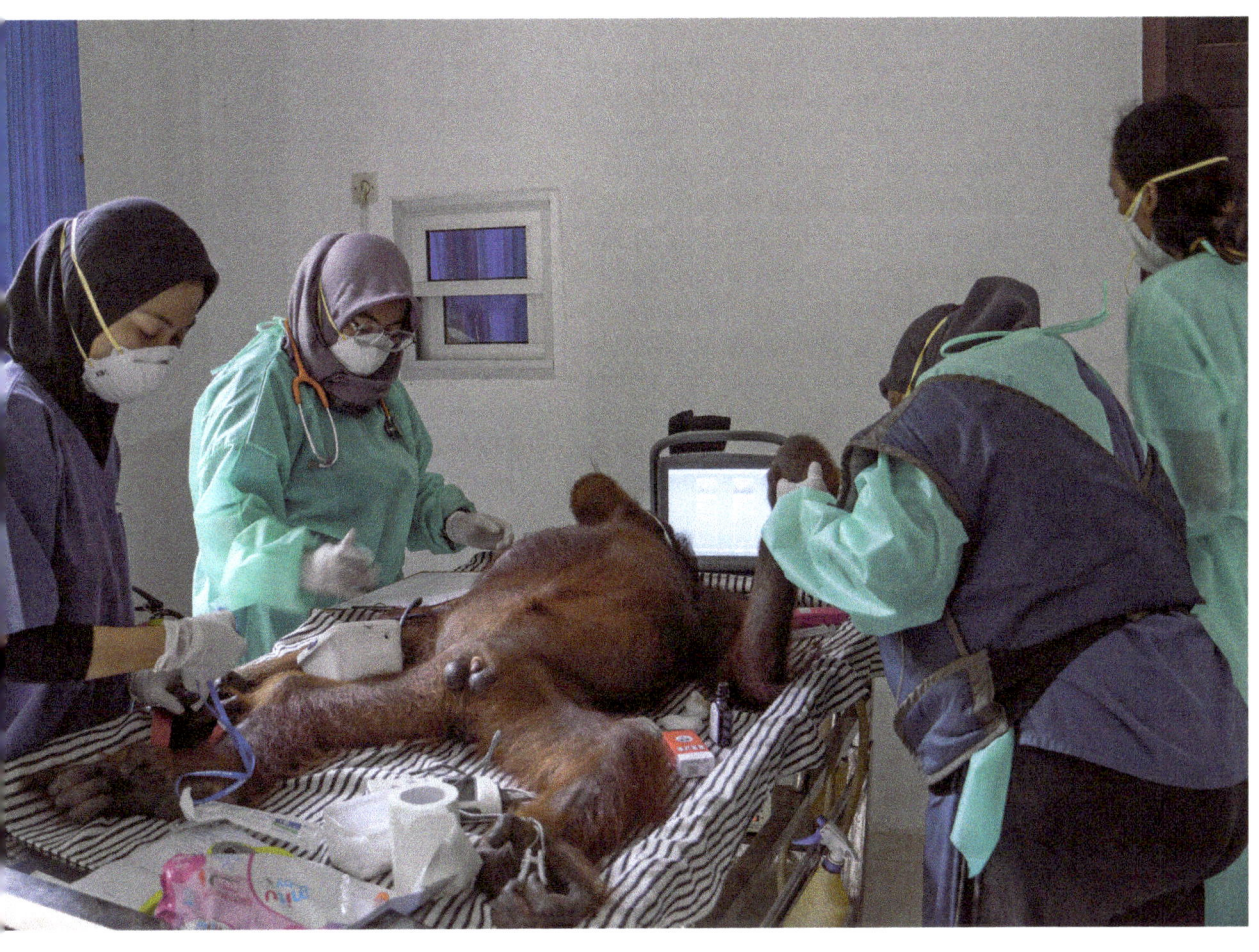

chimpanzees co-occurring with HRSV infections revealed a human origin (Köndgen *et al.*, 2017). Human pneumococci were found in both chimpanzees and orangutans living in closer contact with humans, such as in zoos, rehabilitation centers and wild-living populations (Köndgen *et al.*, 2017; Szentiks *et al.*, 2009). Whereas respiratory viral infections are normally cleared and do not persist, pneumococci can become part of the nasopharyngeal flora upon transmission. Once established in an individual, these infections can be transmitted to other group members and can eventually become endemic in a population, potentially influencing the severity of other diseases.

Another bacterium that has been associated with acute lethal pneumonia (co-occurring with HMPV and *S. pneumoniae*) or air sacculitis in wild chimpanzees is *Pasteurella multocida* (Köndgen *et al.*, 2011). The strain's genetic information and phenotype showed no clear evidence of direct acquisition from other animals or humans. Despite the paucity of data available for RNA viruses other than influenza, pneumoviruses have been shown to favor bacterial colonization in the lung through multiple pathways (McCullers, 2014). Based on the evidence gathered thus far, it seems plausible that infections with members of this viral family caused more overt clinical signs and mortality, often due to co-infections, prompting outbreak investigations and opening up the possibility of obtaining a diagnosis. Infections that cause milder clinical

Photo: To reduce the risk of infection, a facility can ensure that staff members who attend to captive apes are healthy and vaccinated and integrate the use of personal protective equipment into its mitigation strategy. © IAR Indonesia (YIARI)/MoEF of Indonesia

signs may be more difficult to observe and diagnose if continuous behavioral observation and routine sampling are not in place.

Infection with *Mycobacterium tuberculosis* has been reported in wild chimpanzees (Coscollá *et al.*, 2013). Bacterial isolation confirmed the initial pathological diagnosis of tuberculosis. Genomic analyses performed on the strain found in a wild chimpanzee revealed a novel *M. tuberculosis* complex isolate, suggesting that a human origin was unlikely.

Thanks to recent advances, diagnostic tools can now be applied to non-invasive samples to determine which types of pathogens are causing disease or death in wild great apes in various settings, including tourism sites, research areas and forests used by local human populations (see Box 1.3). This knowledge can allow for the design of targeted vaccination strategies for people entering great ape habitats, such as local residents, researchers and tourists. Employee health programs—including routine health checks, mandatory vaccinations against pathogens that also have the potential to cause disease in apes, strict hygiene rules and quarantine based on syndromic surveillance—have proven to be effective measures for reducing the risk of disease transmission (Gilardi *et al.*, 2015; Grützmacher *et al.*, 2018a). The presence of asymptomatic carriers within the human population poses a challenge, however.

The establishment of field laboratories to test all staff and visitors entering great ape habitats may represent another step towards improving prevention measures and maximizing the benefits of conservation actions (Grützmacher *et al.*, 2016). Such testing would only cover a part of the human presence in the forest, however. Additional measures for reducing the risk of disease introduction while improving human health include the broadening of vaccination coverage to the population living around the forest. Programs could make use of commercially available pneumococcal vaccines and, eventually, vaccines against respiratory viral diseases that are currently under development (Leendertz and Kalema-Zikusoka, 2021; see Chapters 2 and 4). This One Health strategy, developed together with local public health authorities, would be an additional way of ensuring direct benefits of conservation activities to local communities (see Chapter 2).

Non-Infectious Causes of Disease

Non-infectious threats to ape health are present both in captive and natural settings. This section discusses some of the human-induced elements that most severely affect ape populations in their natural habitat.

Encroachment into ape habitat by human-caused forest fires, road building, laying of electrical cables, and various farming and mining practices can have immediate effects on animals. These activities can also have long-term impacts on the environment—such as by inducing microclimate change, diminishing food availability and decimating biodiversity—further endangering the survival of great apes (Bettinger *et al.*, 2021; Erb *et al.*, 2018).

Apart from destroying ape habitat, forest fires can cause burns and inhaled smoke can damage the respiratory system, increasing the risk of respiratory infections. These effects can affect many aspects of ape health, as has been documented in humans and orangutans alike (Aguilera *et al.*, 2021; Erb *et al.*, 2018).

Various problems also stem from the construction of roads and associated infrastructure designed to serve the mining or agricultural industries and to connect human settlements. Easier human access to ape environments through roads increases the likelihood of hunting (Laurance *et al.*, 2006). Moreover, roads running through territories directly affect animals by dividing popula-

BOX 1.3

Sample Collection

This box presents sampling options for the study of ape diseases and health. A sample collection approach is best selected based on available methods for subsequent analyses, the markers of health or disease that are being examined, and available resources. The availability of infrastructure—such as liquid nitrogen, a freezer and a refrigerator—may limit the types of samples and storage media, for example. Since methods are continuously improving, it is useful to undertake a careful review of the literature and consultation with experts before developing and following a sampling protocol (Gillespie, Nunn and Leendertz, 2008; Leendertz *et al.*, 2006b; see Chapter 4).

Generally, sampling is either invasive, requiring physical contact with the animals, or non-invasive, in which case it can rely on the collection of samples such as feces, urine, hair or saliva. Many of the techniques for the study of wild animals can also be applied in captivity, but most techniques that are feasible in captivity are not applicable for the study of wild ape health (Gillespie, Nunn and Leendertz, 2008; Leendertz *et al.*, 2006b). This discussion focuses on sample collection in the wild.

Wild apes need chemical immobilization—anesthesia—to allow for invasive sampling. Anesthetizing animals, especially in remote conditions, carries an inherent risk that must be carefully considered against any benefits derived from the procedure (Gillespie, Nunn and Leendertz, 2008; Leendertz *et al.*, 2006b). Non-emergency handling of wild apes is generally considered unethical, so it is only included in management strategies for exceptional circumstances (Gilardi *et al.*, 2015; Gruen, 2018; see Chapter 5). Any proposed invasive sample collection for surveillance purposes needs to go through an extensive review by an ethical committee and secure approval from local and national authorities. To maximize the benefits associated with immobilization, veterinarians can collect a wide range of sample types—including blood, plasma, swabs, biopsies and ectoparasites—for use in a multitude of research programs (Gillespie, Nunn and Leendertz, 2008; Leendertz *et al.*, 2006b).

Post-mortem samples collected after death (during a necropsy) are invaluable for the understanding of disease in wild populations. Since carcasses can contain any number of pathogens that are known (or not yet known) to infect humans, however, the disease risks associated with performing a necropsy are considerable, particularly in remote field settings. Key steps for minimizing risks include restricting post-mortem sampling to veterinarians who have received special training and ensuring procedures are undertaken in consultation with experts and in line with rigorous safety standards (Gillespie, Nunn and Leendertz, 2008; Leendertz *et al.*, 2006b).

Non-invasive sample collection has become an invaluable tool in the diagnosis of disease and the study of behavioral ecology in wild apes. Non-invasively collected samples allow

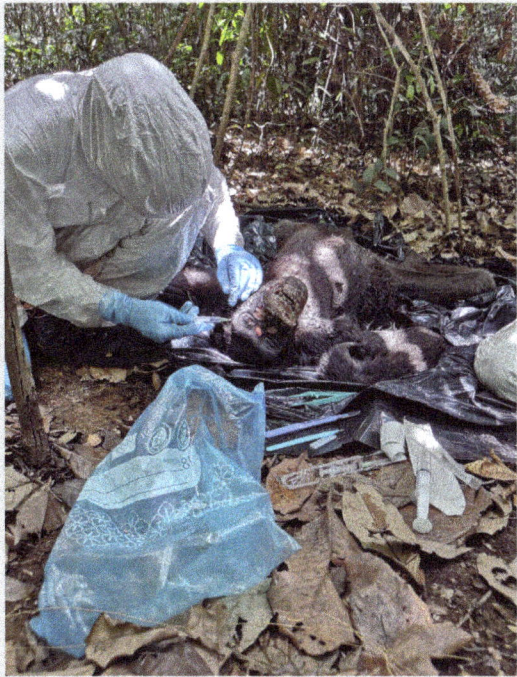

for repeated collection for longitudinal studies without major disturbance (such as chemical immobilization) of the subject of interest (Behringer and Deschner, 2017; Calvignac-Spencer *et al.*, 2021; Smiley Evans *et al.*, 2015, 2016). Molecular analysis of samples has proven fruitful for understanding the ecology of a diversity of pathogens as well as the apes themselves. Many techniques can be used on non-invasive samples to assess a wide range of factors beyond the animal's own nucleic acids, including infection history (through serology), stress and health status (via hormone analysis) and diet (such as by using metabarcoding or isotope ratios) (Gogarten *et al.*, 2018; Patrono *et al.*, 2022; Samuni *et al.*, 2018). As noted above, collection and preservation strategies are selected based on which analyses are planned (Gillespie, Nunn and Leendertz, 2008; Leendertz *et al.*, 2006b).

If samples are to be obtained from captive apes who cannot be released, operant conditioning can be employed to improve their psychological wellbeing and handleability, which can facilitate both non-invasive and minimally invasive sampling (Rasmussen, Newland and Hemmelman, 2020). Non-invasive sample collection under these conditions does not raise stress hormones in bonobos or orangutans (Behringer *et al.*, 2014). Operant conditioning also facilitates routine imaging techniques, such as radiology and ultrasound, including monitoring of pregnancies (Drews *et al.*, 2011). If there is a need to determine which ape was the source of a fecal sample and it is not possible to observe animals defecating, the animals may be fed inert substances such as indigestible grains, food colorants or colored glitter to aid in stool identification (Fuller, Margulis and Santymire, 2011).

tions, cutting them off from food and water supplies and potential mates, and exposing them to the risk of traffic accidents, which are often fatal. One proposal for reducing the number of road accidents is the construction of artificial canopy bridges that allow for safe animal crossings (Chan *et al.*, 2020).

While the mining and agricultural industries drive road construction and deforestation, they can also impact soil and water supplies by overexploiting and poisoning these resources. Gold ore processing often involves the uncontrolled use of mercury, which can potentially lead to neurological or renal malfunctions and even death in primates (Ontl, 2017). Pesticides from agricultural areas also have the potential to have dire effects on primates (Botha *et al.*, 2015). For example, facial dysplasia has provisionally been attributed to pesticides in wild baboons and chimpanzees in Uganda, where DDT/p,p'-DDE, chlorpyrifos and imidacloprid levels in maize exceeded recommended limits in areas used by chimpanzees. Further studies are needed to confirm that pesticides were related to the observed signs (Krief *et al.*, 2017).

Many anthropogenic disturbances can lead to decreases in the food supply, forcing apes to resort to crop-foraging, which further endangers them in several ways. Exposure to crops that are treated with the above-mentioned chemicals can poison apes, while ongoing, sometimes violent conflict between farmers and apes can lead to lethal physical injuries (Humle and Hill, 2016).

Captive Apes

Captivity significantly alters the environmental conditions for apes and their pathogenic organisms. Enclosure designs thus need to meet physical, social and psychological needs, while also incorporating strategies to reduce infection pressure (see Chapter 8).

In general, population density is higher in captivity than in the wild, as animals are confined to a specified space. Measures are therefore required to minimize the possibility that infectious agents will enter the captive population.

In captive settings, close contact with humans can potentially expose apes to pathogens to which they are susceptible, which can lead to serious outbreaks (Kilbourn *et al.*, 2003; Liptovszky *et al.*, 2019). In addition, stressful situations may create stereotypical behaviors (such as repetitive movements without an apparent function) and other psychopathologies that require managing. Prolonged stress can also impair an ape's immune system and the ability to fight off certain infections or regulate microbiomes. The combination of these factors usually results in a higher disease prevalence under captive conditions (Kilbourn *et al.*, 2003). Particular attention to infectious diseases is needed in rehabilitation centers, especially prior to an animal's release into the wild, to minimize the risk of introducing a novel disease into a wild population (Sherman *et al.*, 2021).

Diseases with a Likely Effect on Health

Malaria

Malaria is a potentially deadly disease that is caused by *Plasmodium* parasites transmitted through the bites of infected female *Anopheles* mosquitoes. In rescue centers, chimpanzees and orangutans are often diagnosed with *Plasmodium* infections. In most cases, there are no overt clinical signs, or they are mild and transient and do not require treatment. In rehabilitant orangutans, the severity of clinical signs appears to be correlated with increases in parasitemia, especially in individuals with anemia or persistent fever that is unresponsive to acetaminophen or non-steroidal anti-inflammatory drugs. In these cases, there is some evidence to suggest that antimalarial treatment brings improvement in signs following a treatment-based reduction of parasitemia. The correlation suggests that *Plasmodium* parasites might be clinically relevant and that treatment could be considered when parasitemia is high (J. Philippa, personal observation, 2020).

Altered living conditions in rescue centers could play a part in the ecology of *Plasmodium* infections in orangutans. One set of living conditions relates to population density, which is higher on the ground in captivity than in the more arboreal natural habitat of orangutans. The density of mosquitoes is similarly higher at the ground level than in the canopy. Another set of conditions concerns the proximity to other species, such as humans and wild macaque (*Macaca fascicularis*) populations, which may act as a reservoir or amplifier host of *Plasmodium* parasites (Brant *et al.*, 2016; Siregar *et al.*, 2015). Further studies are needed to elucidate these factors.

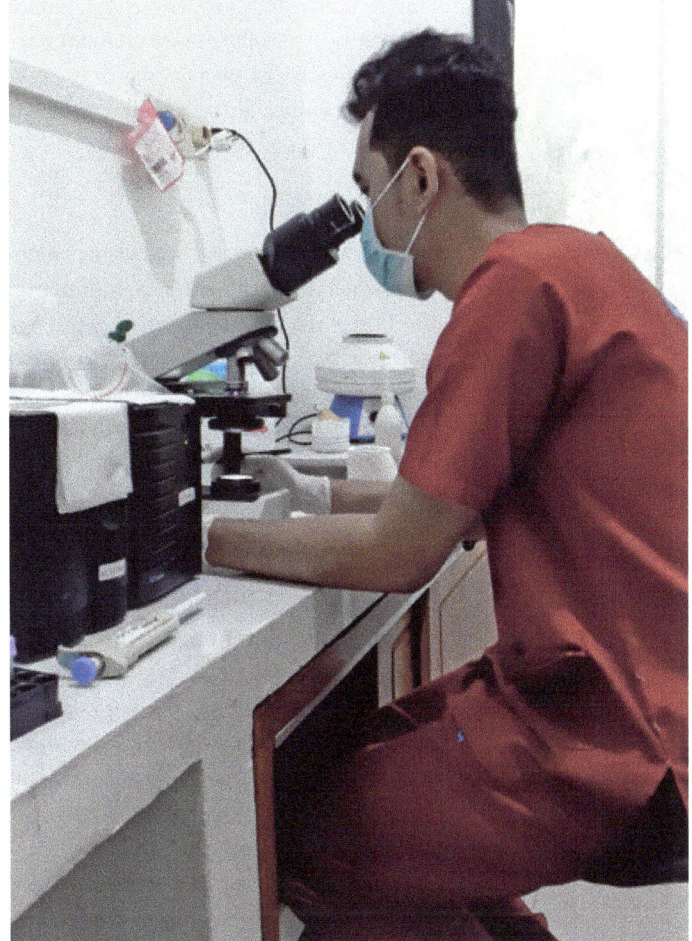

Gastrointestinal Parasites

The gastrointestinal tract comprises all the organs of the digestive system, extending from the mouth to the anus. Despite a high prevalence and variety in wild apes, diseases associated with gastrointestinal parasites (protozoa and helminths) are not well documented in the wild (Medkour *et al.*, 2020). In contrast, changes in the gastrointestinal parasite load and clinical disease in captive apes—in both rehabilitation centers and zoos—have been linked to factors such as increased host population density and infection pressure (due to a small living area or substandard hygiene practices), stressful situations and disturbances of the gastrointestinal microbiome, for example due to oral antibiotics (Labes *et al.*, 2010; Maertens *et al.*, 2021; Nurcahyo, Konstanzová and

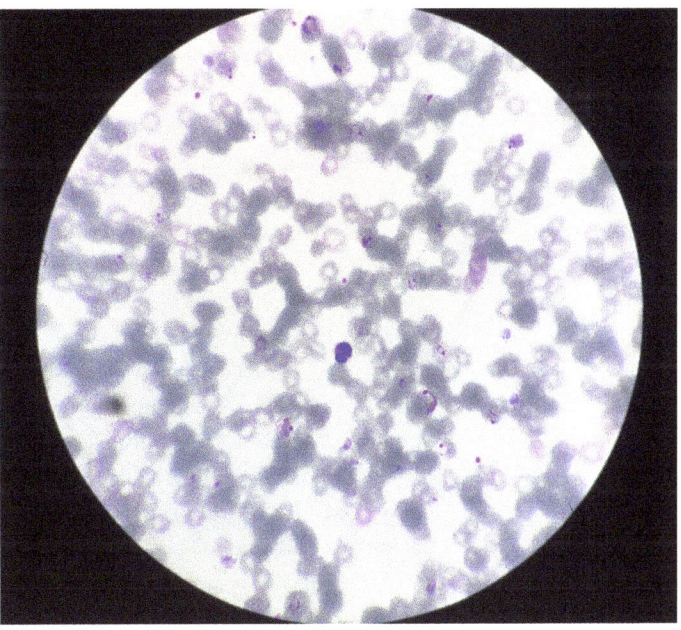

Foitová, 2017). Many captive facilities reduce the severity of parasite infections in the apes by ensuring use of proper biosafety practices and routine anthelmintic treatment (Liptovszky *et al.*, 2019).

Protozoa

Protozoa are single-celled organisms. *Balantidium coli* is a very common commensal infection of wild and captive apes; like other protozoa, it is part of a healthy intestinal microbiome in low to moderate numbers. Its prevalence in captive orangutans is generally higher than in their wild counterparts, however; indeed, observation of clinical disease associated with these infections is restricted to captive animals. Contributing factors to clinical balantidiasis include increased infection pressure in captivity, largely due to higher host population density and stress, and diets rich in easily digestible carbohydrates, or starch (Labes *et al.*, 2010; Schovancová *et al.*, 2013). *Balantidium* infections are usually left untreated in rehabilitation centers and zoos, unless clinical signs accompany increases in numbers. There are case reports of a balantidiasis epidemic in captive western lowland gorillas, including typhlitis requiring surgery and a fatal *B. coli* infection (with a *Salmonella* co-infection) in a captive western lowland gorilla in Cameroon (Lankester *et al.*, 2008; Lee *et al.*, 1990; Teare and Loomis, 1982).

In some situations, other common gastrointestinal protozoa—such as *Cryptosporidium*, *Entamoeba histolytica* and *Giardia*—have caused clinical infections with bloating, cramping or diarrhea in captive apes. In zoos, *Giardia* has been implicated in clinical disease (diarrhea and vomiting). Meanwhile, *Entamoeba* spp. have caused irritable bowel-like signs, ulcerative colitis and diarrhea in gorillas, as well as ulcerative colitis and lung or liver abscesses in chimpanzees. Increased contact with humans has been linked to an increased prevalence of protozoa such as *Entamoeba histolytica* in rehabilitated orangutans (Stuart *et al.*, 2020).

Balamuthia mandrillaris is a recently described, free-living protozoal organism that has caused fatal acute to subacute necrotizing or granulomatous meningo-encephalitis in humans and captive apes. Isolated cases have been reported in the northern white-cheeked crested gibbon (*Nomascus leucogenys*), western lowland gorilla and orangutan in Australia, Europe and North America.[2] Unvalidated immunofluorescence antibody assays, which permit the identification and highlighting of antibodies in a blood sample, have shown promising results in orangutans. A validated test would be extremely useful for preventive screening of captive apes (Ferris, Ali and West, 2021).

Helminths

The group of worms known as helminths comprises nematodes, cestodes and trematodes. Some of the most common nematodes (roundworms) found in captive apes are *Ancylostoma, Ascaris, Capillaria, Enterobius, Oesophagostomum, Strongyloides* and *Trichuris*.[3] As gastrointestinal commensals they generally do not cause significant morbidity. One important exception is *Strongyloides*, a very common nematode in wild and captive apes (Mul *et al.*, 2007; Nurcahyo, Konstanzová and Foitová, 2017; Penner, 1981; Zulfikri, Ridwan and Cahyaningsih, 2018). Although it is not a clinically important parasite while in the gastrointestinal tract, its larval forms travel widely throughout the body and often result in fulminant, fatal verminous pneumonia and peritonitis, which is commonly fatal in juvenile orangutans housed in zoos (Liptovszky *et al.*, 2019). In rehabilitation centers, young orangutans have been found to be more at risk than older animals (Labes *et al.*, 2010). Fatal strongyloidiasis has been described in a Lar gibbon (*Hylobates lar*) colony as the most common cause of death, with erosive and ulcerative enteritis, and multifocal-diffuse hemorrhage associated with migrating larvae (DePaoli and Johnsen, 1978). Disseminated infections have been diagnosed ante-mortem in orangutans and are curable (Kleinschmidt, Kinney and Hanley, 2018).

Chimpanzees and orangutans are natural hosts of the *Enterobius* species (Foitová *et al.*, 2008, 2014; Labes *et al.*, 2010). *Enterobius* infections usually cause asymptomatic to mild clinical disease, but there are reports of fatal hemorrhagic colitis in captive chimpanzees, with the parasite maintained in the population for more than 20 years following introduction—despite attempts at treatment (Hasegawa and Udono, 2007; Murata *et al.*, 2002; Yaguchi *et al.*, 2014). Heavy clinical infections have also been recorded in gibbons kept as pets (Smith *et al.*, 1969).

Cestodes, like the other parasites, generally cause low morbidity in their natural ape hosts. *Echinococcus multilocularis*, the fox tapeworm, is widespread in the northern hemisphere and causes alveolar echinococcosis after infection. Captive gorillas seem to be very susceptible, but infected chimpanzees and orangutans have also been reported in European and Japanese zoos (Federer *et al.*, 2016; Wenker *et al.*, 2019). The infection can remain asymptomatic for years, but clinical disease can be (sub)acute and fatal (Wenker *et al.*, 2019).

Other sporadic cases of severe cysticercosis (an infection caused by larval cysts of the tapeworm) in captive apes include a recent case of fatal disseminated *Versteria mustelae* infection in a captive Bornean orangutan (*Pongo pygmaeus* spp.) with a rapid and severe disease progression (Goldberg *et al.*, 2014). Metabarcoding techniques have the potential to standardize helminth taxonomic identification from ape and other primate fecal samples, while simultaneously allowing for descriptions of

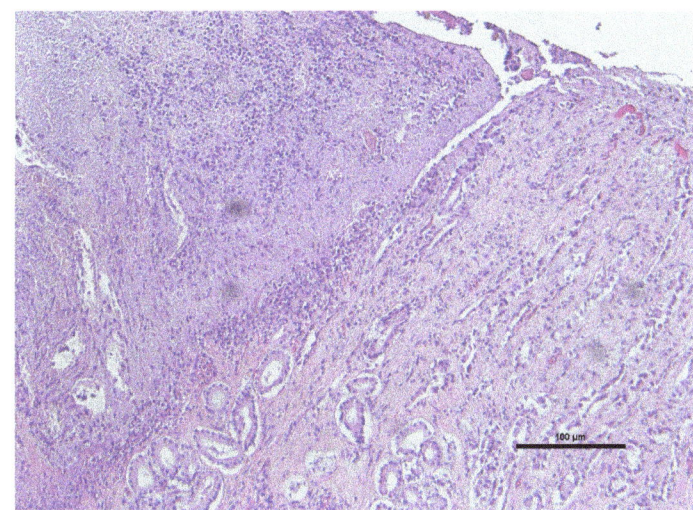

primate-associated parasite communities (Gogarten *et al.*, 2020).

Herpes

Herpes is a group of viral diseases caused by the herpes viruses, which affect the skin (often characterized by blisters or sores) and nervous system. Herpes virus infections have been documented in all apes, and species-specific herpes viruses likely evolved with humans' primate ancestors.[4] Antibodies to human herpes simplex viruses have been reported in rescued gibbons, with a high prevalence likely due to close human contact (Eberle and Jones-Engel, 2017; Sakulwira *et al.*, 2002). Apes are susceptible to other herpes viruses, such as Cytomegalovirus, Epstein-Barr virus and Varicella-zoster (Haberthur and Messaoudi, 2013); mountain gorilla lymphocryptovirus infections have been likened to an Epstein-Barr virus-like epidemiology (Smiley Evans *et al.*, 2017). Manifestation of human herpes simplex virus infections range from stomatitis, or localized signs on the mucous membranes, to systemic infections with encephalitis and fatal outcomes (Gilardi *et al.*, 2014). They have been reported in captive gorilla, orangutan and gibbon populations.[5]

Photo: Inflamed stomach tissue, adult female mountain gorilla, severe acute to subacute ulcerative gastritis. © Gorilla Doctors

Infectious Diseases with a Measured Effect on Health

Candidatus Sarcina troglodytae

Sarcina are bacteria that synthesize and release toxins that cause degeneration in the nervous system (Brown, 2019). Recently, the new, highly virulent *Candidatus Sarcina troglodytae* strain was linked to disease in captive, rehabilitant chimpanzees; the bacteria cause "epizootic neurologic and gastroenteric syndrome," characterized by neurologic and gastrointestinal signs that may result in mortality despite medical treatment (Owens *et al.*, 2021). Further research is warranted to elucidate the exact role of this bacterial strain in the development of the syndrome.

Respiratory Disease

Tuberculosis

Tuberculosis (TB) is the disease caused by infections with *Mycobacterium tuberculosis*, which has a wide host range and is the leading bacterial cause of death for humans worldwide. For these reasons, TB is of specific concern in relation to captive apes. Although prevalence in captive apes is low, an outbreak with environmental shedding could have a disastrous impact considering the large host range and zoonotic aspects (Kock *et al.*, 2021; Lécu and Ball, 2011; Michel *et al.*, 2003; Montali, Mikota and Cheng, 2001). *Mycobacterium tuberculosis* infections have sporadically been reported in captive chimpanzees, orangutans and gibbons in zoos (Michel *et al.*, 2003; Shin *et al.*, 1995; Wilson *et al.*, 1984). Zoo infections are generally thought to have arisen from contact with humans, although animals have been known to carry the mycobacteria into a facility. In one case, an elephant was the source of TB in a chimpanzee and zoo staff (Stephens *et al.*, 2013).

As ape populations are undeniably susceptible to this pathogen, testing for TB is critical before they join captive populations in a rehabilitation center or zoo, especially during the quarantine period (Lécu and Ball, 2011). Orangutan rehabilitation centers appear to be especially vulnerable and affected, as the TB incidence in the human population is very high in range countries where orphaned orangutans are confiscated. Indonesia has a particularly high burden of 312 cases per 100,000 people (WHO, 2020c); Malaysia's TB rate is 92 per 100,000 people (Avoi and Liaw, 2021). Several orangutan centers have had to construct dedicated TB quarantine facilities to house TB-positive animals. These individuals can never be released, as *M. tuberculosis* has never been detected in wild orangutans and the bacteria may be shed years after treatment (Dench *et al.*, 2015). Surveys in wild chimpanzees have not shown the presence of the bacterium (Wolf *et al.*, 2016). Nonetheless, captive, rehabilitant apes cannot be released without a negative TB test and efforts are required to mitigate the risk of transmission from humans and their domestic animals to protect wild populations (Wolf *et al.*, 2014).

Diagnostic challenges may complicate accurate identification of the latent stage of *M. tuberculosis* infection, during which the bacteria remain dormant inside the body, without overt clinical disease or associated shedding of the bacteria. Diagnostics are most accurate when they combine several tests: isolation, culture or molecular detection of the bacteria, chest X-rays and immunological tests that show any previous infection (using antibodies or other immune responses in blood or based on skin tests). Orangutans show a high level of cross-reactivity with non-pathogenic mycobacteria, which can be differentiated by comparative skin tests but may complicate accurate diagnosis (Dench *et al.*, 2015). Furthermore, in its latent stage, TB can remain inside a body for years, capable of escaping stringent therapeutic approaches. These characteristics highlight the risks associated with introducing TB into a captive facility.

Air sacculitis

Air sacculitis is a common inflammatory condition of air sacs. Connected to the laryngeal tubes of apes (and many other animals), these sacs act as resonating chambers that amplify vocalizations and extend the duration of calls (Hewitt, MacLarnon and Jones, 2002; Riede *et al.*, 2008). Air sacculitis is a condition in which pus accumulates within the air sac, with the potential for serious complications, including fatal bronchopneumonia and sepsis. Of all captive ape species, orangutans appear especially susceptible, although cases in captive chimpanzees and bonobos have also been documented.[6] Sinusitis with concurrent pneumonia may play a role in the way this disease develops (Steinmetz and Zimmermann, 2012).

Bacteria isolated from air sacculitis cases in rescue centers often include intestinal bacteria, whose route of entry into the upper respiratory system is facilitated in captive conditions (Philippa and Dench, 2019). The relatively high incidence in captivity may be driven by other conditions as well. Among rehabilitant orangutans, decreased cage space, overcrowded cages, poor ventilation and environmental factors such as smoke appear to increase incidence (J. Philippa, personal observation, 2020).

Other Viral and Bacterial Respiratory Infections

Reports of respiratory infections in captive and semi-captive great apes are common. Human respiratory pathogens have often been involved in outbreaks of respiratory disease in both categories. Infections caused by human pneumoviruses (HMPV and HRSV), often complicated by secondary infections with *Streptococcus pneumoniae*, have been detected in zoo chimpanzees in Europe and the United States, as well as in

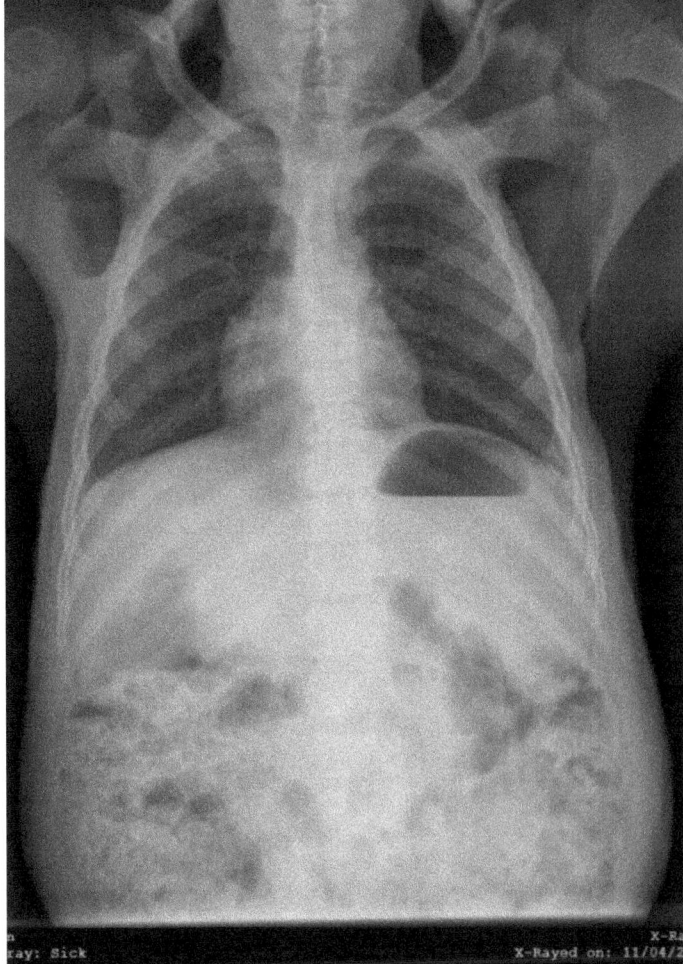

wildlife rescue centers (Köndgen *et al.*, 2017; Slater *et al.*, 2014; Szentiks *et al.*, 2009; Unwin, Chatterton and Chantrey, 2013). Morbidity reached up to 100% and several deaths were reported. Serological investigations have suggested broad exposure to human respiratory pathogens, including influenza A and B viruses of different subtypes (Buitendijk *et al.*, 2014; Kooriyama *et al.*, 2013). These findings were never confirmed by direct pathogen detection methods, however.

The recently emerged severe acute respiratory syndrome coronavirus 2 (SARS-CoV-2), which caused the COVID-19 pandemic, has been transmitted to gorillas

Photo: Reports of respiratory infections in captive and semi-captive great apes are common. Human respiratory pathogens have often been involved in outbreaks of respiratory disease in both categories. © IAR Indonesia (YIARI)/ MoEF of Indonesia

in zoos in Barcelona, Prague, Rotterdam and San Diego. The spread underscores once more the high risk of transmission of human respiratory pathogens and the importance of observing strict hygiene rules when working in proximity to great apes (Gilardi *et al.*, 2015; Reuters and Gorman, 2021; Reuters Staff, 2021). Although there have been no confirmed cases of SARS-CoV-2 in populations of free-ranging apes, the risk is significant given the high prevalence of the disease in surrounding human populations. Measures to reduce the risk of transmission and the likelihood of outbreaks in wild populations include disease risk analysis of apes to be translocated or reintroduced, as well as enhanced pathogen surveillance (Sherman *et al.*, 2021).

Monkeypox

Shortly after the first identification of MPXV in a macaque colony in a research centre in Denmark in 1958, an outbreak was reported by the Rotterdam Zoo in the Netherlands (von Magnus *et al.*, 1959). Among the affected species, chimpanzees, gorillas and orangutans became ill with different degrees of morbidity and mortality (Peters, 1966). Clinical signs included the typical maculopapular rash and nasal discharge.

Subsequently, in 2014 and 2016, two MPXV outbreaks affected semi-captive chimpanzees in sanctuaries in Cameroon (Devaux *et al.*, 2019; Guagliardo *et al.*, 2020). During the first outbreak, at Sanaga-Yong Sanctuary, six animals fell ill and one succumbed to the infection. In the second outbreak, at the Mefou Primate Sanctuary, one out of the two reported cases had a fatal outcome. A serologic survey of the nearby human population showed that farmers had a higher prevalence of MPXV-specific antibodies than sanctuary workers, indicating that contact with rodents was more likely to cause exposure than contact with apes (Guagliardo *et al.*, 2020).

Melioidosis

Also known as Whitmore's disease, melioidosis is a predominantly tropical infectious disease that can infect humans and animals and has a wide range of both symptoms and severity. Melioidosis is a disease of increasing importance in its endemic region of Southeast Asia and northern Australia. It has caused fatal infections in a zoo-kept gibbon and orangutan, as well as in orangutan rescue centers in Malaysia, gibbons in the Singapore Zoo and, more recently, in rehabilitant orangutans in Indonesia (Nathan *et al.*, 2018; Sim *et al.*, 2018; Sprague and Neubauer, 2004; Testamenti *et al.*, 2020). African apes are also susceptible: in the Singapore Zoo, five gorillas and two chimpanzees have had fatal infections (Sim *et al.*, 2018).

The disease is caused by infections with the bacterium *Burkholderia pseudomallei*, which has a broad host range and can have high case fatality rates in animals and humans. Infections tend to coincide with increased rainfall (Cheng and Currie, 2005). Clinically, signs can range from subclinical to subacute, or wasting with subcutaneous and soft tissue abscesses. Melioidosis can be challenging to diagnose and treat because the organism can remain latent for years and is resistant to many antibiotics.

Non-Infectious Causes of Disease

Malnutrition

Malnutrition refers to the effects of a poorly balanced diet, including obesity, but is more commonly associated with undernutrition and starvation. Best practice guidelines on formulations and target nutrient ranges enable careful management of captive ape diets, based on extensive experience and knowledge (Abelló, Rietkerk and Bemment, 2017; AZA Ape TAG, 2010, 2017; Stevens, 2020). Commercial pellets facilitate a balanced dietary composition for captive apes when supplemented with fresh food items that are closer to the natural diet (Nijboer, 2020). In rescue centers in home range countries, however, commercially produced biscuits or pellets may not be available, such that meeting dietary requirements necessarily involves a careful selection of natural foods, based on calculations of their nutritional values.

Despite this progress, nutritional deficiencies still occur in captive situations where unbalanced diets are provided, or in social groups with fierce competition for food, which can lead to the emaciation of certain individuals. To maintain a healthy nutritional state, ape management can include the monitoring of individual food uptake in social groups and a regular weighing schedule and body scoring to monitor body weights (Abelló, Rietkerk and Bemment, 2017; AZA Ape TAG, 2010, 2017; Stevens, 2020).

Deficiencies and Imbalances

Rickets, osteopenia and metabolic bone disease are well documented in captive apes and other primates. These deficiencies are consequences of dietary calcium–phosphorus imbalances, or insufficient calcium or vitamin D intake (Crissey et al., 1998; Farrell, Rando and Garrod, 2015; Junge et al., 2000). They occur when animals—especially infants but also adult females—are insufficiently exposed to natural ultraviolet light, for example because they are housed indoors (Videan et al., 2007). Zoos in regions farther away from the equator require artificial light to supplement the ultraviolet B rays radiated from the sun, which are insufficient at higher and lower latitudes (Nijboer, 2020).

Vitamin C deficiency causes a disease commonly known as "scurvy," which can occur in all primates, as they are unable to synthesize their own vitamin C. To ensure sufficient uptake, most zoos supplement food with commercial primate biscuits containing stable vitamin C, especially if amounts in green vegetables and fruit are insufficient (Lowenstine, McManamon and Terio, 2018).

Obesity

Obesity is the most common form of nutritional disorder observed in zoo apes; orangutans and gorillas appear to be most affected due to the intake of large amounts of easily processable carbohydrates, while physical exercise is limited (Lowenstine, McManamon and Terio, 2018). Obesity, which is difficult to manage in captivity, inherently predisposes animals to diseases such as diabetes and hypertensive heart disease (Gresl, Baum and Kemnitz, 2000; Lowenstine, McManamon and Terio, 2016). As a reduction of calories in the diet usually results in an immediate decrease in activity, a more effective approach to combating obesity involves ensuring that animals engage in foraging-like "work" to access their food, increasing fiber as well as leaves and branches (known as "browse"), and decreasing sugar in their diet. These practices can reduce the frequency of abnormal regurgitation and re-ingestion, while also reversing pre-diabetes in zoo apes (Cabana, Jasmi and Maguire, 2018; Nash et al., 2021).

Photo: Cardiovascular disease, renal disease and osteoarthritis are the most significant age-related or degenerative diseases among all apes. Heart tissue, adult female mountain gorilla, fibrosing cardiomyopathy. © Gorilla Doctors

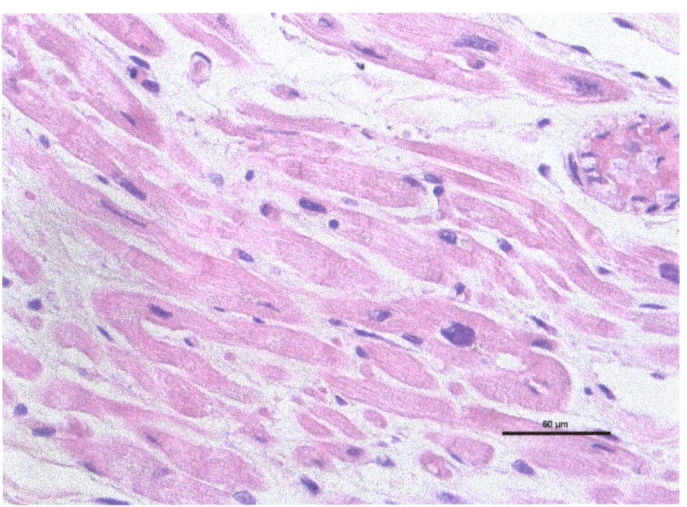

Age-Related Health Issues

Multiple studies examining pathology across captive and free-living apes indicate that cardiovascular disease, renal disease and osteoarthritis are the most significant age-related or degenerative diseases among all apes (Lowenstine, McManamon and Terio, 2018). Other degenerative conditions, including dental disease (dental attrition and tooth loss), ocular conditions (cataracts and retinal disease), and liver disease have also been documented. Pathologic correlates of human brain aging have been reported in chimpanzees, gorillas and orangutans (Lowenstine, McManamon and Terio, 2016). Neoplasms do not seem to be as common in apes as in humans and some other primates, with the exception of benign uterine leiomyomas in female chimpanzees and reproductive malignancies in female lowland gorillas (Brown *et al.*, 2009; Lowenstine, McManamon and Terio, 2016).

Cardiovascular Disease

Cardiovascular disease is an overarching term for conditions that affect the heart and blood vessels. It is a significant contributing factor in the deaths of apes under managed care. Studies indicate that the reported

Photo: Other degenerative conditions include dental disease (dental attrition and tooth loss), ocular conditions (cataracts and retinal disease), and liver disease. © Lwiro Primates Rehabilitation Center

incidence in North American zoos is 45% of bonobos, 41% of western lowland gorillas, 38% of chimpanzees and 29% of orangutans.[7]

Interstitial myocardial fibrosis or fibrosing cardiomyopathy has been the most frequently documented lesion across all great apes, in both zoo and research populations (Munson and Montali, 1990; Schulman *et al.*, 1995). It appears to result in sudden death via malignant arrhythmia or congestive heart failure (Lowenstine, McManamon and Terio, 2016; Murphy *et al.*, 2011). Postmortem data suggest that 41% of gorillas in North America, 81%–100% of chimpanzees living in a research colony and 91% of zoological chimpanzees demonstrate moderate to severe fibrosis (Lammey *et al.*, 2008; Meehan and Lowenstine, 1994; Strong *et al.*, 2018). Left ventricular hypertrophy, combined with coronary arterial arteriosclerosis, is suggestive of systemic hypertension as an underlying pathogenesis (Schulman *et al.*, 1995).

At the time of writing, the only study that had investigated the presence of myocardial fibrosis in sanctuary chimpanzees showed no evidence of the disease in a sample of 23 sanctuary chimpanzees aged 8–27 years (Strong *et al.*, 2020). There are currently no published data examining myocardial fibrosis in sanctuary bonobos, gorillas or orangutans. Further work is required to establish whether myocardial fibrosis presents a similar burden to wild and sanctuary captive apes and, if not, which predisposing factors are responsible for the disease in zoological and research facilities.

Other important cardiovascular lesions among apes are aortic dissection (a major disease in bonobos and lowland gorillas), atherosclerosis and degenerative valvular disease (Lowenstine, McManamon and Terio, 2018). Strokes have been well documented in captive chimpanzees (Jean *et al.*, 2012). Coronary atherosclerosis that was once common in captive apes is now rare, present only in old apes who previously lived under outdated husbandry conditions (Lowenstine, McManamon and Terio, 2016).

Three ongoing projects are specifically examining great ape cardiac disease:

- the International Primate Heart Project (Cardiff Metropolitan University, n.d.);
- the Great Ape Heart Project (Detroit Zoological Society, n.d.); and
- the Ape Heart Project (Twycross Zoo, n.d.; see Case Study 2.4).

Work from these groups has identified specific cardiac conditions, potential risk factors and early markers of cardiac disease, such as multifocal ventricular ectopy as detected through an electrocardiogram, diabetes, renal disease, obesity, hypertension and metabolic syndrome.[8] In time, comprehensive databases of standardized ante-mortem and post-mortem data generated through these projects are expected to improve the understanding of cardiac disease in these endangered species and may help guide improved husbandry and veterinary practices to mitigate and treat this disease.

Renal Disease

The renal system includes the kidneys, ureters, bladder and urethra, which are responsible for the production and excretion of urine. Renal disease occurs commonly in captive apes. The North American species survival plan ape pathology databases list chronic interstitial nephritis as the most common diagnosis, followed by glomerular lesions (Lowenstine, McManamon and Terio, 2018). Aging, laboratory-housed chimpanzees commonly exhibit evidence of clinically declining renal function (Videan, Fritz and Murphy, 2008). Renal disease was also identified as the cause of death in 26% of orangutans over the age of 40 and 15%–18% of 15–40-year-old orangutans, but it was less common in lowland and mountain goril-

las (Lowenstine *et al.*, 2008; Meehan and Lowenstine, 1994; Nutter *et al.*, 2005). There appears to be a statistical association between cardiac and renal disease in zoo-housed orangutans (Lowenstine *et al.*, 2008).

Osteoarthritis

Osteoarthritis is a condition that results in stiff, painful joints and is commonly reported in captive apes, although the overall prevalence in apes under managed care has not been determined. Ape species survival plan pathology advisors report that osteoarthritis typically occurs on the knees, hips, elbows and lower spine. Lesions have been documented in both captive and free-living individuals (Lowenstine, McManamon and Terio, 2016).

Dental Disease

Enamel hypoplasia (thin or missing tooth enamel) of deciduous and permanent teeth occurs in both wild and captive apes. Enamel formation can be disrupted by external stressors, including rainy seasons in which food

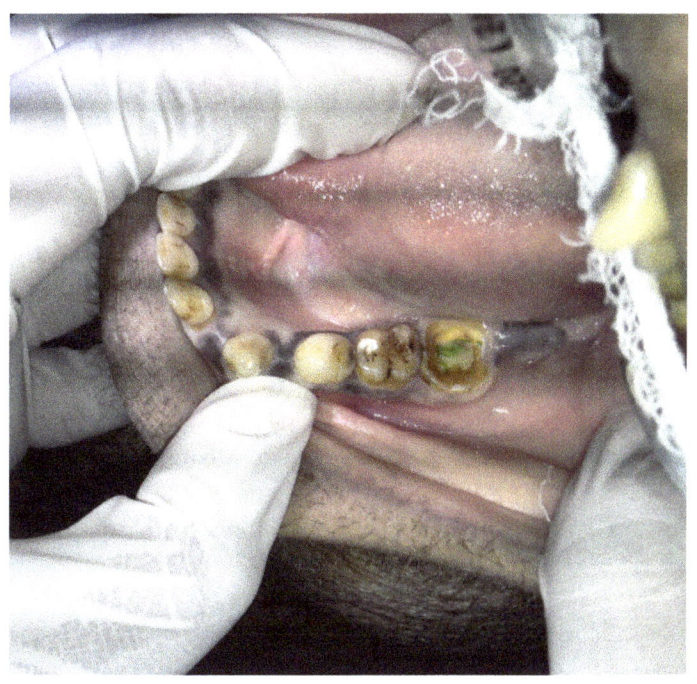

Photo: Enamel hypoplasia (thin or missing tooth enamel) of deciduous and permanent teeth occurs in both wild and captive apes. © IAR Indonesia (YIARI)/ MoEF of Indonesia

availability is limited (Skinner, 1986). Orangutans are most prone to both linear and localized enamel hypoplasia, which also occurs in chimpanzees and gorillas, while gibbons are seldom affected (Guatelli-Steinberg, 2000; Guatelli-Steinberg, Ferrell and Spence, 2012; Guatelli-Steinberg and Skinner, 2000; Hannibal and Guatelli-Steinberg, 2005).

Psychological Disorders

Psychological disorders are also known as psychiatric disorders or mental health problems. Limited opportunity or ability to conduct natural behavior, physical exercise and—most importantly—mental exercise increases the chances of the development of psychological disorders, including stereotypical behaviors, accompanied by increased levels of stress hormones such as cortisol and catecholamines (Jacobson, Ross and Bloomsmith, 2016; Nash *et al.*, 1999; see Chapter 8). Psychological disorders are more likely to develop in captive apes whose history is not considered. Apes exhibit behavioral disturbances similar to post-traumatic stress disorder (PTSD) following traumatic experiences. Carers are advised to take such signs into consideration, particularly when rescuing orphans, translocating "displaced" apes or confining apes in captivity (Ferdowsian *et al.*, 2011).

Illegal Captivity

Illegal captivity combines multiple health threats resulting from poor husbandry. Illegal captivity generally starts at an early age, when young apes are violently separated from their mothers. They are often kept in deplorable living conditions and generally are not provided with an adequate diet. Illegally kept apes tend to exhibit signs of nutritional deficiencies and PTSD; many are malnourished and emaciated, while a smaller number are obese (Ferdowsian *et al.*, 2011). In the best-case scenario, orphaned babies are intro-

> 66 Given the myriad threats faced by apes in the wild, understanding what influences their health and fitness may provide critical knowledge for their long-term conservation. 99

duced to a proper diet and weaned off any inappropriate food they may have been given previously. Among apes whose illegal captivity lasts longer, physical changes can become irreversible, including metabolic bone disease (Farrell, Rando and Garrod, 2015).

In addition to suffering from psychological disorders and malnourishment, some illegally kept apes are used as photo props or tourist attractions. In Thailand, young gibbons are exhibited at beaches, bars and restaurants, where they are given drugs such as amphetamines to keep them awake at night—and alcohol to "perform" (Gray, 2012). It is thus not uncommon for rescued gibbons to have alcohol or drug dependencies (J. Philippa, personal observation, 2021).

Conclusion

This chapter discusses factors that have major and plausible impacts on the health of wild and captive apes. Far from being an exhaustive review of such factors, it provides a preliminary outline. The expanding body of long-term research is likely to reveal new pathogens and non-infectious factors that influence ape health. From a public health perspective, these ongoing research activities may be able to inform disease risk reduction strategies for humans (Calvignac-Spencer *et al.*, 2012). At the same time, studies of humans' fellow hominins can provide insight into the factors that influenced the health of early human societies and their relationship to the microbial world; these findings could further contribute to improvements in human health (Gogarten *et al.*, 2019b; Moeller, 2017). Given the myriad threats faced by apes in the wild, understanding what influences their health and fitness may provide critical knowledge for their long-term conservation.

As this chapter reveals, only a small proportion of the factors that have a demonstrated or suspected influence on ape health impact both wild and captive individuals. This finding may be unsurprising, as bacterial and phage communities in captive apes' guts are completely different from those of their wild conspecifics. Indeed, in captive settings, the components of wild apes' microbiomes appear to undergo a complete replacement by human-associated microbes (Campbell *et al.*, 2020; Gogarten *et al.*, 2021). Just as the microbial world facing captive apes is substantially different from that of their wild counterparts, so too are many of the infectious and non-infectious factors affecting their health.

The threat posed by human respiratory pathogens, which have caused significant mortality in both populations, seems to represent the clearest intersection between wild and captive apes. Given ever-increasing rates of anthropogenic disturbance and the resulting increase in human–wildlife contact, the overlap in health threats faced by wild and captive populations is likely to expand. Nevertheless, this overview indicates that targeted strategies are required for the management of both wild and captive ape populations. Closer collaboration among practitioners and researchers working in both in situ and ex situ situations is key to bridging the data gaps and turning anecdotal clinical data into robust peer-reviewed evidence.

Acknowledgments

Principal authors: Ariane Düx,[9] Fabian H. Leendertz,[10] Jan F. Gogarten,[11] Livia V. Patrono,[12] Kamilla Pléh[13] and Joost Philippa[14]

Contributors: Sébastien Calvignac-Spencer,[15] Aimee Drane,[16] Tim Georoff[17] and Benjamin Mubemba[18]

Endnotes

1 Bermejo *et al.* (2006); Leroy *et al.* (2005); Marí Saéz *et al.* (2015); Olival and Hayman (2014); Pigott *et al.* (2014, 2016).

2 Canfield *et al.* (1997); Gjeltema *et al.* (2016); Hawkins *et al.* (2021); Mätz-Rensing *et al.* (2011); Rideout *et al.* (1997).

3 Labes *et al.* (2011); Mbaya and Udendeye (2011); Mul *et al.* (2007); Panayotova-Pencheva (2013); Tangtrongsup *et al.* (2019); Teo *et al.* (2019); Toft (1982).

4 Eberle, Black Hilliard (1989); Eberle and Jones-Engel (2017); Lavergne *et al.* (2014); Seimon *et al.* (2015); Wertheim *et al.* (2014).

5 Emmons and Lennette (1970); Heldstab *et al.* (1981); Kik *et al.* (2005); Landolfi *et al.* (2005); Mootnick *et al.* (1998); Ramsay *et al.* (1982).

6 Cambre *et al.* (1980); Clifford *et al.* (1977); Kumar *et al.* (2012); Lawson, Garriga and Galdikas (2006); McManamon, Swenson and Lowenstine (1994); Stevens (2020); Zimmermann *et al.* (2011).

7 Gamble *et al.* (2004); Lammey *et al.* (2008); Laurence *et al.* (2017); Lowenstine *et al.* (2008); McManamon and Lowenstine (2012); Meehan and Lowenstine (1994); Seiler *et al.* (2009).

8 Celestino-Soper *et al.* (2018); Doane, Lee and Sleeper (2006); Ely, Zavaskis and Lammey (2013); Lowenstine, McManamon and Terio (2016); Nunamaker, Lee and Lammey (2012); Rosenblum and Coulston (1983); Tong *et al.* (2014).

9 Helmholtz Institute for One Health (www.helmholtz-hzi.de/en) and Robert Koch Institute (www.rki.de).

10 Helmholtz Institute for One Health (www.helmholtz-hzi.de/en) and Robert Koch Institute (www.rki.de).

11 Helmholtz Institute for One Health, Helmholtz-Centre for Infectious Research (www.helmholtz-hzi.de/en), Robert Koch Institute (www.rki.de) and University of Greifswald (zoologie.uni-greifswald.de/en/organization/departments/applied-zoology-and-nature-conservation).

12 Helmholtz Institute for One Health, Helmholtz-Centre for Infectious Research (www.helmholtz-hzi.de/en) and Robert Koch Institute (www.rki.de).

13 Helmholtz Institute for One Health, Helmholtz-Centre for Infectious Research (www.helmholtz-hzi.de/en) and Robert Koch Institute (www.rki.de).

14 At the time of writing: International Animal Rescue (www.internationalanimalrescue.org).

15 Helmholtz Institute for One Health, Helmholtz-Centre for Infectious Research (www.helmholtz-hzi.de/en) and Robert Koch Institute (www.rki.de).

16 International Primate Heart Project, Swansea University (www.swansea.ac.uk and primate-heartproject.co.uk).

17 North Carolina Zoo (www.nczoo.org).

18 Copperbelt University School of Natural Resources (www.cbu.ac.zm/schoolsAndUnits/schoolofnaturalresources).

CHAPTER 2

The Role of One Health at the Human–Ape Interface

Introduction

In the context of multiple interconnected social and environmental crises, addressing health-related threats requires integrated approaches that consider the interconnections among humans, companion animals, livestock and wildlife and their respective social and ecological environments (Zhu *et al.*, 2020; Zinsstag *et al.*, 2011). One Health is a systems-based, transdisciplinary approach to creating and implementing health-based solutions in the face of such complexity.

The definition and implementation of the One Health approach has matured over the past two decades (Capua and Cattoli, 2018; Gibbs, 2014; Lainé and Morand, 2020; see Boxes 2.1 and 2.2). Early efforts recognized the importance of sustainable, healthy, nature–human coupled systems and focused on the need for an expanded conceptual approach to the interconnected health and wellbeing of humans, animals and their environment. During the subsequent period, many groups concurrently defined principles, methods and the practice of One Health. The most recent approach, which is still in its infancy, places the focus on design thinking, which includes measuring and evaluating programmatic impacts.

This chapter provides an overview of the history and core principles of One Health and explores its role in ape conservation, in the context of the challenges of environmental degradation and biodiversity loss. To highlight the diverse applications of One Health in supporting ape conservation, it presents case studies on various themes: community health, ecosystem health methods, capacity building, land use, internal health, translational medicine and ecosystem conservation.

Key findings include:

- The complex challenges of improving ape conservation must be contextualized

---

BOX 2.1

What Is One Health?

The One Health paradigm is:

> a collaborative, multisectoral, and transdisciplinary approach—working at local, regional, national, and global levels—with the goal of achieving optimal health outcomes recognizing the interconnections between people, animals, plants, and their shared environment (CDC, n.d.-b).

The approach reflects the complex, interconnected environmental systems that make up the planet, while also providing an effective means of solving individual, population, species and ecosystem health issues holistically. While One Health has been represented in multiple fashions over the years, this chapter adopts the view of the One Health High-Level Expert Panel to illustrate the paradigm's most up-to-date collaborative nature (see Figure 2.1).

FIGURE 2.1

Representation of One Health

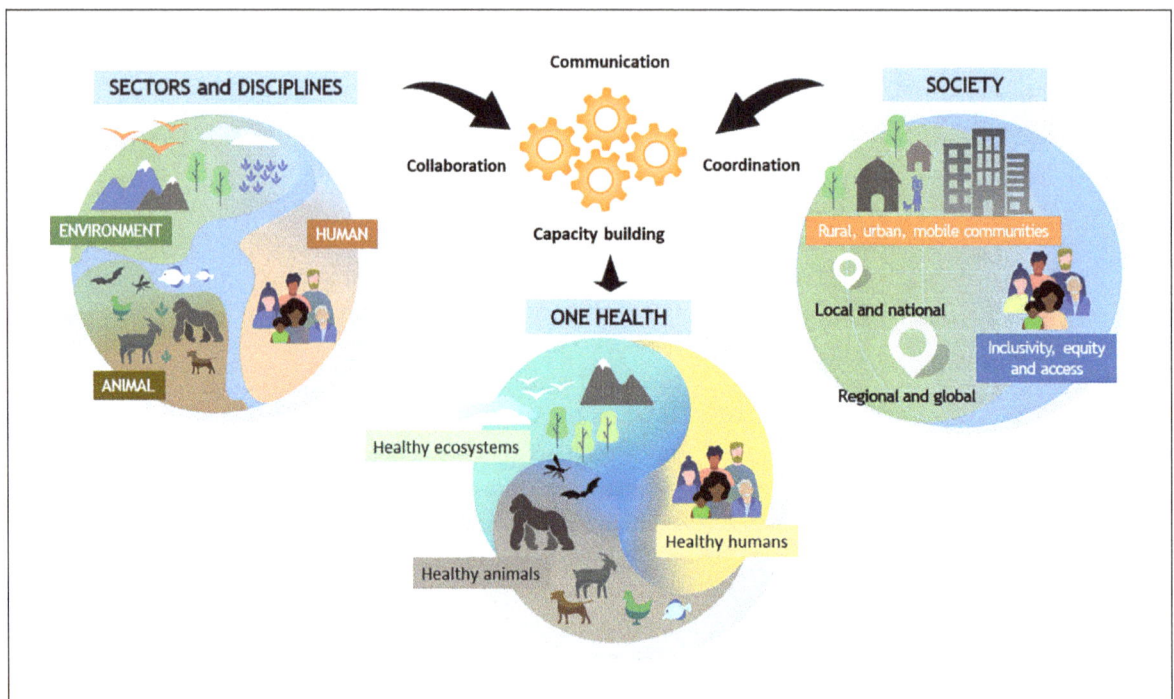

within global sustainability goals. One Health is a mechanism for success here.

- The One Health approach to solving health issues converges with those of related fields, including ecosystem health, conservation medicine and planetary health.

- One Health principles and approaches have a clear role to play in multiple facets of ape health, welfare and conservation (Grützmacher *et al.*, 2021), as evidenced by an expanding set of case studies in areas as diverse as ecology, disaster preparedness, public health, community development, tertiary education, veterinary science, microbiology, science communication, land tenure and environmental law, and resource management.

Ape Conservation, One Health and Sustainable Development

The world changed with the advent of severe acute respiratory syndrome coronavirus 2 (SARS-CoV-2), which caused the COVID-19 pandemic (Guo *et al.*, 2020). No longer can the pieces of the global biosocial system be viewed as disconnected. That recognition can help to support the United Nations (UN) 2030 Agenda for Sustainable Development, which provides a shared blueprint for meeting human needs while protecting the planet. Adopted by all UN member states in 2015, the Agenda's 17 Sustainable Development Goals (SDGs) "recognize that ending poverty and other deprivations must go hand-in-hand with strategies that improve health and education, reduce inequality, and spur economic growth—all while tackling climate change and working to preserve our oceans and forests" (UN DESA, n.d.). Ape conservation intersects with almost every SDG.

Indeed, the conservation and sustainability agendas dovetail with respect to biodiversity protection, climate effects on habitat, migration, health and sustainable natural resource management. The SDGs with the most direct links to the conservation agenda are Goal 13 on climate change, Goal 15 regarding life on land and Goal 16, which focuses on peace, justice and strong institutions. More indirect connections to conservation are evident with respect to poverty and hunger reduction (Goals 1 and 2) and responsible consumption (Goal 12), which relates to human demand for ecosystem services and the trade in wildlife and their products, in the context of supporting economic and nutritional stability.

Overall, growing human contact with apes has negatively affected the "good health and well-being" (Goal 3) of all ape species and humans, particularly through infectious disease "spillover" events and their "spillback" from animal care staff to captive apes. Increased contact does not necessarily result in exclusively negative health impacts, however. While ecotourism can threaten apes' wellbeing by raising their stress levels, for example, it can also positively influence people's mental health and support ape welfare, such as by enhancing protection from hunting or securing revenue for conservation (see Chapter 3).

Education and gender equality (Goals 4 and 5), both of which link to human health, are increasingly part of conservation discussions, especially in relation to tropical areas with suitable ape habitat. On the whole, investments in community capacity and capability in such areas are sorely needed (Razanatsoa *et al.*, 2021; Unwin *et al.*, 2022).

The convergence of ape conservation and sustainability targets has also been promoted at the highest political levels. In 2006, in his keynote address to the International Primatological Society in Entebbe, Uganda, President Yoweri Museveni made a compelling argument for the shared agendas of ape conservation and sustainable development in his country. He highlighted clean water and sanitation (Goal 6), affordable energy (Goal 7), economic growth (Goal 8), innovation and infrastructure (Goal 9) and reduced inequality (Goal 10) as key to sustainability for both apes and humans in Uganda.[1]

While some progress has been made in efforts to achieve both the SDGs and ape conservation goals, increasing contact between humans and apes is ushering in new challenges that call for innovative approaches (Travis, Lonsdorf and Gillespie, 2018). The One Health model provides a foundation for much-needed win–win solutions.

A Short History and Core Principles of One Health

The past two decades have seen milestones in the development of the One Health model. In 2004, the Wildlife Conservation Society and Rockefeller University hosted the "One World, One Health" symposium, which focused on the movement of infectious diseases among humans, domestic animals and wildlife populations. The output, entitled "Manhattan Principles," listed 12 recommendations for establishing a more holistic approach to preventing cross-species infectious disease transmission while maintaining ecosystem integrity and the benefits it bestows on humans and animals (Karesh and Cook, 2009).

In 2009, following growing acceptance and application of this paradigm, the One Health Commission was created to foster consensus—or at least convergence—among practitioners around the globe. In 2019, the Manhattan Principles were updated in Berlin to reframe the unifying approach to human, animal and ecosystem health in an economic and sociopolitical context (WCS, n.d.-a; see Box 2.2). Two years later, in 2021,

the One Health High-Level Expert Panel was created as an advisory body to the Commission's key supporting agencies, the World Health Organization (WHO), the Food and Agriculture Organization (FAO) and the World Organisation for Animal Health (founded as OIE), which were joined by the UN Environment Programme in 2022. Defining One Health was among the Panel's first duties (see Box 2.1).

In both its scope and its approach, One Health intersects—and converges—with several related disciplines and paradigms, including ecosystem health, conservation medicine and planetary health (Lerner and Berg, 2017; Wallace *et al.*, 2015; Wilcox *et al.*, 2019; Xie *et al.*, 2017). All these fields can contribute towards achieving the SDGs. Errecaborde *et al.* (2019) summarize their similarities and differences, while Roger *et al.* (2016) draw out their complementary natures, highlighting that all these paradigms:

- are motivated by the conviction that health concerns must be addressed at the human–animal interface within their broader natural and social environments;

- seek to integrate scientific disciplines through multi- and cross-disciplinary approaches;

BOX 2.2

Berlin Principles of One Health

The 2019 Berlin Principles, presented in their entirety below, are aimed at overcoming systemic policy and social challenges to enable a holistic, global approach to addressing growing health threats.

We urge world leaders, governments, civil society, the global health and conservation communities, academia and scientific institutions, business, finance leaders, and investment holders to:

1) Recognize and take action to retain the essential health links between humans, wildlife, domesticated animals and plants, and all nature; and ensure the conservation and protection of biodiversity which, interwoven with intact and functional ecosystems, provides the critical foundational infrastructure of life, health, and wellbeing on our planet;

2) Take action to develop strong institutions that integrate understanding of human and animal health with the health of the environment, and invest in the translation of robust science-based knowledge into policy and practice;

3) Take action to combat the current climate crisis, which is creating new severe threats to human, animal, and environmental health, and exacerbating existing challenges;

4) Recognize that decisions regarding the use of land, air, sea, and freshwater directly impact health and wellbeing of humans, animals, and ecosystems and that alterations in ecosystems paired with decreased resilience generate shifts in communicable and non-communicable disease emergence, exacerbation and spread; and take action to eliminate or mitigate these impacts;

5) Devise adaptive, holistic, and forward-looking approaches to the detection, prevention, monitoring, control, and mitigation of emerging/resurging diseases and exacerbating communicable and non-communicable diseases, that incorporate the complex interconnections among species, ecosystems, and human society, while accounting fully for harmful economic drivers, and perverse subsidies;

6) Take action to meaningfully integrate biodiversity conservation perspectives and human health and wellbeing when developing solutions for communicable and non-communicable disease threats;

7) Increase cross-sectoral investment in the global human, livestock, wildlife, plant, and ecosystem health infrastructure and international funding mechanisms for the protection of ecosystems, commensurate with the serious nature of emerging/resurging and exacerbating communicable and non-communicable disease threats to life on our planet;

8) Enhance capacity for cross-sectoral and trans-disciplinary health surveillance and clear, timely information-sharing to improve coordination of responses among governments and non-governmental organizations, health, academia and other institutions, the private sector and other stakeholders;

9) Form participatory, collaborative relationships among governments, NGOs, Indigenous Peoples, and local communities while strengthening the public sector to meet the challenges of global health and biodiversity conservation;

10) Invest in educating and raising awareness for global citizenship and holistic planetary health approaches among children and adults in schools, communities, and universities while also influencing policy processes to increase recognition that human health ultimately depends on ecosystem integrity and a healthy planet.

Source: Grützmacher *et al.* (2021, p. 3), reproduced under the Creative Commons Attribution License

- aim to mitigate the risks threatening ecosystems and public health, including veterinary public health;

- address the complexity of diseases and health; and

- struggle to define their boundaries despite their apparent similarities regarding principles and objectives.

Globalization of the Principles of One Health

While the term One Health is the product of a long history of philosophical discussion, the focus of more recent deliberations has been on advancing theories and methods in support of implementation strategies (Mackenzie and Jeggo, 2019; World Bank, 2021). Today, One Health is globally recognized as a theory of change underpinning a renewed focus on systemic approaches to complex problems; however, the conditions under which they should be employed, the modalities for their implementation, and evaluation processes and metrics are still being designed and tested.

As it continues to gain momentum, One Health is serving as a roadmap for connecting collaborative science to policymaking in the context of several international health-related efforts, including:

- the Global Health Security Agenda, a partnership of more than 50 countries, international organizations and non-governmental stakeholders focused on

Photo: One of the 2019 Berlin Principles for One Health is to take action to combat the current climate crisis, which is creating new severe threats to human, animal and environmental health, and exacerbating existing challenges.
© SOCP

combating infectious diseases (Center for Global Health, 2016);

- the WHO–FAO–OIE agreement to combat zoonotic diseases and antimicrobial resistance (Lee and Brumme, 2013; WHO, FAO and OIE, 2019);

- the US Agency for International Development's Emerging Pandemic Threats Program (USAID, n.d.); and

- the global COVID response (Ruckert *et al.*, 2020).

Recognizing the need for more tangible and implementable processes to support this widespread acceptance, in 2018 the EcoHealth Alliance and the World Bank published the *Operational Framework for Strengthening Human, Animal, and Environmental Public Health Systems at Their Interface.* The framework updates the previous operational definition of One Health, which had been in use during the response to the avian and pandemic influenzas in 2005–2014. Specifically, it expands the description of One Health as "a framework for enhanced collaboration in areas of common interests (intersections), with initial concentration on zoonotic diseases, that will reduce risk, improve public health globally and support poverty alleviation and economic growth in developing countries" to include "the discrete disciplinary involvement of human health, animal health, and environmental health, and focus on those infectious disease-related issues (including antimicrobial resistance) that undermine overall health and well-being" (World Bank Group, 2018, p. 3). The intent is to improve health in all the above dimensions, address drivers of processes that threaten health and optimize the effectiveness of systems to achieve these goals.

Evaluations of the One Health framework have led to the development of dedicated monitoring systems. In 2014, a qualitative assessment concluded that "there is no shared conception of health across disciplines and One Health suffers from a lack of strong environmental stakeholders" (Stephen and Karesh, 2014). Subsequent studies of One Health in practice, which found that it lacked a standardized framework and evaluation metrics, called for greater proof of concept and standard indicators (Baum *et al.*, 2017; Häsler *et al.*, 2014). In response to such criticism, the European Union in 2014 supported the Network for the Evaluation of One Health working group, which established, tested and published an open-source text on a comprehensive monitoring and evaluation protocol for One Health (Rüegg, Häsler and Zinsstag, 2018). These tools and indicators have since been incorporated into a One Health package under the Global Health Security Agenda (CDC, n.d.-a; Center for Global Health, 2016; Fasina *et al.*, 2021; Kelly *et al.*, 2020; Rabinowitz *et al.*, 2018).

To produce a comprehensive systems model of One Health, Xie *et al.* (2017) conducted a systematic literature review of 577 One Health articles. Based on the findings, they developed the "One Health Cosmos," which illustrates the fundamental roles of ecology and social aspects such as economy and commerce in the management of health issues at a biosocial system level (see Figure 2.2).

The One Health model was designed by health professionals working at the wildlife–human–livestock interface. It has since been applied in differing contexts the world over and has become an accepted—if not yet standardized—way to approach various complex problems, including human and animal welfare and wellbeing, environmental and ecotoxicological concerns, agricultural (plant or animal) sustainability issues, and challenges related to the aquatic environment. Nevertheless, conservation efforts remain largely focused on infectious disease and translational and

FIGURE 2.2

The One Health Cosmos

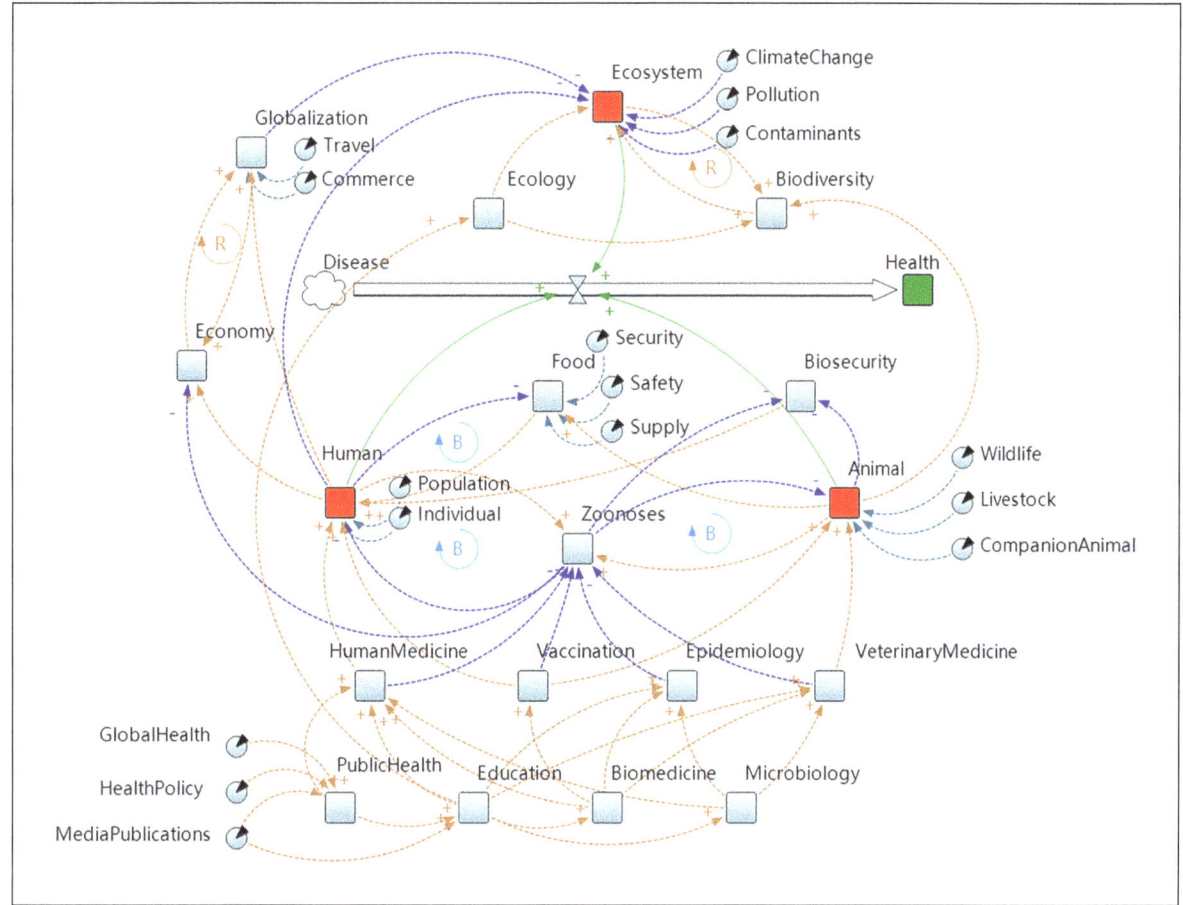

Notes: As specified by Xie *et al.* (2017): "One Health Cosmos shows the relationships between the various disciplines and complex problem descriptors that are reported to fall within the One Health concept. Squares and circles represent nodes, and the arrows connecting nodes represent causal links. Brown color is used to show positive causal link which also includes a '+' sign beside each arrowhead. Negative causal link is portrayed with a blue color and '-' sign beside each arrowhead. A positive causal link means that both the causative and the resultant factors increase or decrease in the same direction. A negative causal link indicates that the two linked factors change in opposite directions. The positive reinforcing loop has a 'R' in the clockwise cycle. A negative reinforcing loop opposite has a 'B' in the counter-clockwise cycle. A big arrow shows the direction of this relationship between disease and health through One Health."

Source: Xie *et al.* (2017, fig. 2), reproduced under the Creative Commons Attribution License

comparative medicine, perhaps reflecting the dearth of guidance on employing the One Health approach in real-world settings. The World Bank's abovementioned *Operational Framework* begins to fill this gap by providing detailed guidance on multiple aspects of the model. Box 2.3 presents basic steps and questions that can guide any group in developing a One Health project or consortium (Waltner-Toews, Kay and Lister, 2008).

One Health and the Human–Ape Interface

Complex problems often require multifocal solutions and teamwork. One Health supports the inclusion of all relevant stakeholders in solution-based processes, using the theory that inclusion leads to more robust, implementable and sustainable solutions, both in-situ and ex-situ (Xie, 2021).

BOX 2.3

How to Develop a One Health Project

The following steps and questions are designed to assist groups in establishing a One Health project or consortium.

1. **Problem formulation**. Why is a potential One Health approach needed or useful? Discuss which systems are involved and how they interact. Do they involve complex issues that involve the human–animal–environmental interface and that cannot be solved using traditional methods and approaches?

2. **Stakeholders**. What stakeholders are involved or likely to be affected? One Health approaches are suited to problems that require engagement from multiple stakeholders, potentially with differing agendas or preferred outcomes. Identify principal stakeholders, their conflicts of interest and related power dynamics.

3. **Problem mapping**. Map out the problem, including all interconnecting systems. Highlight the intersections and how they influence each other. Think about the historical development of systems: what has stabilized or destabilized them?

4. **Access to information**. Identify critical barriers to a better understanding of the problem, assess what information is available and determine what further information is needed to understand or address the problem.

5. **Solutions**. Explore whether and how potential solutions can be made acceptable, affordable, sustainable and equitable for all stakeholders.

Source: Waltner-Toews, Kay and Lister (2008)

The model also allows for an assessment of the role socioeconomic factors play in decisions and behavior that increase health risks, including disease emergence (Dobson *et al.*, 2020; Wallace *et al.*, 2015). The likelihood of such risks has grown during the current geological epoch—commonly termed the Anthropocene—as human values, choices and activities have increasingly driven both physical and biological global processes (Crutzen, 2006). An estimated 75% of the planet's terrestrial landscapes is currently human-modified (Venter *et al.*, 2016). As highlighted in other volumes of *State of the Apes*, agricultural intensification (of crops and livestock) is the dominant form of human-driven landscape modification. Often undercutting the resilience and sustainability of natural systems, the process has resulted in increased contact between

humans, livestock and wildlife, including apes (Arcus Foundation, 2015; IPBES, 2020; UNEP and ILRI, 2020; Williams *et al.*, 2021). This increased interaction is associated with an estimated 25% of all infectious disease emergence and 50% of all zoonotic disease emergence (Rohr *et al.*, 2019).

Efforts to minimize health risks associated with the interface between humans and captive apes can also benefit from the One Health approach, particularly as it interacts with the fields of animal welfare science and translational medicine research (Pinillos *et al.*, 2016; see Chapter 8). The model is applicable in all captive situations, no matter whether apes are kept as pets, for exhibition or educational purposes, as part of a conservation program or for research. An example of One Health under the One Welfare paradigm is the design and validation of the Enclosure Design Tool, which allows caregivers to compare captive apes' behaviors to those in the wild and to adjust welfare plans accordingly (see Case Study 8.1).

Apes, Humans and Infectious Disease

As evidenced by the COVID-19 pandemic, human health and the environment are intimately intertwined. Although it is well established that human disturbance of ecosystems can create disease-related threats for apes and other endangered species, predicting the direction, magnitude and mechanisms of disease emergence remains a challenge (Gillespie and Chapman, 2006; Gillespie, Chapman and Greiner, 2005). For instance, wild mammals are frequently the primary source of novel pathogens found in humans, but making related predictions and assessing risk remain difficult due to the lack of basic presence and absence data for zoonotic viruses for almost 90% of wild mammal species that could serve as reservoirs (Calvignac-Spencer *et al.*, 2012; Johnson *et al.*, 2020; Zhu *et al.*, 2020). This example highlights the desperate need for well-designed empirical studies that integrate animal and human pathogen surveillance as well as robust ecological data on natural and anthropogenic systems, as a detailed understanding of population and community dynamics is central to solving these problems (Gillespie, Nunn and Leendertz, 2008; Lonsdorf *et al.*, 2022).

The risk of disease spillover is directly related to human behaviors, including values and choices. In fact, evidence suggests it is most strongly associated with agricultural-driven decisions such as tropical forest conversion to monoculture plantations and industrial livestock production (Rohr *et al.*, 2019). This association reflects three key factors, namely that pathogen diversity correlates with host diversity; that the highest host diversity occurs in tropical forests; and that commercial agriculture is expanding rapidly in tropical forest regions (Gillespie *et al.*, 2021). Forest conversion increases the risk of pathogen spillover in two ways. First, it increases the interface between wild mammals and people—and thus the number of interactions that facilitate disease transmission directly or indirectly. Second, it promotes novel behaviors by wild mammals who seek new food sources as their long-standing food supply becomes less dependable (Faust *et al.*, 2018). In addition to increasing the risk of spillover from wildlife to humans, as has been well documented, anthropogenic disturbance may also heighten wild apes' exposure to human pathogens (Grützmacher *et al.*, 2018b; Köndgen *et al.*, 2008; Parsons *et al.*, 2015; Rwego *et al.*, 2008).

Given the influence of humans on disease transmission dynamics at the human–animal interface, connecting human socio-behavioral contexts to models that address threats to ape survival is vital. The use of ethnography and mixed methods approaches to support human wellbeing and animal welfare promise insights into disease spill-over risk and control at the human–ape interface (Dore, Riley and Fuentes, 2017; see Chapter 8). In addition, enhanced human socio-behavioral understanding is likely to help identify drivers that contribute to ape decline, such as wildlife trade, consumption and cohabitation. Creating integrated approaches that treat human values, decisions and their impacts on the physical world as one meta-system is key to future efforts in this area (Wallace *et al.*, 2015). In this context, support for effective (human) community health programs can benefit both ex-situ and in-situ ape projects.

As discussed below, people around the world are already implementing the One Health model at the human–ape interface. On the whole, however, there are limited avenues for sharing successes, failures and lessons learned. Fostering a global community of practice is essential for the creation and implementation of new, effective solutions for sustainability and resilience at the human–ape interface.

Community Health Initiatives as Drivers of Improved Ape Conservation

As highlighted in Chapter 1, many diseases that affect humans are also a threat to apes, especially those who have never had any contact with humans. A solid understanding of human health is thus necessary for assessing the risks they pose to apes. Factors that perpetuate a health risk to apes include poor human health, especially in remote rural areas where people share a common habitat or ecosystem with ape populations. In such areas, community health practices and interventions can affect human and great ape health, both positively and negatively.

A community is often described as a social unit that is organized around a

geographic area and shares common norms, customs and a sense of place.[2] Unlike clinical approaches, which focus primarily on the individual, community health considers the extent to which shared beliefs, norms and practices affect risk factors for human disease, including ill health. It seeks ways to leverage shared beliefs to ensure good health, including for the community as a whole.[3] The community health approach requires an understanding of disease processes that are relevant to the local community—as well as local health beliefs and how the community functions as a social unit (Goodman, Bunnell and Posner, 2014). Ecosystem approaches to health widen the lens to include the broader environment surrounding human and animal communities (CBD, 2020).

For the purposes of this chapter, the Human Development Index (HDI) can be seen as a quantitative measure of the environmental setting in which humans and apes interact. It is a composite index of life expectancy, education and per capita income indicators, used to rank countries in four tiers of human development. While the HDI is inadequate in terms of measuring inequality, it does recognize people and their capabilities—rather than economic growth in isolation—as the ultimate criteria for assessing the development of a country (Giannetti *et al.*, 2015; UNDP, n.d.). Many ape range states rank low on the HDI, largely due to relatively high levels of infant and child mortality, low levels of immunization for childhood communicable diseases and poor access to safe water sources, sanitation and health care services, especially in rural areas where contact with wild ape populations is most likely.

Particularly in range states with low HDI scores, prevention strategies employed in the human population can have a critical protective effect for humans and apes (Deem, 2016). For instance, vaccines against childhood communicable diseases routinely

Photo: Evidence suggests that the risk of disease spillover is most strongly associated with agricultural-driven decisions such as tropical forest conversion to monoculture plantations and industrial livestock production. © Alison White

offered at health facilities throughout range states reduce the circulation of disease in environments that humans and apes share. Prevention strategies are not universally accessible, however. To combat barriers to access, including logistical challenges and systemic inequality, ministries of health and local non-governmental organizations periodically organize mobile vaccination campaigns, often in collaboration with community leaders and other decision-makers. Still, households that are located on the outskirts of a community—and potentially closer to great ape habitats—may be far from areas where campaigns are organized, which can reduce their access to preventive care.

In addition, some children may not receive vaccines if their mothers or families mistrust immunizations or the intentions of workers providing these services. The exclusion of these or other community members not only reduces individual protection and the potential achievement of herd immunity, but it may also weaken community cohesiveness, a key factor affecting community health, with unknown consequences. A successful community health approach begins with a clear understanding of what different community members know about how immunizations work, and why they may or may not want or be able to access these services (Wiysonge, 2019). Clear communication about each immunized individual's value to the whole community can be part of a comprehensive strategy to increase understanding and acceptance of, as well as access to, the service.

Food and Nutritional Security

Food and nutritional insecurity is defined as a lack of secure access to sufficient amounts of safe and nutritious food for people's normal growth, development, and an active and healthy life (FAO, 2018; Ingram, 2020). Malnutrition affects people's immune system

and their ability to ward off infection, one of a range of negative outcomes that, in turn, pose indirect health threats to ape species. Solutions include nutritional education, particularly maternal and prenatal education and food supplementation, given women's vulnerability to malnutrition. As discussed below, promoting and supporting diversified food production systems is also key to food and nutritional security.

Food security programs aimed at addressing human malnutrition often focus on ensuring adequate calorie consumption, especially during droughts or extreme weather events. Nutrition-based programs do not typically emphasize broader environmental risks, such as those associated with the use of genetically modified seeds or chemical fertilizer. Such campaigns do not generally cover the risks of wild meat consumption either, even though unsustainable hunting for wild meat threatens apes (Arcus Foundation, 2020). Since the global spread of COVID-19 in 2020, however, some of these programs have begun to take on more holistic approaches (Kumareswaran and Jayasinghe, 2022).

The Community Health Approach: Beyond Food Security

Diversified, sustainable food production systems—such as agroecology, permaculture and sustainable agriculture—can help address food and nutritional insecurity, as well as unsustainable wild meat consumption. A host of local factors affect agricultural practices and a community's potential to adequately feed itself; these include soil quality, climatic conditions and socioeconomic dynamics, such as land tenure and food sovereignty. In this context, industrial agriculture, infrastructure and poor soil management are community-wide challenges. Traditional laws about the use of community land and resources, cultural food taboos and food preferences also play a role in community food security and nutritional health.

A community health approach considers all these factors and potential interventions to promote food security and nutritional health for all community members and the health of the community at large. For example, it would take into consideration whether land is being used unsustainably by some members and whether it is being polluted by harmful chemicals that directly or indirectly affect other community members (Ostrom and Cox, 2010).

The community health approach also leverages interventions provided by health facilities, community health workers and community-wide public health groups, while taking into account the social, economic and environmental determinants of health of a particular community. Globally, this idea has gained traction under the label

Photo: Health in Harmony and their local partner, Alam Sehat Lestari (ASRI), support human health and wellbeing by providing discounted health care services, education and alternative livelihood programs to communities that are protecting their surrounding forests. Treatment at ASRI's clinic can be paid for with cash, handicrafts, tree seedlings, grain, manure/compost or work.
© Alison White

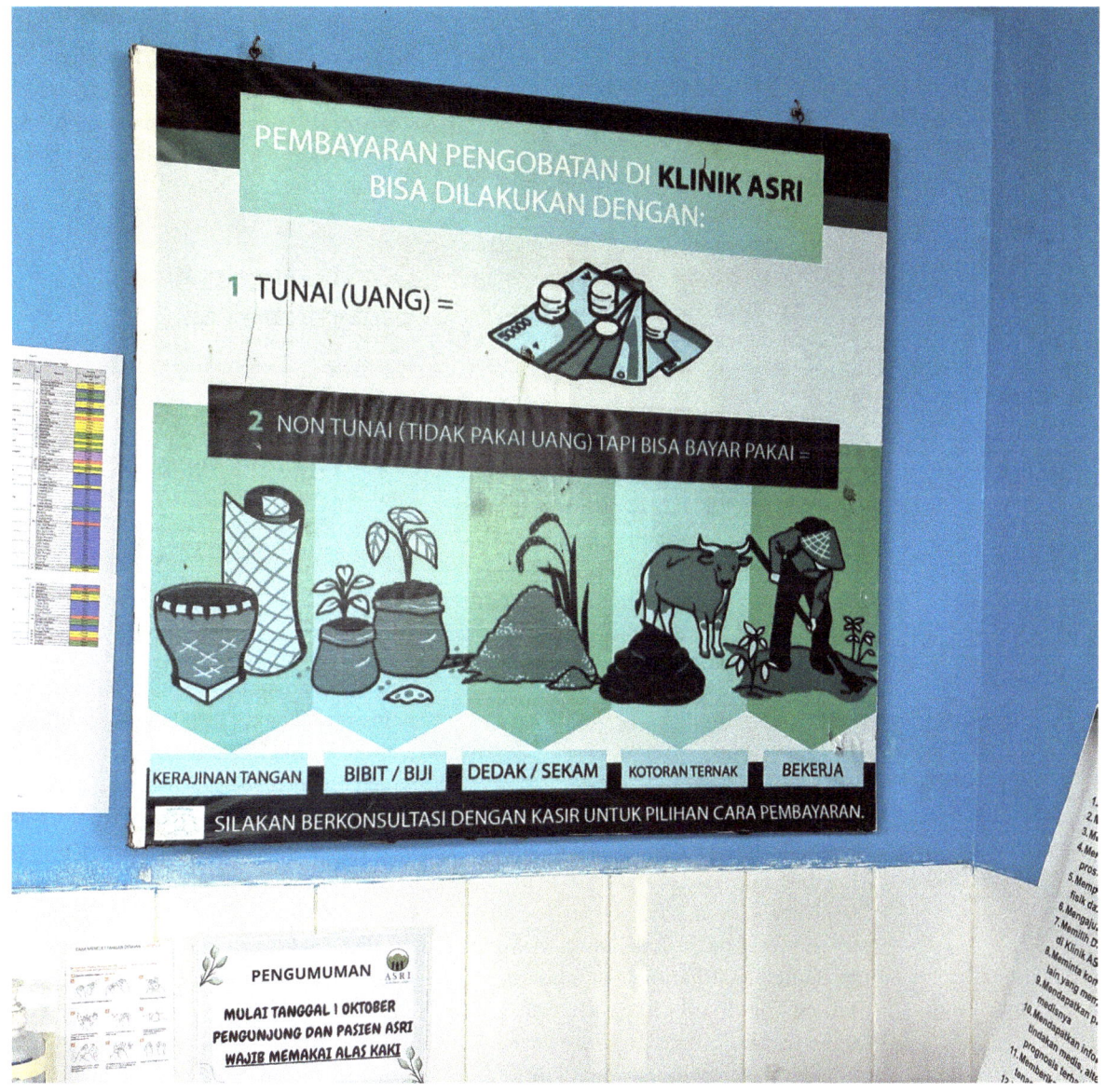

"social and environmental determinants of health" (WHO, 2012).

As discussed in the next section, the community health approach has had positive impacts around ape habitats, including when applied by conservation organizations. This volume does not examine beneficial community health outcomes outside of ape ranges, although the methodologies may be widely applicable. Relevant examples include:

- **The Konashen Community-Owned Conservation Area in Guyana**. Using ethnoprimatological techniques, this area is being studied with the aim of identifying links between sustainable primate hunting by Indigenous Peoples and cultural identity (Shaffer *et al.*, 2018). The results are being integrated into the understanding of zoonotic disease in the area (Milstein *et al.*, 2020). This research combines multiple knowledge systems from earlier work with advanced genomics to better understand and prevent emerging zoonotic diseases.

- **The PIVOT program in Madagascar**. This program serves as a framework for a model district in Madagascar, where national policies are implemented along with additional health system interventions to allow for bottom-up adaptation. The aim is to protect human health and the health and restoration of the ecosystem on which the people of Madagascar depend (Rakotonanahary *et al.*, 2021; Roberts, 2019).

One Health in Action

This section presents case studies of One Health applications at the human–ape interface, authored by the individuals who built the models. As there is no standardized method for discussing real-world One Health examples, this section uses the Berlin Principles as a unifying foundation (see

Box 2.2). By highlighting relevant Principles, the case studies indicate to what degree system-based health paradigms, including ecosystem health and planetary health, converge with One Health practices.

Each case study begins with an introduction to the problem or challenge and then appraises the solutions in progress. All employ systems-based approaches and recognize that progress is incremental, accruing over iterations and, ideally, resulting in continuously improved outcomes. The authors highlight successes and failures, data gaps that could hamper decision-making, and attempts to fill them. They also explain how solutions were implemented or improved through a One Health approach. Connections between cases help to draw out the core themes.

Community Health and Ecosystem Health

Most ape species are found in the tropics, in areas that are also home to some of the world's lowest-income communities. The land is often rich in natural resources, which tend to be exploited unsustainably, to the detriment of local people and wildlife. Rapid population growth combined with increased demand for resources has led to significant anthropogenic interference in ape habitats globally (Estrada, 2013; Junker *et al.*, 2012). The consequences include habitat loss, human–wildlife conflict, increased hunting and the spread of zoonotic diseases, all of which threaten ape survival. The critically endangered mountain gorilla (*Gorilla beringei beringei*) is in particular danger (Dunay *et al.*, 2018; Hockings and Humle, 2009; Kalema-Zikusoka, Kock and Macfie, 2002).

Case Study 2.1 highlights the role of community health—which is central to the planetary health concept—in efforts to increase positive ape conservation outcomes. Case Study 2.2 is focused on ecosystem health.

CASE STUDY 2.1

Conservation through Public Health: Towards Shared Community and Gorilla Health[4]

Theme: Incorporating community health into conservation at the human–ape interface.

Applicable Berlin Principles:

☑ 1. Conservation   ☐ 2. Strong institutions
☐ 3. Climate crisis   ☐ 4. Ecosystems
☐ 5. Disease control   ☑ 6. Biodiversity integration
☐ 7. Investment   ☑ 8. Enhanced capacity
☑ 9. Multilevel collaboration   ☑ 10. Awareness raising

Disciplines relevant to One Health: Public health, community development, veterinary science, ecology, ape conservation and welfare

Conservation through Public Health (CTPH)—a Ugandan grassroots non-governmental organization (NGO) and US-registered non-profit organization—was founded in 2003, after two outbreaks of skin disease occurred in mountain gorillas in 1996 and 2001/2002 (CTPH, n.d.-c; Graczyk *et al.*, 2001; Kalema-Zikusoka, Kock and Macfie, 2002). The causative agent—scabies—was traced to people with inadequate access to basic health services living around Bwindi Impenetrable National Park, Uganda (see Figure 2.3 and Chapter 1). Gorillas were probably infected when they foraged on community land and touched scarecrows made with infected clothing.

CTPH promotes biodiversity conservation by enabling people to coexist with wildlife through integrated One Health programs that improve animal health, community health and livelihoods in and around Africa's protected areas and wildlife-rich habitats. The organization had already implemented integrated wildlife and community health programs for ten years when it added "improving livelihoods" to its activities. It did so with the aim of addressing poverty that was exacerbating poor health among local communities (CTPH, n.d.-a). CTPH delivers its One Health activities through three integrated programs:

- wildlife conservation, including wildlife health and habitat conservation;
- community health with a focus on zoonotic disease prevention and control; and
- alternative livelihoods through a social enterprise called Gorilla Conservation Coffee, which provides local farmers with above-market prices for good coffee.

The majority of CTPH efforts target the area around Bwindi Impenetrable National Park and the surrounding communities. The ecosystem showcases the current challenges for great ape conservation, protected area management and human health and development. About 100,000 people live within 5 km of the park, which covers 331 km² (33,100 hectares) and is home to 459 (43%) of the world's mountain gorillas (*Gorilla beringei beringei*) (CTPH, n.d.-b). The main income-earning

FIGURE 2.3

CTPH Work in and around Bwindi Impenetrable National Park

Sources: Protected areas—UNEP-WCMC (2021c, 2021f, 2021i); country boundaries—GADM (n.d.); other base map detail—OpenStreetMap (n.d., © OpenStreetMap contributors, published under Creative Commons Attribution License CC BY; for more information see http://creativecommons.org)

activity in the area is subsistence farming, and up to one-quarter of all smallholder households live on less than US$ 1.25 per day (World Bank, 2018).

There is anecdotal evidence of frequent interactions between mountain gorillas and local people, as the apes forage outside the park and community members engage in unauthorized resource extraction activities in the park (Harrison *et al.*, 2015). The increasing direct and indirect contact through these interactions facilitates the transmission of commensals and pathogens, including scabies, respiratory and diarrheal diseases across the human–ape interface (Guerrera *et al.*, 2003; Rwego *et al.*, 2008; see Chapter 1).

In 2007, CTPH established Village Health and Conservation Teams comprised of community volunteers from the government-supported Village Health Team network, whose members are trained to promote health and conservation at the household and community levels. The teams promote good hygiene and sanitation, infectious disease prevention and control, family planning, proper nutrition, sustainable agriculture, and gorilla and forest conservation, including by educating their communities on the dangers of hunting and deforestation. They also document gorilla encounters near homes, enabling Human and Gorilla Conflict Resolution Teams to herd gorillas back to the park (Kalema-Zikusoka and Rwego, 2016). The Village Health and Conservation Teams reach 30,000 people in 6,000 households across 6 front-line parishes where there is conflict between people and gorillas (CTPH, n.d.-d).

The One Health approach has contributed to a reduction in human-related disease outbreaks in the gorillas. No scabies outbreaks have been recorded since 2002, and gorilla exposure to human and livestock diseases such as giardiasis has dropped (Kalema-Zikusoka *et al.*, 2018). This outcome is attributed to a combination of increased referrals of people with infectious diseases and improved health and hygiene in homes bordering park boundaries. Data collected by CTPH reveal that the percentage of pit latrines with external hand-washing facilities rose from 10% to between 30% and 75%, depending on the parish. There was also a significant increase in the proportion of women who use modern contraceptives, from 22% to 67% in the first two parishes, Mukono and Bujengwe (Ainerukundo, Gaffikin and Kalema-Zikusoka, 2019). The rate is above the national average for rural areas, which did not exceed 47% in the same period (June to December 2016) (UBOS and ICF, 2018).

Gorillas are at risk from diseases of global pandemic significance, including severe acute respiratory syndrome, or SARS, and COVID-19 (Gillespie and Leendertz, 2020; see Chapter 1). CTPH's long-term presence and established relationships meant it was well placed, starting in 2020, to assist in developing activities to mitigate the impacts of the COVID-19 pandemic on mountain gorillas.

CTPH worked with the Uganda Wildlife Authority and local partners—including the Mountain Gorilla Veterinary Project/ Gorilla Doctors, the International Gorilla Conservation Programme (IGCP) and the Max Planck Institute—to strengthen great ape viewing guidelines to prevent transmission of COVID-19 and other respiratory diseases between people and from people to gorillas (see Case Study 2.2). Park staff members were required to wear protective face masks and were trained to enforce hand hygiene and a 7-meter great ape viewing distance in 2014, which the Uganda Wildlife Authority increased to 10 meters in 2020. The same training was provided to Human and Gorilla Conflict Resolution Teams and Village Health and Conservation Teams (Kalema-Zikusoka *et al.*, 2021).

In the absence of tourism revenue for park edge communities, incidents of wild meat hunting increased and contributed to the killing of a gorilla by a community member who was hunting duiker of various species (*Cephalophinae*) and bushpigs (*Potamochoerus larvatus*). As a result, CTPH resumed efforts to support reformed hunters with group livestock projects as a way of discouraging them from going back to the park, while encouraging other community members not to hunt. CTPH also started a new emergency food relief program to provide fast-growing seedlings to vulnerable community members, mainly to address hunger brought about by the lack of tourists and other factors affecting the economy during the pandemic. In addition, CTPH encouraged them to go back to sustainable farming methods that they had abandoned to earn a living through the tourism industry.

The COVID-19 pandemic provided an opportunity for CTPH and IGCP to advocate for more responsible tourism involving great apes in Africa. They called for the adoption of International Union for Conservation of Nature guidelines on viewing distances to reduce the risk that tourists might get too close to gorillas (Hanes *et al.*, 2018; Mbayahi and Kalema-Zikusoka, 2020; Weber, Kalema-Zikusoka and Stevens, 2020). They also emphasized the need to support community health and hygiene and non-tourism-dependent livelihoods of people who share habitats with great apes. Further, CTPH is advocating for wildlife trade policies to help prevent future pandemics.

CTPH's overall experience of implementing activities through a One Health approach has been positive. The organization's conservation efforts have complemented government programs and initiatives by other NGOs, including ones with a focus on improving law enforcement, veterinary care, monitoring, research, and community engagement through education, health service provision and livelihoods linked to tourism. Together, these activities have helped mountain gorillas to be the only gorilla subspecies whose population is showing a positive growth trend over the past 25 years (Hickey *et al.*, 2019b). A recent study based on population monitoring estimates that the population in the entire Virunga Massif is growing at a rate of about 3% per year (Granjon *et al.*, 2020a).

Yayasan Ekosistem Lestari and Gorilla Doctors: Ecosystem Health in Africa and Asia[5]

Theme: Applying ecosystem health methods within a One Health approach to promote ape-focused conservation and welfare.

Applicable Berlin Principles:

☑ 1. Conservation   ☑ 2. Strong institutions
☐ 3. Climate crisis   ☑ 4. Ecosystems
☑ 5. Disease control   ☑ 6. Biodiversity integration
☐ 7. Investment   ☐ 8. Enhanced capacity
☐ 9. Multilevel collaboration   ☑ 10. Awareness raising

Disciplines relevant to One Health: Community development, veterinary science, public health, ecosystem management, reintroduction biology, ecosystem health

Yayasan Ekosistem Lestari

Yayasan Ekosistem Lestari (YEL)—the Sustainable Ecosystem Foundation—was founded in 2000 with the vision "to preserve the environment with sustainable benefits for the whole community" (YEL, n.d.-b). One of YEL's main programs is the Sumatran Orangutan Conservation Programme (SOCP), a collaboration with the Swiss PanEco Foundation and Indonesia's Ministry of Environment and Forestry. Another program is the Environmental Education Centre Bohorok, in Bukit Lawang, North Sumatra (YEL, n.d.-a, n.d.-b; see Figure 2.4). Both were established in response to the massive impact of large-scale deforestation on Sumatra's rainforest and the

FIGURE 2.4

Yayasan Ekosistem Lestari Work in North Sumatra, Indonesia

Sources: Protected areas—UNEP-WCMC (2021d); country boundaries—GADM (n.d.); other base map detail—OpenStreetMap (n.d., © OpenStreetMap contributors, published under Creative Commons Attribution License CC BY; for more information see http://creativecommons.org)

endemic fauna it supports (YEL, n.d.-b). Deforestation has a direct impact on ecosystem health because it reduces biodiversity and thus weakens the ecosystem's ability to cope with challenges.

The SOCP employs the One Health paradigm by focusing on the ecology, health and welfare of orangutans in its ex-situ conservation program, which covers the rescue, rehabilitation and reintroduction of ex-captive orangutans to establish new viable wild populations for its in-situ conservation program. This program utilizes knowledge on disease spread between orangutans and other species, including humans, to provide risk-based data on disease and health issues (SOCP, n.d.-d). To enable rehabilitated orangutans to return to a life in Sumatra's tropical rainforests, the YEL team uses the data as part of a One Health approach that combines community development in ecosystem health and support for law enforcement.

Since 2002, more than 350 orangutans have arrived at the SOCP's Orangutan Quarantine and Rehabilitation Centre in North Sumatra for the health screening they need prior to joining the rehabilitation program. In the same time, nearly half (more than 170) of the orangutans have been released to Bukit Tigapuluh National Park in Jambi (with the help of the Frankfurt Zoological Society) and around 100 have been transferred to the Jantho Pine Forest Nature Reserve in Aceh, in release operations wholly managed by YEL (SOCP, n.d.-a, n.d.-d).

At the YEL reintroduction center in Jantho, the SOCP team conducts remote surveys to monitor habitat and identify threats, assess the health and welfare of the released orangutans, and evaluate dispersal of the reintroduced orangutan population—the latter as an indicator of the ecosystem health service the species provides as a seed disperser (McConkey, 2018; SOCP, n.d.-b). The post-release monitoring team consists of YEL staff members from local villages, who are trained in monitoring orangutan behavior and conducting phenology surveys of forest composition and fruit availability. The habitat monitoring team initially comprised only YEL staff, but by mid-2019 all the habitat monitoring team members were local residents working part-time. At the end of 2019 the habitat monitoring team was fully handed over to the local community under what is called the Jantho Community Ranger program, with members representing seven local villages around the periphery of the Jantho Nature Reserve. Related data analysis is still carried out using the expertise of YEL personnel.

The plan is to transfer knowledge and the capacity for data analysis to the Jantho Community Rangers, to enable the team to carry out habitat monitoring independently, wherever and whenever needed. Via the Jantho program, a sustainable habitat monitoring team is gradually being established, with members from the adjacent communities trained in species management, habitat monitoring and protection methods, and data analysis. This work also helps the local population develop an understanding of a healthy ecosystem.

The key for orangutan conservation is protection of their remaining habitat. Rescue and rehabilitation projects cannot guarantee the survival of orangutans in the wild without the protection of the rainforests and the existing wild populations.

The SOCP's in-situ conservation focuses on wild orangutan populations in Sumatra in several ways. The SOCP manages three field monitoring stations, two in the Leuser Ecosystem, home to the Sumatran orangutan (*Pongo abelii*), and one in the Batang Toru Ecosystem, home to the recently described Tapanuli orangutan (*Pongo tapanuliensis*) (SOCP, n.d.-c). In these landscapes, students and field assistants can study orangutan behavior and ecology as well as their interaction with their forest habitat. Their research helps to inform the One Health approach by providing vital behavioral ecology information, integrated with disease surveillance data gathered during rehabilitation. These data are critical for assessing disease spread risk through the orangutan population.

YEL and the SOCP are taking a holistic approach to tackling the diverse challenges facing the health of the Leuser Ecosystem. They do so by working on the interconnected issues of orangutan conservation, habitat protection and monitoring, by promoting and encouraging sustainability, and by actively supporting direct community involvement in preserving healthy ecosystems.

Gorilla Doctors' Health Programs

Numerous international and local non-governmental organizations are working together to conserve the mountain gorillas (*Gorilla beringei beringei*) and Grauer's gorillas (*Gorilla beringei graueri*), as well as their habitat in the Democratic Republic of Congo (DRC), Rwanda and Uganda (see Figure 2.5). Part of the focus is on disease control, as described in Case Study 2.1, with different projects targeting the community, rangers, researchers and tourists. This case study focuses on Gorilla Doctors' employee health program.

Michael Cranfield, who served as Gorilla Doctors' project director from 1998 to 2019, ranked disease as the biggest threat to mountain gorillas:

> Bush meat and logging were for a long period the two leading causes of decreasing gorilla numbers in general, with disease being third. Because the mountain gorillas are in protected areas, the first two factors take a backseat to disease (Nolen, 2006).

Conservation workers and tourists spend time with habituated gorilla groups on a daily basis. Gorillas can also venture outside protected areas, where they may come into contact with local people and domestic animals. The long-term health of mountain gorillas cannot be ensured without addressing human and domestic animal health. As a result, Gorilla Doctors approaches gorilla conservation medicine from a One Health perspective (Gorilla Doctors, n.d.-e). Specifically, the organization undertakes activities as part of a suite of One Health initiatives, including:

- Facilitating annual health screenings, follow-up care and health education for people who work in the national parks through an employee health program (see below).

FIGURE 2.5

Gorilla Doctors Work in the DRC, Rwanda and Uganda

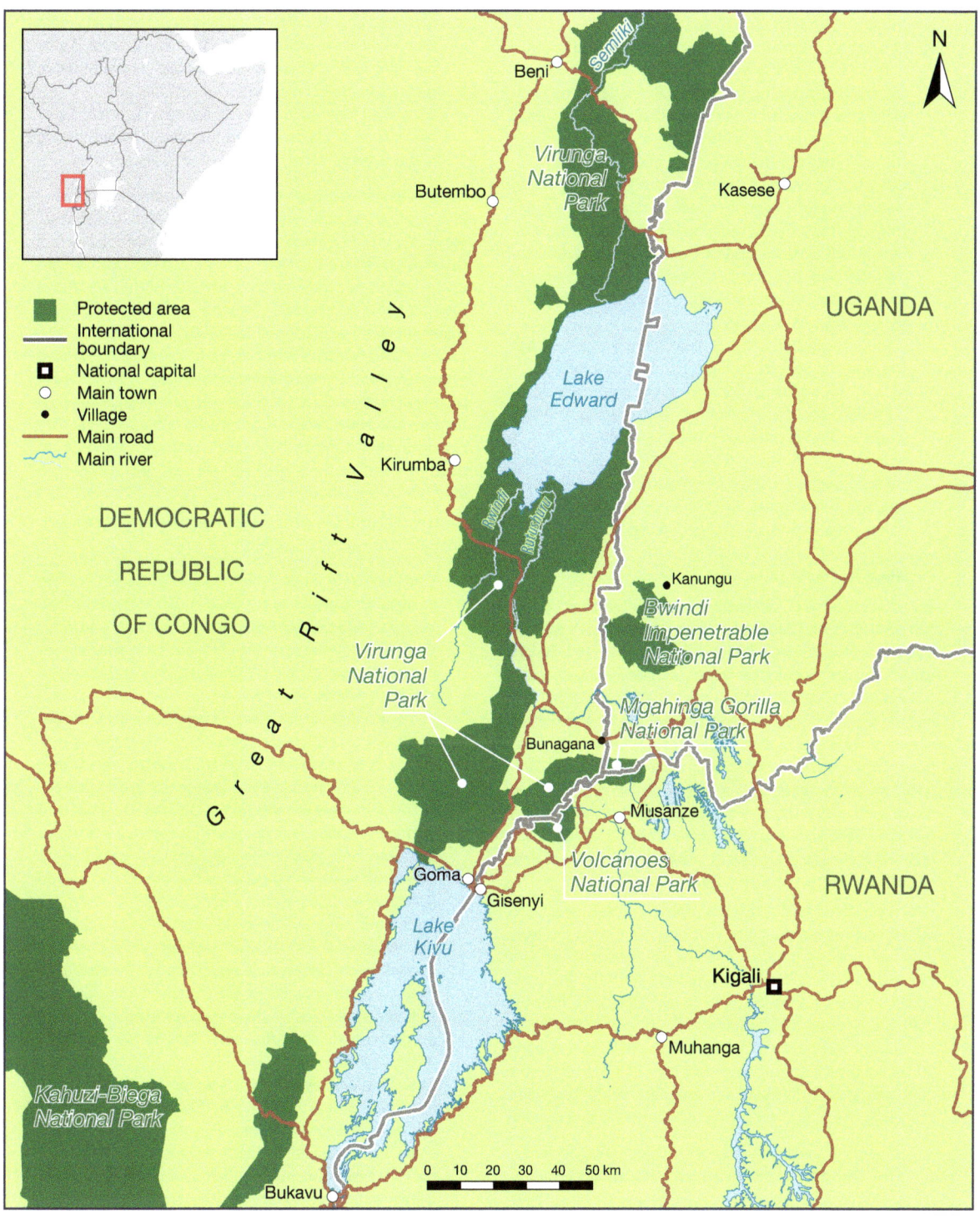

Sources: Protected areas—UNEP-WCMC (2021c, 2021f, 2021i); country boundaries—GADM (n.d.); other base map detail—OpenStreetMap (n.d., © OpenStreetMap contributors, published under Creative Commons Attribution License CC BY; for more information see http://creativecommons.org)

- Providing preventive health programs for livestock and pets living near the gorilla habitat and educating their owners about best animal care practices.
- Facilitating a feral dog control program near the gorilla habitat to reduce the number of feral dogs and cats.
- Conducting pathological examinations of tissues collected during post-mortem examinations of deceased gorillas and other wild animals with whom gorillas share the ecosystem, to determine the causes of death.
- Carrying out targeted sampling and testing of priority wildlife taxa for emerging infectious disease pathogens that pose the greatest health risk to people and gorillas, largely through collaborative research projects with universities and other research institutions, as a way of informing best practices for highly infectious disease prevention and preparedness.
- Conducting research to investigate the linkages between the health of gorillas, humans and other animals.
- Making recommendations to government and local authorities about best practices for managing a healthy gorilla population based on research findings.
- Providing local, regional and international veterinary, medical, public health, epidemiology and environment students and professionals with opportunities for internships, volunteering, training and research.

Employee Health Program

To reduce the risk of disease transmission between people and gorillas, Gorilla Doctors facilitates annual health screenings and follow-up care for people who come in contact with gorillas as part of their work to protect these populations in national parks (Gorilla Doctors, n.d.-c). Great ape site staff and tourism support workers such as rangers, guides and porters spend many hours in the forest tracking habituated great apes. They are also in indirect contact with unhabituated apes, simply by virtue of their shared use of habitat. While researchers and veterinarians may be fewer in number and perhaps spend less time in the forest, they have close contact with individual animals and groups as they collect data or treat ill or injured animals (Gilardi et al., 2015).

Each year, hundreds of rangers, trackers, researchers and others who work in the parks participate in Gorilla Doctors' employee health program. Currently, this program is offered to people who work in Volcanoes National Park, Rwanda, and Virunga and Kahuzi-Biega National Parks, DRC. The program is administered in conjunction with local hospitals, which provide the necessary facilities and medical staff. Nearly 300 national park workers and more than 2,000 family members participate in the health screening and health education programs. Through this program, conservation employees:

- undergo physical examinations and screening for infectious and chronic diseases such as tuberculosis;
- are immunized against infectious diseases such as measles, polio and rabies;
- receive treatment or referrals for additional care if they are sick;
- are linked to government programs that provide treatment for chronic diseases, such as antiretroviral drugs if they are living with the human immunodeficiency virus (HIV);
- participate in health education programs; and
- bring their families to receive treatment for tropical diseases such as intestinal parasites, as well as counselling about hygiene, family planning and HIV prevention.

Due to the close phylogenetic relationship between humans and apes, great apes are susceptible to human diseases, especially those to which they have no natural immunity due to a lack of previous exposure, such as measles, polio, severe acute respiratory syndrome coronavirus 2 (SARS-CoV-2) and tuberculosis (Ferber, 2000; Gillespie and Leendertz, 2020; Gillespie, Nunn and Leendertz, 2008). Until recently, evidence of direct disease transmission to wild apes was limited to bacterial and parasitic infections; however, several new studies provide evidence of direct viral transmission between humans and apes.[6] Also of concern are pathogens that may remain dormant in the environment for extended periods, such as some intestinal parasites (Gillespie et al., 2010; Zommers et al., 2013; see Chapter 1).

Best practices for any great ape conservation employee health program include the following:

- Examination of employees before or at the time of hire to determine eligibility for specific job responsibilities, such as fieldwork, but with permission from the employee to prevent employment discrimination against sick individuals, protect patient confidentiality and adhere to ethics guidelines.
- Fecal testing for gastrointestinal parasites and immediate treatment of employees if needed, for the following reasons: employee infection rates tend to be high; employees are the people most likely to defecate in great ape habitat because they are often in the forest all day; treatment regimens are standard; drugs are inexpensive and widely available; and gastrointestinal pathogen transmission between people and great apes has been documented (Gilardi et al., 2015; Parsons et al., 2015).
- Recommending and directing employees to appropriate hospitals and clinics for treatment of any medical conditions or infections detected through the employee health program, combined with the referral of employees with confirmed chronic disease cases to national health programs or to other local institutions for more complex or emergency care.
- Extending employee health program services to family members to help achieve overall objectives—wherever the requisite financial, human and other resources are in place. In Rwanda, for example, there is a comprehensive national health insurance scheme for all citizens, which allows for referrals for chronic and complicated medical conditions detected through the employee health program.

In contrast, the bulk of medical bills associated with such referrals in the DRC can only be covered through fundraising.

Partnerships for Community and Animal Health Programs

Gorilla Doctors understands the interconnectedness between humans, domestic animals, wildlife and other elements in the ecosystem as determinants of health, while also recognizing the need for a multisectoral approach (Gorilla Doctors, n.d.-c). Consequently, the organization has partnered with local human and animal health care providers to participate in One Health programs in the gorilla conservation area. Over the past few years, this work has involved raising community awareness, vaccination, disease surveillance, research and advocacy, including through the following initiatives:

- surveillance for zoonotic diseases of public health importance;
- veterinary support to rural communities living within or close to protected areas;
- support for capacity building of local animal health professionals, along with veterinary student training and internship opportunities;
- massive dog and cat vaccination campaigns to try to eradicate human and animal rabies;
- vector-borne disease control through spraying of mosquitos and other vectors; and
- advocacy for community action in One Health and engagement of local government and local communities on improved domestic waste management, especially around conservation areas.

The Role of Capacity Development

The Global Health Security Agenda identifies a critical lack of adequately trained wildlife health professionals around the world (GHSA, 2020). For projects such as those illustrated above to be successful, effective capacity development programs linking One Health principles to veterinary health and conservation management practice are thus needed.

The Pan African Sanctuary Alliance in Africa and the Orangutan Veterinary Advisory Group in Southeast Asia have facilitated ape-focused capacity building programs for several years (OVAG, n.d.; PASA, n.d.-b; see Chapter 4). These programs focus on practitioner education and empowerment through the creation of health-focused networks to increase the capacity to improve ape health. They are designed to provide veterinary-led training that advocates the inclusion of One Health principles in ape-focused conservation action plans (Unwin *et al.*, 2022; see Case Study 2.3). They pair community-based domestic animal, human and wildlife health practitioners with academics from all continents and practitioners within the International Union for Conservation of

CASE STUDY 2.3

A Team-Based Approach to Ape Health Networks to Develop an Ape-Focused Global Health Initiative

Theme: Building capacity to allow for the application of One Health approaches.

Applicable Berlin Principles:

- ☐ 1. Conservation
- ☑ 2. Strong institutions
- ☐ 3. Climate crisis
- ☐ 4. Ecosystems
- ☑ 5. Disease control
- ☐ 6. Biodiversity integration
- ☐ 7. Investment
- ☑ 8. Enhanced capacity
- ☑ 9. Multilevel collaboration
- ☑ 10. Awareness raising

Disciplines relevant to One Health: Veterinary science, public health, tertiary education practice, science communication, succession planning

Together, the Pan African Sanctuary Alliance and the Orangutan Veterinary Advisory Group (OVAG) represent an ape-focused global health initiative. They are capacity-strengthening and expertise networks that synergize efforts from multiple organizations, all of which aim to improve their individual and collective impact on ape welfare and conservation.

To achieve maximum impact, the organizations involved in this network use interdisciplinary One Health principles whenever approaches or aims overlap. For example, practitioner mental health and resilience are vital to maintaining a community of practice that can sustain One Health programs. In nearly all ape health contexts, practitioners who inform welfare assessments of ape populations can benefit from an understanding of their own mental states and behaviour that could drive disease risk, including in relation to the transmission of zoonotic disease (see Chapter 8).

Network members build capacity via didactic and problem-based workshops, practical laboratory sessions, online advisories, and as advocates between participants and their organizations. As a collective, the network provides direct input into conservation management at the organizational and government level. Capability is enhanced through a forum, which empowers ape health practitioners and academics to formulate plans relevant to all wildlife health management needs. As a rule, these plans are respected and utilized by decision-makers. The assumption—based on global One Health guidance from the World Organisation for Animal Health and the One Health High-Level Expert Panel—is that this consolidation of experts from a wide variety of organizations improves individual impact in conservation and welfare efforts (OHHLEP *et al.*, 2022; WOAH, 2021).

In evaluations carried out over the past decade, the OVAG program has been found to help improve the quality of participant outputs on an iterative basis (Unwin *et al.*, 2022). OVAG's integration of One Health programs into welfare and conservation efforts has led to improved disease mitigation strategies in both in-situ and ex-situ populations, particularly by linking public health and environmental disease issues that contribute to the protection of apes, their habitat and human health. Utilizing community engagement platforms such as WhatsApp is a core approach of the United Nations strategy and is embedded in the Sustainable Development Goals and the Global Health Security Agenda (Armstrong-Mensah and Ndiaye, 2018; UN, 2019).

Every network participant has a responsibility and role to play, including wildlife center administration and participating staff, local public health representatives, and community leadership and members. Key goals of the network's internet information hub, which allows open access to all participants simultaneously, are the longevity of its hardware, software and content (capacity), as well as human participant capability long into the future. With respect to the transfer of knowledge, the hub's features are aimed at minimizing inequalities among staff members from different stakeholder organizations and among local communities. OVAG has already registered evidence of at least partial achievement of these desired outcomes, based on increased staff retention, successful train-the-trainer initiatives and positive reviews from network participants, over 80% of whom said that their participation was either critical or very important in shaping their decision-making processes (Unwin *et al.*, 2022).

Nature, the American Association of Zoo Veterinarians and the European Association of Zoo and Wildlife Veterinarians networks. In 2020, these networks were further integrated via a combined concern over the SARS-CoV-2 threat to all apes, via a new online technical service, the Non Human Primate COVID-19 Information Hub (University of Minnesota, n.d.-b).

Disease risks can only be mitigated effectively if capacity building on One Health extends beyond conservation to cover the extractive industries and industrial agriculture sectors. As discussed in the next section,

ape habitats face significant threats from land use changes associated with resource extraction and agricultural expansion (Arcus Foundation, 2014, 2015).

Resource Management and Land Use

Agriculture is the biggest cause of habitat loss in ape range countries, particularly in Asia, given the global demand for oil palm products (Estrada *et al.*, 2017; Williams *et al.*, 2021). Significant damage is also done by the extractive industries, such as large-scale logging and mining, as well as clearance for large-scale infrastructure. Ape habitats are also threatened by unsustainable local-level exploitation and extraction of resources, both of which are driven by a range of actors.

The links between threats driven by various land uses are complex. For instance, the expansion of industrial agriculture in one area may drive communities from their land and push them to farm within a protected area. Similarly, the establishment of a national park may prevent communities from accessing traditional food sources within the forest and push them into other economic activities, which ultimately threaten ape habitat. Many people who live around protected areas—including some who may have been evicted when such areas were created—still rely on accessing protected ape habitats for non-timber forest products such as firewood, charcoal, food, fodder, thatch and medicines. Effective implementation of the One Health approach requires consideration of the diverse land uses driven by different sectors at different scales, as well as the complex links between them.

Land use changes not only risk degrading environmental health, but they can also impact ape health through loss of habitat, loss of preferred diets, heavy metal contamination and the introduction of disease (Estrada *et al.*, 2017). In addition, apes may be attracted to newly disturbed sites due to the availability of food (cultivated or herbaceous secondary growth), which can further increase their exposure to hazards such as human and livestock excrement, as well as mining pollution (Ontl, 2017). Moreover, increased contact between apes, people and domestic livestock can lead to the transmission of diseases (Parsons *et al.*, 2014; Spelman *et al.*, 2013). One Health approaches have enabled positive ape health and conservation outcomes in the context of land use changes, however. Case Study 2.4 introduces collaborative approaches that successfully integrate land use practices with ape conservation programs.

CASE STUDY 2.4

Using One Health to Link Land Use to Ape Health

Theme: Balancing approaches to land use with ape health and conservation needs.

Applicable Berlin Principles:

☑ 1. Conservation   ☑ 2. Strong institutions
☑ 3. Climate crisis   ☑ 4. Ecosystems
☐ 5. Disease control   ☑ 6. Biodiversity integration
☑ 7. Investment   ☐ 8. Enhanced capacity
☐ 9. Multilevel collaboration   ☑ 10. Awareness raising

Disciplines relevant to One Health: Community health, land tenure and environmental law, community conservation, resource management

Conservation organizations have developed different ways to mitigate the impacts of land use changes caused by agriculture and resource extraction on ape habitats and other wildlife areas. These include improved land tenure, land use planning, resource management, sustainable use and legislation for conservation (TAWIRI, 2018). Other initiatives involve working with local communities to improve their health, education and livelihoods, as well as advocacy and campaigns against destructive industrial activities and their financing. These approaches—which often focus on addressing human rights and land use issues—tend to entail collaboration between conservation organizations, local communities, private-sector firms and governments. Multisectoral and multidisciplinary approaches are required to coordinate holistic and sustainable solutions to ecosystem, human and animal health.

Improving Human Health and Wellbeing

A successful approach to reducing illegal logging and forest loss on Borneo is implemented by the US-based planetary health organization Health in Harmony. The organization focuses on improving human health and wellbeing by providing discounted health care services, education and alternative livelihood programs to communities that are protecting their surrounding forests (HiH, n.d.). A ten-year study of the Gunung Palung National Park indicates that this approach

FIGURE 2.6

Health in Harmony Work in Central and West Kalimantan, Indonesia

Sources: Protected areas—UNEP-WCMC (2021d); country boundaries—GADM (n.d.); other base map detail—OpenStreetMap (n.d., © OpenStreetMap contributors, published under Creative Commons Attribution License CC BY; for more information see http://creativecommons.org)

reduced illegal logging and simultaneously improved health care access and health outcomes (Jones *et al.*, 2020; see Figure 2.6). The focus on improving human health expanded to the Bukit Baka Bukit Raya National Park in Indonesia, where Health in Harmony partnered with the UK based wildlife welfare group International Animal Rescue, which coordinates orangutan rescue and rehabilitation (Finley, 2019).

Buffer Zones

Buffer zones are areas of controlled resource extraction around protected areas. They provide benefits to local communities while reducing pressures associated with human encroachment on wild spaces (Nepal and Weber, 1994). In Bwindi Impenetrable National Park in Uganda, the International Union for Conservation of Nature and UNESCO support the demarcation of multiple-use zones for regulated harvest of certain resources (such as honey and medicinal plants) by authorized resource users (Harrison *et al.*, 2015). Despite these allowances, however, illegal resource extraction continues in some areas. These incursions increase the risk of human–wildlife conflict and disease transmission to gorillas. Drivers of unauthorized resource extraction include poverty and resentment linked to historic evictions from the park, "crop-raiding" by wildlife and perceived unfairness in benefit sharing (Harrison *et al.*, 2015). Local knowledge and engagement in the establishment of buffer zones is necessary for their success (Nepal and Weber, 1994). Alternative approaches to buffer zones include community-managed forests in land-sharing arrangements (Estrada *et al.*, 2017). In Bwindi, conservation organizations such as Gorilla Doctors, the International Gorilla Conservation Programme and the World Wide Fund for Nature foster local engagement to minimize encroachment into the forest.

Agricultural Innovation and Transformation

Another approach to reduce habitat destruction and resource extraction in ape-range countries is to improve livelihoods for small- to medium-scale farmers. Agriculture is the most common source of livelihoods for these communities (Plumptre *et al.*, 2004). Improvements to farming practices can be achieved through conservation agriculture and revitalization of some of the traditional farming techniques that were eroded by the green revolution (FAO, n.d.-b; John and Babu, 2021). These approaches, utilised by stakeholders such as the International Livestock Research Institute and the UN's Food and Agriculture Organization, advocate reduced tillage, permanent soil coverage, reduced use of agrochemicals, increased plant diversity and supporting ecosystem services with trees and other natural features (Arcus Foundation, 2015; FAO, n.d.-a; see Box 1.4). Improvements in livestock farming can be secured through enhanced nutrient cycling based on maintaining association between crops and livestock and controlling livestock disease through biosecurity,

vaccination and disease-resistant breeds (ILRI, 2019). Smallholder farmers can benefit from new technologies to improve production, including mobile phone innovations; access to markets and reliable value chains for produce; and education in new business models (ILRI, 2019). The combined benefits of improving environmental and livestock health through a holistic approach include enhanced livelihoods and hence better human health and nutrition, as well as a reduction in the pressure on apes and their habitats.

Community-led Conservation

It is widely recognized that Indigenous Peoples and Local Communities (IPLCs) are central to effective conservation of nature. Ranging from community conservancies and community forestry to integrated conservation and development, a wealth of holistic approaches attempt to reconcile the needs of species and ecosystems with the cultural, social and economic wellbeing of IPLCs. "Territories of life"—a term used by IPLCs that is gaining traction within the conservation sector and among international and local NGOs—refers to areas governed and conserved by IPLCs according to their particular cultures, governance systems and practices. The *Territories of Life* report presents a series of case studies that showcase holistic approaches used by IPLCs to protect their territories and the life they harbor. These approaches involve conducting participatory mapping and securing land rights; revitalizing environmental knowledge and language, including about how to live alongside and care for species and their habitats; strengthening governance systems and developing conservation by-laws; and supporting nature-friendly food production systems and livelihoods (ICCA Consortium, 2021). The report links these approaches to the improved protection of ecosystems and the species they support, as well as to improved health and wellbeing of communities. Several case studies focus on territories of life that are home to apes, including Hkolo Tamutaku K'rer (the Salween Peace Park) in Myanmar, Kisimbosa in the DRC and Yogbouo in Guinea (ICCA Consortium, 2021).

Internal Health Systems: Microbiomes and Ape Health

Biodiversity loss can both promote and be affected by emerging infectious diseases, with great potential for shaping human history through epidemics and pandemics (Keesing *et al.*, 2010; Nicholson, 2016). Managing these threats depends as much on effective internal microcosms that can interpret changes in the world and respond appropriately, as it does on external ecosystems. In this context, One Health practice is relevant to successful immune and microbiome system operation.

A significant part of the internal microcosm is the gut microbiome. Evidence suggests that, in humans, health status and particularly gut microbiome dysbiosis (imbalance) may be driving clinical signs. Confounders of this relationship include lifestyle factors such as diet, including fiber intake, which is driven by socioeconomic status in humans; behavior, including periods of stress or conduct risky to health; demographic factors such as race, host genetics and geography; and the use of antibiotics (Clayton *et al.*, 2016). Dysbiosis can occur for several reasons (which may apply concurrently), including a reduction in certain gut microbes; an increase in harmful infectious pathogens; increases in the prevalence of an ordinarily commensal bacterium; and a decrease in microbial diversity (Gagliardi *et al.*, 2018). Further, the status of the human gut microbiome has been linked to chronic diseases such as autoimmune and inflammatory conditions that affect the gastrointestinal system, such as ulcerative colitis and Crohn's disease, colorectal cancer, diabetes, Kwashiorkor, non-alcoholic fatty liver disease and obesity (Clayton *et al.*, 2016; Gevers *et al.*, 2014; Turnbaugh *et al.*, 2008; Yang and Jobin, 2014).

While investigations into ape microbiomes remain a nascent research area, studies in monkeys show convergence of

CASE STUDY 2.5

The Primate Microbiome Project: One Health and the Individual Microcosm

Theme: Linking internal health to One Health.

Applicable Berlin Principles:

☑ 1. Conservation   ☑ 2. Strong institutions
☐ 3. Climate crisis   ☐ 4. Ecosystems
☑ 5. Disease control   ☐ 6. Biodiversity integration
☐ 7. Investment   ☑ 8. Enhanced capacity
☐ 9. Multilevel collaboration   ☑ 10. Awareness raising

Disciplines relevant to One Health: Microbiology, veterinary science, evolution, medical science, public health

There are currently no data on the effect of gut microbiome changes on non-infectious and infectious gastrointestinal disease in apes — nor is there even knowledge on whether changes occur. The Primate Microbiome Project was established to develop a systematic map of variation in microbiome structure and function across all primates and to relate the findings to primate behavior, conservation, evolution and health (PMP, n.d.).

For rehabilitant orangutans, gastrointestinal illness represents a significant barrier to successful release into the wild. Indeed, anecdotal and peer-reviewed evidence indicates that clinical gastrointestinal disease is an important issue in all captive orangutans (Strong *et al.*, 2016). Unpublished data highlight that gastrointestinal upset in orangutan rehabilitants, without evidence of a confirmed pathogenic cause or origin, appears to worsen during the translocation process and following release into the wild (Y.S. Saraswati and C. Nente, personal communication, 2019). Endoparasites in clinically healthy wild and semi-captive orangutans have also been reported; these may exacerbate clinical signs due to a dysbiosis — an imbalance of the gut microbiome (Labes *et al.*, 2010; Mul *et al.*, 2007).

Since orangutans live in captive, semi-captive and wild settings and some are (re-)released into the wild, there is an opportunity to examine microbial transmission and determine how resilient or susceptible a microbiome is based on lifestyle. Studies can consider whether and under what circumstances orangutans in captivity acquire human microbes, given that such transfers have been documented in sanctuary chimpanzees (Schaumburg *et al.*, 2012). Daily record-keeping in such environments allows for monitoring of the effects of diet and other lifestyle factors on microbiome composition. One lifestyle factor is antibiotic exposure, which is prevalent in captive settings both in and outside of range countries and has been demonstrated to be a risk factor for wild chimpanzees (Parsons *et al.*, 2021).

In 2019, a collaborative One Health mapping project was launched by the Primate Microbiome Project, the British and Irish Association of Zoos and Aquariums, the University of Birmingham and the Orangutan Veterinary Advisory Group (OVAG) to investigate this issue from an integrated perspective, in both in-situ and ex-situ scenarios. The aim was to begin answering the question, "What does the gut microbiome look like in a healthy orangutan?" In 2022, following delays linked to COVID-19, a pilot study began to assess the microbiome of orangutans housed in a UK zoo. Perhaps more importantly, this project has resulted in the training of OVAG participants in DNA extraction, polymerase chain reaction and gene sequencing to continue investigations in the field.[7]

In the future, this project aims to examine gut microbiome interactions between orangutans and humans at points of stress in a captive orangutan's life—for example, after confiscation and the initial rehabilitation process; during transportation between zoos; and during release back into the wild—and compare those to the wild or "normal" microbiome. The gut microbiome of wild populations has not yet been investigated either; one of the initial tasks of this study is to examine the "normal" microbiome of Bornean orangutans (*Pongo pygmaeus*) in Sabah and Sumatran orangutans (*Pongo abelii*) in the Leuser Ecosystem of Indonesia. Data gathered from the following perspectives are to inform orangutan conservation management decisions:

- **Biological perspective:** How is gut health affected by microbial diversity, dietary traits and behavior? What are the effects of different habitats on the orangutan microbiome (such as disturbed forests compared to intact habitats)? Does the microbiome of infants who drink breast milk differ from those of infants who are fed formula?
- **Health perspective:** Is gut microbial diversity species-specific and linked to a healthy gut? Is diversity a good indicator and possible driver of susceptibility to gastrointestinal pathogens?
- **Evolutionary perspective:** What are the effects of host genetics versus lifestyle factors on shaping the gut microbiome? What impact does acquisition of human microbes have on orangutans in captive and semi-captive settings?
- **Conservation perspective:** Are humans introducing microbiome changes in wild populations that could lead to increased disease risk? Are humans creating an antibiotic resistance issue through conservation reintroductions? Do the microbiomes in captive and semi-captive orangutans differ from those of wild orangutans? If so, what are the potential negative consequences for populations in the long term? Moreover, what could be the reasons for the differences—could they be linked to changes in diet, exposure to humans, artificial circumstances, antibiotic usage, changes in social structure or the reduction in opportunities to acquire microbes from other individuals?

their gut microbiome with that of their human caregivers (Clayton *et al.*, 2016, 2018). This trend cannot just be due to diet, as zoo-based diets for primates are arguably much better than those of many humans in the developed world (Muegge *et al.*, 2011; Nagpal *et al.*, 2018). What, then, are the environmental factors that drive this convergence? Are they the same drivers that have changed the human microbiome over the past several thousand years—water quality, pollution

and behavior? Case Study 2.5 explores a One Health project that examines potential links between external drivers, the orangutan microbiome, and orangutan health at the individual and population levels.

Translational Medicine and Ape Health

One Health can be interdisciplinary, multidisciplinary and translational in its approach. Translational medicine expedites the discovery of new diagnostic tools and treatments by using a multidisciplinary, highly collaborative approach. It links medical research, medical practice and community to produce a holistic approach to medicine, linking patient to environment. Case Study 2.6 illustrates how this holistic approach is applied to ape cardiac health research to improve both clinical and animal management decisions. Translational medicine encourages using methodologies across species to confirm cardiovascular "normals" and abnormalities.

Conservation Projects and the Impact of Human Activity on Ecosystem Health

The real and perceived risks of wildlife-associated diseases for human and domestic animal health can potentially erode public support for wildlife conservation itself (Buttke, Decker and Wild, 2015). Wildlife conservation projects are therefore ideally placed to inform public perceptions of disease risks, particularly through community engagement and related efforts in environmental protection. Conservation practice itself can become more effective at mitigating harm by recognizing and understanding the complexity of social impacts on the environment and on wild animal populations and individuals, despite the lack of

CASE STUDY 2.6

International Primate Heart Project: Translational Medicine in Ape Health

Theme: Indicating the importance of the translational medicine approach to One Health.

Applicable Berlin Principles:

☐ 1. Conservation ☑ 2. Strong institutions
☐ 3. Climate crisis ☐ 4. Ecosystems
☑ 5. Disease control ☐ 6. Biodiversity integration
☐ 7. Investment ☑ 8. Enhanced capacity
☐ 9. Multilevel collaboration ☐ 10. Awareness raising

Disciplines relevant to One Health: Medical science, veterinary science, capacity building, tertiary education

Historically, cardiac disease has been shown to be a condition of concern in ex-situ primate populations, particularly in North American and European zoos (Lowenstine, McManamon and Terio, 2016; Strong et al., 2016). In captive apes, idiopathic myocardial fibrosis and cardiomyopathy were found to predominate. In humans, idiopathic myocardial fibrosis is abnormal and pathogenic; in chimpanzees, however, it may be normal. Similar cardiomyopathy lesions are rarely found in wild apes. Vascular changes in the heart and kidneys and aortic dissections in captive gorillas and bonobos suggest that hypertension may be involved in pathogenesis (Lowenstine, McManamon and Terio, 2016). Anecdotal evidence indicates that all bonobos in American Zoo Association collections are on some sort of cardiovascular medication (N. Lung, personal communication, 2020). If that is the case, is it due to misdiagnosis? Or is something in ex-situ environments causing captive apes to suffer from similar organ issues to humans?

Analysis of cardiac disease in apes often uses inferences based on data from their closest genetic relatives—humans. The Great Ape Heart Project was formally established in 2010 to study and understand cardiac disease in great apes in captivity, using data from American Zoo Association collections (Detroit Zoological Society, n.d.). The International Primate Heart Project (IPHP) followed in 2012, with the aim of creating improved understanding of cardiac disease in great apes, initially as a collaboration between veterinary practitioners, cardiac physiologists and cardiologists (Cardiff Metropolitan University, n.d.). In 2016, the Ape Heart Project was launched by Twycross Zoo, the University of Nottingham and the University of Birmingham to "develop a collaborative and co-operative approach to the investigation of heart diseases among the European great ape population" (BBC, 2016; Twycross Zoo, n.d.).

The IPHP methodology highlights consistency of data gathering—using a multidisciplinary team of cardiologists, echocardiographers, physiologists and veterinarians, who travel internationally to gather consistent, and therefore comparable, cardiac data and provide bespoke training to field practitioners. The team's first publication was a how-to guide on conducting a cardiac assessment (Shave et al., 2014). IPHP postulates that veterinary professionals may be better able to diagnose, treat and manage captive great apes with, or at risk of developing, heart disease by applying a thorough, systematic "animal in the environment" approach, rather than treating the cardiovascular system as a separate entity.

Given widespread data deficiencies, it is common across the wildlife veterinary profession to use "similar" species' clinical values as a proxy for species on which information is limited. IPHP results suggest that this proxy approach is not ideal for cardiovascular disease. The project encourages professionals to use a comprehensive approach to cardiac assessments—one that employs various ultrasound modalities to provide a thorough description of overall cardiac structure and function, which can then be used to inform clinical opinion (Shave et al., 2014). IPHP data indicate that translational medicine approaches could have profound impacts on both the welfare and conservation management of captive apes as well as wild populations. Evidence for this position is growing, supported by the IPHP network's scientific studies on ape cardiovascular health, cardiac structure and function, and electrocardiogram assessments (Curry et al., 2023; Drane et al., 2019, 2020).

This work could have far-reaching implications for medical research. If consensus is reached that the human being is not a good model for ape cardiac health, questions may similarly be raised with respect to the widespread use of domestic species as physiological models for the health of wildlife species. Moreover, in addition to improving decision-making processes in apes' and other animals' cardiac health, this understanding and use of translational medicine could inform the evolution of and approach to cardiovascular disease in humans themselves (Drane et al., 2019, 2020).

The Borneo Nature Foundation: Ecosystem Conservation and One Health

Theme: Integrating the One Health approach into ecosystem conservation.

Applicable Berlin Principles:

☑ 1. Conservation   ☑ 2. Strong institutions

☑ 3. Climate crisis   ☑ 4. Ecosystems

☐ 5. Disease control   ☐ 6. Biodiversity integration

☑ 7. Investment   ☐ 8. Enhanced capacity

☐ 9. Multilevel collaboration   ☑ 10. Awareness raising

Disciplines relevant to One Health: Ecology, disaster preparedness, community development, capacity building

The Borneo Nature Foundation and One Health

Apes have a role to play in maintaining a healthy ecosystem. Borneo Nature Foundation (BNF) is a non-profit wildlife and biodiversity conservation and research organization that protects and safeguards tropical rainforests and the environment in Borneo (see Figure 2.7). BNF collaborates with the Central Kalimantan Department of Environmental Services, the provincial division of the Indonesian Ministry of Environment and Forestry, and with the Sebangau National Park authorities. As a landscape conservation-focused organization, BNF has successfully merged ape population health surveillance with landscape conservation goals. Data gaps remain, particularly regarding disease investigation within the system, but the research infrastructure is already in place to include that target in future operations.

BNF's research-focused, iterative approach has provided a robust framework for tackling future One Health concerns in the region (BNF, n.d.-d). Compared to great ape research, the study of wild gibbons is particularly difficult due to the speed with which they move through the canopy and the difficulty of habituating them. These factors render disease surveillance in the field challenging. Nevertheless, BNF is the only project to have more than seven years' worth of accumulated data on the behavior, health and wellbeing of wild, individually identified animals in three populations, which provides useful ecological and health data as a foundation for One Health-focused research and messaging (S. Cheyne, personal communication, 2021).

Recent publications from the BNF research team and collaborators reveal the importance of incorporating the social sciences into orangutan conservation (Chua *et al.*, 2020; Palmer, 2020; Sherman *et al.*, 2021). These studies explored the hypothesis that greater conservation benefits could be achieved if ape practitioners deliberately and consciously reduced the pace of decision-making in the face of environmental crises (real and potential) and thought more reflexively and creatively about how the work they do can be improved (Chua *et al.*,

FIGURE 2.7

Borneo Nature Foundation Work Area

Sources: Protected areas—UNEP-WCMC (2021d); country boundaries—GADM (n.d.); other base map detail—OpenStreetMap (n.d., © OpenStreetMap contributors, published under Creative Commons Attribution License CC BY; for more information see http://creativecommons.org)

2020). This openness to changing their paradigm and improving preparedness can produce more impactful responses.

BNF's Use of a Systems Approach to One Health Challenges

BNF's projects integrate wildlife, landscape and Indigenous culture via community-led initiatives. These include monitoring the distribution, population status, behavior and ecology of both the Bornean orangutan (*Pongo pygmaeus*) and Bornean gibbons (*Hylobates funereus* and *Hylobates albibarbis*) (BNF, n.d.-d). The systems approach allows BNF to target the drivers of ape population decline and disease in the area. It also enhances the local human population's understanding of their contribution to maintaining healthy primate populations—in terms of the size of the population, its genetic potential and protection against disease. As discussed below, the approach include projects focused on fire prevention, youth, social forestry schemes, community engagement, and orangutan and gibbon health.

Preventing Forest Fires: Peatland Protection and Restoration in Sebangau National Park

Whereas BNF previously ran a program on reacting to forest fires, today its focus is on prevention—a central tenet of One Health practice (BNF, n.d.-c). This initiative involves:

- forest restoration, particularly in the area bordering the major city of Palangka Raya;
- community patrol teams whose aim is to prevent illegal logging, illegal hunting and electric fishing;
- community-based fire-prevention units that patrol and extinguish any identified fires;
- the detection, mapping, reporting and blocking of illegally built drainage channels, which dry out the peatland forest and make the area more susceptible to fire; and
- new technologies, including handheld data collection devices; aerial drones fitted with thermal imagery software to spot and map fires; and automated data loggers to collect hydrological data (see Chapter 6).

All of these activities are coordinated with the Sebangau National Park Authority, the regional Disaster Management Agency and the Peatland Restoration Agency to ensure an integrated, collaborative approach, which supports a network of fire-fighting teams.

Connecting Young People to Nature

The Sebangau research camp, managed by BNF partners at the University of Palangka Raya, is located on the edge of the forest. Less than an hour from the city center, the camp includes a network of forest trails and an environment rich in wildlife. The camp allows young people to experience the rainforest and learn to care for and protect it. One aspect of this initiative is the Gibbon Goes to School education program for 6–8-year-olds.

Social Forestry Schemes and Habitat Protection through Industry Engagement

The Rungan River Landscape contains 1,474 km^2 (147,357 hectares) of forest, a significant proportion of which is in industrial tree concessions, with between 2,220 and 3,275 orangutans living in a mosaic of habitats (Ancrenaz *et al.*, 2021; Jong, 2022). The landscape represents one of the largest unprotected areas of forest in the Bornean lowlands and is critically important not only for biodiversity conservation, but also for supporting the livelihoods of Indigenous Dayak people (Liswanti *et al.*, 2004). To protect this forest, BNF is working together with government partners to promote the expansion of a local social forestry initiative, through which villages can claim management rights over their forest, to ensure its protection as a village resource for the future (S. Cheyne, personal communication, 2021).

Community Engagement

Having identified that wider community engagement was urgently needed, BNF prioritized the construction of a new education and conservation hub in 2020–2021 (BNF, n.d.-e). The hub supports all community projects in the village of Kereng Bangkerai, the gateway to the Sebangau National Park, including:

- community-based fire-prevention units;
- children's education activities;
- research teams engaged in data processing and the writing of reports; and
- the Sebangau National Park Authority, which plans to construct a visitor's center to showcase the park to visitors.

Promoting Healthy Populations of Orangutans and Gibbons

Aerial surveys of orangutans. BNF was founded by scientists studying the density and distribution of orangutans and collecting field-based nest data (BNF, n.d.-a). Plans are in development to supplement field-based data collection with drone-based surveys, to increase both the scale and efficiency of BNF's surveys. In partnership with Liverpool John Moores University, researchers are planning to use drones to locate nests and attempt to locate apes in the forest using thermal imaging cameras (BNF, n.d.-b).

Assessing the status of key endangered species in Kalimantan. In addition to studying orangutans, BNF's scientists observe many other endangered species in Kalimantan. These include gibbons, wild cats, birds and sun bears (*Helarctos malayanus*). In 2020, they also began to conduct riverine surveys of crocodilians and proboscis monkeys (*Nasalis larvatus*). Their aim is to improve ecosystem health by presenting information on important hotbeds of biodiversity and encouraging a rounded approach to biodiversity conservation in Central and West Kalimantan.

clear or singular solutions to some problems (Bennett *et al.*, 2017; Game *et al.*, 2014). A case in point is the Borneo Nature Foundation, a conservation project whose approach integrates consideration of the human impact on ape health (see Case Study 2.7).

Conclusion

The concept of One Health has matured since the term was first used around 2003 (Mackenzie and Jeggo, 2019). Today it is acknowledged as a way of thinking about,

approaching and solving ecosystem-level health problems. The case studies presented in this chapter—and their links to the Berlin Principles of One Health—illustrate the breadth of ape conservation endeavors that take a One Health approach. Nevertheless, there is a need for supporting data and evaluation of this approach in ape-specific situations going forward.

The threats to apes, ape habitats and ape health are cross-sectoral and multifaceted—hence the need for collaborative and interdisciplinary solutions. The One Health approach has traditionally been driven from the animal health perspective, but the process of considering and designing ape conservation programs requires expertise from a wide range of disciplines. The solutions are many, varied and complex, and they may be implemented at an individual or population level. They also come in a variety of forms, including specific individual treatments translated from human internal medicine; community-level changes to land management; and regional, national and international policy interventions. Since ape survival is inextricably linked to human development, best practice is to factor consideration of the SDGs into any One Health approach to ape wellbeing. The key elements for success are capacity, collaboration, resources and motivation.

Acknowledgments

Principal authors: Steve Unwin,[8] Dominic Travis,[9] Thomas Gillespie[10] and Elizabeth Cook[11]

Contributors: Lynne Gaffikin,[12] Gladys Kalema-Zikusoka,[13] Citra Nente[14] and Benard Ssebide[15]

Box 2.1: Steve Unwin

Box 2.2: Steve Unwin

Box 2.3: Dominic Travis

Case Study 2.1: Gladys Kalema-Zikusoka

Case Study 2.2: Citra Nente and Benard Ssebide

Case Study 2.3: Steve Unwin

Case study 2.4: Elizabeth Cook

Case studies 2.5–2.7: Steve Unwin

Endnotes

1   From a presentation at the XXI Congress of the International Primatological Society, Entebbe, Uganda, June 25–30, 2006 and noted by D. Travis.

2   The concept of community has expanded to include like-minded groups, communities of practice and others that may communicate virtually (MacQueen *et al.*, 2001).

3   These approaches have been referred to as "whole-of-society" efforts (Warren *et al.*, 2021).

4   Unless otherwise cited, material in Case Study 2.1 is based on the author's extensive knowledge of this situation, as founder and leader of CTPH and as a veterinarian whose work has contributed to ape health in Uganda since 1996.

5   Unless otherwise cited, the material presented in Case Study 2.2 is based on the authors' extensive knowledge of these situations. Citra Nente has been head of ex-situ conservation for YEL since 2017; prior to that, she was a veterinary clinician at the Borneo Orangutan Survival Foundation for 20 years. Benard Ssebide has worked for Gorilla Doctors as a Uganda-based clinician and manager for almost 15 years.

6   Deere *et al.* (2019); Grützmacher *et al.* (2018b); Köndgen *et al.* (2008); Negrey *et al.* (2019); Palacios *et al.* (2011); Parsons *et al.* (2015); Rwego *et al.* (2008); Scully *et al.* (2018).

7   Pilot research was conducted by Steve Unwin (at the time: University of Birmingham) and Yenny Saraswati (Sumatran Orangutan Conservation Programme), with funding from both the University of Birmingham and the Primate Microbiome Project.

8   University of Birmingham (https://www.birmingham.ac.uk/schools/biosciences/index.aspx) then Wildlife Health Australia (https://wildlifehealthaustralia.com.au).

9   University of Minnesota (https://vetmed.umn.edu).

10 Emory University (http://envs.emory.edu/home/index.html).

11 Consultative Group on International Agricultural Research (https://www.cgiar.org).

12 Stanford University (https://www.stanford.edu).

13 Conservation through Public Health (https://ctph.org).

14 Sumatran Orangutan Conservation Programme (https://www.sumatranorangutan.org).

15 Gorilla Doctors (https://www.gorilladoctors.org).

Photo: Buffer zones are areas of controlled resource extraction around protected areas. They provide benefits to local communities while reducing pressures associated with human encroachment on wild spaces. Village on the edge of Gunung Palung National Park, Indonesia. © Alison White

CHAPTER 3

The Impact of Tourism and Research Activity on Ape Health

Introduction

Apes have fascinated humans since antiquity: several ancient myths and legends mention their existence (Russon, 2004). Over time, scientists, philosophers and others have made comparisons between humans and apes in efforts to define precisely what characterizes "humans" and what is unique about "humanity." Such investigations spurred research into ape behavior, communication, tool use, self-awareness, social structure, culture and social learning. As behavioral and ecological research was made available to the public, people's fascination with the apes grew, stimulating the desire to see them in captivity and in the wild. In 1925, Parc Albert (renamed Virunga National Park in 1969), the first national park in Africa, was created to protect mountain gorillas (*Gorilla beringei beringei*) (Virunga National Park, n.d.-a). With the growth of the leisure industry, ex-situ and in-situ tourism operators have increasingly sought to capitalize on the fascination with apes.

Behavioral research is based on direct observation, which brings people close to apes in captivity and in their natural habitat. In response to observation in the wild, apes typically become habituated to the presence of field researchers. In turn, habituation opens the possibility of revenue creation from paying visitors and, potentially, the development of ape tourism projects.

As a primary global economic sector, tourism contributes substantially to most countries' incomes. In 2019, before the onset of the COVID-19 pandemic, the travel and tourism industry accounted for approximately 10% of the global gross domestic product (GDP), fluctuating between 2% and 15% for the African and Asian ape range states. The sector—a significant component of which is nature-based tourism—contributed about 10% of the global workforce (WTTC, 2020).

While ecotourism is a sustainable subset of nature-based tourism, the overall sector is a double-edged sword that can bring both significant economic and conservation benefits but also threats to great ape populations and their habitat (Wood, 2002). For example, the industry has played a critical role in the movement of pathogens and disease transmission, particularly since the 1970s, with the boom in national and international air travel.[1] Today, many pathogens can travel the world in less time than the incubation period of the diseases they cause (UNEP and ILRI, 2020).

Apes—humans' closest living relatives—are intelligent, sentient beings with complex social lives. As such, they attract local and international scientists, students, tourists, filmmakers and other visitors in the wild

and in captivity (Carr, 2016; Nielsen and Spenceley, 2011; Rose, 2011). Governments, tour operators, local communities and conservation organizations increasingly view wildlife and ape tourism as a potential source of funding to support national and regional economies, sustain local development and employment, contribute to biodiversity conservation and raise awareness about wildlife and nature.[2] The ape tourism industry has grown considerably since the 1950s and is expected to expand even more in the future (Macfie and Williamson, 2010; Russon and Susilo, 2014; Russon and Wallis, 2014a).

Tourism also poses a significant risk to the apes, however. Behavioral disturbance, disease transmission, overhabituation and a higher risk of aggression, conflict and stress are among the documented impacts of tourism (Ampumuza and Driessen, 2021; Macfie and Williamson, 2010). In addition, unregulated and poorly planned ape tourism activities can lead to human–ape conflicts in adjacent communities, particularly if apes lose their fear of humans and enter cultivated fields, where they may consume or damage crops and engage in aggressive interactions with people (Ampumuza and Driessen, 2021). Such conflicts can negatively impact ape behavior and culture, as well as their chances of survival in areas where communities retaliate (Kühl *et al.*, 2019; Macfie and Williamson, 2010).

As detailed in Chapter 1, owing to their close genetic relationship to humans, apes are at risk of disease transmission from people (see Chapter 1). Pathogens of human origin can easily be transmitted to apes who are in close and repeated contact with people, both in the wild and in captive settings within and outside ape range countries (Dunay *et al.*, 2018; Hosey, Melfi and Ward, 2020). Studies have documented pathogen transfers from scientific researchers to free-ranging apes, for example (Köndgen *et al.*,

2008; Köster *et al.*, 2022; Nuno *et al.*, 2022). In captivity, the sheer number of visitors and daily close contact between apes and zookeepers, sanctuary managers or other professionals pose a threat to apes kept in these confined environments (Liptovszky *et al.*, 2019).

In their natural habitat, apes tend to avoid humans. Only habituated apes let people approach and observe them, whether for research or tourism (Knight, 2009; McLennan and Hockings, 2016). Habituation of wild apes for tourism and research in their natural habitat is mostly a planned process, unlike habituation of captive (and semi-captive) apes. The aim of a habituation process is to decrease the flight distance of apes when they encounter humans. The removal of apes' fear and need to flee effectively reduces any significant anthropogenic effect on their natural behavior, although some degree of human influence on their behavior is inevitable (Tutin and Fernandez, 1991; Williamson and Feistner, 2011). Moreover, habituation directly heightens the risks of disease spill-over to apes, as they tolerate closer proximity to people (Köster *et al.*, 2022; Russon and Wallis, 2014a). One way of minimizing these risks is to ensure that habituation and

Photo: Apes—humans' closest living relatives—are intelligent, sentient beings with complex social lives. As such, they attract local and international scientists, students, tourists, filmmakers and other visitors in the wild and in captivity. © Paul Hilton/ Earth Tree Images

other activities related to tourism and field research have a minimum negative impact on ape health and behavior, as well as on ecosystems (Friend *et al.*, 2006; Muehlenbein and Ancrenaz, 2009; Williamson, 2001).

In some ape range states, sanctuaries have been established to care for apes who have been rescued from the illegal pet and wild meat trades or displaced by forest conversion (Farmer, 2002). Many sanctuaries, zoos and other wildlife collections aim to serve as vehicles for education and awareness (Ferrie *et al.*, 2014). As a result, hundreds of thousands of local and foreign visitors per year travel to these ex-situ facilities in ape range countries. In so doing, they create a considerable risk of disease transmission to captive and semi-captive apes (Muehlenbein and Wallis, 2014).

This chapter assesses the risks of disease transmission by humans who come in close contact with wild habituated and captive apes. It examines these risks from a health perspective, recognizing habituation as a risk factor and visitors, carers and scientists as potential hazards to apes. The chapter identifies the costs and benefits of habituating apes and keeping them in captivity for research and tourism. Lastly, it identifies current knowledge gaps and ways to address the sanitary risks linked to ape research and tourism activities.

Key findings include:

- In Africa and Asia, stakeholders increasingly perceive wild apes as opportunities for socioeconomic development, spurring growth in the number of habituated groups across their ranges.

- A growing body of evidence shows that habituated apes—including those in captive facilities—are at risk of disease spillover from humans. The inverse is also true, with apes posing a threat to humans.

- A dearth of information on the risks of disease transmission between humans

and apes—especially from Asia—hampers the design of effective management strategies that could minimize the risks linked to habituation for research and tourism, including in entertainment facilities.

- Although best management practices are available for research and tourism, poor enforcement, a lack of awareness, insufficient resources and inadequate capacity are hindering implementation.

- Disease surveillance, epidemiology and health studies at the human–ape interface are critical priorities for the prevention of disease transmission between humans and apes and vice-versa.

- Reducing the risks of disease transmission between people and apes is a conservation priority. The ape–human interface could become more secure for these threatened species through collaboration among stakeholders, including academics, businesses, conservationists, government authorities, local communities, scientists, tour operators, tourism facilities and tourists.

Habituation: A Prerequisite for Ape Research and Tourism

Habituating Wild Apes

Studying or visiting unhabituated wild apes is difficult. Apes tend to be suspicious of humans and generally flee; they may also display unnatural behaviors or become aggressive (Gruen, Fultz and Pruetz, 2013). For scientists who document ape behaviors and tourists who want to see and photograph the animals, however, being in close visual range is essential (Williamson and Feistner, 2011). Consequently, the first step towards developing and sustaining long-term scientific research and tourism is the

> A growing body of evidence shows that habituated apes are at risk of disease spillover from humans.

habituation of wild apes (Schaller, 1963; Tutin and Fernandez, 1991).

Habituation is the process by which animals gradually experience a loss of fear and become desensitized to the presence of human observers (Thorpe, 1963; Whittaker and Knight, 1998; Williamson and Feistner, 2011). It is a relatively new concept, triggered during the second half of the 20th century, following modern commercial tourism and scientific research on great apes (Gruen, Fultz and Pruetz, 2013; Russon and Wallis, 2014a). While humans can approach and observe habituated apes more easily than unhabituated ones, higher levels and frequencies of close contact with people result in increased stress, reduced resting and feeding times, and considerable disease transmission risks (Homsy, 1999; Knight, 2009; Köster et al., 2022; Woodford, Butynski and Karesh, 2002).

In the past, scientists conditioned apes with food to approach them easily, but the artificial presence of food modified the animals' behavior and ranging patterns, while also creating a risk of food contamination (Goodall, 1986). Therefore, current best management practices for habituation strongly discourage the provisioning of food to apes and other primates (Macfie and Williamson, 2010; Power, 1986; Wrangham, 1974). Not all tour operators take the guidance on board, however. Hand feeding has become a popular means by which to increase the likelihood of close sightings of free-ranging apes, particularly in areas where they have been released following captivity (Orams, 2002). During feeding activity, humans and apes are in close contact, and the risk of disease transmission increases, compounded by a heightened risk of aggression towards humans and conspecifics, as well as conflict (Lappan et al., 2020).

Habituation usually involves regular visual tracking of the target group until animals ignore the observers (Blom et al., 2004; Doran-Sheehy et al., 2007). During habituation, the animals' reactions fluctuate between aggression (especially for terrestrial African apes), avoidance (hiding or fleeing), curiosity and indifference to the observer's presence (Shutt, 2014). The duration of the habituation process depends on the species, the nature of prior encounters with humans, the frequency and type of contact, the social structure of the group and personalities of the individuals, and the habitat (Bertolani and Boesch, 2008; Morgan and Sanz, 2003; Werdenich et al., 2003). In all cases, ape habituation is a long and challenging process. It can take more than 6 months for gibbons, 1–2 years for mountain gorillas, up to 4 years for orangutans, 2–5 years for bonobos (Pan paniscus) and more than 5 years for western lowland gorillas (Gorilla gorilla gorilla) or chimpanzees (Pan troglodytes).[3] In some instances, individuals may never become habituated and may remain wary of human presence (Oram, 2018).

In the scientific literature, habituation and visitation are often presented as conservation tools that support the protection of animals and their habitat—or diversify and improve the livelihoods of local communities (Butynski and Kalina, 1998; Köster et al., 2022; Robbins and Boesch, 2011; Spenceley et al., 2010). In practice, however, the risks to the apes may outweigh the conservation benefits in the long run (Butynski and Kalina, 1998; Ferber, 2000; Shutt et al., 2014). More information is needed to understand the risk balance at the individual and species levels (Russon and Wallis, 2014b). In the meantime, best management practices (BMPs) are key to guiding both habituation and visitation in ways that minimize costs and maximize benefits (Macfie and Williamson, 2010). Such BMPs ensure that new habituation attempts consider the latest scientific evidence on disease transmission and animal welfare (Gruen, Fultz and Pruetz, 2013; Laurance, 2013).

" Reducing the risks of disease transmission between people and apes is a conservation priority. "

Habituation of Captive and Semi-captive Apes

Captive apes may be held permanently at research facilities, zoos or sanctuaries, or temporarily at rescue and rehabilitation centers. Their level of fearlessness towards humans depends on the duration and degree of human intervention while they are in captivity, as well as their temperament and experience.[4] Habituation of captive and semi-captive apes is often an unintentional consequence of repeated exposure to human carers (Chelluri, Ross and Wagner, 2013).

Carers tend to condition captive apes to facilitate animal compliance, as well as individual or group behavior management more generally (Bloomsmith *et al.*, 1994; Leeds, Elsner and Lukas, 2016). Such conditioning also allows scientists and medical professionals to carry out research and routine veterinary health screenings without having to use physical restraint or tranquilization, ensuring both human and animal

safety and welfare.[5] Apes receive positive reinforcement, often in the form of a food reward (known as a "conditioned reinforcer"), at the end of each completed task to ensure their continued compliance.

In captive settings, wild-caught and semi-captive apes are typically exposed to stress factors such as environmental novelty, a new diet, an unfamiliar social group structure and the presence of humans. An individual's failure to cope with these changes may result in physiological strain (Morgan and Tromborg, 2007). This stress may lead to suppressed immune function, increasing susceptibility to various conditions and diseases, growth impairment and reproductive failure.[6] While some individuals may adapt before permanent physiological damage sets in, those who do not can continue to deteriorate and suffer premature death (Fischer and Romero, 2019; Špinka and Wemelsfelder, 2018).

Familiar and unfamiliar human presence can significantly impact the movement and spatial dispersion within an enclosure, aggressive behaviors and interaction among group members at captive facilities (Hosey and Druck, 1987; Lee, 2012). The presence of unfamiliar visitors and researchers has caused a decrease in grooming, foraging and tool-use behaviors in outdoor-housed captive chimpanzees (Wood, 1998). In contrast, indoor-housed animals demonstrate an increase in agonistic (combative) behaviors, such as aggression, biting, attacking and injuring group members (Lambeth, Bloomsmith and Alford, 1997; Maki, Alford and Bramblett, 1987). When some indoor-housed chimpanzees were given access to an outdoor area, there was a significant decline in aggressive behavior among group members and an increased interest in human visitors (Stevens *et al.*, 2008). Conversely, outdoor-housed captive apes such as orangutans were generally unaffected by the presence of unfamiliar zoo visitors (Choo, 2011).

However, whenever visitor numbers have been high and and there is nowhere for the animals to hide, captive orangutans have been observed to use paper sacks to hide their heads (Birke, 2002).

The habituation process among captive apes may be reversible, as these animals can develop new normal or abnormal behaviors to cope with the presence of familiar and unfamiliar humans in their captive environment (Hosey and Druck, 1987). The response to the presence of humans may vary across arboreal and terrestrial captive ape species and may also be influenced by a host of other factors, including enclosure design and size, group size, available space per animal, the visual distance between apes and visitors, the ability of the captive apes to hide from humans, and visitors' activity level and associated noise (S. Sumita, personal observation, 2022).

Some conservationists argue that the cost of ape habituation for tourism and research may ultimately outweigh the benefits (Ferber, 2000; Litchfield, 2008; Shutt *et al.*, 2014). Others contend that without the economic incentive of ape tourism, it is unlikely that mountain gorillas and their habitat would have been protected, especially during prolonged periods of armed conflict in the region (Maekawa *et al.*, 2013). The following sections discuss the benefits and costs of habituating apes.

Actual and Potential Benefits of Habituating Apes for Research and Tourism

Owing to the global interest in apes and their corresponding economic value, many ape range countries use these animals as icons of national identity and pride, featuring them on passports, postage stamps, banknotes, statues and posters (Williamson, 2001). Some also use apes as major attractions for international and national tourists (Digun-Aweto, 2020; Shutt, 2014). In the Democratic Republic of Congo (DRC), Rwanda and Uganda, for instance, mountain gorilla tourism was established to prevent the extinction of the species. Tourism was a tool used for the conservation of apes, and its protective effects have been a source of pride in the three countries (Harcourt and Stewart, 2007; Mukanjari *et al.*, 2013; Robbins *et al.*, 2011b; Sabuhoro *et al.*, 2017). In Indonesia and Malaysia, increasing interest in conserving the only Asian great ape species—orangutans—has led to research activities, support for rehabilitation programs, and

ape tourism projects (Rijksen, 1978; Rijksen and Meijaard, 1999). Recently, other Asian countries have started to encourage gibbon tourism to promote their conservation and support local livelihoods. These include Cambodia, India and the Lao People's Democratic Republic (Williams and Behie, 2020).

Apes represent a significant economic asset so long as best management practices are implemented and enforced (English and Ahebwa, 2018; Litchfield, 2008; Macfie and Williamson, 2010; Munanura *et al.*, 2016). Mountain gorilla tourism, for example, is among the most important sources of foreign exchange income for Rwanda and

Uganda, where it accounted for 15% and 7.7% of national GDP in 2018, respectively. After coffee and tea exports, the sector was the greatest foreign exchange earner in both countries that year.[7] In 2005, in the Bwindi, Mgahinga, Virunga and Volcanoes National Parks—which range across the DRC, Rwanda and Uganda—mountain gorilla tourism activities generated US$20.6 million per year in direct benefits and much more in indirect benefits. About 53% of that direct income accrued at the national level and 41% at the international level, but only 6% at the local level (Maekawa *et al.*, 2013). In 2010 alone, wildlife tourism, driven mainly by orangutan viewing, generated between US$13 and US$23 million for the local economy of Sarawak, Malaysian Borneo (Zander *et al.*, 2014).

Ape tourism activities provide local, national, regional and international employment opportunities. Increased income, more secure livelihoods and awareness among communities that live adjacent to the parks can contribute to poverty reduction and a better appreciation of the apes and other wildlife, as recently shown with gorillas in Uganda's Bwindi Impenetrable National Park and in Gabon's Loango National Park (Robbins, 2021). In the Kinabatangan floodplain in Sabah, Malaysian Borneo, tourism receipts surpassed an estimated MYR100 million (US$24 million) in 2019, although fewer than half of the local people are involved in the sector (Chan, Marzuki and Mohtar, 2021; Wong, 2020). Tourism has created a source of income for local residents, who work as tour guides, drivers and skippers or operate their own tourism businesses, such as homestays or bed and breakfasts. At the same time, work remains to be done to overcome barriers to local participation in ecotourism and to prevent the exploitation of local Indigenous communities (Chan, Marzuki and Mohtar, 2021; Latip *et al.*, 2015).

With the right policies and planning, ape tourism can theoretically benefit local and national economies through multiplier effects. It is considered successful when local communities around an ape tourism site derive tangible benefits, have an active say in how they are involved and impacted, and enjoy respect for their social and economic systems and values, rather than being overwhelmed by tourism (Dawson, 2008; Litchfield, 2008). In practice, however, these objectives have not generally been met at the community level, which tends to benefit the least.[8] Inequalities in ape tourism benefit-sharing are significant concerns for ape conservation and could be among the reasons why hunting and poaching persist in many sites used for tourism (Munanura *et al.*, 2020; Tolbert *et al.*, 2019). Other reasons may relate to social, cultural and economic factors, which financial benefits from tourism alone cannot address (Munanura *et al.*, 2016; Plumptre *et al.*, 2004).

Concepts such as "pro-poor tourism" and "sustainable tourism" were developed in part to help address the unfair distribution of benefits and to ensure that poverty alleviation and reduction were key objectives addressed through tourism activities (Chok, Macbeth and Warren, 2007; Goodwin, 2007, 2014, 2016; Roe and Urquhart, 2001). In range states, the focus of ape tourism is on the reduction of poverty among communities living in the vicinity of ape habitats (Maekawa *et al.*, 2013). Most of these countries have developed legal frameworks to ensure the revenue from tourism in national parks is shared with adjacent local communities (Ahebwa, van der Duim and Sandbrook, 2012; Archabald and Naughton-Treves, 2001; Zander *et al.*, 2014). In Uganda, for example, 20% of park entrance fees and US$10 per US$700 gorilla permit are provided to people who live in the vicinity of the park. In Rwanda, in 2017, the government increased its revenue sharing allocation for community development projects from 5% to 10% of each gorilla permit (Maekawa *et al.*, 2013; Mukanjari *et al.*, 2013; Plumptre and Williamson, 2001).

In the past 25 years, the price of gorilla tracking permits has increased substantially. In Uganda, tracking fees rose from US$175 in 1997 to US$700 in 2022. In Rwanda, permit costs increased from US$250 in 1999 to US$1,500 in 2017 (see Case Study 6.2). No comparable or even significant increase in revenue has been accorded to the communities of the gorilla parks.[9] Despite some examples of success at the local level, ape tourism has yet to maximize benefits for economically deprived rural communities across range countries (Baker, Milner-Gulland and Leader-Williams, 2012; Maekawa *et al.*, 2013).

Tourism revenue arguably benefits governments, elites, foreign investors and other professionals disproportionately, while local communities face opportunity costs in land they cannot cultivate or exploit and the challenges of "crop-raiding" wildlife or conflicts with animals venturing out of the forests (Odhiambo, 2021; Scherl *et al.*, 2004; Tumusiime and Vedeld, 2012). Tourism has yet to fulfill its potential to contribute to poverty alleviation in a meaningful way.

Nevertheless, answering some of the basic needs of communities that share habitats with apes forms an integral part of the conservation agenda to protect habituated apes (Munanura *et al.*, 2016; Tolbert *et al.*, 2019). Tourism activities can provide direct or indirect support for schools and health facilities, for example. In line with the One Health approach—which recognizes that ape health and the wellbeing of human communities sharing the same habitat are interconnected—they can also allow for family planning, vaccinations and other prophylactic campaigns (see Chapter 2). Benefits can include improved health and hygiene among communities and a lower risk of disease transmission between the local people and habituated apes (Cranfield and Minnis, 2007; Kalema-Zikusoka and Byonanebye, 2019).

Ape research and tourism may also lead to a better understanding and acceptance of the biodiversity conservation agenda, which can result in fewer poaching or snaring incidents (Ancrenaz, Dabek and O'Neil, 2007; Robbins, 2021). In addition, the daily presence of researchers and rangers researching and monitoring the apes offers increased protection from hunting and snaring (Köster *et al.*, 2022; see Box 3.1). By following habituated groups, researchers also facilitate rapid veterinary interventions when necessary (Robbins *et al.*, 2011b). Indeed, studies show that habituated mountain gorillas exhibit a higher population growth rate (4.1%) than wild groups (0.7%) (Gray *et al.*, 2010; Robbins *et al.*, 2011b). Taken together, research and tourism activities, positive engagement from communities around the mountain gorilla habitat and other cross-boundary collaborative efforts probably contributed to the downlisting of *Gorilla beringei beringei* from critically endangered to endangered. While they are still threatened with extinction, mountain gorillas are the only great ape species whose conservation status has improved (Hickey *et al.*, 2019b; Robbins *et al.*, 2011b).

In addition to fostering a better understanding and acceptance of conservation goals, tourism revenue can help to safeguard protected areas. In Uganda, mountain gorilla tourism generates more revenue than other protected areas and contributes more than half of the funds used to run other national parks managed by the Uganda Wildlife Authority (Ahebwa, van der Duim and Sandbrook, 2012; English and Ahebwa, 2018; Walaga and Mashoo, 2009). Without these funds, most of the country's protected areas would be in peril and perhaps destroyed by land-use conversion or illegal resource extraction (English and Ahebwa,

2018). Indeed, Litchfield (2008) notes that without mountain gorilla tourism in Uganda, it is unlikely that the tiny Mgahinga Gorilla National Park (about 40 km² or 4,000 hectares) would exist today.

Limitations and Costs of Habituating Apes for Research and Tourism

Ape tourism is associated with some benefits, as discussed above, yet it also poses real risks to apes (Buckley, Morrison and Castley, 2016; Krüger, 2005). Ape visitation, even with fully habituated apes, often results in individual and group behavioral changes (time budget and daily ranges), higher vigilance levels and various stress-related signs.[10]

Compared to wild orangutans who are not subjected to tourist visitation, those who are visited are characterized by range restriction, reduced arboreal behavior, increased infant mortality and inter-female aggression, both in the wild and in semi-captive conditions (Kuze et al., 2012; van Noordwijk et al., 2018). At the group level, ape tourism may harm intergroup dynamics by discouraging female or male immigration or preventing unhabituated animals from coming close to habituated groups and individuals (Goldsmith, 2000; Morton et al., 2013). In rehabilitation sites such as Indonesia's Bukit Lawang in Sumatra and Tanjung Puting in Kalimantan, as well as in Sepilok in Sabah (Malaysian Borneo), former rehabilitants not only stopped their usual foraging activities during tourist visitation and food provisioning, but also increased their vigilance and self-directed behaviors (Dellatore, Waitt and Foitovà, 2014).

Stress triggered during habituation may negatively impact apes' welfare and reproductive success (Moberg, 1985). Chronic and repeated exposure to stress can eventually suppress immune function, increasing

BOX 3.1

Impact of the COVID-19 Pandemic on Ape Research and Tourism Activities

On 30 January 2020, the World Health Organization declared the outbreak of COVID-19 a public health emergency of international concern (WHO, 2020b). As a result, all countries were expected to curb the spread of the disease among human populations. Some immediately instituted measures such as the closure of national borders, nationwide lockdowns and travel restrictions. By April 2020, protective measures against COVID 19 were instituted in all African and Asian ape range states. Other protective measures included a ban on tourism and research activities in national parks and other sites (Orangutan Foundation, 2020; Richardson, 2021; UWA, 2020a).

Research and tourism activities were probably among the first and most severely affected of all conservation-related activities (Henseler, Maisonnave and Maskaeva, 2022; Huynh et al., 2021; Reuter et al., 2022). By mid-March 2020, the fear of COVID-19 transmission by travelers resulted in travel bans and entry refusals in ape range countries, which drastically reduced and eventually halted the influx of tourists and researchers into ape sites. In most range countries where ape visitation sites are located, much of the funding for running these sites comes from tourism activities, international donor agencies and foundations (Maekawa et al., 2013; Tumusiime and Vedeld, 2012). National governments provide a relatively small amount of funding, most of which is earmarked for staff salaries at ape facilities and sites.

The COVID-19 pandemic resulted in a drastic reduction or halt of revenue from ape tourism. Consequently, ape monitoring patrols were reduced in number, intensity and quality at most sites, or stopped altogether. In the Virunga Massif and the Bwindi Impenetrable National Park, the cuts severely affected mountain gorilla patrols, while the monitoring of park boundaries came to a halt (N. Guma, personal communication, 2020). In Bwindi alone, the number of snares and illegal park entries substantially increased in 2020, due to the measures instituted to curb COVID-19 (IUCN, 2020a; UWA, 2020b). Furthermore, in June 2020, a male silverback gorilla was killed by poachers who took advantage of the reduced presence of rangers, trackers and tourists (BBC, 2020).

The decline in the number of tourists significantly affected most countries' economies, including the local economy of communities engaged directly or indirectly in ape tourism, as well as local residents who were involved in income-generating activities in captive ape facilities (Henseler, Maisonnave and Maskaeva, 2022; Huynh et al., 2021). This situation also resulted in reduced revenue to support parks and locally active conservation organizations, which were forced to cut down or pause most field activities, such as community conservation programs. In addition, travel restrictions placed on researchers caused a drop in conservation research fee revenues at ape facilities in range countries.

Between June and July 2020, most ape sites reopened for local and international research and tourism activities—with strict standard operating procedures (SOPs) in place (UNCST, 2020). These SOPs required tourists and researchers to be tested for COVID-19 before they entered the ape range states, quarantine for 14 days before visiting apes, wear face masks and use hand sanitizer, and maintain a physical distance of at least 10–15 meters from the apes (UWA, 2020a). Despite the reopening of ape research and tourism sites and a decline in COVID-19 cases, researcher and tourist numbers have remained low at most ape sites.[11]

Photo: In rehabilitation sites such as Indonesia's Tanjung Puting in Kalimantan and Malaysian Borneo's Sepilok in Sabah, former rehabilitants not only stopped their usual foraging activities during tourist visitation and food provisioning, but also increased their vigilance and self-directed behaviors. Tanjung Puting National Park, Borneo, Indonesia. © Suzi Eszterhas / naturepl.com

susceptibility to diseases (Sapolsky *et al.*, 1990; Shutt *et al.*, 2014; Wasser, Sewall and Soules, 1993; Woodford, Butynski and Karesh, 2002). Studies undertaken during ape habituation processes have documented clinical signs of infectious diseases in chimpanzees, as well as higher parasitic loads in mountain gorillas, although the latter could be related to them living in close proximity to humans, near the park boundary (Fujita, 2011; Morton *et al.*, 2013). In contrast, analysis of fecal and hair cortisol concentrations shows that wild chimpanzees habituated to ecotourism are not chronically stressed, unlike orangutans and western lowland gorillas (Carlitz *et al.*, 2016; Muehlenbein *et al.*, 2012; Shutt *et al.*, 2014).

Tourists in ape habitats create additional pressure on the environment, including trampling, habitat damage, noise and waste pollution (Plumptre and Williamson, 2001). The construction of facilities and infrastructure necessary for tourism also has a negative impact on wildlife habitats as it directly alters the landscape and forest connectivity. In addition, deforestation and habitat fragmentation can lead to an increase in illegal activities such as poaching, while also contributing to environmental degradation (Arcus Foundation, 2018).

As noted above, habituation reduces apes' natural fear of humans. In some cases, however, apes may become overhabituated —or completely unafraid of humans. These apes may seek out physical interaction with humans, which can increase the risk of conflict, aggression and disease spillover (Ampumuza and Driessen, 2021; Williamson and Feistner, 2011). Habituated individuals occasionally leave the safety of protected areas, approach tourist facilities and enter local residential areas, gardens and orchards, where they may engage in crop-foraging or find themselves in conflict with local communities.[12] People may commit retribution killings in response to crop damage or human–animal conflict, particularly if they perceive apes as life-threatening or want to express their anger to park authorities (Davis *et al.*, 2013; McLennan and Hockings, 2016).

In Bwindi, Uganda, where several habituated mountain gorillas have regularly entered local people's gardens and damaged crops, human–ape conflict is not uncommon (Seiler and Robbins, 2016). Indeed, conflicts between humans and gorillas are a concern for most protected area managers in gorilla range countries (Hockings and Humle, 2009). At sites where habituated apes cross over into human-dominated landscapes, establishing buffer zones between community lands and ape habitats is a potential solution, as is the cultivation of crops that are non-palatable to gorillas (such as tea) or the creation of barriers. In Asia, conflicts with habituated apes are far less frequent, mainly because apes there are primarily or fully arboreal and live in smaller groups (see the Apes Overview).

Increased proximity between habituated apes and humans heightens the risk of direct disease transmission via contact with aerosols and droplets as well as indirect transmission via fecal deposits, contaminated fomites or substrates.[13] Before reaching ape visitation sites, international tourists and researchers often pass through several countries and continents. From an epidemiological perspective, they present a very effective means of transporting and disseminating pathogens to apes (Litchfield, 2008). In the past few decades, apes have been victims of several disease outbreaks of human origin, such as anthrax, measles, parasites, respiratory viruses, scabies and yaws (see Chapter 1).[14] Meanwhile, people involved in ape research and tourism are also susceptible to possible zoonoses of viral, bacterial or parasitic origin as they come near apes.[15] Box 3.2 and Chapter 1 focus on disease risks from tourists, or researchers.

As range states realize that habituated apes are a potential source of foreign exchange earnings, managers of protected areas are increasingly under pressure to habituate more ape groups for tourism (Munanura *et al.*, 2020; Nielsen and Spenceley, 2011). For example, in Uganda's Bwindi Impenetrable National Park, only three groups of gorillas were habituated in 1994 for research or tourism activities (Kabano, Arinaitwe and Robbins, 2014). Today, 17 groups are regularly used for research and tourism (Hickey *et al.*, 2019b). Indeed, about 43% of Uganda's mountain gorilla population has been habituated to tourism and research (Hickey *et al.*, 2019b). In the Virunga Massif, which spans Rwanda and the DRC, nearly three-quarters (73%) of the total gorilla population is habituated (Gray *et al.*, 2013). Furthermore, several groups of mountain gorillas specifically habituated for research are also concurrently used for tourism. This additional pressure increases all the human-induced negative impacts on the target groups and individuals.

As noted above, ape habituation is a long and challenging process. It is very costly and exceeds the budget of most national parks, typically requiring support from the international donor community. In the Central African Republic, for example, habituating one group of western lowland gorillas at Dzanga-Sangha took more than two years and cost at least US$250,000 (Blom, 2001b). This figure excludes the budget for the health monitoring programs that are conducted prior to, during and after the habituation processes (Blom, 2001a). In view of the resources necessary for habituating apes, carrying out the process would be nearly impossible without international agencies' support.

Risks to Habituated Wild and Captive Apes

Diseases of Human Origin

Apes have been regular victims of zoonotic disease outbreaks since people have shared their range. However, data from health monitoring have only been collected since humans started visiting apes in the wild for tourism or research.

The longest-running research program on wild chimpanzees, the Gombe Stream Research Centre, is located in Gombe National Park, which was established in 1968 (Collins and Goodall, 2008; Figure 3.1). Before the early 2000s, many Gombe chimpanzees died of flu-like diseases that were probably of human origin (Lonsdorf et al., 2006; Wallis and Lee, 1999; Williams et al., 2008). In Mahale, Tanzania, human respiratory viruses were identified in habituated groups of chimpanzees, who exhibited morbidity rates between 34% and 98% and mortality rates between 3% and 7% (Hanamura et al., 2008; Kaur et al., 2008). In Ivory Coast, five distinct outbreaks of respiratory diseases occurred in groups of wild chimpanzees habituated for research; their morbidity rate reached 90% and their mortality rate fluctuated between 3% and 19% (Köndgen et al., 2008). More recently, several outbreaks of respiratory diseases have affected chimpanzee and bonobo communities across their range (Grützmacher et al., 2018b; Negrey et al., 2019; see Box 3.2 and Chapter 1).

In the past 20 years, habituated mountain gorilla groups experienced 18 documented outbreaks of respiratory disease, possibly of human origin; nearly every group that was habituated for research or tourism suffered from at least one outbreak (Spelman et al., 2013). Health monitoring of mountain gorillas between 2018 and 2019 showed that habituated individuals had a higher infection rate than their unhabituated counterparts, particularly with reference to intestinal parasites of livestock or human origin (Hickey et al., 2019b). Similar results were found among gibbons but not orangutans (Ancrenaz, 2015; Hilser, 2011). No large-scale epidemics of human origin have yet been documented in Asian apes, although disease transmission between humans and orangutans or gibbons has been reported in the region (Kilbourn et al., 1997, 2003; Mul et al., 2007; Rijksen, 1978; Smith et al., 1969).

Risks of Exposure in the Wild

In the wild, differences in diet, social structure and ranging behavior influence the risks of exposure and disease spread between wildlife populations and groups (Herrera and Nunn, 2019; see Chapter 1). For example, parasite richness depends on host body size, social group size, diet and individual ranging patterns (Freeland, 1976; Nunn et al., 2003; Vitone, Altizer and Nunn, 2004).

The spread of a disease within an infected population depends on the nature of the pathogen, its infectiousness and the host's recovery time (Masi et al., 2012; Rushmore et al., 2013). It is also influenced by the social organization of the host species and the frequency of contact between individuals (Altizer et al., 2003; see the Apes Overview).

Immunocompetence is an essential element of an individual's survival. As immunity and stress levels are correlated, increasing stress levels due to habitat fragmentation, climate change and regular close exposure to humans make animals more susceptible to pathogens (Acevedo-

BOX 3.2

Disease Spillover at the Oldest Chimpanzee Research Program in Africa: Gombe

Since the inception of standardized data collection on the Gombe chimpanzees, the primary cause of death has been infectious disease, often of suspected human origin (Williams *et al.*, 2008). In the 1970s and 1980s, during the initial decades of research at Gombe National Park, close interaction between researchers and chimpanzees was not unusual. Such interactions were encouraged through supplemental food provisioning of the wild chimpanzee population (Goodall, 1986). By the mid-1970s, tourists began to visit the park—without any specific regulation protocol regarding group size or safety distance (Collins and Goodall, 2008).

The 1990s witnessed repeated fatal epidemics of flu-like disease in chimpanzees. While the pathogens responsible for these outbreaks remained unknown, it was recognized that exposure to humans increased the risk of illness for chimpanzees (Wallis and Lee, 1999). By 2000, researchers stopped regular provisioning of chimpanzees and worked with park staff to codify and implement best management practices for disease monitoring and prevention, targeting scientists and tourists (Collins, 2003; Pusey, Wilson and Collins, 2008). Following the recommendations of Homsy (1999), these protocols stipulated a minimum distance from chimpanzees of 7.5 m for researchers and 10 m for tourists, who are more likely to carry unfamiliar disease strains. Tourist visits were restricted to no more than six individuals for no more than one hour. Visiting researchers were asked to supply proof of vaccinations and to complete a seven-day quarantine prior to following wild habituated chimpanzees (Collins, 2003). For more on the chimpanzees of Gombe, see *State of the Apes: Industrial Agriculture and Ape Conservation* (Arcus Foundation, 2015, pp. 207–15).

Additional measures were introduced in view of chimpanzees' frequent movement through areas of the park where staff members lived. Researchers moved families of personnel out of the park, built wire mesh cages around the front of staff houses to prevent chimpanzees from accessing cooking and cleaning implements, and introduced a shift system to reduce the numbers of staff members present at any given time. Latrines and garbage pits were also modified to prevent access. In 2012, latrine facilities were upgraded with secure structures, plumbing and flush toilets. In addition, since 2017, observers (researchers and tourists) have been required to wear face masks in the presence of chimpanzees (Lonsdorf *et al.*, 2022).

Most recently, in 2020, outbreak response protocols were implemented, which include a threshold for additional fecal and fruit wadge sampling from infected chimpanzees and their social contacts. While flu-like outbreaks have occurred once or twice a year in the two decades since the best management practices were implemented, only four have resulted in mortality in the chimpanzee community (T.R. Gillespie, personal communication, 2021).

FIGURE 3.1

Gombe National Park and Vicinity

Sources: Protected area—UNEP-WCMC (2021h); country boundaries—GADM (n.d.); other base map detail—OpenStreetMap (n.d., © OpenStreetMap contributors, published under Creative Commons Attribution License CC BY; for more information see http://creativecommons.org)

Whitehouse and Duffus, 2009; Lochmiller, 1996). In two separate studies, orangutans and western lowland gorillas used for tourism had higher glucocorticoid metabolite concentrations than unhabituated conspecifics, indicating a certain level of chronic stress (Muehlenbein *et al.*, 2012; Shutt *et al.*, 2014). No visible symptoms were associated with these findings, however. Little is known about these mechanisms, and more research is needed to explore their long-term impact on ape survival.

Anthropogenic habitat fragmentation exacerbates the risks of disease spillover from humans to both Asian and African apes by causing ecological stress, temporary or permanent increases in population densities, confinement of pathogens that can disseminate more rapidly and efficiently, and exposure to pathogens from people or domesticated animals (Daszak, Cunningham and Hyatt, 2001; Gillespie and Chapman, 2008; Nunn and Altizer, 2006). Compared to large populations living in vast protected areas, fragmented populations are more vulnerable to stochastic events, including disease outbreaks (Acevedo-Whitehouse and Duffus, 2009). However, the paucity of data on ape adaptation and survival in highly fragmented landscapes precludes accurate assessments of the real risks (Ancrenaz, 2015).

Risks of Exposure in Captivity

The risk of disease transmission between people and captive apes depends on the facility design and the nature and management of the captive environment. Apes at rescue and rehabilitation centers may have limited or no contact with visitors. Nevertheless, they are subjected to close contact with their regular carers and staff working at these ex-situ facilities. Occasionally, short-term visitors, such as interns, researchers or documentary film crews, may be present.

Before arriving at rehabilitation centers and approaching apes who are expected to be released into the wild, visitors are required to follow strict standard operating procedures (SOPs), including stringent health screenings, which are drawn from best management practices (see Box 3.3).

Captive apes are frequently housed with conspecifics in enclosures designed to facilitate the expression of natural behaviors. Carers monitor their health and welfare closely, with a certain degree of intervention, to ensure that the animals are provided with appropriate nutrition and optimum veterinary care (see Chapter 8). Captive apes may experience nutritional, metabolic or degenerative conditions that are rarely reported in wild ape populations.[16] Such conditions can be exacerbated by a pronounced sedentary life and an increase in lifespan.

The risk of zoonosis is high in captive facilities such as zoos, where a constant stream of carers and visitors come close to the apes. The risk is especially high when tourists are encouraged to have close encounters with habituated apes during photography sessions or similar experiences. If people who come close to apes fail to abide by SOPs, they effectively place both themselves and the apes at risk of contracting a transmissible disease. A recent review has identified a minimum of seven occurrences of respiratory anthroponosis among captive apes in zoos and rehabilitation centers (Dunay *et al.*, 2018).

Poorly designed and non-species-specific enclosures, in tandem with a lack of biosecurity measures, pose a health risk to captive apes by enabling the transmission of diseases from visitors. However, funding gaps and inadequate knowledge of appropriate enclosure design, infrastructure materials, landscaping and enrichment requirements can make it challenging to provide a suitable captive environment for apes. In addition, visitors often throw food

and other items into outdoor exhibits in efforts to attract attention or provoke a reaction from apes. These items may be contaminated with pathogens of human origin that can potentially cause disease in naive captive apes. Too often, signage discouraging feeding of captive wild animals at zoological facilities is ignored by zoo visitors (S. Sumita, personal observation, 2021).

Apes held as pets by private owners are generally at greater risk of contracting and transmitting infectious diseases, as they live in closer proximity to humans. Illegally kept apes often suffer from various degrees of malnutrition and malabsorption. They are also susceptible to various zoonotic diseases, trauma and mental health issues due to their experience and abnormal living conditions (see Case Study 4.3).[17]

Serological surveys of rehabilitant orangutans show that captive apes visited by tourists have been exposed to human diseases, including typhoid, hepatitis (A, B, C), tuberculosis, scabies, measles, conjunctivitis and meningitis, as well as various parasites. Mortality among rehabilitant apes has been linked to respiratory ailments, tuberculosis, hepatitis B and scabies (Rijksen, 1978; Warren, 2001; Yeager, 1997). In Sabah, rehabilitant orangutans who were exposed to humans produced antibodies in response to common human respiratory viruses, unlike their wild conspecifics with no human exposure (Kilbourn *et al.*, 1997, 2003). Gilardi *et al.* (2014) report a case of vesicular stomatitis caused by human herpes simplex virus type 1 in a confiscated juvenile Grauer's gorilla (*Gorilla beringei graueri*) who had been hand-reared by humans in the DRC.

Some rehabilitation centers continue to promote tourism for educational and financial reasons, thereby heightening the risk of spreading human diseases to rehabilitant apes and wild ape populations following the translocation and release of rehabilitated individuals (Rijksen, 1978; Russon and Susilo,

> Apes held as pets by private owners are generally at greater risk of contracting and transmitting infectious diseases.

2014). Due to the frequent movement of related government agency personnel involved in implementing and enforcing established policies, practices and processes related to translocation and release of rehabilitants, best management practices (BMPs) are not always observed. Appropriate handover documentation can minimize risks associated with staff turnover and ensure continuity and adherence to SOPs and BMPs.

Managing Disease-Related Risks in Apes

It is not possible to prevent new pathogens from emerging or to eliminate the occurrence of zoonoses. Reducing the risks of pathogen spillover from humans to wildlife and vice versa requires the implementation of adequate legislation and the vigorous enforcement of practices regulating human–wildlife interactions and contacts, especially for people who come in regular or close contact with apes, such as tourists, researchers and local community members.

Disease-Related Risks Associated with Visitors

People who visit apes can be divided into two main categories: short-term visitors (mainly tourists) and long-term visitors (mainly researchers). To date, most documented cases of zoonotic disease in wild habituated apes have been linked to local communities, park staff and researchers—rather than tourists (Muehlenbein and Ancrenaz, 2009; Wallis and Lee, 1999). This finding is not surprising, as tourists spend less time on site than it takes most diseases to incubate and infected animals to display clinical signs, making it very challenging to identify the source of infection. The risk of disease transmission depends on several

factors, such as the distance between people and apes, the duration of contact and visitor numbers. Tourism exposes certain groups of habituated apes to more people in a single year than average people let into their homes throughout a lifetime (Homsy, 1999).

Short-term visitation typically involves proximity to apes for less than a few hours; however, visitors can potentially spend several hours or days in the visited animals' range. Short-term visitors include national and international tourists, interns, camera crews and reporters, VIPs, veterinarians and medical staff, and general workers (in the case of captive or semi-captive facilities). Overall, tourists have a poor understanding of the risks they pose to the places and animals they are visiting, and they rarely adhere to a preventive health strategy before visiting (Hamer and Connor, 2004; Van Herck *et al.*, 2004). Many travelers are unaware of their vaccination status and are not protected against vaccine-preventable diseases that may spread to apes (Van Herck *et al.*, 2004). A key step in mitigating these threats to ape health is enhancing communication about transmission risks (see Chapter 2).

A study conducted in Sabah shows that the vaccination status of nearly half of the visitors at the Sepilok Orangutan Rehabilitation Center was unknown or not up to date before the introduction of COVID-19 regulations. In addition, more than two-thirds of visitors with medical occupations who were aware of the risks of influenza were not vaccinated when they visited the orangutan rehabilitation center (Muehlenbein *et al.*, 2008). About 15% of the tourists reported symptoms of respiratory or gastrointestinal diseases during their visit, meaning that their presence represented a real risk of disease transmission to the apes (Muehlenbein *et al.*, 2010). More recently, similar findings were reported with respect to mountain gorillas (Hanes *et al.*, 2018). Future research

❝ Due to the frequent movement of related government agency personnel, best management practices (BMPs) are not always observed. **❞**

could usefully assess whether tourists and short-term visitors have become more aware of the risks since the COVID-19 pandemic (Anthes, 2022; BES Press Office, 2022; Gilardi and Uwingeli, 2022).

International tourists are of particular concern because they are often victims of respiratory infections or gastrointestinal ailments (due to unfamiliar diets and tropical intestinal pathogens) that can potentially be passed on to apes (Rack *et al.*, 2005). While traveling, they spend hours in enclosed spaces such as airplanes and are exposed to thousands of other people when transiting. Many are under physiological stress resulting from their journey, such as from a lack of sleep or jetlag, or in response to an unfamiliar environment (Gilardi *et al.*, 2015). Tourists often visit apes shortly after arrival or explore several ape sites in succession (Muehlenbein and Wallis, 2014).

The scientific literature has not yet documented contamination of apes by short-term tourists. Nevertheless, several factors could combine to devastating effect for entire groups of apes, including the sheer number of people visiting apes each year, the proximity they seek with the animals, their overall lack of consideration for health issues, the transmissibility of some diseases even before the first symptoms are detected and the presence of many people in a restricted environment (Russon and Wallis, 2014a).

On the whole, tourists fail to adhere to rules at many wild and captive ape ecotourism sites (Russon and Wallis, 2014a). As they typically spend a considerable amount of money in order to see apes in their natural habitats or in captive settings, they may be reluctant to declare illness for fear of being barred from visiting them. Moreover, park and sanctuary staff, who depend on tourists' financial contributions, may feel uncomfortable about challenging visitors who seem unwell or are not adhering to standard operating procedures.

Long-term visitors include researchers, documentary film crews, rangers and park personnel, local community members, carers and volunteers for captive and semi-captive

Photo: Apes held as pets by private owners are generally at greater risk of contracting and transmitting infectious diseases, as they live in closer proximity to humans. Illegally kept apes often suffer from various degrees of malnutrition and malabsorption. Juvenile gorilla at a hotel in Gabon. He was later sent to a sanctuary. © Alison White

apes. These visitors are more likely to engage in close and repeated contact, including physical contact with apes. The risk of disease transmission is thus high, and the consequences can be devastating. The number of long-term visitors at any single location is generally low, however, which may facilitate enforcement of SOPs to reduce the likelihood of disease spillover. Strict enforcement can help ensure adherence to precautionary practices. Targeted strategies to minimize the risk of disease transmission can usefully be developed, adapted and implemented among local communities that share the same habitat as apes, given that ape health is intimately linked with human health (see Chapter 2).

Many short- and long-term visitors—including primatologists, conservationists, volunteers, carers and workers in the entertainment industry—share photographs and videos showing close interaction with captive or wild apes. Photos of apes in close contact with humans can promote the view that these animals are suitable pets or that they are not endangered (Leighty *et al.*, 2015; Ross *et al.*, 2008; Ross, Vreeman and Lonsdorf, 2011). By giving the false impression that touching apes is acceptable, such images play down sanitary risks associated with these situations and undermine conservation objectives (Ross, Vreeman and Lonsdorf, 2011).

A recent analysis of holiday photographs taken with wild animals shows that many pictures portraying close encounters with great apes are extremely popular on various social media sites, including Instagram, Facebook and personal blog sites (Otsuka and Yamakoshi, 2020; Waters *et al.*, 2021). The popularity of shared human–animal close contact in photographs and videos on social media platforms encourages tourists to engage in risky behaviors (Van Hamme *et al.*, 2021). As noted above, many tourists spend a significant amount of money and

time traveling in the hopes of getting as close as possible to apes; they want to be able to get their fill of "memories" from their once-in-a-lifetime encounters, irrespective of the potential sanitary risks such situations create (Cox *et al.*, 2009).

In response to the growing popularity of human–animal images, an increasing number of conservationists are calling for an end to the posting of images of physical and close contact between people and wildlife (Sherman, Brent and Farmer, 2016). Relevant guidance, recently published by the International Union for Conservation of Nature (IUCN) Primate Specialist Group Section for Human Primate Interaction, is entitled *Best Practice Guidelines for Responsible Images* (Waters *et al.*, 2021).

Best Management Practices for Disease Prevention in Apes

The IUCN Red List classifies all ape species and subspecies as vulnerable, endangered or critically endangered. All three categorizations signal the need to minimize the risks created by human proximity to habituated and captive apes and to adopt precautionary measures regarding the use of apes in research and tourism activities (Macfie and Williamson, 2010). As it is impossible to eliminate the emergence of zoonoses and anthroponoses, the focus is on minimizing the risks of disease transmission to apes. Indeed, it is far easier, more affordable and more efficient to prevent the introduction of a pathogenic agent to a population than to control, treat or eradicate a disease outbreak (Macfie and Williamson, 2010; Santos, Guiraldi and Lucheis, 2020). Therefore, protected area authorities, ape researchers and tourism projects emphasize the implementation of vital disease prevention programs and the adoption of BMPs for disease prevention.

The first regulations to target mountain gorilla tourism were developed in the Virungas in the 1970s (Williamson, 2001). They subsequently underwent scientific review and were later revised based on field experience and impact studies (Homsy, 1999). These regulations limited each habituated gorilla group to a single one-hour visit per day by no more than eight tourists at a time, with a minimum of 7 meters between gorillas and humans (Weber, Kalema-Zikusoka and Stevens, 2020).

In view of the increasing number of ape tourism sites, the IUCN Species Survival Commission Primate Specialist Group developed the *Best Practice Guidelines for Great Ape Tourism* (Macfie and Williamson, 2010). The guidelines aimed to manage various risks, including diseases, to ensure that tourism contributed positively to ape conservation. Five years later, the IUCN produced the *Best Practice Guidelines for Health Monitoring and Disease Control in Great Ape Populations* (Gilardi *et al.*, 2015). Separate guidelines have been developed for

managing the risk of disease transmission to gibbons in rehabilitation centers and to great apes (Beck *et al.*, 2007; Campbell, Cheyne and Rawson, 2015; PASA, 2009).

Practical tools found in BMPs include "dos and don'ts" that park managers and authorities can easily implement. They were translated into simple, straightforward SOPs to offer practical guidance about how to implement the BMPs. They were produced in local languages and adapted to the local ape species, habitat conditions and socioeconomic contexts, as well as the type of human interventions at each site (Gilardi *et al.*, 2015; Macfie and Williamson, 2010). BMPs aim to cover the various scenarios in which apes come into close contact with humans (see Box 3.3). Habituated mountain gorillas, for example, are exposed to more than 2,000 hours of visitation by tourists every year and thus face far more risks than groups that are followed by a limited number of scientists (Homsy, 1999; Litchfield, 2008).

In response to the COVID-19 pandemic, guidelines were produced to minimize risks

of human transmission of the disease to captive and wild apes (Gillespie and Leendertz, 2020). Following the precautionary principle, many conservationists advocated halting ape-related fieldwork in the wild (Reid, 2020). However, the sudden cancellation of these activities had detrimental effects on habituated apes, as field workers had afforded them some degree of protection against poaching; local communities also suffered, largely due to loss of employment and income (Lappan *et al.*, 2020; see Box 3.1). This experience has highlighted the need to adapt and modify field protocols and develop more robust occupational health policies to make fieldwork safer for both people and apes (Lappan *et al.*, 2020; Trivedy, 2020).

Despite BMPs and localized site-specific SOPs, one of the significant weaknesses at most sites is poor enforcement, which can lead to an overall lack of adherence by tourists, researchers and park staff (Daud, 2019; Hanes *et al.*, 2018; Sandbrook and Semple, 2006; Weber, Kalema-Zikusoka and Stevens, 2020). On several occasions in Bwindi, for example, tourists and researchers did not maintain the 7-meter minimum distance from gorillas, while others stayed with gorillas for more than an hour (Hanes *et al.*, 2018; Sandbrook and Semple, 2006; Weber, Kalema-Zikusoka and Stevens, 2020).

If tourists and researchers are well informed and understand the risks they pose to the animals they encounter, they are more likely to follow best practice guidelines (Russon and Wallis, 2014a). Clear punitive and incentive measures are prerequisites for the enforcement of proper sanitary guidelines (Sandbrook and Semple, 2006). The dissemination of such guidelines requires the development of targeted awareness raising materials for various audiences, especially on travel websites (Horvath, Murray and DuPont, 2003; Muehlenbein and Ancrenaz, 2009). Park personnel, tour operators, surrounding local communities and anyone who

BOX 3.3

Health Best Management Practices for Ape Visitation: A Summary

The health best management practices summarized below are designed to minimize the risk of disease transmission from people—including personnel, researchers, filmmakers, tourists and veterinarians—to great apes and gibbons in captive facilities and in their habitats.

- People who are feeling unwell or exhibiting signs of illness may not visit the apes.
- Before visiting the apes, people who have been ill must undergo a quarantine period of at least seven days after the cessation of clinical signs of illness.
- People who may have acquired infectious agents prior to or during international travels must undergo a seven-day quarantine prior to visiting the apes.
- People who visit the apes must be at least 15 years old.
- People who are likely to come within 10 m of apes must wear a surgical mask.
- People must maintain a minimum distance of 7 m from the apes.
- No individual ape or ape group may be visited by more than one tourist group per day.
- The number of tourists in such groups may not exceed the maximum deemed appropriate for the relevant ape species.
- A visit to an ape group may not last more than one hour.
- People must sanitize their hands before and after entering ape habitat and captive ape facilities.
- Before and after visiting apes, people must clean (and then, if possible, disinfect) their clothing and footwear, including between visits to different ape groups.
- People who need to sneeze or cough while visiting apes must keep their mask on, turn away from the animals and cover their mouth and nose with the crook of the elbow or clothing, rather than with the hand.
- People who need to urinate while in ape habitat must move away from and out of sight of the apes and dig a hole at least 30 cm deep.
- Defecation is not permitted in ape habitat. People who need to defecate while in ape habitat must bag and then dispose of any solid waste, such as feces and toilet tissue, outside of the forest.
- Cigarette smoking is prohibited and cigarette butts may not be disposed of in ape habitat.
- Artificial items, such as plastic bags and containers, may not be discarded in ape habitat.
- People who are likely to come into frequent or close proximity to apes must be immunized according to local government recommendations. At a minimum, long-term visitors—including site staff, researchers and veterinarians—must be immunized against measles and other highly infectious diseases that can affect apes.
- Site staff and longer-term visitors must be tested for tuberculosis annually and show negative results before commencing ape visits (Gilardi *et al.*, 2015; Johnson *et al.*, 2009; Jones and Brosseau, 2015; Macfie and Williamson, 2010; Monto, 2002; Muehlenbein *et al.*, 2012; Shutt *et al.*, 2014; Xie *et al.*, 2007).

could come near habituated apes (including poachers) also need to be aware of the risks of disease transmission between people and apes (Filippone *et al.*, 2015). Park authorities, tour operators and site managers can ensure that all tourists and other people who visit apes have read and understood these recommendations—and that they adhere to them.

Another concern is the disposal of non-biodegradable personal protective equipment (PPE), such as face masks, gloves and hand sanitizer dispensers. Most PPE items contain plastic or microplastic ingredients, which negatively impact the environment, especially if they are not disposed of properly. Plastics can act as fomites for pathogen transmission and are potentially dangerous pollutants of ape habitats such as Bwindi (Bitariho, Akampurira and Mugerwa, 2020). While current BMPs recommend the use of PPE items, they do not specify how to dispose of them after use.

Over the past 15 years, a growing number of zoological facilities issued plastic bans to prohibit the provision, sale, distribution and introduction of plastics on their premises. One of the first was Nepal's Central Zoo in Kathmandu, which educates members of the public on the environmental concerns related to plastic. In 2009, the zoo banned plastic bags and began to offer its visitors environmentally friendly alternatives. Its messaging included information on the health implications of accidental ingestion of plastic material by the zoo animals (Himalayan News Service, 2009). Since then, many other captive facilities worldwide have launched on-site campaigns about plastic waste and the danger it poses to both terrestrial and aquatic animals and it has become a global focus with a number of ape range states banning single use plastics (Cerdán and Kirk-Cohen, 2020; Greenpeace Africa, 2020; Inclean Magazine, 2019; Rivas *et al.*, 2022).

The Role of Legislation and Regulatory Frameworks

Most sanitary guidelines and BMPs for habituated ape populations were developed voluntarily by conservationists or practitioners. On the whole, BMPs are not legally binding, although a few research sites implement these guidelines strictly and captive facilities are required to adhere to legislation about animal welfare or fulfill health requirements (see Chapter 8). Overall, the implementation and effectiveness of non-binding BMPs are poor, and the legal framework for addressing the risk of disease transmission to habituated apes is still fragile. The lessons from the COVID-19 pandemic could help to bridge these gaps. Adequate legislation is required to support the enforcement and implementation of BMPs so that they can be applied across all ape sites, rather than at exceptional facilities. Incentives and punitive measures would help managers of protected areas to ensure that people who come into contact with apes comply with the guidelines.

Researchers must adhere to national and international laws concerning veterinary medicine, biological sample collection and interactions with both wild and captive apes. Depending on the range country, researchers must submit a "wildlife animal use protocol" under the Institutional Animal Care and Use Committee and undergo health screenings before arriving at the field research site. On-site animal welfare and ethics committees further scrutinize the research topic, especially in the case of captive apes, before permission is granted to proceed with the proposed research.

The Role of Evidence-Based Assessments

In view of the growing demand for ape visitation, various stakeholders perceive the habituation of additional ape groups as a way to promote conservation while diluting sanitary and other risks, as this approach could provide more options for research and tourism activities (Ancrenaz, 2018). Disease-related and other risks associated with habituation are barely understood, however. Before any additional ape habituation processes for research or tourism could begin, extensive site- and species-specific risk assessments and feasibility studies would need to be carried out. Such evaluations are most useful when they consider the environmental, welfare and socioeconomic characteristics of a situation, as well as the vulnerability and long-term protection of the apes selected for habituation (Russon and Wallis, 2014a).

Moreover, the governance of tourist sites that offer visits to wild, habituated ape populations would benefit from the joint input of professionals, including conservationists, ecologists, ape managers, travel medicine specialists and social scientists (Muehlenbein and Ancrenaz, 2009; Munanura, Backman and Sabuhoro, 2013; Russon and Wallis, 2014a). The first step could be to conduct an in-depth assessment of ape visitation sites, including through a cost–benefit analysis of current ape tourism projects and analysis of their contributions to ape conservation. The assessment could give rise to recommendations for improving governance and decision-making processes that guide ape habituation and ape-related tourism.

Photo: Habitat destruction leads to disease emergence, but the underlying mechanisms, the prediction and the prevention of possible outbreaks are still poorly understood. Knowing and cataloging pathogens that affect ape species may benefit human medicine while supporting conservation efforts. Habitat conversion along the edge of Volcanoes National Park.
© Ronan Donovan

Understanding Disease Ecology in Natural Habitats

Addressing health threats requires an understanding of what can potentially infect or kill apes, how and under which circumstances. There is an urgent need to collect reliable baseline data to quantify the impact of habituation on the health of ape populations that are used for tourism and research (Leendertz *et al.*, 2006b; see Chapter 2). Although a fair amount of research is being carried out to identify pathogens that occur in the wild, very little is known about their actual impact and the health risks they pose. The lack of information about what could be considered "normal" in a population severely impedes the identification of sanitary issues that require attention. Baseline data on unhabituated apes in Asia and other fast-changing environments is particularly scarce (Calvignac-Spencer *et al.*, 2012).

Over the past decade, several African sites have reported a significant increase in respiratory disease outbreaks in wild, habituated chimpanzees (Desmond and Desmond, 2014; Fujita, 2011; Negrey *et al.*, 2019; Scully *et al.*, 2018). Scientists are still trying to determine whether this observation reflects ecological changes—such as climate change or increased contact with human beings or domestic animals—or improvements in the ability to detect outbreaks. For now, the real threats to the survival of these populations remain unknown.

The COVID-19 pandemic illustrates how little is known about the dynamics between hosts, reservoirs and pathogens on the one hand, and the impacts of deforestation, habitat fragmentation and climate change on the other (Lappan *et al.*, 2020). Habitat destruction leads to disease emergence, but the underlying mechanisms, the prediction and the prevention of possible outbreaks are still poorly understood. Knowing

and cataloging pathogens that affect ape species may benefit human medicine while supporting conservation efforts. For example, in Southeast Asia, ticks are increasingly found on newly captured wild orangutans (Sabah Wildlife Department, personal communication, 2019). However, scientists still do not know whether this increase is a result of the apes' close contact with cattle or people, habitat fragmentation, or an adaptation of the parasites to new environmental conditions resulting from climate change. Such examples illustrate the knowledge gaps in disease ecology, highlighting the need for epidemiological and holistic studies that investigate the underlying variation of infectious disease pathogenicity under different environmental conditions (see Chapter 1).

Early warning systems based on monitoring protocols prioritize risks to habituated apes and support immediate intervention to prevent catastrophic outbreaks (Leendertz *et al.*, 2006b; see Chapters 4 and 6). Direct and visual monitoring can easily be implemented at all sites without expensive equipment (Knott *et al.*, 2021; Shutt, 2014). Although the lack of rapid diagnostic tests is still a challenge in the field, recent technological advances in molecular diagnostics, in tandem with improved sequencing techniques and mobile diagnostic laboratories, can improve the current understanding of disease ecology and complement the arsenal already used to monitor animal health, pathogen burden and physiological status (Calvignac-Spencer *et al.*, 2012; Knott *et al.*, 2021; Quick *et al.*, 2016). Combined with human disease monitoring, such early warning mechanisms can contribute to effective One Health approaches (see Chapter 2).

Recent technological improvements provide tools for studying apes in their natural habitat without habituation and without the need for human observers to be in close proximity to them. Genetic sampling is a practical and effective non-invasive approach

> " Recent technological improvements provide tools for studying apes in their natural habitat without habituation or the need for human observers to be in close proximity to them. "

for studying apes in their natural habitat (Arandjelovic *et al.*, 2010, 2011; McCarthy *et al.*, 2015). Over the past few years, scientists have used terrestrial environmental DNA sampling for ecosystem and biodiversity surveys. Animals are shedding DNA in the environment: hair and skin, feces and urine, saliva and blood. By testing contaminated water or soil, scientists can identify the shedding individual's species (Deiner *et al.*, 2017; Leempoel, Hebert and Hadly, 2020). In practice, however, such non-invasive techniques are challenging, and genetic material is often difficult to analyze due to DNA fragmentation and degradation, as well as allelic dropout.

Camera-trapping is increasingly used to study the impacts of conservation threats and management, socio-demographics, behavior and feeding ecology, disease screening, mapping of habitat use and ranging patterns of wild habituated and unhabituated apes (Boyer-Ontl and Pruetz, 2014; Head *et al.*, 2013; Klailova *et al.*, 2013; Steinmetz *et al.*, 2014). One major limitation of using camera-trapping, smartphones or conventional cameras to collect images is the time-consuming nature of processing hundreds or thousands (or even more) pictures. However, the emerging field of animal biometrics and ape facial recognition can overcome some obstacles (Crunchant *et al.*, 2017; Loos and Ernst, 2013; Loos and Kalyanasundaram, 2015). Devices such as drones and other unmanned tools can potentially serve to minimize the proximity of tourists and researchers to the apes they are observing. Passive acoustic monitoring has been used to monitor long calls of wild orangutans, as well as vocalizations of chimpanzees (especially "pan-hoots" and "drumming") and gibbons; this method allows for the monitoring of spatio-temporal patterns of habitat use by unhabituated groups of apes (Clink, Crofoot and Marshall, 2019; Kalan *et al.*, 2016; Kaplan and Rogers, 2000; Spillmann *et al.*, 2015).

Despite these tools, effective health monitoring of apes (both habituated and unhabituated) is still rare. The number of sites where health monitoring and disease prevention are implemented is small compared to the number of habituated ape populations in Africa and Asia (Calvignac-Spencer *et al.*, 2012; Knott *et al.*, 2021; Morton *et al.*, 2013). As long as governments and other stakeholders perceive wildlife health as a low priority, funding for sanitary monitoring is likely to remain insufficient.

Conclusion

Although ape research and tourism have often been viewed as potential tools for conserving wild populations and supporting captive care, they are associated with significant risks to ape health. Evidence at most ape sites shows that research and tourism are managed with little adherence to sanitary best management practices (Russon and Wallis, 2014a). Given that research and tourism activities pose enormous health risks to habituated apes and their environments, a clear assessment of the actual benefits and costs of habituation is necessary, particularly to guide the use of tourism as an ape conservation tool.[18]

A key health concern is the distance between people and habituated apes (see photograph on pp. 76–77). While some argue that proximity to the apes is necessary to satisfy tourists' expectations, better management of such expectations, greater public awareness of health risks and more responsible behavior by tour operators and guides would go a long way to protecting the apes. Similarly, researchers, captive care staff and park employees could prioritize methods that do not require close contact with apes (Knight, 2009; Russon and Wallis, 2014a; Tapper, 2006). In addition, social

> " Most of the time, a relatively small portion of the funds from research and tourism are channeled directly into ape conservation programs or to the communities living near ape habitats. "

media could be utilized to make it "shameful" for people to be close to apes. In this context, conservationists and carers have an essential role to play.

Enforcing simple BMPs such as those calling on visitors to wash their hands, wear a mask or maintain a minimum distance from apes can significantly reduce the risk of disease transmission to the animals (Macfie and Williamson, 2010). Before the COVID-19 outbreak, tourists, scientists and park personnel were reluctant to wear masks, thereby posing a severe risk to habituated apes (Van Hamme *et al.*, 2021). Future research could examine public perceptions of mask wearing since the start of the coronavirus pandemic, as well as its impact on the health of habituated apes and the financial value of the ape tourism experience (Anthes, 2022; BES Press Office, 2022).

Guidelines and BMPs have been used to minimize the risk of disease transmission to many habituated wild ape populations. However, laws and regulations in ape range countries generally do not require the implementation of BMPs. As is already the case in accredited rescue centers, sanctuaries and zoos, robust legal frameworks that assist practitioners in both captive and wild settings to enforce BMPs would ensure best practice for all apes, especially with the addition of punitive measures for BMP offenders. Simultaneously, government departments and conservation practitioners, including park managers, could secure more resources and incentives to support the enforcement of BMPs.

In theory, research and tourism can secure much-needed revenue for the protection of wild ape populations, yet how the money is actually invested depends on a range country's priorities. Most of the time, a relatively small portion of these funds is channeled directly into ape conservation programs or to the communities living near ape habitats. Moreover, a significant

Photo: In theory, research and tourism can secure much-needed revenue for the protection of wild ape populations, yet how the money is actually invested depends on a range country's priorities. Most of the time, a relatively small portion of these funds is channeled directly into ape conservation programs or to the communities living near ape habitats. Ngamba Island Chimpanzee Sanctuary, Uganda. © Friends of Chimps

Photo: In captive settings, an individual ape's failure to cope may result in severe stress leading to suppressed immune function, increasing susceptibility to various conditions and diseases, growth impairment and reproductive failure. While some individuals may adapt before permanent physiological damage sets in, those who do not can continue to deteriorate and suffer premature death. © Jo-Anne McArthur/ Born Free Foundation/ We Animals Media

proportion of great ape and gibbon conservation funding still comes from international donors (Macfie and Williamson, 2010). A greater proportion of the proceeds from ape research and tourism is required for the long-term protection of habituated ape populations, as well as for related law enforcement, research, veterinary and local community health care, education campaigns and interventions through One Health programs (see Chapter 2).

By integrating long-term observational health data with non-invasive diagnostics,

future studies could fill knowledge gaps regarding the epizoology and biology of pathogens in apes. They could also allow for impact assessments of the habituation process and people's presence near apes, especially those whose populations are dwindling and whose habitats are declining because of development projects and hunting (Calvignac-Spencer *et al.*, 2012; Devaux *et al.*, 2019).

Finally, addressing the risks of spillover diseases caused by ape-related research and tourism requires a multipronged approach

—one that combines the development and implementation of stringent biosafety protocols, the adoption of appropriate practices during ape encounters, the development of adequate outreach and social media campaigns, the collection of long-term data on the sanitary status of apes living at the interface with people, and the prioritization of sanitary risk as a significant conservation threat to the survival of great apes and gibbons (Lappan *et al.*, 2020).

Acknowledgments

Principal authors: Sumita Sugnaseelan,[19] Marc Ancrenaz[20] and Robert Bitariho[21]

Contributors: Tom Gillespie[22] and Elizabeth Lonsdorf[23]

Box 3.1: Sumita Sugnaseelan and Robert Bitariho

Box 3.2: Tom Gillespie and Elizabeth Lonsdorf

Box 3.3: Marc Ancrenaz and Robert Bitariho

Endnotes

1   Chomel, Belotto and Meslin (2007); Hall, Scott and Gössling (2020); Koeppel et al. (2018); Lyra (2006); Rodriguez-Morales and Schlagenhauf (2014).

2   Buckley, Morrison and Castley (2016); Hvenegaard (2014); Nielsen and Spenceley (2011); Ringer (2002); Russon and Wallis (2014a).

3   Ando, Iwata and Yamagiwa (2008); Bertolani and Boesch (2008); Chivers (1974); Doran-Sheehy *et al.* (2007); Macfie and Williamson (2010); Oram (2018); Schaller (1963); Susman (1984).

4   Choo, Todd and Li (2011); Hosey (2008); Hosey, Melfi and Pankhurst (2013); Mitchell *et al.* (1992); Pedersen *et al.* (2019).

5   Bloomsmith *et al.* (2015); Laule, Bloomsmith and Schapiro (2003); Pomerantz and Terkel (2009); Schapiro, Bloomsmith and Laule (2003); Westlund (2015); Whittaker and Laule (2012).

6   Berga (2008); Fischer and Romero (2019); Morgan and Tromborg (2007); Sapolsky, Romero and Munck (2000); Špinka and Wemelsfelder (2018).

7   English and Ahebwa (2018); Macfie and Williamson (2010); Maekawa *et al.* (2013); Trogisch and Fletcher (2022); NPA (2020).

8   Maekawa *et al.* (2013; 2015); Mazimhaka (2006); Sandbrook (2010); Spenceley *et al.* (2010); Spencer, Amony and Dube (2020); Tolbert *et al.* (2019).

9   Adams and Infield (2003); Nkuringo Safaris (2021); Rwanda Development Board (2017); UWA (2022); Visit Rwanda (n.d.); H. Goodwin, personal communication, 2022.

10   Cipolletta (2003); Doran-Sheehy *et al.* (2007); Johns (1996); Mabano (2013); Muyambi (2005); Oram (2018); Shutt (2014); Williams and Behie (2020).

11   Information gathered by authors from access to internal documents showing visitor numbers, and conversations with representatives of national park authorities in 2022.

12   Goldsmith (2014); Knight (2009); Macfie and Williamson (2010); Seiler and Robbins (2016); Shutt *et al.* (2014).

13   Dawson (2008); Ferber (2000); Litchfield (2008); Lonsdorf *et al.* (2006); Williams *et al.* (2008); Woodford, Butynski and Karesh (2002).

14   Dunay *et al.* (2018); Leendertz *et al.* (2006a; 2006b); Litchfield (2008); Nizeyi *et al.* (2001); Patrono *et al.* (2018).

15   Calvignac-Spencer *et al.* (2012); Hahn *et al.* (2000); Keele *et al.* (2006); Keita, Hamad and Bittar (2014); Krief *et al.* (2010); Mutombo, Arita and Jezek (1983).

16   Cabana, Jasmi and Maguire (2018); Ely *et al.* (2010); Gresl, Baum and Kemnitz (2000); Kumar *et al.* (2017); McTighe *et al.* (2011); Nunamaker, Lee and Lammey (2012).

17   Information obtained by the authors in conversations with those involved in rescuing apes 2021–22.

18   Desmond and Desmond (2014); Goldsmith (2014); Hingham (2007); Russon and Susilo (2014); Russon and Wallis (2014a).

19   Universiti Putra Malaysia (www.upm.edu.my).

20   HUTAN–Kinabatangan Orangutan Conservation Programme (www.hutan.org.my).

21   Institute of Tropical Forest Conservation, Mbarara University of Science and Technology (https://itfc.must.ac.ug).

22   Rollins School of Public Health, Emory University (https://sph.emory.edu/index.html).

23   Emory University (www.emory.edu).

CHAPTER 4

Managing Ape Health: Informing Interventions

Introduction

Ape populations no longer reside in remote strongholds, away from human interaction. Even if some are inaccessible to tourists and researchers, they are certain to experience impacts of climate change and other anthropogenic effects (Kühl *et al.*, 2019). As ape health declines, the need to intervene at the individual, population and ecosystem levels thus becomes more acute.

The veterinary profession is framed by codes of Good Veterinary Practice, which promote intervention in animal health via the use of tools and approaches that ensure the dignity and humane treatment of all animals (FVE, n.d.; Martinsen and Jukes, 2005). Kiran, Sander and Duncan (2022) successfully link veterinarians, as public health practitioners, to climate change impacts, promoting interdisciplinary policy and adding a layer to the process employed to decide whether an intervention is warranted (see Figure 4.1).

FIGURE 4.1

Aspects of Climate Change That Impact Veterinary and Public Health

Climate change issues are interconnected and fall within the One Health and planetary health frameworks, as well as the public health model, which is within the scope of veterinary practice. These issues include:

1. rising temperatures;
2. extreme weather events;
3. air quality;
4. vector-borne disease;
5. food safety and security;
6. water-related health issues; and
7. mental health.

By explicitly framing them as climate change issues, policy can serve as a transcendent tool that connects all domains, thereby fostering veterinary capacity and empowering veterinarians to be climate stewards and protect planetary health.

Source: Kiran, Sander and Duncan (2022, fig. 1). Reproduced under the terms of the Creative Commons Attribution License (CC BY).

Although this model is useful from a One Health perspective, it does not provide for a decision-making process in wildlife health interventions (see Chapter 2). Carver, Peters and Richards (2022) offer a model to support evidence for wildlife disease control solutions. They concentrate on the need for improved integration of in situ wildlife disease management and modeling to guide and assess disease management actions (see Figure 4.2). As shown in Figure 4.3, their model has been used to manage health interventions for wombats (*Vombatus ursinus*); it promises to help identify sustainable disease management solutions for all wildlife species, including apes.

In focusing on the reasons and processes that lead to decisions to intervene—or not to intervene—in ape health issues, this chapter acknowledges the wider system needs that are illustrated in Figures 4.1–4.3. In the context of ape health, an intervention is a clearly defined action taken to improve the health of an individual, group, population or ecosystem. The decision-making process that considers intervention in response to a given injury or health issue is usually based on the local environmental situation (in situ or ex situ), accessibility of the individual animals (captive, habituated or fully wild), and the potential for improvement in either welfare or conservation of the individual, species or ecosystem that is the subject of the intervention. Constraints that can influence decisions include data gaps and a lack of resources. As this chapter shows, effective decision-making is based on reliable risk assessments and entails the formulation of a justification for any decision—be it to intervene or not to intervene—regardless of the type of intervention or context.

Veterinarians, like professionals in human health, have long been taught: "first, do no harm." The desire to do good can sometimes get in the way of an appropriate decision. This chapter focuses on the need

FIGURE 4.2

A Framework for Integrated Disease Management for Wildlife

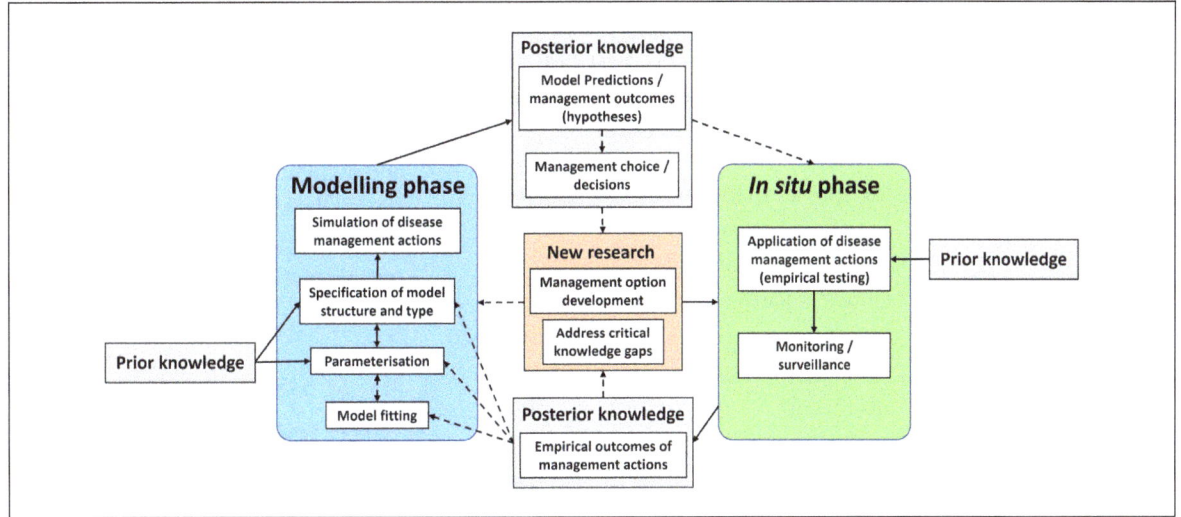

Notes: Solid arrows symbolize established research pathways; dashed arrows represent common gaps that often limit integration between modeling and in situ phases of wildlife disease management, such as culling, therapeutic interventions, host movement restriction and combinations of actions. Programs begin with prior knowledge and lead to posterior knowledge, which can inform additional research; in turn, new research results can advance modeling and in situ phases. Key personnel involved include practitioners and modelers tasked with finding effective and sustainable management solutions for wildlife disease issues, as well as a broader array of stakeholders, such as landholders, Indigenous and community groups, governments and students.

Source: Carver, Peters and Richards (2022). Reproduced under the terms of the Creative Commons Attribution License.

FIGURE 4.3

Integrated Disease Management of Wombat Mange

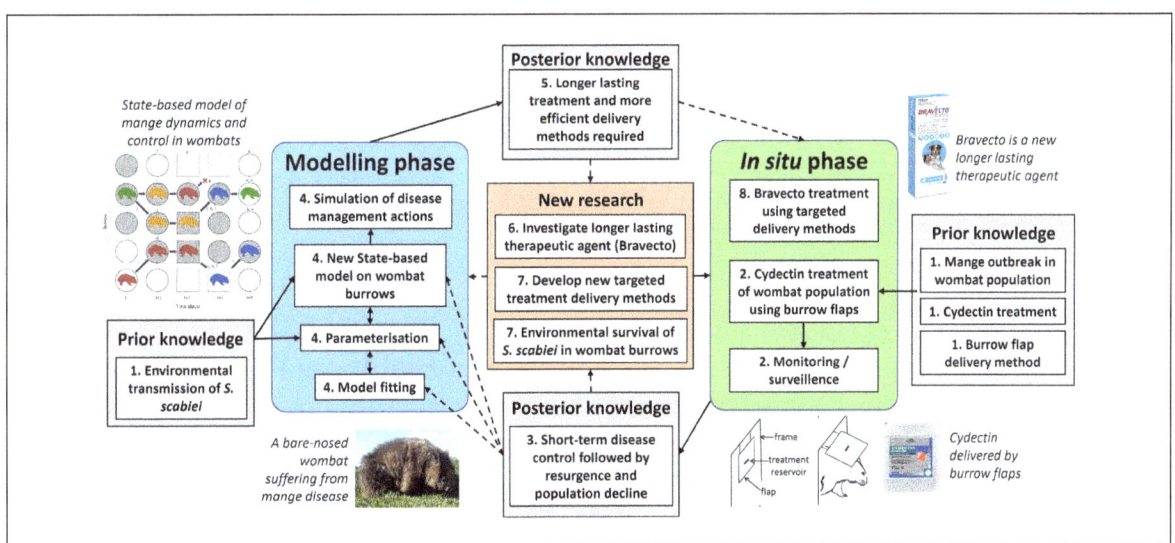

Notes: Solid arrows symbolize established research pathways; dashed arrows represent common gaps that often limit integration between modeling and in situ phases of wildlife disease management. This research commenced with the in situ phase and looped back to the in situ phase. Numbers indicate the sequence of learning. The experience informed additional research into the development of a longer-lasting treatment and more effective delivery of the treatment. While this application of the model is focused on managing sarcoptic mange disease—which is caused by the parasitic mite *Sarcoptes scabiei*—in bare-nosed wombats (*Vombatus ursinus*), the approach is the first to offer a potential solution for intervention decisions regarding the health of all wildlife, including apes.

Source: Carver, Peters and Richards (2022). Reproduced under the terms of the Creative Commons Attribution License.

for a decision-making process for intervention in each situation—and on the importance of using available information and knowledge to inform and guide that process. The following ape-specific best practice guidelines can assist decision-makers in considering potential interventions:

- for apes in their natural habitat: *Best Practice Guidelines for Health Monitoring and Disease Control in Great Ape Populations* (Gilardi *et al.*, 2015);

- for apes in range state sanctuaries: *Primate Veterinary Health Manual* (PASA, 2009); and

- for captive apes in non-range countries: "Is Your Facility Prepared?" (ZAHP, n.d.).

This chapter begins with a historical recap of the evolution of ape health-focused interventions. It continues by examining reasons to intervene, skills required for effective interventions, the ethical implications of vaccination and the factors that inform interventions at the systems level, such as capacity building, technological advances and the availability of relevant toolboxes and approaches. Through a variety of case studies—on topics as diverse as intervening in unregulated settings and improving diagnostics—the chapter delves into real-world scenarios that are rarely covered in ape health management resources.

Key findings include:

- Decisions to intervene are necessarily context-specific and potentially differ based on whether targeted apes live in ex situ or in situ settings and whether they are captive, habituated or wild.

- Concerns regarding the health of both the individual and the population require consideration in intervention decisions.

- Best practice in the consideration of potential interventions involves a risk-based approach designed to inform the decision-making process based on assessments of the consequences of both intervening and not intervening.

- Ape health intervention teams that possess requisite qualifications—such as diagnostic, veterinary and communication skills—are more likely to secure and maintain positive health outcomes, especially if they arrange for independent auditing of their welfare and health management processes.

A Brief History of Ape Health Interventions

When it comes to the health of captive apes, the duty-of-care concept emphasizes the need to intervene (Blackett *et al.*, 2017; Deem, 2007; Hernandez *et al.*, 2018). With respect to apes in their natural habitat, however, decision-making on whether to intervene for health reasons is a more ambivalent process, as the animals are more difficult to access and diverse ethical frameworks apply. This section presents two conservation-centered perspectives on ape health interventions for captive orangutans and gorillas in their natural habitat, as well as a historical overview of the evolution of ape health interventions since the middle of the 20th century.

Orangutan Health Interventions in Historical Perspective

Orangutan conservation efforts were initiated in the 1960s and 1970s in response to the high number of individuals, especially young orphans, caught and sold in wildlife markets. During those decades, four rescue and rehabilitation centers were established in Sumatra and Borneo, with the understanding that these species were decreasing in numbers and that displaced individuals needed to be returned to the wild to prevent them from becoming extinct (Rijksen, 1978; Smits, Heriyanto and Ramono, 1995). Systems approaches to multispecies health issues, such as One Health, were not common practice at that time (see Chapter 2). Disease transmission between wildlife and humans was overlooked, especially during the early attempts of rescued orangutan release, which involved minimal health examinations and pathogen screening. During this period, practitioners released rehabilitants in sites that were home to wild orangutan populations, thereby increasing the risks of species-specific disease transmission and spillover into other species, including humans.

In the 1990s, a new approach to rehabilitation was initiated in East Kalimantan (Smits, Heriyanto and Ramono, 1995). This new Borneo Orangutan Survival Foundation (BOSF) program based the rehabilitation and reintroduction methods on creating social bonds among rehabilitated orphans; rigorous disease screening, especially targeting zoonotic pathogens; and release into sites without a resident population. Today, most orangutan centers in Indonesia follow this approach, under the supervision of the Indonesian Ministry of Environment and Forestry.

Indonesia saw vast increases in forest exploitation and conversion to monoculture plantations throughout the 1990s. Sumatra and Borneo were especially affected by intensive oil palm cultivation (Arcus Foundation, 2014, 2015; Tsujino et al., 2016). This massive change in land use displaced hundreds of orangutans and other wildlife (Russon, 2009; Spehar et al., 2018). In response, a number of orangutan centers ramped up their rescue activities, which resulted in many taking in hundreds of additional apes, e.g., one in Kalimantan, Indonesia had rescued nearly 700 orangutans by 2009 and by 2019, that number had increased to more than 1,000.

These numbers raised concern over disease spread, especially zoonoses from humans, including human and orangutan-specific hepadnavirus (a group of DNA viruses that can cause liver damage, such as the hepatitis B virus), *Mycobacterium tuberculosis* complex (a genetically related group of bacteria that cause tuberculosis), *Plasmodium* spp. (single-celled parasites that cause malaria) and *Strongyloides stercoralis* (a parasitic roundworm, known as threadworm in the United States).[1] Centers that house large ape populations witness increased outbreak risk of these and other pathogens. The outbreaks represent a significant additional burden on the centers' operations, staff and community health around the centers, and they can potentially jeopardize the success of entire reintroduction programs (S. Unwin, personal observation, 2021).

Research projects have been undertaken to investigate these pathogens. Until 1999, many practitioners had assumed that seroconversion to human hepatitis B occurred in many orangutans in rehabilitation centers. This view was initially revised when Warren *et al.* (1999) and Warren (2001) confirmed that a wild, endemic orangutan hepadnavirus cross-reacted in the human hepatitis B serology test, thereby effectively eliminating this specific infection as a barrier to most reintroductions. It was not until 2010, however, that this information was widely acted upon by those working with orangutans.

Tuberculosis remains one of the most worrisome, confirmed pathogens in orangutan rehabilitation programs (S. Unwin, personal observation, 2021; see Case Study 4.6). The development of a robust diagnostic protocol for this challenging pathogen is crucial to successful disease screening in both rescued individuals entering captivity and rehabilitants released into the wild. A combination of polymerase chain reaction and tuberculin skin tests is often used in parallel with further tests, with the aim of improving diagnostic reliability and effectiveness. Ongoing research on field-based tuberculosis diagnostics for African great apes is expected to yield results that can also be applied to Asian apes.[2]

Indonesia and Malaysia have a combined total of 13 orangutan facilities, all of which have at least one full-time veterinarian (Unwin *et al.*, 2022). All orangutans who are to be reintroduced into the wild undergo a thorough health examination and disease screening to ensure that they will not harm wild populations or impact the health of human communities living close to the release site. Since its creation in 2009, the Orangutan Veterinary Advisory Group (OVAG)—a network of orangutan veterinarians and related professionals—has utilized One Health principles to help centers communicate with each other and share best practices in the health management of orangutans (and gibbons) (see Case Study 4.4).

Gorilla Health Interventions in Historical Perspective[3]

By the mid-1980s, Dian Fossey's research indicated that the mountain gorilla (*Gorilla beringei beringei*) population was rapidly declining and that fewer than 300 known individuals remained in the world. Gorillas were being killed through hunting, suffering life-threatening injuries caused by snares and succumbing to illnesses that Fossey suspected were being transmitted by humans. As no health system was in place to treat sick or injured gorillas at the time, Fossey envisioned a veterinary program to meet those needs. James Foster, a veterinarian at the Seattle Zoo, agreed to move to Rwanda to run this program and arrived in 1986, just months after Fossey's death. The same year saw the establishment of the Virunga Veterinary Center in Rwanda. Funded by the Morris Animal Foundation, the Center

> **Tuberculosis remains one of the most worrisome, confirmed pathogens in orangutan rehabilitation programs.**

aimed to take care of injured and critically ill gorillas and to provide medical treatment and quarantine for orphans (Gorilla Doctors, n.d.-d). In 2006, the Mountain Gorilla Veterinary Project was created and three years later it partnered with the School of Veterinary Medicine at the University of California, Davis, to take over the funding of the Virunga Veterinary Center, which was renamed Gorilla Doctors (Gorilla Doctors, n.d.-d, n.d.-f).

While the Gorilla Doctors began with a single veterinarian, the group now employs 16 and operates across three countries—the Democratic Republic of Congo (DRC), Rwanda and Uganda (Gorilla Doctors, n.d.-a, n.d.-b, n.d.-d). In the early years, the group's work focused on rescuing gorillas from hunters' snares; over time, the intervention philosophy developed to include treatment when gorillas had been exposed to potentially fatal human diseases and darting the animals with antibiotics, as well as anesthetizing them and operating on the jungle floor.

Today the group undertakes dozens of medical interventions annually on both subspecies of eastern gorilla—mountain gorilla and Grauer's gorilla (*Gorilla b. graueri*)—for the treatment of diseases and conditions caused by humans, as well as life-threatening injuries not caused by humans. Interventions include darting with antibiotics, anthelmintics, vaccinations, and anesthesia of mothers and babies for extensive diagnostics and treatment, including surgery. The veterinarians carry the necessary equipment—including an X-ray machine, gas anesthesia and ultrasound—and perform all interventions in the field. They also complete post-mortems on all recovered carcasses; this process reveals a considerable amount of data, not only on the cause of death, but also on underlying morbidities in the population (M. Cranfield, personal communication, 2021).

Gorilla Doctors has carried out more than 200 medical interventions on wild habituated gorillas and has cared for more than 20 orphans, many of whom needed 24-hour care to address dehydration, mental distress or wounds and would not have survived if left in the wild (Robbins *et al.*, 2011b; B. Ssebide, personal observation, 2021). Aside from contributing to the growth of the mountain gorilla population, the work has helped to build the capacity of African veterinarians to address apes' veterinary needs. While veterinary care for the mountain gorillas is expensive, its benefits arguably outweigh the costs, particularly with respect to population viability. Moreover, the positive impacts of veterinary care for mountain gorillas can serve as a symbol of positive conservation outcomes that help to maintain the balance of fragile ecosystem services to host countries.

> " Veterinary intervention used to be criticized as interference with the natural course of an ape's life. "

The Evolution of Great Ape Health Interventions

For more than 60 years, primatologists have been carrying out behavioral studies of great apes in the wild. Their research has been critical in informing the decision-making process for health-related interventions:

- In 1959, George Schaller began to study mountain gorillas in the Virunga Mountains of East Africa (Nicholls, 2015).

- Dian Fossey observed mountain gorillas for 18 years, from 1967 until her murder in 1985. Through her work, mountain gorillas became well known (Erdős, 2019).

- Jane Goodall and colleagues have studied chimpanzees at Gombe National Park, Tanzania, since 1960. Some Gombe chimpanzee communities have been habituated since the mid-1960s (Lonsdorf *et al.*, 2014). Theirs constitutes the longest continuous study of any great ape population.

- In the early 1960s, primatologist Toshisada Nishida began studying chimpanzees in Mahale, Tanzania, at what is now a long-term, research-productive field site (Nishida, 1968; Nishida, Matsusaka and McGrew, 2009).

- In 1971, Birute Galdikas began studying the now critically endangered orangutans in Indonesia (Gruen, Fultz and Pruetz, 2013).

- In 1973, Takayoshi Kano established a field site at Wamba in the DRC to study bonobos (Furuichi et al., 1999).

- More recently, numerous long-term and short-term study sites for bonobos, chimpanzees and gorillas were established (Kappeler and Watts, 2012).

During the early years of great ape research, health interventions were rare. When they did occur, their focus was on diagnosis and treatment to prevent animal suffering, such as rescue from hunting snares (Lonsdorf et al., 2014). In Gombe, for example, direct veterinary intervention in the form of anesthesia for disease investigation and treatment occurred only three times prior to 2005, despite severe disease outbreaks such as suspected polio in 1966, respiratory syndromes in 1968, 1987, 1996, 2000 and 2002, and sarcoptic mange in 1997 (Goodall, 1983, 1986; Mlengeya, 2000; Nutter, 1996; Williams et al., 2008).

Veterinary intervention used to be criticized as interference with the natural course of an ape's life. Indeed, environmental philosophers and conservationists have long debated the ethics of human intervention in nature, including with reference to the eradication of invasive species that threaten native species; the prevention of suffering that accompanies predation; and the release of captive rehabilitants into the wild (Gruen, Jamieson and Schlottmann, 2012; see Chapter 5). As discussed below, some interventions, such as preventive vaccination, remain controversial (Ryan and Walsh, 2011; see Chapter 5).

On the whole, however, support for interventions—especially ones designed to save apes' lives—has been growing, especially where care quality has improved. This shift may be partly due to a recognition that some "wild" apes live in circumstances that do not necessarily qualify as "natural." Mountain gorillas, for instance, draw more than 60,000 tourists annually and are thus exposed to a high risk of disease transmission from humans. Another reason for the shift may be linked to the growing impact of certain human activities on great apes, such as the use of indiscriminate snares and steel traps, or development-induced habitat loss that triggers aggression between individual apes or rival communities. In such cases, conservationists and others recognize an ethical duty to act if there is a safe and ready way of reversing an illness or injury (Gilardi *et al.*, 2015; Gruen, Fultz and Pruetz, 2013; Hockings *et al.*, 2015). At the same time, primatologists increasingly find themselves navigating the blurred line between illnesses and injuries that are directly caused by humans and those for which humans may be indirectly responsible (Fedigan, 2010).

In contrast to the earlier interventions, more recent ones have been conducted for both welfare and investigative purposes—to determine the cause of disease or suffering (Lonsdorf *et al.*, 2014). Veterinarians who conduct interventions are encouraged to take full advantage of the opportunity to undertake extensive sampling, not only for the patient, but also to build a biobank collection of biological samples for future research.

Some interventions can be perilous for both the humans and the ailing or injured apes. While interactions with wild gorillas can be somewhat risky for people, the dangers are more pronounced in interventions involving chimpanzees, as they tend to be more aggressive. To separate infected chimpanzees from their groups, veterinarians often need to wait until they are sick enough to be handled safely; such interventions are just as psychologically distressing for the ape, however, and the chances of a positive result may be significantly lower by that point.[4] Trapped apes, who typically struggle to free themselves, can die or develop gangrene, infections or deformities unless they are swiftly released. They cannot be released until they are anesthetized, which is usually dangerous, especially if other apes stand in the way. It is easier to administer an anesthetic dart to gorillas, who do not climb trees, than to chimpanzees, who may flee into the trees—only to fall to their death or sustain greater injury once the anesthesia

takes effect. Nevertheless, many interventions to remove snares and traps from chimpanzees have been performed successfully (ASP, n.d.; JGI, n.d.; Ohashi and Matsuzawa, 2011; B. Ssebide, personal observation, 2021).

In rare cases, researchers have intervened to treat disease outbreaks in ape communities. During the suspected polio outbreak at Gombe in 1966, for example, Jane Goodall's team administered the polio vaccine non-invasively to chimpanzees, by placing it into provisioned bananas. Goodall defends the procedure, which prevented the spread of paralysis and death among the apes (Greene, 2005). In the mid-1980s in Rwanda, an intervention designed to prevent death from a measles outbreak involved the vaccination of habituated gorillas by darting (Webber and Vedder, 2001). As public awareness of the transmission of human diseases to great apes has grown, so has the interest in inoculating the animals against diseases for which vaccines have been developed (Gruen, Fultz and Pruetz, 2013).

In contrast, preventive vaccination remains controversial, partly because it is experimental rather than reactive (C. Walzer, personal observation, 2021). Another concern is the expense involved in vaccinating great apes, particularly if the local human population has limited resources for health care and disease prevention (see Chapter 5). A third worry pertains to the lack of coordinated oversight for this sort of experimentation. Addressing some of the skepticism about preventive vaccination requires assessing the safety and efficacy of a potential vaccine delivery (Gruen, Fultz and Pruetz, 2013). As conservationists are often reacting to an immediate deadly infectious disease outbreak, there usually is no time for an intervention protocol to be developed, validated and approved through an oversight process. Preparedness is thus key to averting inappropriate intervention decisions, which

can be made in the heat of the moment (see Chapter 6). The ethical dimensions of ape vaccination are discussed in greater detail below.

Health interventions also include rescue operations, which can involve seizing apes from people who keep them as pets or entertainment props, generally with the aim of rehabilitating them for release into the wild. In some cases, apes are captured for translocation, typically to decrease the risk of human–wildlife conflict (see Case Study 4.1). While translocations may be undertaken as preventive care measures, they carry their own health risks, including

that released apes may transmit diseases to resident wild apes (Schaumburg *et al.*, 2012). Moreover, habituated chimpanzees can prove dangerous following translocation and their release is sometimes opposed by local communities (Hockings *et al.*, 2010; Sherman, Ancrenaz and Meijaard, 2020).

Reasons to Intervene and Skills Needed to Maximize Effectiveness

This section examines the complexity of the decision-making process that informs individual ape health interventions. It features an example of the Gorilla Doctors' context-specific decision tree, which is continually revised in line with emerging information (Decision Tree Writing Group, 2006; see Figure 4.4). The group relies on experienced professionals with clinical skills to maximize intervention effectiveness (B. Ssebide, personal observation, 2021). As discussed below, the administration of anesthesia is among the skills that are often required to secure intervention success.

The two case studies below explore decisions to intervene in response to human–orangutan conflict in Indonesia and in relation to gorilla snare and fight-induced injuries (see Case Studies 4.1 and 4.2). In such situations, the decision not to intervene can be among the most important a wildlife veterinarian can make under the duty of care, in terms of both animal welfare and conservation (Gray and Favre, 2022). Case Study 4.3 considers the role of veterinarians in protecting gibbon health in an unregulated setting, the United Arab Emirates. This section may be read alongside Chapter 2, which explores One Health and focuses on the need for multi-disciplinary collaborations in complex systems to improve intervention outcomes, and Chapter 5, which debates the ethics of health interventions.

FIGURE 4.4

Flow Chart of the Clinical Response Decision Tree for Mountain Gorillas

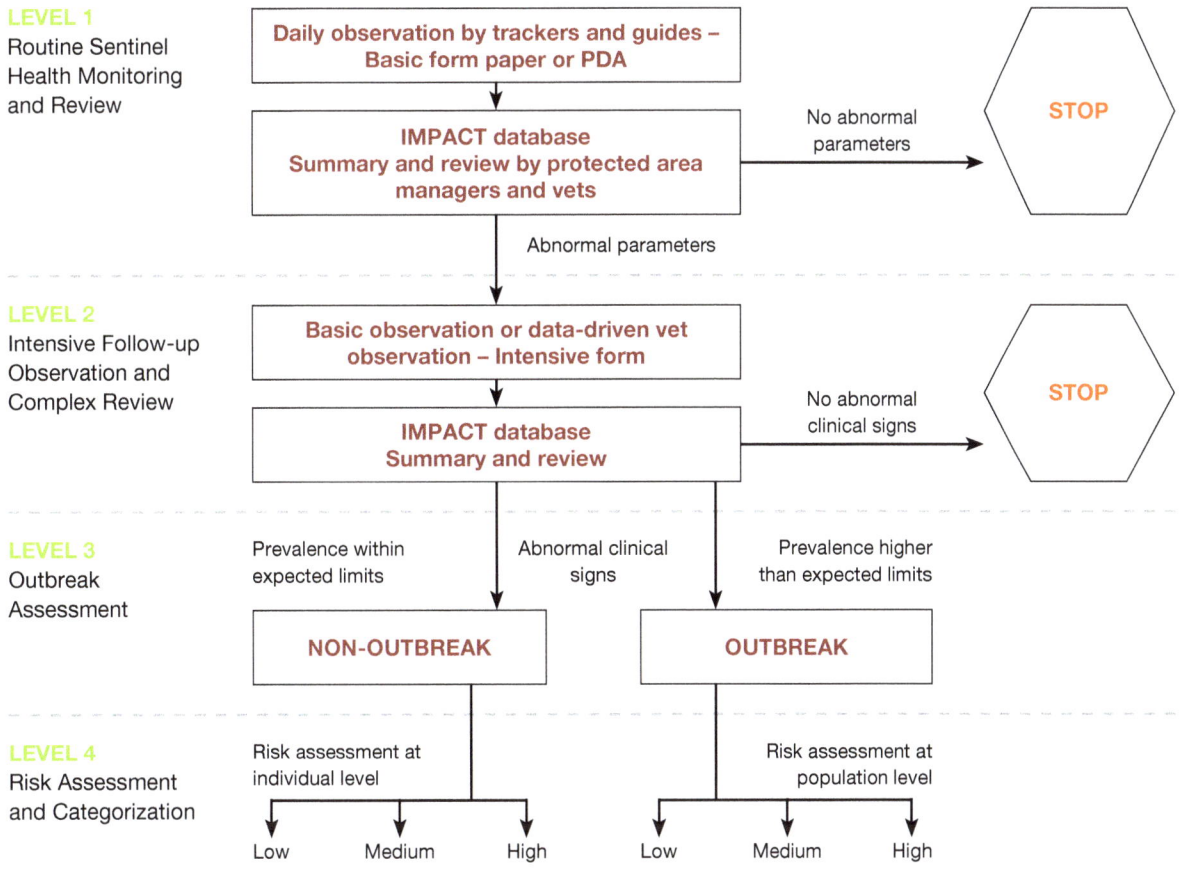

LEVEL 1
Routine Sentinel Health Monitoring and Review

LEVEL 2
Intensive Follow-up Observation and Complex Review

LEVEL 3
Outbreak Assessment

LEVEL 4
Risk Assessment and Categorization

LEVEL 5
Risk Management

Action	Low	Medium	High	Low	Medium	High
Immobilization	N/A	±	+	N/A	±	+
Diagnostics a) Invasive b) Non-invasive	N/A ±	± ±	+ +	N/A ±	± ±	+ +
Treatment	N/A	±	+	±	±	+
Continued observation	±	+	+	±	+	+
Outside help	N/A	N/A	±	N/A	± reg	reg ± inter
Action plan	N/A	N/A	+	±	+	+
Preventive action	N/A	±	±	±	±	+
Reporting	PA + PD	PA + PD	PA + PD	PA + PD	S + SH + PH	S + Approp inst.
Export permits	N/A	N/A	N/A	N/A	±	+

Notes: Approp inst.: appropriate institution (such as National Institutes of Health or Centers for Disease Control and Prevention); **inter:** international help needed; **N/A:** not applicable; **PA:** protected area authority; **PD:** Mountain Gorilla Veterinary Project director; **PDA:** personal data assistant; **PH:** public health official; **reg:** regional or in-country veterinarians can handle situation; **S:** subsequent groups; **SH:** stakeholders; **±:** decision on individual case basis.

Source: Decision Tree Writing Group (2006, fig. 1) © 2006 Wiley-Liss, Inc.[5]

Deciding Whether to Translocate Orangutans in Response to Human–Wildlife Conflict[6]

Background

Humans and wildlife have come into conflict since before recorded history (Dickman and Hazzah, 2016; IUCN SSC Human–Wildlife Conflict & Coexistence Specialist Group, n.d.; Nyhus, 2016). A recent study that focused on human–ape conflict in Borneo found that respondents said they had killed orangutans in self-defense or retaliation; anecdotal field reports corroborate these results (Davis *et al.*, 2013). In contrast, no reliable evidence or published literature indicates that wild orangutans attack or injure people (McLennan and Hockings, 2016).

Resource conflict between humans and orangutans is the leading cause of translocations. Typically, the result of complex decision-making processes, such translocations involve the human movement of orangutans between habitats or from captive facilities into natural habitats (Sherman *et al.*, 2021). Orangutans have been translocated to protect conservation needs while mitigating risks associated with agricultural and infrastructure development (Humle, 2015). Translocation can protect orangutan health, yet it also has inherent health risks as it brings orangutans and humans in ever closer contact (Sherman *et al.*, 2021).

In Sumatra, Indonesia, orangutans tend to be translocated to remote areas, usually in an attempt to avoid further conflict and to minimize the risk that local people will kill apes in retaliation for crop damage or if they fear for their personal safety. Transporting orangutans is costly and effective post-release monitoring is rare due to a lack of resources and capacity; meanwhile, the apes themselves suffer health and welfare impacts as a result of translocation (Meijaard *et al.*, 2012; Robins *et al.*, 2019; Sherman *et al.*, 2021).

Anesthesia and Unintended Consequences of Translocation

In deciding whether to translocate an orangutan, conservationists and veterinarians weigh potential benefits—such as the prospect of enhanced disease management—against a host of potential harms, such as the risk of transferring undiagnosed disease or of upsetting population genetics in severely degraded habitat (Ancrenaz *et al.*, 2021; Kock, Woodford and Rossiter, 2010).

The capture itself can be life-threatening for both the human participants and the orangutan.[7] Orangutans generally require chemical immobilization via remote dart delivery (using a rifle). Since it is rare to get closer than 20 m to an orangutan, veterinary professionals generally calculate the dose of a species-safe anesthetic based on visual estimations of an individual's age, sex and body weight. Under such circumstances, it is difficult to identify—and impossible to confirm the presence of—any underlying medical conditions, such as congenital heart issues, respiratory problems or drug allergies. As a result, reactions to an anesthetic are unpredictable, particularly in agitated or stressed orangutans.

A safe capture depends on the team members' expertise and resources, and their ability to adapt quickly to a rapidly changing situation. In many cases, darted, sedated orangutans remain hanging up to 20 m from the ground and need to be caught safely in a net. At this critical point, an anesthetized orangutan can shift and drop outside of a predicted falling spot, which can lead to injury or death. Over the past 15 years, translocation operations in Sumatra have led to a few orangutan deaths and serious injuries, such as broken bones. While exact figures of injuries have not been recorded, research suggests there may be a significant association between darting with a rifle and serious or fatal injuries in primates globally (Cunningham, Unwin and Setchell, 2015).

Most translocated orangutans are in good physical condition and are sent straight to a translocation site to be released as soon as possible. Although this process always includes a physical exam, teams generally do not undertake further diagnostics for disease. They make exceptions for orangutans who exhibit abnormal wild behavior, a major injury (such as a bone fracture) or a debilitating condition (such as blindness) that may impair their survival. In such cases, orangutans are sent to rehabilitation centers.

The public tends to view translocation as a positive action for orangutan conservation, perhaps because it is often highlighted by non-governmental organizations in campaign materials showcasing animals in poor condition, appealing for public support. As the reality is usually more nuanced, a discussion is needed between policy-makers and practitioners on the conservation benefits of the translocations process, given that it can lead to the death of orangutans or the disturbance of population genetics in fragmented habitats (Ancrenaz *et al.*, 2021).

CASE STUDY 4.2

Deciding Whether to Intervene in Response to Gorilla Snare and Fight-Related Injuries

In the early 1980s Dian Fossey noted that gorillas were dying from treatable wounds caused by snares, as well as from inter- and intragroup aggression (Harcourt, Fossey and Sabater-Pi, 1981; Hassell et al., 2017). Veterinary interventions in response to non-infectious health issues in mountain gorillas (*Gorilla beringei beringei*) are not uncommon (Barone, 2015; Burt *et al.*, 2017). Every year, Gorilla Doctors rescues dozens of gorillas from hunters' wire snares and many silverbacks are treated for severe fighting-related trauma.

Gorillas can be accidental victims of wire snares, which may be set by hunters who legally target wildlife such as forest antelopes (Haggblade *et al.*, 2019). In many cases gorillas cannot free themselves from the snare loops, which can tighten as they fight to remove them. In the absence of veterinary intervention, snaring can potentially result in the loss of limbs, infection, sepsis or death. Since the impacts of snares and other traps on gorillas are clearly caused by humans, veterinary intervention is a duty-of-care obligation.

Rationales for intervening in response to injuries from inter- and intragroup fighting are more complex. Although aggression between silverbacks from different groups is rare, the apes do occasionally fight to protect their core range and group members; within the same group, silverbacks fight for dominance (B. Ssebide, personal observation, 2021). In both cases, the fighting individuals—as well as some young gorillas—can suffer mild to life-threatening injuries.

While fighting among gorillas is natural, the species' low numbers render every individual's genetic input critical to the health of the population. Efforts to save individuals are thus made even if veterinarians can only deliver a guarded prognosis that a gorilla may suffer or die in the absence of an intervention. The tourism industry also plays a role in pushing for interventions, as tourists and protected area authorities consider some wounds sustained by habituated gorillas unsightly and upsetting.

Such cases can present ethical dilemmas for veterinarians, who have a duty of care to ailing or injured animals, not an obligation to consider tourists' perceptions. In practice, however, these perceptions may not just be difficult to ignore—they may prove to be a factor in deciding whether to intervene. For example, all efforts are made to save a silverback from a single-silverback group, as that individual's death may lead to group disintegration and thereby reduce the number of groups available for tourism. From a conservation perspective, the decision to save the silverback also makes sense, given their importance to the genetic health of the population.

Photos: Rationales for intervening in response to injuries from inter- and intragroup fighting are more complex. Although aggression between silverbacks from different groups is rare, the apes do occasionally fight to protect their core range. Silverback gorilla whose lower lip was torn apart during an interaction with another group; on this occasion there was no intervention and the injury healed entirely on its own. Left – pre-injury. Right – during healing. © Gorilla Doctors

Administering Anesthesia: A Requisite Skill for Successful Ape Health Interventions

Veterinarians are regularly required to administer anesthetics to enable intensive diagnostic examinations, therapeutics, surgical procedures, and safe transport and translocation for ape conservation purposes. During anesthesia, continuous monitoring of vital signs is essential, as is maintaining the airway, which can require the provision of oxygen. For any procedure involving painful stimuli, the anesthesia protocol includes the provision of analgesia. The growth of conservation management of in situ great ape populations has led to development of field anesthesia techniques for translocation, reintroduction into the wild and clinical interventions (Cerveny and Sleeman, 2014).

Ketamine is frequently used to immobilize apes, with or without an adjunct sedative (such as midazolam or other benzodiazepine). Alternatives include a proprietary mixture of tiletamine and zolazepam (Telazol™ or Zoletil®), as well as an alpha-2 agonist such as medetomidine, in combination with either tiletamine/zolazepam or ketamine. While these drugs and drug combinations generally provide safe and effective immobilizations, alpha-2 agonists may present a significant risk to apes who are predisposed to or already have cardiovascular disease (GAHP, n.d.). This brief enumeration does not cover all the drug combinations that can be used for ape anesthesia, nor does it identify all the corresponding evidence-based concerns.

In administering anesthesia, site staff and area veterinarians develop the most effective anesthetic plan relative to their context. They have access to numerous guidelines on the use of anesthetics in primates (Abelló, Rietkerk and Bemment, 2017; PASA, 2009; Research Animal Resources, n.d.). In all jurisdictions, anesthetics are classified as veterinary-only medicines and therapeutics, meaning that it is illegal for non-veterinarians to administer (and often to handle) these drugs without direct veterinary supervision (Cunningham, Unwin and Setchell, 2015). General considerations for anesthesia in apes include the following:

- **Intravenous catheterization**: Following sedation, an indwelling venous catheter is placed in a vein so that it can serve as a port for the administration of anesthetic drugs, emergency drugs and intravenous fluid support. The most common sites for catheter placement are the saphenous vein (in the hindlimb) and cephalic vein (in the forelimb).

- **Fluid support**: The provision of supplemental fluid support is recommended for animals under anesthesia for longer than 30 minutes. Appropriate fluid rates range from 5 to 10 ml/kg per hour and may vary based on the anesthetic combination used.

- **Monitoring**: Standard mammalian monitoring techniques apply to apes. The goal of monitoring is to maintain cardiovascular homeostasis and core body temperature. An understanding of the basic physiologic effects of anesthetics is key to the correct interpretation of monitoring parameters in anesthetized apes, including anesthetic depth, heart rate, respiratory rate, oxygen saturation (SpO_2), expired carbon dioxide ($EtCO_2$), temperature, blood pressure and mucous membrane color.

- **Heat support**: As most anesthetic drugs cause hypotension and hypothermia, the provision of supplemental heat (such as circulating water blankets) to apes under anesthesia is recommended. Regardless of the heat source, animals are never placed directly on the heat.

Ape Health Interventions in Unregulated Settings

Settings with limited regulatory control and governance mechanisms to support practitioners represent a challenge to successful ape health interventions and the sustainable implementation of the duty of care. Unregulated wildlife trade can exacerbate the problem by facilitating ape ownership by individuals who are ill-equipped to care for them (Arcus Foundation, 2020). Under these conditions, interventions are highly likely to be reactive, rather than preventive.

Such is the case for gibbon health in the United Arab Emirates (UAE), where veterinary care for trafficked apes is unregulated and ape health data are scarce. In Case Study 4.3, a veterinarian who provides care for privately kept gibbons offers expert opinion on local barriers to gibbon health and welfare in the UAE (see Chapter 8). The previous volume in this series, *State of the Apes: Killing, Capture, Trade and Conservation*, provides further information on ape trafficking (Arcus Foundation, 2020).

Ethical Complexities Related to Vaccination

In addition to a regulatory framework, an ethical framework is key to the success of health interventions. This section explores ethical considerations of vaccination as an intervention that can impact the health of an entire population. Ethical considerations are examined further in Chapter 5 of this volume.

In human and companion animal medicine, vaccines are a mainstay. They are among the most efficacious and cost-effective prevention tools used to promote population health (Orenstein and Ahmed, 2017; Sánchez-Vizcaíno *et al.*, 2018). Vaccination is also used to prevent the spread of a variety of

CASE STUDY 4.3

Veterinary Interventions for Privately Held Gibbons in the United Arab Emirates[8]

Drivers of the Wildlife Trade

There is often a disconnect between wildlife law and practice (Roe and Booker, 2019; Runhovde, 2022). In the United Arab Emirates (UAE), information on wildlife trade is tightly controlled. There is no evidence of prosecutions of UAE citizens in relation to illegal wildlife trade or welfare neglect.

Documentation filed under the Convention on International Trade in Endangered Species of Wild Fauna and Flora (CITES) regarding wildlife movements into and out of the UAE is sparse, not just for apes, but also for many other endangered species, especially birds (Morocco World News, 2018; Soorae *et al.*, 2008). No one questions the origin (or the destination) of animals (see Figure 4.5). Veterinarians are under immense pressure to sign off on incomplete official documentation without question, or risk losing their jobs and being banned from the country. There may be reason for optimism, however, as the UAE and UK governments launched a toolkit to support financial institutions in tackling illicit money flows associated with the illegal wildlife trade in March 2022 (TRAFFIC, 2022).

In the UAE, apes account for a smaller proportion of the illegally traded and held primates than do monkeys and other species—including baboons, slow lorises and vervet monkeys (*Chlorocebus pygerythrus*). Wildlife veterinarians working in the UAE are generally aware that many gibbons are smuggled into the country via Oman, as the border can easily be crossed by vehicle, with the apes hidden in car trunks or under seats (which is also a common method for smuggling cheetahs). One prominent figure who has utilized veterinary services claims to have "rescued" more than 70 gibbons of various species.

For the elite and influential individuals in the UAE, ape ownership is a status symbol. While reports of ape ownership are invariably linked to the royal family, information is compartmentalized, making total numbers impossible to estimate. What can be verified is that charities and non-governmental organizations that might monitor and publicize the situation are very restricted in the UAE. Wildlife trade in the region was much more obvious in the early 2000s; there is little indication that trade levels have declined since then, but more deals may be occurring behind closed doors. The experience related below is representative of only a small proportion of ape-related veterinary issues in the UAE, as many ape owners rely on veterinary clinics that belong to the royal family.

Reliance on Diagnostic Tests

For those who are involved in the illegal ape trade, the top priority is neither health nor welfare, but cash return. Given concerns about personal health and the spread of zoonotic disease, however, pre-purchase testing has become the norm, posing additional risks to animal welfare. Many gibbons who arrive on the market in the UAE test positive for hepatitis B and thus end up in limbo, typically going from vet to vet before being returned to the dealer. What happens to these animals in the long term is unknown. Since they are already in the country, it is unlikely that they would be repatriated to their country of origin—at least there is no evidence thereof. In all likelihood, dealers then try to find a more naïve buyer and sell for less, to secure some return on investment.

FIGURE 4.5

Routes Used for Trafficking Gibbons to the United Arab Emirates

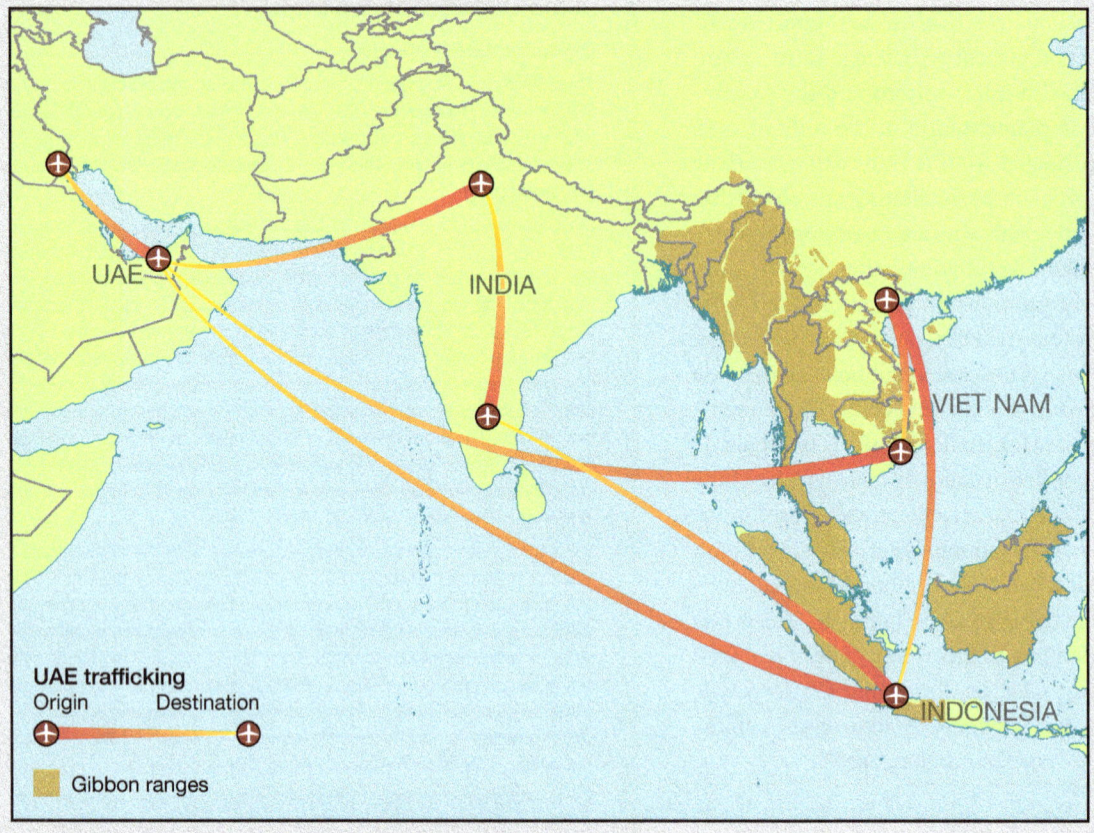

UAE trafficking

Origin   Destination

⊕————————⊕

▨ Gibbon ranges

Source: Adapted from Utermohlen and Baine (2018, fig. 90)

Not all diagnostic tests available in the UAE are reliable, however. For example, they cannot differentiate the zoonosis hepatitis B from the endemic gibbon-specific hepatitis, which does not appear to be a clinical issue in these apes (Norder *et al.*, 1996; Robertson and Margolis, 2002). In a sanctuary or zoo, standard biosafety protocols prevent the spread of the disease between gibbons and humans. But many of these apes are kept in peoples' homes, close to children, so owners insist on their own version of the precautionary principle. Perhaps unsurprisingly, clinical evidence shows that recently imported, underweight gibbons have had problems seroconverting to hepatitis B vaccination, indicating that the vaccination is not protective. Indeed, since the animals are immunosuppressed due to their general poor condition, vaccine effectiveness of any sort is likely to be reduced.

Prospective purchasers can also be unnecessarily thorough in checking for diseases, including cytomegalovirus and infections that may not be major health concerns. If a gibbon tests positive for any pathogens, they will not buy them.

Health Concerns Regarding Privately Held Gibbons in the UAE

Privately held gibbons in the UAE commonly exhibit evidence of infection and stress, such as raised white blood cell counts, septicemia and high parasite burdens, as well as evidence of malnutrition that could lead to growth defects, such as abnormal calcium-to-phosphorus ratios in juveniles and increased susceptibility to pathogen infection. *Strongyloides* spp. — a type of parasitic worm seen in many species, including humans—kills many trafficked gibbons in the region. Although this parasite is part of the normal gut flora in many species, it regularly causes superinfections in gibbons who were separated from their mothers before they could be weaned. Complicating matters, these infants often get shuttled from carer to carer, who provide them with varying milk formulas, which can attack overall immunity and gut microbiome balance.

To prevent fatalities, *Strongyloides* in preweaned gibbons is treated as a matter of urgency. If caught early enough, the

FIGURE 4.6

Strongyloides in Privately Held Gibbons in the UAE

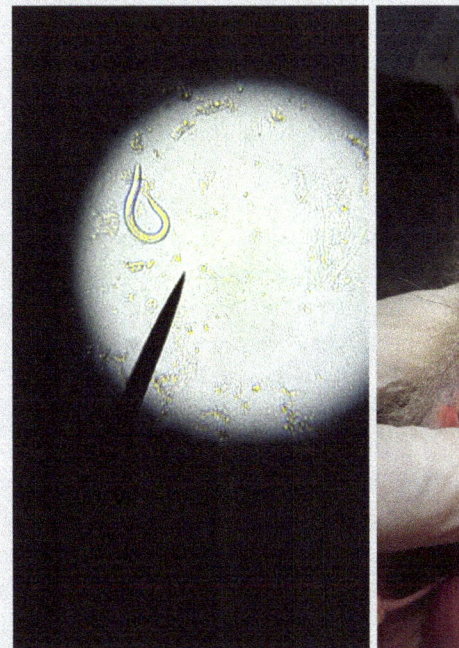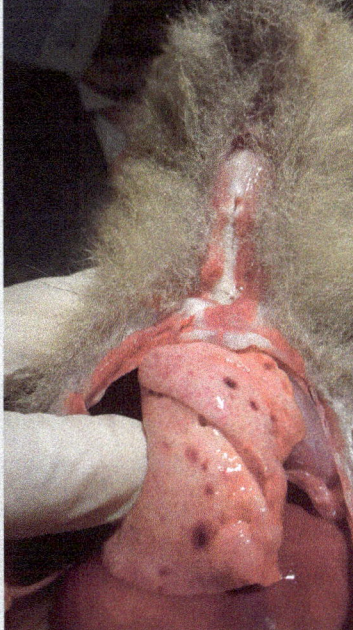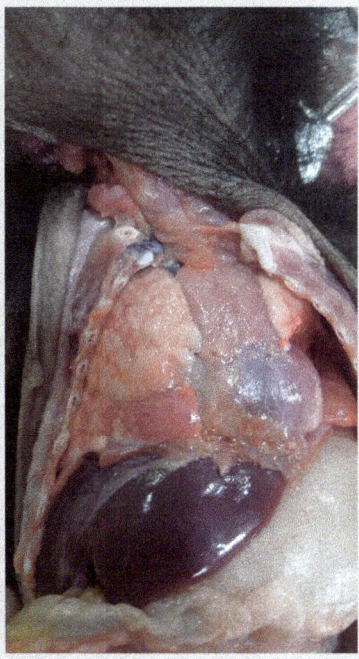

Notes: Left: *Strongyloides* L1 larvae from a gibbon's fecal sample, indicating an overwhelming parasite burden. Center and right: Gross lesions on a gibbon's gut and lungs, as revealed in a post-mortem examination.

Photos courtesy of the author.

infection can be managed with anthelmintic treatment, so long as there is good owner engagement.

The most common clinical signs in *Strongyloides* cases are anemia, vomiting and diarrhea. At this stage, most gibbons are unresponsive to anthelmintic therapy. The parasite lifecycle is such that most fecal examinations do not reveal eggs, but rather dead or live larval forms in fecal samples (see Figure 4.6). Related risk factors include chronic weight loss due to inappropriate nutrition, which allows infections to spread throughout the body, as also seen in immunocompromised human patients. For both gibbons and humans, treatment consists of repeat doses of avermectin until the infection is cleared.

Other easily preventable diseases and conditions that cause clinical concern with respect to the privately held gibbon population in the UAE include *Giardia* spp., *Balantidium coli* (when there is a heavy burden), *Trichuris* spp. (also when there is a heavy burden), dermatophytes, fractures, hernias and umbilical infections. Intestinal yeasts such as *Candida* spp. can be problematic, especially when linked to the overuse of antibiotics (a common and chronic issue in the country) or when young gibbons are on high-fruit diets, which, in the absence of correct treatment, can allow yeast overgrowth with the potential for lethal diarrhea (see Figure 4.7).

FIGURE 4.7

Candida in Privately Held Gibbons in the UAE

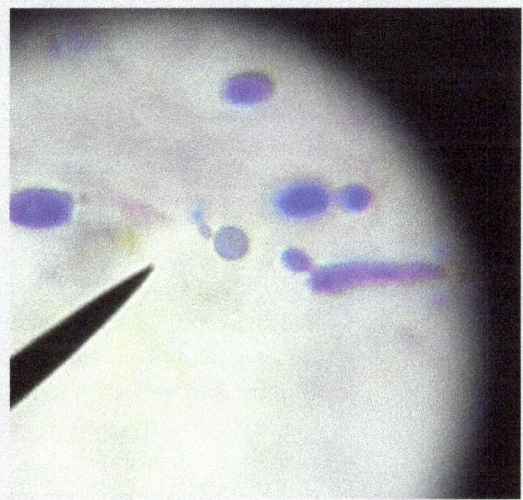

Note: *Candida* spp. found on gibbon feces.

Photo courtesy of the author.

pathogens to production animal species—for welfare, population health and food security reasons (Richeson *et al.*, 2019; Roth, 2011). In wild animals, however, vaccinations are far less common and tend to be used in specific instances. One example is the use of oral rabies vaccines in red foxes (*Vulpes vulpes*) in Europe, where the fox is seen as a wildlife reservoir for rabies (Cliquet *et al.*, 2003). Indeed, this disease prevention tool is often prohibited, either because diagnostic tests cannot differentiate between genuine infection and vaccination, or based on a widespread—but perhaps waning—assumption that it would not be feasible (Abbott, 2020; Buddle *et al.*, 2018; Edwards, Chatterjee and Santini, 2021).

With respect to the vaccination of apes and other species, a single rationale applies: disease prevention in individuals as a tool for disease reduction in populations. In the case of apes, this approach fulfills both welfare and conservation aims, yet few vaccines have been specifically manufactured for apes. One notable exception was the development of a customized encephalomyocarditis virus vaccine, after a multispecies outbreak in Australia that included zoo-housed orangutans (Reddacliff *et al.*, 1997; L. Vogelnest, personal communication, 2021).

Evidence of antibody response to vaccination in apes is scant, in part due to extremely small sample sizes in studies to date, including on Ebola, hepatitis B and *Streptococcus pneumoniae* (Solleveld *et al.*, 1984; Thornton, Walker and Zuckerman, 2001; Walsh *et al.*, 2017). Thornton, Walker and Zuckerman (2001) immunized London Zoo gorillas and gibbons against hepatitis B using the standard human protocol of the day: one dose every month for three months and then at 12 months. They found that gorillas required an extra dose of the vaccine at three months to become protected—that is, to have a serum antibody level of more than 100 milli-international

units per liter, matching protective levels in humans.

In zoo and non-release sanctuary settings, a decision to vaccinate is usually based on the risk of exposure, while vaccine regimes follow protocols designed for humans (Mugisha *et al.*, 2010; Weston-Murphy, 2015). Decisions to vaccinate in such settings are relatively easy to justify; they are typically based on welfare grounds and on risk analysis of a situation, in relation to protecting an ape from human infection. Deciding whether to vaccinate free-living and rehabilitant apes is far more complicated.

In past decades, wild mountain gorillas were successfully vaccinated against measles (Hutchins, Foose and Seal, 1991). As was the case then, identifying potential "super spreaders" can still inform conservation measures aimed at limiting the spread of epidemics, including via vaccination programs (Carne *et al.*, 2013). The vaccination of wild animals is associated with a multitude of disadvantages, however. These include the expense; the difficulty of implementation—that is, vaccinating enough individuals for the population to develop herd immunity; disruption and stress to populations, which can further lower immunity; and the potential reduction of selection pressure for natural resistance to diseases (allowing colonization of the host by a strain of pathogen against which the vaccine does not provide protection) (Cabezas, Calvete and Moreno, 2006; Carne *et al.*, 2013).

Disease susceptibility and vaccine efficacy vary across species, in part due to their behavioral ecology. For orangutans, targeted vaccinations may be a valuable preventive measure for any pathogen (Carne *et al.*, 2013). In contrast, vaccinating targeted chimpanzees would not necessarily qualify as a useful preventive measure, even though these apes appear to be far more susceptible to disease spread than orangutans. Given the severe risk that human diseases could spread

> " Disease susceptibility and vaccine efficacy vary across species, in part due to their behavioral ecology. "

to chimpanzees, alternative preventive measures are required, particularly since it is difficult to stop the spread of disease once it has penetrated a chimpanzee community.

Experts generally agree that a vaccine must be shown to be efficacious, deliverable and safe for both the target (ape) and non-target species (domesticated animals, feral animals, other wildlife and humans) before deployment (Cameron and Reed, 2019). In response to the increased frequency of Ebola epidemics, however, some researchers have proposed new approaches that stretch the ethical and moral boundaries for intervention (see Chapter 5).

Overall, fighting the spread of Ebola among great apes requires an understanding of virus biology and ecology, vaccine composition and vaccination dosing requirements for effectiveness. The selection of a vaccine and a cost-effective vaccination strategy are mainly determined by the accessibility of the apes—that is, whether they are habituated to human presence or not—and by the aim of vaccination. The aim could be to prevent the introduction of Ebola from the natural source into the ape population, or to stop the spread of infection within populations once an outbreak has started. Since great apes are endangered species, the potential use of experimental vaccines on them gives rise to ethical concerns, regardless of whether they are wild, habituated or captive (Leendertz *et al.*, 2017).

In a "conservation-oriented" vaccine trial on captive chimpanzees, however, Warfield *et al.* (2014) tested an experimental virus-like particle as a vaccine against the Ebola virus. Without the mitigating effects of a vaccine, they postulated, the virus would impact ape conservation due to mortality in the wild. This approach arguably failed to consider the welfare and population significance of the captive chimpanzees on whom the trial was conducted. It also appears to have overlooked practical considerations

for wild ape populations, such as mode of delivery (oral or injectable) and ease of access (to habituated vs. non-habituated apes) (Cameron and Reed, 2019).

Some medical professionals have put forward the potentially polarizing view that researchers should test the safety of novel Ebola vaccines in wild apes by employing a pluralistic approach to evidence, which diverges from the traditional reliance on a single method to measure the effects of one intervention at a time (Edwards *et al.*, 2018). They provide two reasons to test vaccines in wild populations of apes: to protect apes and to reduce Ebola transmission from wild animals to humans, a transmission route now known to be highly exaggerated (Kuisma *et al.*, 2019).

Other experts highlight the benefits of human vaccination campaigns in protecting great ape populations, noting that such interventions have been successful in eradicating extremely damaging diseases (Capps and Lederman, 2016). This approach has limitations, however. It is only effective with respect to habituated ape populations, which make up a small subsection of wild populations, and it can jeopardize the habituation process.

Practitioners of human and veterinary medicine recognize that any medical intervention carries risks. The principle guiding whether to vaccinate apes arguably needs to reflect that realization, perhaps by placing the greatest emphasis on ensuring that potential benefits outweigh any risks, rather than on prioritizing the aim to do no harm (Varkey, 2021).

Informing Interventions at the Systems Level

Health interventions on individual apes or populations can impact entire ecological systems. To be successful, such interventions

> " Since great apes are endangered species, the potential use of experimental vaccines on them gives rise to ethical concerns, regardless of whether they are wild, habituated or captive. "

require access to resources, human capacity, and appropriate techniques and approaches.

Building Human Capacity for Successful Ape Health Interventions

Insufficient human capacity can represent a major limiting factor on ape health-focused interventions. This problem is generally linked to an absence of empowerment rather than a lack of will, as conservation and welfare endeavors begin with a duty of care, not just for individual patients, but also for the environment (Kelly, Osburn and Salman, 2014; Lyons, Smuts and Stephens, 2001). Capacity building is thus critical to the success of ape health interventions. Case Study 4.4 discusses the creation and facilitation of an international capacity-building network for the promotion of orangutan and gibbon health. Case Study 4.5 presents a preventive health assessment framework that can be used for interdisciplinary capacity building in both wild and captive settings.

Translocation-Related Risks and Pathways to Solutions

IUCN has published best practice guidelines for wildlife translocations as well as for great ape disease risk management, including in the context of translocations (Beck *et al.*, 2007; Gilardi *et al.*, 2015). A great ape release that is consistent with the precautionary principle is one that does not endanger resident wild populations via communicable disease, hybridization, excessive social disruption or exacerbated competition for resources. IUCN guidelines further stipulate that individual welfare benefits alone are not a valid rationale for a conservation release and that conservation of a taxon and wild conspecifics takes precedence over the welfare of captive individuals. Since IUCN is not a

regulatory body, its guidelines are only enforceable in countries or areas where national or local decision-makers mandate compliance with them.

As IUCN guidelines are not legally binding in some range states, ape translocations and releases throughout these countries

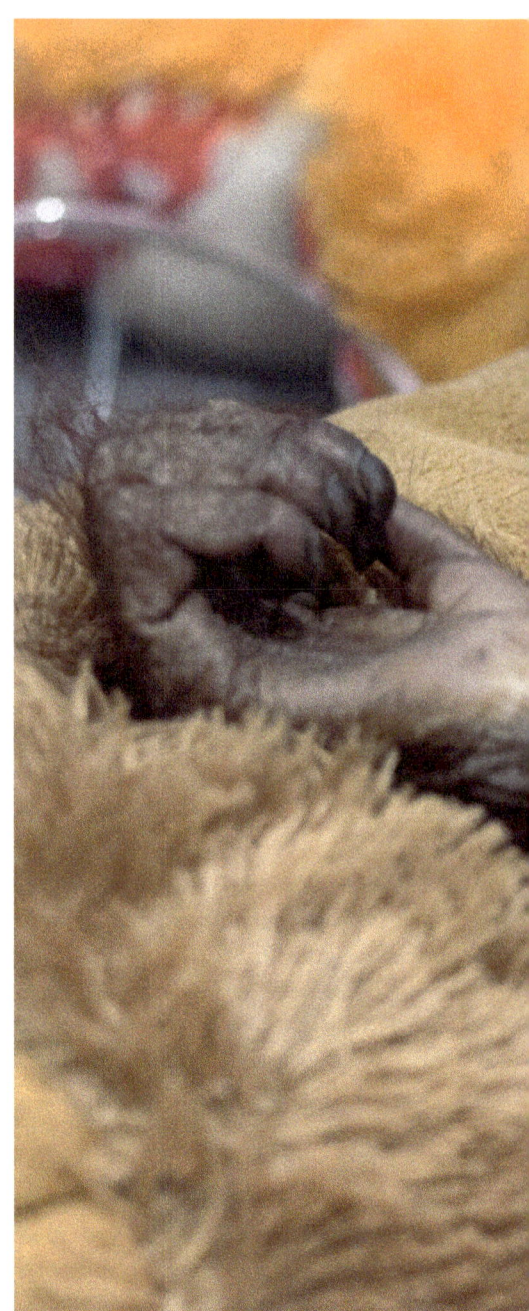

are carried out on an unregulated basis. Moreover, some governments may actually be driving ape releases into the wild. In August 2019, for example, the Ministry of Environment and Forestry of Indonesia issued a draft national plan for orangutan conservation for 2019–29, which called for all releasable orangutans to be translocated by 2024 (Scorpion, 2019). Such political pressure for translocation interventions can lead to avoidable health dilemmas (Sherman, Ancrenaz and Meijaard, 2020; Sherman *et al.*, 2021). Following review, the ministry withdrew the plan a few months after its release.

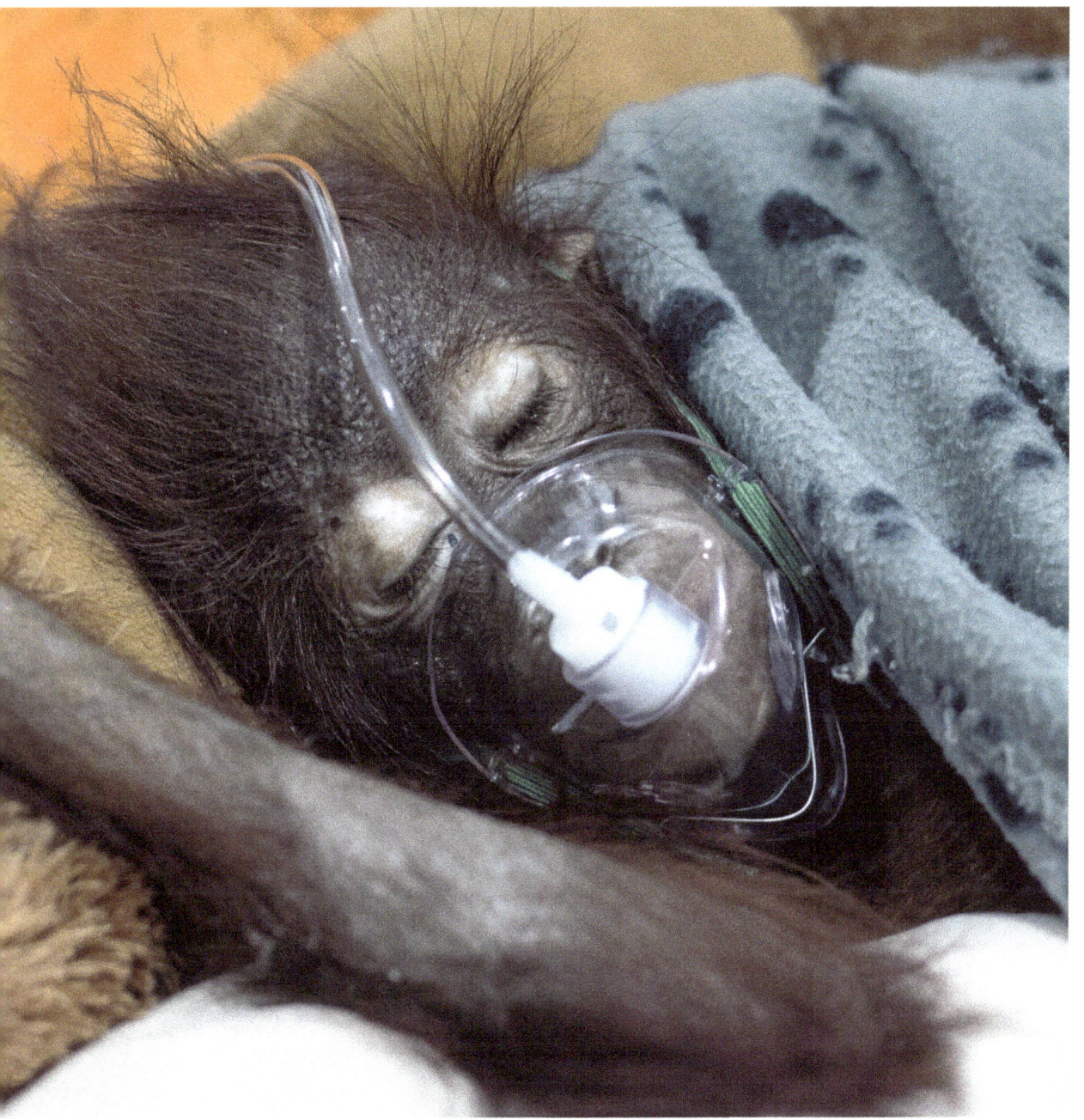

The Orangutan Veterinary Advisory Group

Created in 2009, the Orangutan Veterinary Advisory Group (OVAG) is a capacity-building and expertise network that brings together experts from a wide variety of organizations in an effort to ensure orangutan health (Unwin *et al.*, 2022). OVAG is linked to a global network of practitioners, researchers and specialists, including, since 2015, a gibbon practitioner network operating under the Section on Small Apes of the International Union for Conservation of Nature (IUCN) and, intermittently, professionals working with other species. OVAG programs are based on the One Health concept and practice; they focus on orangutans as a model for providing participants with skills in wildlife clinical needs and an understanding of veterinary, public and ecosystem health (see Chapter 2).

The OVAG forum is designed to empower Indonesian and Malaysian ape health practitioners and academics to formulate practicable policies and plans relevant to all wildlife health management needs. In so doing, the network is developing a community of practice. OVAG's key objective is to develop a sustainable regional cadre of professionals who are able to provide capacity building, advice, guidance and management of One Health matters with a wildlife focus in Indonesia and Malaysia.

Impact and Influence on Capacity-Building Practices

OVAG supports a multimodal approach aimed at ensuring resilience, for example by engaging in succession planning, promoting sustainable outcomes and avoiding methodology drift or variation. The OVAG program, which is evaluated annually, offers members and participants a series of benefits and activities provided by global and local experts and, more importantly, the participants themselves (Unwin *et al.*, 2022):

- Access to materials.
- Access to colleagues.
- In-country and onsite annual workshops on technical skills such as anesthesia and surgery.
- A forum for needs-based discussions, case studies and role-playing in areas such as systems operations, behavior, nutrition, contingency planning, biosafety and disease outbreak.
- Regular back-to-basics field programs in veterinary skills and lab skills.
- Online workshops with limited places (to enhance the experience for participants) and with access to all materials via OVAG's online members-only area and the Canvas (invite-only) tertiary education platform, in partnership with the University of Minnesota and others. This online presence has enhanced network accessibility and allowed OVAG to provide participants with detailed COVID-19 guidance and rapid access to customized information. An active WhatsApp group led by workshop participants helps to mitigate feelings of isolation in the field, serves as a notice board for continuing professional development opportunities and allows for peer-to-peer advice on health issues.
- A trusted environment in which to train the trainers and promote peer-to-peer learning.

- Internships in the UK and the United States, in partnership with zoos, universities and non-governmental organizations (NGOs). One partner is Orangutan Veterinary Aid, which:
  - provides customized clinical training in orangutan centers;
  - coordinates funding and logistics for internships in the UK for OVAG participants who have both clinical and managerial responsibilities, and who are identified as potential leaders in the field as part of OVAG's succession plan;
  - provides expertise in brokering veterinary equipment for orangutan field sites, from syringes to digital radiography systems, along with the required technical backup and operator training; and
  - supplies materials that are not available or prohibitively expensive in-country (other such suppliers include Worldwide Veterinary Services and individual zoos).

OVAG Participants

The network includes more than 300 participants, from the core of wildlife veterinarians working at the human–ape interface to academics, researchers, project managers and government authorities from Indonesia, Malaysia and elsewhere. Courses take the form of workshops and webinars in One Health, veterinary science, ecology, primatology and environmental science. Through its affiliated professionals, staff and participants, the network provides peer-to-peer support as well as direct input into conservation management at the NGO and government levels. Indonesian undergraduates in veterinary science and forestry have participated in OVAG's summer school program.

Impact and Influence

In 2020, the OVAG committee developed COVID-19 guidance for OVAG participants, in addition to guidance on the practical aspects of pandemic protection (University of Minnesota, n.d.-a). NGOs made use of this material to inform government decision-makers in environmental policy about impacts of the pandemic on wildlife centers. OVAG participants are also "road-testing" COVID-19 preparedness and response protocols in relation to ape-specific guidance from IUCN, the leading international body on wildlife conservation (IUCN SSC PSG SGA, n.d.-a).

Future Directions

OVAG intends to integrate its online presence into a blended learning experience for clinical and conservation practitioners who are participating in other programs in Africa and Southeast Asia. Teaching materials are being linked to IUCN to enhance participant influence on national environmental policies. They will also be used as the basis of an MSc program in conservation medicine at the University of Gadjah Mada, the first of its kind in Indonesia. Another goal is to expand current research programs — including investigations into Orangutan Respiratory Disease Syndrome and reviews of the effects of captivity on the gut microbiome — to provide answers to health-related questions from the participating projects.

A Framework for Preventive Health Programming and Interventions

Several zoos in the UK run internal–external audit systems that, implicitly or explicitly, integrate general welfare with health management and animal husbandry. The resulting preventive health framework recognizes a duty-of-care responsibility to both in situ and ex situ ape populations and can be adapted to multiple situations, as demonstrated by the Orangutan Veterinary Advisory Group and Gorilla Doctors approach to clinical advice and programs. Under the framework, preventive health programming seeks to:

- prevent disease from entering the animal population and assess the physical and psychological wellbeing of new arrivals;
- maintain the health and welfare of the animal population; and
- prevent dissemination of disease to other institutions, release programs, populations and ecosystems.

Many pathogens are difficult to eliminate once established in a population—be it wild or captive. Due to the lack of access to individuals in wild populations, it is often too late to intervene in their psychological and physical health, particularly if animals are showing signs of overt disease. In captivity, access is considerably easier, but diagnosis and treatment remain complicated. The framework thus takes a preventive medicine approach, rather than a reactive one.

From a welfare perspective, the robust nature of the framework facilitates diagnosis and mitigation of issues in social groups with complex, adaptable behavioral ecologies (see Chapter 8). The framework is also designed to promote positive conservation outcomes, as healthy individual apes increase the likelihood of healthy populations.

When applied through a preventive health program, the framework follows an evidence- and risk-based approach that can:

- highlight data gaps;
- generate accurately cost–benefit assessments; and
- improve communication through enhanced teamwork, which can further practitioner understanding and compliance with disease management guidance.

Data Gathering, Communications and Preventive Health Procedures

Data collection processes differ across settings. Before deciding whether to acquire an animal, many zoos request pre-import testing and husbandry records via the Zoological Information Management Software (ZIMS) database. This sort of data is rarely available to sanctuaries or rescue centers, which fall back on the precautionary principle or more stringent quarantine requirements for new arrivals. Practitioners who care for in situ ape populations can refer to research on interpopulation or interspecies interactions in analyzing what pathogen issues may be on the horizon.

Regardless of the setting, clear communication is key to ensuring the health and wellbeing of ape populations. In zoos, animal managers typically meet on a weekly basis to discuss potential imports and exports. The decision-making process requires trust between parties regarding import disease surveillance, quarantine requirements and a zoo's prerogative to refuse imports based on pre-import health findings.

The gold standard for when an ape arrives at an ex situ facility, or when a resident animal is sick, is a quarantine combined with behavior assessments and a biosecurity program. Veterinarians and senior animal husbandry staff agree on such procedures in advance, lead quarantine implementation and biosecurity enforcement, and ensure communication with relevant staff members. These procedures are integrated with zoonotic disease control, an employee health program and the use of ZIMS (or a similar database) to facilitate the open flow of husbandry and veterinary information among all stakeholders. In situ preventive health programs generally focus on human, domestic animal and wildlife movement around the ape species of interest, typically in cooperation with government, private sector, industry and community stakeholders.

Health Surveillance and Assessments

Depending on the setting, keepers, rangers or researchers monitor the health and welfare of a population. In zoos and sanctuaries, these observations are usually recorded in daily reports and emailed to animal health and management teams. Any concerns are discussed in regular meetings, such as weekly health and welfare briefings on current and potential issues between animal health and senior animal management staff.

Infectious disease surveillance programs tend to be based on species susceptibility and informed by regional and onsite pathogen findings. They include health screening protocols that cover post-mortem and clinical pathology data, and particularly the gathering and investigation of parasitology data. In ex situ situations, effective protocols prescribe thorough post-mortems as standard procedures to be followed after all ape deaths and in any clinical pathology investigations conducted prior to an individual's death. The findings and data gathered through such procedures can also be applied to in situ situations, especially relating to species' pathogen susceptibility.

Health Review and Welfare Audits

Quarterly husbandry, health and welfare audits can help to determine whether health interventions were successful and how approaches and procedures might be improved. UK zoos conduct such audits with external academic experts in health and welfare, as well as veterinarians who specialize in exotic and zoo animals. Internally, they rely on a committee of directorial, scientific, curatorial and animal health teams that prioritizes ongoing actions to enhance health and welfare conditions, for example through changes in husbandry practices or physical alterations in facilities. The committee

maps clinical, pathological, dietary and behavioral trends to highlight health and welfare issues; it also issues recommendations on the risk management of diseases of concern, as required. The quarterly summary and minutes of committee meetings form a written record of the health, welfare and husbandry status for the archives. Such records also form the basis for reports on any health issues for keepers. Regardless of the situation, external review of any ape health program is necessary to ensure the expediency of intervention parameters.

These processes allow zoo staff to spot potential health and welfare issues in species and enclosures early and to respond in a coordinated, timely, multi-departmental fashion. The approach also benefits the organizational culture, given that:

- reviews become second nature;
- health and welfare assessments are facilitated;
- disparate issues that occur over years can be assessed systematically;
- multi-disciplinary expertise is at hand; and
- protocols allow for risk assessments, the filling of data gaps and the quantification of areas of most concern.

Photo: Depending on the setting, keepers, rangers or researchers monitor the health and welfare of a population. In zoos and sanctuaries, these observations are usually recorded in daily reports and emailed to animal health and management teams. Electrocardiogram, Fauna Foundation. © Justin Taus/Fauna Foundation

One Health pathway solutions can be applied to manage health issues related to translocation (Sherman, Ancrenaz and Meijaard, 2020; Sherman *et al.*, 2021). Figure 4.8 presents a risk matrix that allows practitioners to compare levels of risk associated with proposed and used translocation options for orangutans in range countries (Sherman *et al.*, 2021). While the risk analysis process is complex, it yields One Health management pathways that are evidence-based and generally embedded in good biosafety practice (Jakob-Hoff *et al.*, 2014; see Chapter 2).

Based on evidence presented by Sherman *et al.* (2021), a One Health approach is needed for successful disease risk management in orangutan translocations. Pathways to solutions include:

- Increasing effectiveness by:
  - promoting a systems approach to health surveillance that involves coordinated monitoring of wild, captive and released orangutans, as well as transparent information sharing among all stakeholders;
  - using strategies that have proven to be effective, such as tying health care services to reductions in illegal logging and community forest management initiatives; and
  - engaging local community members as "orangutan guardians," by providing indirect incentives such as infrastructure and civic facilities, or by offering direct financial incentives.

- Increasing feasibility by:
  - collaborating with government bodies, local communities and non-governmental organizations (NGOs) to address health and biodiversity conservation;

  - calling for investment in education and policy that recognizes the direct dependence of human health on functional ecosystems and biodiversity; and
  - promoting improvements to law enforcement.

- Increasing both effectiveness and feasibility by:
  - conducting disease risk analysis with orangutan conservation and translocation stakeholders and wildlife health experts; and
  - involving rescue centers, research centers and local NGOs in the development of solutions, as they have long-standing relationships with surrounding communities.

The Increasing Imbalance between Apes and Parasites

Health systems need to be considered at the micro level as well as the macro level. The internal system between host and parasite, for example, can be brought into imbalance due to changing environmental conditions.

Until recently, mountain gorillas were not treated for helminths (parasitic worms). In 2017, veterinarians started noticing individuals who were losing condition and becoming debilitated without obvious cause (B. Ssebide, personal observation, 2021). Due to limitations in obtaining biological diagnostic samples, non-invasive fecal samples were collected. Analysis revealed high helminth infestation rates. Gorillas who were subsequently treated for parasitic helminth infestation by remote delivery of dewormers have shown marked improvement.[9] At the time of writing, a study was being conducted utilizing molecular techniques to determine the origin of these parasites.

FIGURE 4.8

Orangutan Translocation Risk Matrix

Qualitative rank definitions

■ Severe risk: Translocation is not advisable; other conservation solutions should be pursued

■ High risk: Translocation into wild populations is not advisable; extreme caution should be used for reintroduction

■ Moderate risk: Translocation may not be advisable; additional information is needed before proceeding

■ Negligible or low risk: Translocation is not expected to have marked negative impacts

☐ COVID-19 transmission and morbidity/mortality risk rating for orangutan rehabilitation and translocation scenarios

			Consequences				
			Insignificant	Minor	Moderate	Significant	Catastrophic
	Species & ecosystem conservation consequences and likelihood		No expected risk to conspecifics, other taxa, or ecosystem	Low risk to conspecifics or ecosystem; possible risks to other taxa	Some risks for conspecifics and/or other taxa	Significant risk of possibly lethal effects in conspecifics and/or other local taxa; possible ecosystem effects	Lethal effects pose population or species risk; likely negative effects on other taxa or ecosystem
		Health & biosecurity consequences and likelihood	No health effect; little or no transmission risk	No long term health effect; little or no transmission risk	Some health effects, moderate transmission risk	Moderate risk of transmission and/or morbidity and mortality	High risk of transmission, morbidity/mortality, disease spillover
Near certain	Species or ecosystem effects often occur in OU releases	Transmission or disease in OU happens regularly					
Likely	Species or ecosystem effects have occurred multiple times in GA or OU releases	Transmission or disease in GA or humans working with GA has occurred multiple times			**Reintroduction:** Lower initial risk of disease presence due to mitigation. Opportunities for infection through captivity, release and post-release human proximity. Released OU populations susceptible and non-immune; other taxa may be also	**Wild-to-wild translocation and reinforcement:** Many people in contact/proximity to OU. Confirmed human–GA transmissibility; all wild OU susceptible and non-immune; other taxa may be susceptible	**Tapanuli translocation:** Infection, death and transmission could pose catastrophic species impact and effect ecosystem; disease spillover to other taxa and local human populations possible
Possible	Species or ecosystem effects have occurred at least once in OU or other primate releases	Has happened at least once before in GA or other primates, or in humans involved in GA care		**Captive OU:** Lowered risk of disease presence due to mitigation; any active infection poses high risk to OU which are susceptible and non-immune			
Unlikely	Species or ecosystem effects have occurred but not in primate releases	Has not happened in GA but has in other animals					
Rare	Species or ecosystem effects have not been recorded in wildlife releases	Possible; has not been detected in wildlife					

(Likelihood — left axis label)

Source: Sherman *et al.* (2021, fig. 4)

Fatalities and histopathological variations—due to severe chronic gastritis and colitis—in mountain gorillas have been linked to the increasing impact of metazoan parasites. This association suggests a change in the epidemiology of parasitic infections, which may be related to the rapid increase in mountain gorilla population density, particularly in areas of the Virunga Park (Caillaud *et al.*, 2020). Since understanding the epidemiology and impact of parasites in the context of conservation medicine is an important part of population management, several ongoing studies are aiming to bridge knowledge gaps in this domain. In recognition of emerging parasitic infections and their potential link to fatalities, Virunga-based management teams deworm suspected and confirmed helminth infestations as part of the routine treatment of individual mountain gorillas with debilitating conditions.

Launched in 1988, the Gorilla Doctors pathology database contains records of more than 100 gorillas on whom at least a partial histopathology has been performed. In several cases, including two involving gastric cancer, strongylid nematodes (which resemble hookworms) were associated with chronic gastric mucosal proliferation. Since no other pathogens were detected in examined adult mountain gorillas, these unidentified strongylid nematodes are thought to be the cause of chronic gastritis, a condition that can cause debilitation and contribute to mortality (Muhangi *et al.*, 2021).

Situations in which multiple indistinguishable parasite species with presumed differences in pathogenicity occur together call for the use of tools that can uncover quantitative and qualitative parameters of parasite communities. Practitioners may be able to overcome related obstacles thanks to recent advances in the development of next-generation sequencing (NGS), as discussed below (Hu *et al.*, 2021).

Improving Diagnostic Accuracy, Precision and Capacity

Molecular diagnostic methods that employ polymerase chain reaction (PCR) can detect and identify genetic material (DNA or RNA) using highly specific primers (a short nucleic acid sequence that provides a starting point for DNA synthesis). These methods are indispensable in biomedical research, as they can confirm species and identify pathogenic microbes—even when pathogens are present in minute amounts. They also allow for rapid, animal-side individualized treatment.

PCR detects known genetic sequences (very specific targets), although the use of multiplex primers can broaden the range to target multiple DNA or RNA sequences. The development of reverse transcription PCR has made it possible not only to detect the targeted genetic sequence, but also to quantify the number of copies in the sample (Kralik and Ricchi, 2017). An alternative is gene sequencing, which uses similar techniques but is more useful for discovering new pathogens, as it does not require prior knowledge of genetic sequences.

New procedures, such as NGS long-read sequencing, have greatly reduced the time necessary to perform sequencing. In addition, technological advances have allowed for a significant reduction in the size of required machines; the production of highly portable field-based solutions, such as the MinION; and exponential improvements in accuracy (Lu, Giordano and Ning, 2016; Srivathsan *et al.*, 2021).

Molecular techniques used to be the exclusive purview of specialized laboratories or institutes that had the capacity to use them and could bear the costs of procuring the technology, which were prohibitively high for most organizations working with wild apes. Field-based practitioners thus had to arrange for samples from the field to be transported to such labs—a process that typically required permits, involved costs and entailed prolonged periods of time between sample collection and testing. Reverse transcription PCR and gene sequencing machines have since become more affordable, highly portable and battery-powered, enhancing the possibilities for use in the field (Marx, 2015; Tyler *et al.*, 2018). Several organizations that focus on ape health have started using them in the field in Africa and Asia, allowing for rapid diagnosis and real-time patient management, as well as new opportunities for in-country research (Schubert *et al.*, 2021).

Box 4.1 and Case Study 4.6 highlight the need for a One Health approach and a thorough understanding of the local context in decision-making processes that aim to identify the most appropriate use of molecular techniques in the field.

The Conservation Physiology Toolbox

Conservation physiology contributes to conservation solutions by identifying biotic and abiotic environmental and anthropogenic drivers and their impact on the performance and persistence of organisms (Wikelski and Cooke, 2006). The physiological approach to assessing these multi-factorial responses is central to comprehending cause-and-effect relationships and mechanistic processes beyond correlations, as well as to informing predictive models and concepts to conserve animal populations under threat (Cooke *et al.*, 2020).

The relevant physiological subdisciplines include bioenergetics, nutritional and cardiorespiratory physiology, neuro- and endocrinology, immunology, epidemiology, genomics and proteomics, reproductive physiology and toxicology (Madliger *et al.*, 2018). Within these disciplines, researchers have tested and applied numerous physiological parameters and techniques for conservation purposes, using sample media such as saliva, urine, feces, tissue biopsies and blood, which differ in terms of the invasiveness of sample collection techniques.

Despite the increasing diversity and availability of practical tools, stress physiology metrics—particularly the change of glucocorticoid (GC) levels in different sample media—are the dominant tool in the conservation physiology toolbox. Elevated GCs do not necessarily indicate a state of stress or discomfort, however, as baseline and stress GC levels fluctuate among individual life-history stages (Romero and Wingfield, 2015). Moreover, stress responses are context-specific and can be triggered by a variety of stressors. Therefore, the use of GCs as a single metric to gain a comprehensive understanding of individual conditions is limited, as stress responses involve several physiological processes in parallel.

The evaluation of multiple interacting stressors in complex systems is difficult but essential, as stress contributes to chronic degenerative diseases, particularly in ape species, and may have deleterious and long-lasting implications on animal welfare (Edes, 2018). The allostatic load concept allows biomarkers from multiple physiological systems to be combined into an allostatic load index (ALI). As they represent overall physiological dysregulation, ALIs can be applied as a risk assessment tool to monitor health and welfare in captive and free-ranging wildlife species, including apes (Edes *et al.*, 2020; Edes, Wolfe and Crews, 2018). The allostasis concept was extended within the reactive scope model, integrating species' developmental strategies and their potential long-lasting impact on later-life stress responses (Romero, Dickens and Cyr, 2009; Scheffer *et al.*, 2018). These two promising concepts allow for the targeted and combined use of tools from the conservation

BOX 4.1

The Wildlife Conservation Society Community-Based Ebola Program in the Republic of Congo

Over the past 20 years, the Ebola virus appeared several times in Central Africa, affecting both human and wildlife populations. Research indicates its impact on the endangered gorilla and chimpanzee populations in the region was considerable, with mortality rates reaching 90% or more (Fontseré et al., 2021). The most recent Ebola virus disease outbreak in the Republic of Congo occurred in 2004. The initial spillover events that led to outbreaks among people in the region involved direct contact through the consumption of primate and other wildlife carcasses—a finding that establishes a strong link between humans and wildlife health.

To enable rapid detection of Ebola virus epizootics (outbreaks among wildlife), the Wildlife Conservation Society (WCS) set up a community-based, One Health-aligned early warning surveillance system. On detection of an epizootic, program staff disseminate information to public health authorities and local communities to help spur the implementation of prevention and control measures. The main goal is to limit spillover opportunities that could result in a possible human Ebola epidemic.

The surveillance system enables both wildlife conservation and public health organizations to respond with mitigation measures to protect threatened human and great ape populations. Working extensively with hunters and villagers, the program team raises awareness about zoonotic diseases, explaining the risks and the actions to take when a carcass is found in the forest. The messages are simple and clear: People must not touch, move or bury carcasses; rather, they should immediately inform the local authorities and the nearest WCS program or WCS management so that it can be dealt with safely.

Between April 2008 and September 2018, WCS conducted a total of 520 visits to 268 villages spread over 26 separate missions in four departments in the north of the Republic of Congo. The team delivered the educational message to a total of 6,658 hunters and to thousands of women and children who frequently visit the forest to gather food. Many villages are revisited each year. These communities now form a surveillance network covering nearly 30,000 km² (3 million ha) across the Congo Forest; they monitor, locate and report carcasses found in the forest. Between November 2006 and March 2018, WCS responded to 58 reports of carcasses. Community members submitted 21 (36%) of these reports, demonstrating the value of this One Health approach (Kuisma et al., 2019; Seifert et al., 2022). Now that this network is in place, WCS is utilizing patient-side PCR and gene sequencing to confirm the presence and origin of the Ebola virus.

CASE STUDY 4.6

Improving Diagnostics of Tuberculosis in Great Apes[10]

Tuberculosis, a chronic bacterial disease caused by the *Mycobacterium tuberculosis* complex, is one of the main diseases of concern in great ape conservation at the human–ape interface (Zimmerman et al., 2022; S. Unwin, personal observation, 2021). The disease is currently the "world's top infectious killer," claiming 1.5 million human lives every year (Adefuye et al., 2022; WHO, n.d.).

Infections of great apes have been recorded in captive facilities in Africa and Asia and are thought to be predominantly of human origin (Molyneaux et al., 2021; PASA, 2009; Sanchez and Hidalgo-Hermoso, 2022; Zimmerman et al., 2022). Like humans, great apes can carry a latent form of tuberculosis that may eventually become active, complicating diagnosis (Sanchez and Hidalgo-Hermoso, 2022). A failure to treat active cases leads to fatal consequences for both the individual and the population. Infected great apes are a potential source of transmission to other primates and spillback into the human population.

Confirming diagnosis is challenging due to the biology of the *Mycobacterium*, the low specificity and sensitivity of test systems and the general lack of testing modalities. To date, no single test has been validated to detect tuberculosis in great apes; however, the increasing use of molecular techniques in the field, in combination with other tests, such as X-rays, is improving diagnostic accuracy, which can enhance disease management efforts. Mycobacterial culture, the gold standard test if positive, has a high rate of false negatives due to poor sensitivity, although the use of liquid media (for example, in a Mycobacteria Growth Indicator Tube) has shown improvement in both the time to culture and the sensitivity of the result (Thangavelu et al., 2021).

Polymerase chain reaction (PCR) to detect DNA material from *Mycobacterium* has been used widely as a diagnostic test in humans and great apes. Experts recently raised concerns about some PCR protocols for tuberculosis, noting that because they are non-specific within *Mycobacteria* spp., they can lead to inconclusive results (G. Omondi, personal communication, 2021; P. Sudharmono, personal communication, 2021). The Pan African Sanctuary Alliance is investigating interferon–gamma assays and, potentially, gene sequencing for use in captive ape populations in the care of its members. From 2022 onwards, this research has been shared with those working with orangutans in similar situations in Indonesia and Malaysia, as part of a wider capacity development program.

In practice, ape health practitioners face a set of challenges with respect to diagnostics. First, they need to identify the most appropriate (combination of) diagnostic tests or surveillance methods for a given situation. The Borneo Orangutan Survival Foundation, for example, has chosen to use tuberculin skin testing and PCR, following a thorough cost–benefit analysis. Second, many captive facilities are forced to rely on human-focused laboratories, some of which cannot elucidate which PCR primers or protocols are in use, let alone accept a request for a specific PCR method. Third, tuberculin skin test reagents may not be readily available, particularly in Indonesia, where neither tuberculin purified protein (avian/bovine) nor mammalian old tuberculin is obtainable. Fourth, as some veterinarians have not received sufficient training to perform bronchoalveolar lavage (BAL) for sample collection, sample quality may suffer, as may the accuracy of culture and PCR lab results.

By taking the following steps, captive ape facilities can overcome some of the above challenges and enhance tuberculosis diagnostics in great apes:

- Provide training in BAL sample collection to improve the accuracy of laboratory tests using BAL samples.
- To increase the sensitivity and specificity of the screening protocol, choose and use a combination of diagnostic tests that detects both the pathogen (for example, a culture, PCR or acid fast staining) and the host immune response (for example, a tuberculin skin test or interferon–gamma assay).
- Include X-rays in the decision tree for tuberculosis diagnostics, so long as facility staff is properly trained in obtaining and reading radiographs for tuberculous lesions.
- Improve communications with human tuberculosis diagnostic labs to be able to discuss and secure the best interpretation of the results.
- Improve all onsite testing modalities as much as possible; if resources are limited, step up communications with lab facilities that could offer the appropriate services.
- Collaborate with other facilities to establish a pathogen surveillance program in great ape reintroduction sites to measure the effectiveness of preventive health management in rehabilitation facilities and, most importantly, to ensure that the reintroduction process does not contribute to pathogen transmission that could harm the natural ecosystem.

Photo: Confirming a tuberculosis diagnosis is challenging. No single test has been validated to detect tuberculosis in great apes; however, the increasing use of molecular techniques, in combination with other tests, such as X-rays, is improving diagnostic accuracy, which can enhance disease management efforts.
© Lwiro Primates Rehabilitation Center

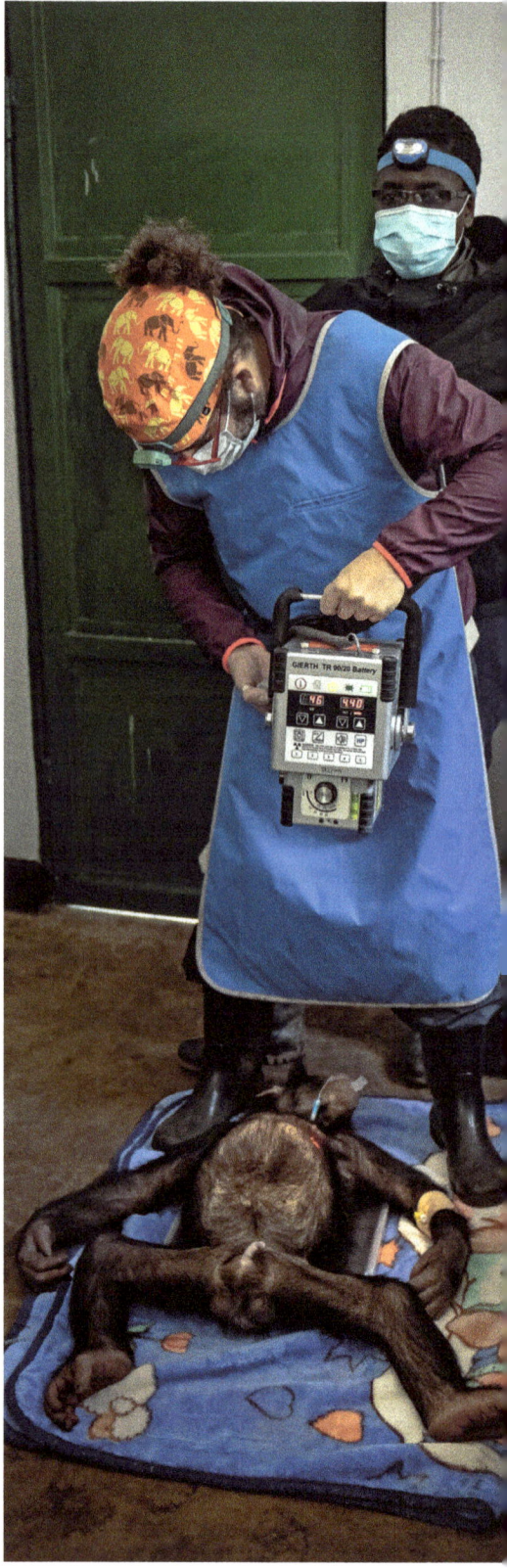

physiology toolbox and help categorize how individuals respond to changes and challenges in the context of conservation physiology.

Post-Release Monitoring: Telemetry

Radio telemetry enhances the ability to conduct state-of-the-art monitoring and data collection, through the development of technology and methods specifically designed to locate individual apes after release into the wild. It allows for the unequivocal identification of individuals, facilitates data collection and permits reintroduction specialists to intervene to promote welfare or prevent potential conflict situations involving released animals (Juarez *et al.*, 2011). With respect to apes, however, telemetry's biggest drawback has been the absence of appropriate species-specific attachment systems (King, Chamberlan and Courage, 2006; Russon, 2009). Traditional radio collars have been successfully employed in monitoring prosimians, some monkeys and reintroduced chimpanzees.[11]

In 2009, in response to these issues, the Research Institute of Wildlife Ecology in Vienna developed subcutaneous, very high frequency (VHF) radio telemetry transmitters and a corresponding surgical implantation method (Robins *et al.*, 2019). Since then, numerous ape reintroduction projects have adopted implanted radio telemetry. The small circular transmitters are available in two sizes: small (d = 28 mm, h = 10 mm, 14 g), with a 280 mAh battery, and large (d = 28 mm, h = 12 mm, 17 g), with a 540 mAh battery (Robins *et al.*, 2019). Post-release detection ranges vary from a few hundred meters to greater distances, as when elevated reception from hilltops is possible.

Before a transmitter can be implanted, a surgeon creates a subcutaneous pouch high up on an ape's back, between the shoulder blades. The transmitter is then inserted with the plane of the transmission facing the suture line to maximize detection by the receiver. The most significant disadvantages of transmitter implantation are that anesthesia and surgery are required to place the devices and again if anything goes wrong, such as the battery failing, and that surgery is followed by a post-operative recovery period during which wound healing is monitored (Robins *et al.*, 2019).

Both rehabilitated and wild translocated apes are most vulnerable immediately following release (Strum, 2005; Tutin *et al.*, 2001). As radio telemetry allows ape health practitioners to relocate them during this phase and beyond, it can help to improve the long-term survival of released individuals, as long as the transmitters are reliable.

Ecoimmunology: The Host Side of the Equation

In apes, infectious diseases are the leading cause of morbidity and mortality (Kuisma *et al.*, 2019). In evaluating animal health and emerging diseases that can threaten both wildlife and humans, however, practitioners often overlook the extent to which environmental and biological contexts modulate physiological processes in vertebrate species (Hing *et al.*, 2016; Phelps and Kingston, 2018; Plowright *et al.*, 2008, 2016; Subudhi, Rapin and Misra, 2019).

The field of ecoimmunology highlights the necessity of a multimodal and integrative physiological approach to immunity in the context of the whole organism, including genetics, developmental environment and individual traits that drive variations in immune function, such as sex, age, body condition and reproductive status (Schoenle, Downs and Martin, 2018). Ecoimmunologists emphasize the importance of these factors in shaping individual immune phenotypes, including resistance and tolerance against pathogens and the concomitant biological costs and consequences on, within and among individuals and populations (Kernbach *et al.*, 2019; Schoenle, Downs and Martin, 2018). These immunological variations can influence host–parasite eco-evolutionary dynamics within populations and communities, which play a central role in the conservation of threatened species (Becker *et al.*, 2020).

By determining urinary neopterin (a catabolite from macrophages that is used as a marker of cellular immune system activation), for example, ecoimmunologists can compare general immune system activation and existing disease dynamics, while also identifying risk factors within and across primate populations (Löhrich *et al.*, 2018). An increasing number of studies provide concrete evidence of a strong and reciprocal interaction between the neuro-endocrine and immune systems (such as during stress responses), which points to an integrated and evolutionary highly conserved element of physiology across phyla (Adamo, 2012; Verburg-van Kemenade, Cohen and Chadzinska, 2017).

Future work in the field of ecoimmunology could provide insight into the environmental drivers of host defense, fill related knowledge gaps and facilitate more accurate risk assessments regarding potential infections in the context of climate and landscape change, such as habitat destruction and loss (Becker *et al.*, 2020).

Conclusion

There is no doubt that human contact facilitates disease transmission to apes (Whittier *et al.*, 2022). For wildlife health professionals, both ethically and morally, there is a requirement to understand when it is best to do nothing (if the guiding ethos is first "do no harm", often the most appropriate way of not doing harm is not to intervene). But this is a dynamic and iterative decision-making process informed by evidence. The decision to mitigate harm by intervening in an ape health or welfare situation—or to stem potential negative health consequences from any other sort of intervention, such as ecotourism—is ultimately informed by the availability of resources, staff capacity and contingency planning. From an ethical perspective, a decision not to intervene must be justified as much as a decision to intervene. Of paramount importance is recognizing where an intervention framework is deficient and for projects to concentrate on building capacity in these areas before any interventions are attempted.

This chapter presents examples of decision-making processes, practical solutions and emerging toolkits that help to inform the intervention decision process, provide opportunities to gain much needed intelligence in a non-invasive way and could change how interventions happen. It is up to legislators, researchers and practitioners to work together to ensure interventions improve ape health, not only by reducing injury and disease transmission, but also by adding to health knowledge and good welfare practice in health-specific interventions.

Acknowledgments

Principal authors: Steve Unwin,[12] Benard Jasper Ssebide[13] and Chris Walzer[14]

Contributors: Mike Cranfield,[15] Nikolaus Huber,[16] Alain Ondzie,[17] Ricko Jaya,[18] Yenny Saraswati[19] and Fransiska Sulistyo[20]

Box 4.1: Alain Ondzie

Case Study 4.1: Ricko Jaya

Case Study 4.2: Benard Ssebide

Case Study 4.3: Steve Unwin[21]

Case Study 4.4: Steve Unwin

Case Study 4.5: Steve Unwin

Case Study 4.6: Fransiska Sulistyo

The Conservation Physiology Toolbox and **Ecoimmunology: The Host Side of the Equation**: Nikolaus Huber

The editors would like to acknowledge the enormous contribution Dr Mike Cranfield made to the science, conservation and care of wildlife, and express our deep respect for his work. He will be dearly missed.

Endnotes

1. These findings are based on 14 years of research conducted through the Orangutan Veterinary Advisory Group (OVAG). They are captured on the OVAG Continuing Professional Development website for practitioners, which is maintained in collaboration with Wildlife Health Australia, the Orangutan Conservancy, the Arcus Foundation and the University of Minnesota.

2. The research is funded by the Arcus Foundation and led by George Omondi of the University of Minnesota, in collaboration with clinicians across the Pan African Sanctuary Alliance and academics based in Africa, Australia and the UK. Chapter co-author Steve Unwin is part of this consortium.

3. Unless otherwise cited, the information in this section is based on B. Ssebide's knowledge and experience, gained from working for Gorilla Doctors for 25 years.

4. The opposite is true with respect to captive apes: medical interventions involving chimpanzees are far simpler than those targeting gorillas, who are much more fragile when it comes to anaesthetics (S. Unwin, personal observation, 2022).

5. This article was made freely available as part of the COVID-19 public health emergency response; to be used for unrestricted research, re-use and analysis in any form or by any means with acknowledgment of the original source.

6. This case study is primarily based on the author's knowledge and experience from ten years as a veterinarian with the Human Orangutan Conflict Response Unit, which is run by the Orangutan Information Center in Sumatra.

7. The capture of an orangutan by government or sanctuary staff is sometimes referred to as a "rescue." Here we have used the term "capture" as, in many cases, the orangutans are healthy individuals who could potentially have been given the opportunity to remain where they were while solutions were found for their coexistence with local communities and/or companies.

8. This case study is based on author interviews with veterinary clinicians in 2018 and 2019, and on author observations of the gibbon health situation in the United Arab Emirates in 2019, as a working veterinarian. The author verified the situations discussed, including by reviewing the clinical records of the cases. All veterinary care for wild-born captive apes falls into the category of intervention, as it necessarily affects an animal psychologically (on those rare occasions conscious examination may be possible) and/or physiologically where anesthetics would be needed.

9. Information contained in internal Gorilla Doctor clinical records, seen by the author.

10. Unless otherwise cited, the information provided in Case Study 4.6 is based on the author's knowledge and 15 years of experience working as a veterinarian in orangutan rescue and rehabilitation.

11. Bearder and Martin (1980); Campbell and Sussman (1994); Charles-Dominique (1977); Fernandez-Duque and Rotundo (2003); Goossens *et al.* (2005); Humle *et al.* (2011); Tutin *et al.* (2001).

12. University of Birmingham (https://www.birmingham.ac.uk/schools/biosciences/index.aspx) then Wildlife Health Australia (https://wildlifehealth australia.com.au).

13. Gorilla Doctors (www.gorilladoctors.org).

14. Wildlife Conservation Society (www.wcs.org) and Research Institute of Wildlife Ecology (www.vetmeduni.ac.at/en/research-institute-of-wildlife-ecology).

15. Gorilla Doctors (www.gorilladoctors.org).

16. University of Veterinary Medicine, Vienna (www.vetmeduni.ac.at/en).

17. Wildlife Conservation Society (www.wcs.org).

18. University of Birmingham (www.birmingham.ac.uk/schools/biosciences/index.aspx).

19. Sumatran Orangutan Conservation Programme (www.sumatranorangutan.org).

20. Orangutan Veterinary Advisory Group (www.ovag.org) and independent consultant.

21. Using information from interviews with a veterinarian in the UAE.

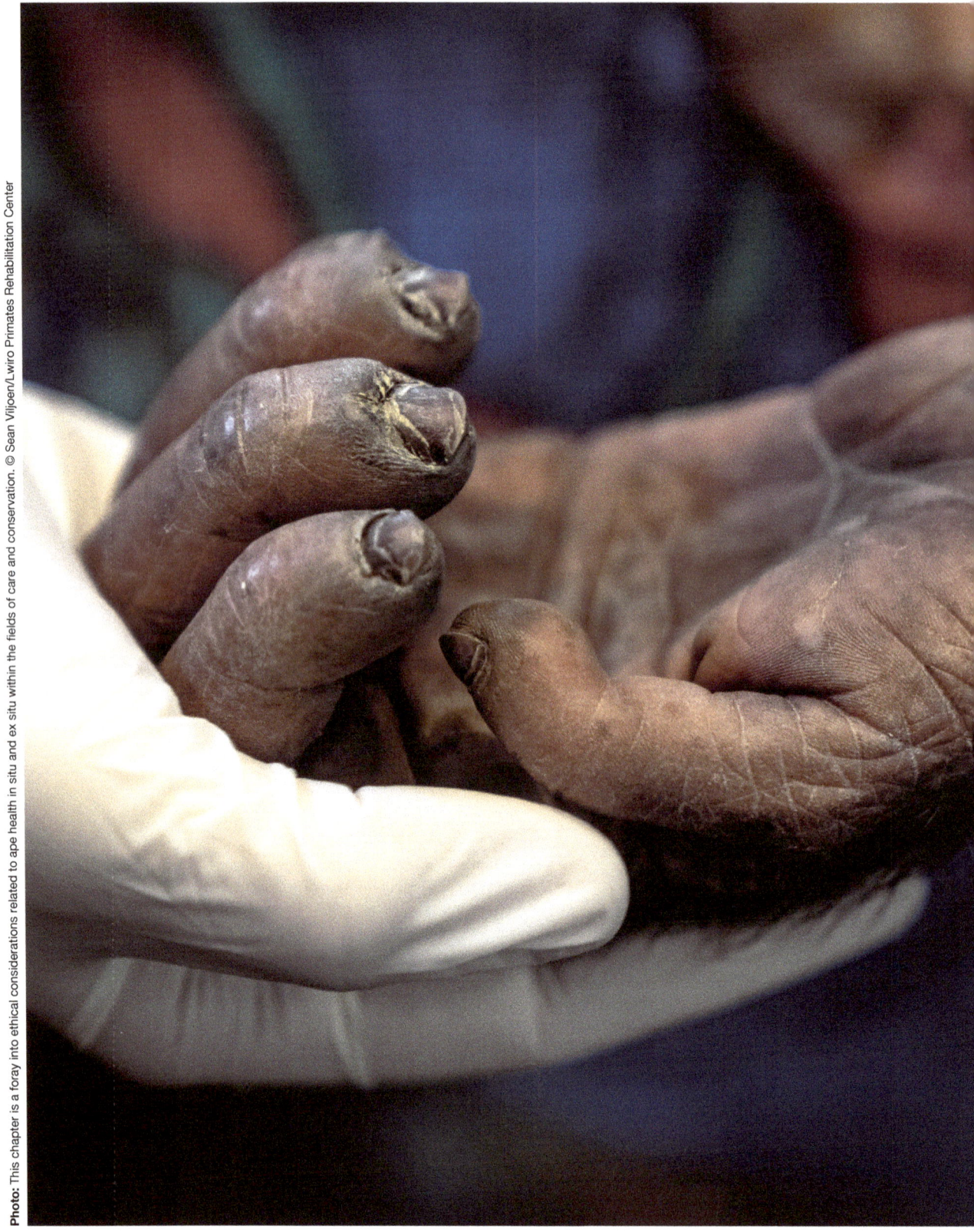

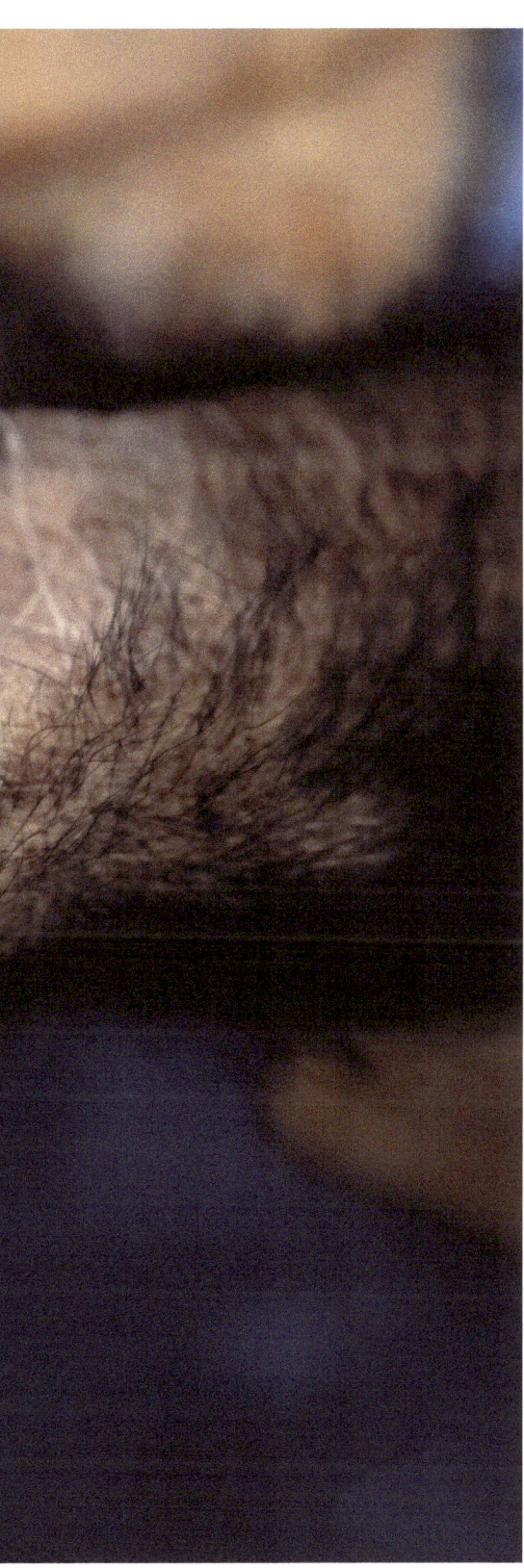

CHAPTER 5

Ape Health and Ethics

Introduction

Ape care and protection give rise to many moral problems. As with other areas of ethics, it is impossible to enumerate the full range of moral dilemmas that can occur in the context of protecting and caring for apes (see Box 5.1). Nonetheless, exploring these issues in general—and specific quandaries in particular—can provide insight into practical methods for ensuring the health and wellbeing of individuals, communities and populations.

Human interventions regarding apes can take the form of primary, secondary or tertiary prevention of injuries, infectious diseases, physical illnesses, psychological disorders, acute or chronic suffering or death.[1] Nearly all such interventions raise moral issues, some of them complex. In primary prevention, for example, technological advancements in vaccine development raise ethical questions about the justifiability of human interference to prevent ape illness (see Case Study 5.1). Other ethical challenges may arise when animals are injured in their natural habitats, such as in connection with treatment strategies or secondary prevention (see Case Study 5.2). Further issues may come to the fore in the context of captive care, such as rehabilitation in sanctuaries or tertiary prevention (see Case Study 5.3).

Since apes live in diverse regions around the world, human deliberations about their protection and care often involve collaboration across the borders of disciplines, countries, ecosystems and cultures (see the Apes Overview). As part of these processes, decision-makers may encounter normative perspectives that converge with or diverge from their own.

This chapter is a foray into ethical considerations related to ape health in situ and ex situ within the fields of care and conservation. It begins by exploring general ethical questions that arise in ape conservation; ethical foundations of ape moral standing; and the role of compassionate conservation in addressing tensions between individual and population health outcomes. Next, it examines particular moral dilemmas with a view to proposing general considerations and approaches for resolving key moral dilemmas of ape care and protection. Finally, the chapter highlights ways in which caregivers and other decision-makers can remain morally courageous and resilient in the face of significant challenges (see Box 5.2).

The aim of this chapter is to support ethical decision-making in practice by offering new perspectives, facilitating critical reflection, and furthering capacity for ethical decision-making across organizations and institutions.

Key findings include:

- Moral dilemmas regularly emerge during the course of ape protection and care.
- Approaches such as compassionate conservation address tensions between individual and population health strategies.
- Interventions are ethically sound if they are attempted by qualified personnel (such as field-trained veterinarians) and if anticipated benefits outweigh identified risks as they relate to individual, environmental and social conditions.
- Retaining an on-site veterinarian has ethical implications in that doing so improves emergency response time and increases the probability of success, thus reducing suffering and improving wellbeing.
- The provision of health care to apes presents ethical challenges that require

BOX 5.1

Ethical Considerations Beyond the Scope of this Chapter

To unpack all the ethical considerations relating to disease and zoonosis prevention would fill an entire book. Among the ones that are not explored in this chapter are:

- providing medical care and disease prevention equipment for the protection and treatment of apes in places where they are not available or accessible for human communities, as was the case when researchers, tourists and veterinarians wore masks when tracking the mountain gorillas (*Gorilla beringei beringei*) during the COVID-19 pandemic, and when life jackets were provided for orangutan transport in Indonesian Borneo (Chua *et al.*, 2021);
- taking and storing samples from humans who live alongside wild and captive apes, for example as part of a One Health approach (Tindana *et al.*, 2014; Vaz, Sridhar and Pai, 2016);
- medically intervening in the case of non-life-threatening injuries and diseases or intervening in response to injuries that resulted from group interactions, which could potentially alter the natural dynamics in a social group setting (Gruen, Fultz and Pruetz, 2013);
- imposing Western or external health and hygiene ideals and standards on local communities. The ethics of such approaches have been explored to some degree in the context of mountain gorilla conservation in Rwanda (Scholfield, 2013); and
- administering euthanasia to seriously sick or injured apes in sanctuaries or in the wild, with the aim of reducing suffering or protracted deterioration of health.

balancing the complex and often conflicting values that define wellbeing.

- To foster moral courage and resilience, organizations can provide support for caregivers and others, for example by developing peer networks, fostering positive coping strategies and ensuring institutional responsiveness.

General Ethical Considerations in Ape Care and Conservation

Moral conflicts in ape conservation often occur along two axes: the individual and the collective. Along one axis are individual humans and individual apes. Along the other axis are groups of humans and groups of apes, who are also members of families, communities, populations, societies, species and ecosystems.

In conservation circles, emphasis is often placed on protecting systems, ecosystems and the biosphere, as well as endangered species and their habitats (Vucetich *et al.*, 2018). Some conservation policies implicitly treat individual animals as irrelevant or expendable, while some frame them only in terms of their contribution to the species overall or to other conservation goals. These are examples of the "holism" of environmental ethics at work—the grounding of values in wholes such as species and ecosystems, to the disadvantage of individuals (Varner, 1998). As a result, the interests of specific individuals may be discounted in favor of conservation policies that are intended to preserve species and their habitats. The ensuing moral question is how to consider individual apes in the ethical decision-making process in the context of conservation.

Ethical Considerations Regarding Apes

There are many ways to include animals in ethical decision-making. One way is to consider who or what matters morally, and

Photo: Some conservation policies implicitly treat individual animals as irrelevant or expendable, while some frame them only in terms of their contribution to the species overall or to other conservation goals. Skull of a western lowland gorilla. © Jabruson/naturepl.com

how much (Goodpaster, 1978). From a utilitarian perspective, for instance, sentient beings matter since they have experiential welfare, but any consideration of their interests may be curtailed by an aim to achieve the best possible outcome for all sentient beings in question (Singer, 2011). Approaches that strive for a collective good, like utilitarianism, may therefore disadvantage individuals. In contrast, rights-based approaches protect individuals from being disadvantaged by collective outcomes, for example by arguing for moral rights not to be harmed, killed or held captive, or for moral rights to the safeguarding of social, environmental and other determinants of health and well-being (Cochrane, 2012; Shue, 1996). The interests of sentient beings can also serve as a basis for rights. Interests in continuing to live, enjoying bodily sovereignty and not being made to suffer, for instance, can serve as grounds for recognizing moral rights for individual humans and animals (Cochrane, 2012; Feinberg, 1974; Ferdowsian, 2020).

Other ethical perspectives highlight the qualities of the decision-maker and their attunement to, or relationship with, the individual or collective in question. In this context, enquiries may look into what would qualify a person as virtuous with respect to animals and nature; how empathy morally enriches relations between humans and other animals; or which moral obligations arise from relations between humans and animals (Gruen, 2015; Hursthouse, 2011; Palmer, 2010; Yu and Fan, 2007).

Across and within cultures and traditions, many ethical perspectives align with this relational approach. For instance, seeing oneself and the rest of reality as thoroughly relational and even interdependent—a key element of Buddhism as well as other traditions—can spark compassion (Halifax, 2011). Similarly, some African cultural traditions emphasize a relational approach. The Ubuntu philosophy of personhood states, "I am because we are." Personhood

arises from participating in the social life of a community of persons (Eze, 2010). The relational understanding of human beings and their ethical commitments in terms of Ubuntu has also been explored in relation to animals and nature (Etieyibo, 2017). While ethical approaches exhibit moral diversity across and within cultures and traditions, they tend to strive for a particular balance between individuals and the collective of which they are part (Prinz, 2007).

The Individual and the Collective

Ape conservationists and health professionals are regularly confronted by the tension between care for the individual and the collective. As discussed below, compassionate conservation may help to integrate not only the individual and the collective, but also the question of moral consideration and the question of how to relate to others.

That the individual matters as part of collectives is further indicated by several initiatives on the moral status of apes. The Great Ape Project, the Nonhuman Rights Project and the Philosophers' Brief, for example, point out that there are good reasons to think individual apes deserve robust protection, perhaps even a basic set of rights (Andrews et al., 2018; Cavalieri and Singer, 1996; Wise, 2010; Wise, Durham and Banes, 2020). Like human rights, a set of ape rights arguably includes the right to be free of unnecessary harm, coercion and restraint; the right to have their autonomy respected; and rights to pursue their interests individually and through their associations with their families and communities, as well as with other animals (Andrews et al., 2018).

While there are good reasons to acknowledge the individual, fostering collectives is vital in view of the ongoing breakdown of the ecological fabric that supports life (O'Riordan and Lenton, 2013). All apes live

> Ape conservationists and health professionals are regularly confronted by the tension between care for the individual and the collective.

under threats posed by human development and encroachment, including agricultural expansion, infrastructure construction, logging and mining, as well as killing, capture and trade (Arcus Foundation, 2014, 2015, 2018, 2020). Activities such as the clear-cutting of rainforest also destroy the habitats of countless other animal species, rob Indigenous Peoples of their homes and livelihoods, and endanger vital ecosystems that affect the climate (Lovejoy and Nobre, 2019).

In considering how best to safeguard species and the habitats on which they depend, conservationists not only consider the collective, but take into account the interests—or rights—of individuals (Bruskotter *et al.*, 2019; Palmer, 2020). When species conservation goals conflict with the interests of human individuals and collectives, however, they can generate difficult moral dilemmas whose resolution requires careful consideration and respect. Flexibility and creative solutions can advance mutual respect and mutual interests. Compassionate conservation may help to prompt moral deliberation and to navigate the moral complexities inherent in fostering collectives and promoting individual flourishing within and across species.

Compassionate Conservation and Managing Ape Health

Compassionate conservation emerged over the past decade as a novel perspective on moral decision-making in conservation practices (Wallach *et al.*, 2018). The idea was born at the nexus of animal welfare science and conservation biology, where recognition of the wellbeing of individual, free-living animals was seen as integral to sound conservation practice (Baker, 2017; Fraser, 2010).

In more recent conceptualizations, compassionate conservation challenges three assumptions in traditional conservation:

collectivism, instrumentalism and nativism (Wallach *et al.*, 2018). A common interpretation of collectivism is that it presumes the primacy of collectives—species, populations and ecosystems—over individuals, rather than seeing individuals as social beings in relation to members of their environment (Baker and Winkler, 2020; Santiago-Ávila and Lynn, 2020). While conservationists are not likely to deny the intrinsic value of individuals, some may base decisions about species preservation exclusively on individuals' instrumental value. The notion of nativism can disadvantage individual animals and their groups for the sake of historical, geographical and attitudinal ideals about the presence of species in a particular ecosystem—ideals that invite ethical reflection on their merit (Wallach *et al.*, 2018).

As discussed below, compassionate conservation is based on four general principles: first, do no harm; individuals matter; inclusivity; and peaceful coexistence (Draper, Baker and Ramp, 2015). Foundational to these four principles is that compassion is a critical moral capacity in ethical decision-making. In other words, for decisions to be ethically robust, they need to be informed by compassion. Why compassion? Compassion, as generally defined, involves a recognition of the suffering of others paired with a motivational response to be helpful in resolving or alleviating the suffering (Singer and Klimecki, 2014). As such, it provides a way to relate to the experience of other sentient beings. While empathy has an important role to play in moral agency, on its own it can introduce bias, such as for those near and dear. Instead, compassion attunes one's perceptions to the suffering of others, regardless of whether they are familiar (Bloom, 2017; Halifax, 2011). Compassion thus serves a highly relevant moral purpose: considering the experience of other sentient beings in as unbiased a way as reasonably possible and thereby providing a basis for an ethics

> Compassionate conservation is based on four general principles: first, do no harm; individuals matter; inclusivity; and peaceful coexistence.

Photo: When species conservation goals conflict with the interests of human individuals and collectives, they can generate difficult moral dilemmas whose resolution requires careful consideration and respect. Orangutan in an oil palm plantation.
© HUTAN-Kinabatangan Orang-utan Conservation Project

of interspecies interdependence, attuned to the flourishing of both humans and animals as members of their ecological communities (Batavia *et al.*, 2021; Kirby, Steindl and Doty, 2017; Nieuwland, 2020).

Although highlighting compassion cannot solve all moral problems in the field, in part because tragedy is sometimes inescapable, compassionate conservation encourages conservationists to reflect on their goals as well as on their practice before exploring possibilities for a compassionate approach to safeguard and promote the protection of animals and biodiversity (Batavia, Nelson and Wallach, 2020; Wallach *et al.*, 2018). Compassionate conservation is a topic of ongoing intellectual and practical debate, such that employing it and discussing its merits within a specific conservation context can help bring out different points of view and individual moral commitments (Batavia *et al.*, 2021). In that sense, the approach can be used to navigate the moral complexities of including the health and wellbeing of individual apes in the context of conservation. It invites those with an affinity for collectives to explore moral deliberation that highlights individuals as morally valuable participants of those collectives, while calling on those with sympathy for the individual to bear witness to the complexities of moral decision-making that extend beyond mere individualism.

First, Do No Harm

With respect to ape health, compassionate conservation calls for the application of the "first, do no harm" principle in any given context. In settings where humans are or could be present near apes, for instance, a critical evaluation can establish whether associated risks of disease transmission from humans to apes are too high (Woodford, Butynski and Karesh, 2002). Such critical evaluation may also be applied to the practice of habituation, which may be considered harmful given the level of stress caused over a long period of time when unhabituated apes are confronted by frequent human presence (Williamson and Feistner, 2011).

Individuals Matter

Given the immense and assorted pressures on ape species and populations, the principle of individuals matter is perhaps already engrained in much of ape conservation. The interests of apes, especially great apes, are increasingly recognized in terms of individual moral and legal rights (Andrews *et al.*, 2018; Cavalieri and Singer, 1996). Still, when individual apes cannot reproduce or be reintroduced into their natural habitats, or when the reintroduction of, for example, an orangutan appears contrary to his or her interests, the interest of the individual may be obfuscated in deference to the promotion of species or population sustainability (Palmer, 2020).

Emphasizing the individual in conservation works in at least two ways. First, compassionate conservation aims to promote the flourishing of collectives, such as species

and populations, in a way that aligns with individual flourishing. As part of this strategy, seeing populations and their individuals as thoroughly interdependent encourages efforts to protect the social fabric. Second, when individual interests become detached from the collective goals of conservation, compassionate conservation highlights the moral relevance of individuals in their own right (Wallach *et al.*, 2018).

Inclusivity

The principle of inclusivity counters biases towards apes in conservation practices, especially when these biases disadvantage others. This principle is pertinent to ape conservation, given that various cultures ascribe a distinctive status to apes, as compared to other animals (Corbey, 2005). It points towards a recognition of apes as living in multispecies collectives, with other species not merely instrumental to human or ape interests. Compassionate conservation works to undo biases in ethical decision-making. It remedies unfair distribution of resources and brings attention to how certain conservation practices that benefit particular species, such as apes, could potentially marginalize and overlook other species, communities and individuals (Santiago-Ávila and Lynn, 2020; Wallach *et al.*, 2018).

Peaceful Coexistence

The principle of peaceful coexistence primarily aims to mediate human–ape conflict through the exploration of potential changes in the behavior of both humans and apes. Rather than placing more emphasis on the competing interests of humans and apes, it promotes creative inquiry into the possibilities of living peacefully together (Wallach *et al.*, 2018). As part of this inquiry, compassionate conservation underlines the critical importance of human behavioral change

(Hockings *et al.*, 2015). Moreover, compassion necessarily attunes all conservation efforts to the needs and involvement of local communities (Santiago-Ávila and Lynn, 2020).

Primary Prevention: Conservation and Inter-species Health Policy

Attuning conservation efforts to local communities and every individual animal requires unwavering moral resolve. Complicating matters, conservation is confronted with an increase in emerging infectious diseases, which can threaten the health and wellbeing of individual animals and wild-

life communities (Capps and Lederman, 2015; Jones *et al.*, 2008). These diseases, and some of the underlying drivers of their increased emergence—such as deforestation and human encroachment—reflect the interdependence of human, animal and ecosystem health (Daszak, Cunningham and Hyatt, 2000; Patz *et al.*, 2004). The One Health initiative has captured the push for an interspecies health policy, providing a framework for conservation to align with public health goals (Nieuwland, 2020; see Chapter 2).

Ebola virus disease has made the international community critically aware of the dangers of emerging infectious disease. It has also sparked debate about human intervention in the wild (Capps and Lederman, 2015). This debate can inform the ethics of intervention: Should humans interfere in the lives of apes who are relatively independent of humans? More recently, the COVID-19 pandemic and its potentially devastating impact on ape health served as a reminder of the critical importance of understanding human and ape health within an interspecies health policy perspective (Gillespie and Leendertz, 2020). Exploring whether it is appropriate to intervene in the lives of apes in the wild is a central aspect of One Health ethics (Edwards *et al.*, 2018; Gruen, 2018; Nieuwland, 2020; see Case Study 5.1).

Secondary Prevention: Responding to Injury

The ethical and practical question of whether humans should interfere in situ is also commonly raised in response to the detection of injuries in apes. While opportunities to vaccinate apes in situ against a disease may be rare, practitioners often find themselves in a position to intervene in cases of human-caused injuries to apes (see Case Study 5.2).

Ebola Virus Disease and Vaccinating Apes[2]

In view of the potential impact of Ebola virus disease on apes in the wild and the ongoing development of vaccines against the disease, some ape health experts have suggested immunizations of African apes in situ, with the aim of diminishing the threat of infection among ape populations (Leendertz et al., 2017; Ryan and Walsh, 2011; Walsh et al., 2017; Warfield et al., 2014; see Chapters 1, 4 and 6). These contributions to the literature have triggered ethical debate and raised a range of distinct ethical concerns. Among these concerns is the question of whether humans should interfere with the lives of apes in their natural habitats (see also Case Study 5.2). This question has elicited a range of responses, from principled objections to any human meddling in the ecological systems of which apes are a part, to arguments for intervention. The ethical question of whether humans should intervene in situ is limited in part by various empirical questions related to the possibility and consequences of interventions. While these concerns may suffice to steer clear of any intervention, they do not eliminate the need to consider various ethical and empirical questions (Nieuwland, 2020).

Questions also emerge with reference to settings where there are no principled objections to vaccinating apes in situ. In the case of Ebola virus disease, one justification for pursuing vaccination is that the disease is generally considered a major risk to the survival of ape populations in Africa, yet other considerations could also prove relevant. For instance, individual apes arguably have an interest in being protected against Ebola virus disease (Capps and Lederman, 2016; Nieuwland, 2020; Ryan and Walsh, 2011). Moreover, vaccinating apes in situ has been suggested as a One Health approach for protecting ape populations while simultaneously decreasing the risk of disease spillover into human communities, reflecting a concern for public health in wildlife interventions (Capps and Lederman, 2015; Edwards et al., 2018; see Chapter 2).

Ethical Considerations Regarding Vaccine Development

Other concerns relate to the development of a vaccine against the Ebola virus and particularly whether the allocated effort and resources are fair and reasonable in the light of other moral demands. Is it justifiable, for instance, to spend significant resources on a potentially unachievable goal of protecting apes against Ebola virus disease (or another threat to their health) in situ while the health needs of neighboring human communities remain unmet due to a lack of funds? Apes may be better protected against disease if resources are spent on the prevention of hunting, other forms of habitat encroachment, and ecosystem fragmentation and destruction (Addison and Malone, 2018; Gruen, 2018). Alternatively, humans should arguably allocate resources to vaccine development for apes precisely because of large-scale anthropogenic intrusion into ecological systems on which apes rely for their health, well-being and survival (Osofsky, 2016).

In addition to concerns about resources that flow into pharmaceutical development, distinct scientific questions relate to the development of an Ebola vaccine. Whether health knowledge can be readily transferred from animal experiments to human biology is unclear; the same applies to transference between animal species, such as gorillas and chimpanzees (Addison and Malone, 2018; Gruen, 2018; Nieuwland, 2020). While chimpanzees have historically stood in as models of human biology due to observable similarities between the two species, Jones and Greek (2014) demonstrate that complex, systemic differences between and within species consistently prove to be biologically meaningful. Separately, changes to the individual due to environmental conditions—especially if such conditions affect multiple generations—can hinder the transfer of knowledge related to different cases from a single species, such as captive apes and their conspecifics in situ (Gruen, 2018). Another concern relates to differences across Ebola virus species, which could limit the cross-species immunizing potential of a vaccine (Feldmann and Geisbert, 2011; Leendertz et al., 2017). Furthermore, the knowledge base regarding Ebola vaccines in apes is limited, as only two vaccine trial studies have been undertaken, both on captive great apes (Gruen, 2018; Walsh et al., 2017; Warfield et al., 2014).

Since vaccine development has until now been inextricably tied to the use of animals in research, moral questions have been raised about the justification of the harms involved (Nieuwland, 2020). Do the presumed benefits really outweigh the harms inherent to such research (Barnhill, Joffe and Miller, 2016; DeGrazia, 2016; Ferdowsian and Fuentes, 2014; Ferdowsian et al., 2020)? Is it morally acceptable to harm apes in captivity for the benefit of their conspecifics elsewhere (Capps and Lederman, 2015; Nieuwland, 2020; Wendler, 2014)? Moreover, if the use of apes in medical research were to be deemed unacceptable because of the suffering and restriction of freedom it would involve, then additional moral problems would emerge with respect to the use of monkeys or, more broadly, any sentient animals. Once the human–animal distinction and, subsequently, the species barrier lose (much of) their moral relevance in justifying harms to animals, any invasive research that imposes harm on sentient beings to benefit other sentient beings may become morally problematic (DeGrazia, 2016).

Vaccine Implementation

Vaccination is seldom used to protect ape health in situ, because of both practical challenges and ethical concerns. In large part, the practical challenges depend on how the vaccine is to be administered, which is determined based on whether apes are habituated to human presence. One notable case in which habituated apes were vaccinated occurred in 1966, when Jane Goodall detected severely ill chimpanzees and immunized these apes to protect them against infection with poliovirus, using bananas as bait (Goodall, 2000). In 1989–1990, the Gorilla Doctors observed an outbreak of respiratory disease, to which they responded by vaccinating

Photo: Is it justifiable to spend significant resources on a potentially unachievable goal of protecting apes against Ebola virus disease (or another threat to their health) in situ while the health needs of neighboring human communities remain unmet due to a lack of funds? © Pete Oxford/Minden/naturepl.com

60 mountain gorillas (*Gorilla beringei beringei*) to protect them against what they presumed was measles, although the diagnosis remained unconfirmed (Cranfield and Minnis, 2007).

Evaluations of vaccines developed or in development for protection against Ebola virus disease consider a host of factors that determine their applicability for use in situ (Leendertz *et al.*, 2017; Nieuwland, 2020). For instance, while a single injection of a vector-based vaccine such as cAd3-EBO-Z or rVSV-EBOV proves sufficient to protect an individual ape, virus-like particle-based vaccines instead require multiple injections to protect individual apes, making them much less realistic for use in situ (De Santis *et al.*, 2016; Henao-Restrepo *et al.*, 2015; Leendertz *et al.*, 2017; Warfield *et al.*, 2014). The cytomegalovirus-based vaccine developed specifically for great apes in situ disseminates throughout a population when an individual is vaccinated (Marzi *et al.*, 2016). For each of these vaccines—but especially the cytomegalovirus-based vaccine, given that it self-disperses—it is ethically crucial to ensure that the effects in both apes and non-target species are not disadvantageous (Gruen, 2018; Leendertz *et al.*, 2017; Osofsky, 2016).

Do No Harm

Precaution appears warranted in addressing the unforeseen and unforeseeable consequences of introducing genetically engineered vaccines within ape populations (Gruen, 2018). The many scientific and ethical concerns that pertain to the safety of introducing vaccinations into wild populations seem to outweigh any prospects of expeditious implementation. These concerns also support the application of the principle central to both medicine and compassionate conservation—*first, do no harm*—in any consideration of implementing vaccines in situ.

Still, certain situations may call for a carefully coordinated reactive vaccination strategy with (habituated) apes, as did the abovementioned disease outbreaks witnessed by Goodall and the Gorilla Doctors (Leendertz *et al.*, 2017). Coordination is key since Ebola outbreaks occur randomly, which makes it practically impossible to assess the risk to any single population with any degree of accuracy. Complicating matters, outbreaks leave little to no time for planning a medical response. Practitioners who are tasked with managing ape health thus prepare situational assessments and develop protocols for potential medical intervention in advance (Leendertz *et al.*, 2017; see Chapter 6). To ensure robust ethical decision-making in the face of future disease outbreaks requires that such preparations include investment in ethical oversight and best practices to guide veterinarians and others charged with protecting ape health (Gilardi *et al.*, 2015; Gruen, 2018; Gruen, Fultz and Pruetz, 2013; Osofsky, 2016).

Addressing Ape Health in Natural Habitats[3]

Uganda's Rationale for Health Interventions in Situ

In 2009, at Uganda's great ape health workshop, a collective of invited national and international researchers, veterinarians and wildlife experts, agreed to intervene in response to all human-caused injuries to great apes and other primates in the country. This landmark decision was made in part because approximately one-third of Uganda's chimpanzees were estimated to have snare injuries (Plumptre et al., 2010).

While chimpanzees and other primates are not necessarily the intended targets of Ugandan hunters, many snares are set in forests and gardens, increasing the risk to chimpanzees almost everywhere. Snares can cause prolonged suffering and, when left untreated, snare injuries can result in permanent disfigurement or death (Hartel et al., 2020). For survivors, severe snare wounds impose long-term damage through their effects on behavior, social status and reproductive success.[4] Interventions provide an opportunity to alleviate human-induced suffering, mitigate permanent injury and preserve behavioral and social integrity (Gruen, Fultz and Pruetz, 2013; Hartel et al., 2020; Hyeroba, Apell and Otali, 2011).

Conditions Necessary for Snare Removal Interventions

Snare removal interventions are logistically challenging and accompanied by inherent risks, such as those related to dart deployment, drug administration, potential falls and responses by the target or their conspecifics. These risks are factored into any decision-making process. Standardized protocols can help to evaluate each situation and to determine the probability of success.

In deciding whether to intervene, practitioners and other stakeholders focus on one main question: Do the potential benefits of intervention outweigh the inherent risks? Qualified personnel—including veterinarians, field teams, project managers and directors, and wildlife authorities—are best placed to make that assessment. Similarly, interventions are best carried out by the most qualified experts, such as highly trained wildlife veterinarians who have the requisite equipment (including dart guns, controlled drugs and medical supplies) and are accompanied and advised by field assistants, rangers or guides who have intimate knowledge of the target apes, their conspecifics and the forest (Gruen, Fultz and Pruetz, 2013).

Figure 5.1 outlines the individual, environmental and social prerequisites for an intervention attempt. If all criteria are met,

FIGURE 5.1

Requirements for Successful Snare Removal Intervention

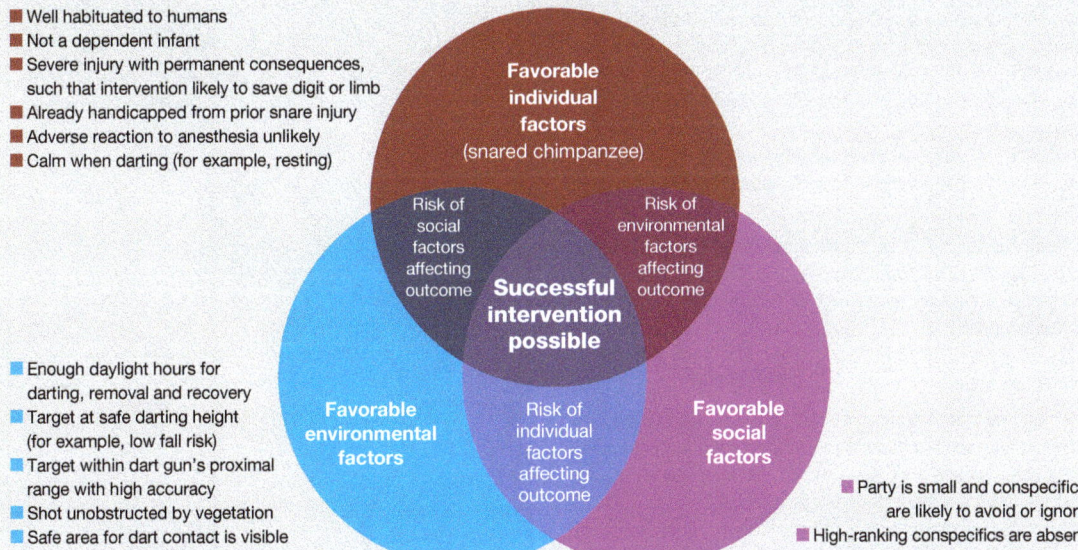

- ■ Well habituated to humans
- ■ Not a dependent infant
- ■ Severe injury with permanent consequences, such that intervention likely to save digit or limb
- ■ Already handicapped from prior snare injury
- ■ Adverse reaction to anesthesia unlikely
- ■ Calm when darting (for example, resting)

- ■ Enough daylight hours for darting, removal and recovery
- ■ Target at safe darting height (for example, low fall risk)
- ■ Target within dart gun's proximal range with high accuracy
- ■ Shot unobstructed by vegetation
- ■ Safe area for dart contact is visible (for example, target's thigh or arm)

- ■ Party is small and conspecifics are likely to avoid or ignore
- ■ High-ranking conspecifics are absent
- ■ Mother is absent or likely to avoid

Notes: This figure shows that individual, environmental and social factors must all be favorable for a snare removal intervention to be successful. If only two of these three sets of conditions can be met, the absence of the third risks undermining the outcome of the intervention. In deciding whether an experienced veterinarian should plan and perform a snare removal intervention, qualified personnel evaluate the presence of all three sets of conditions. Deciding to proceed with a possible intervention is not the same thing as choosing to go ahead with dart deployment. These options require separate assessments; the diagram includes conditions relating to both. Since 2006, intervention attempts have not been possible in 27% of snare cases in Uganda due to challenging individual, environmental and social factors.

Source: Based on Hartel et al. (2020)

the probability of success is expected to outweigh potential risks and an intervention may be attempted. If it is not possible to meet all the criteria, then the risks may exceed the potential for success, which suggests that the intervention should be pursued with extreme caution, postponed until all conditions can be met or abandoned entirely.

Chimpanzee Snare Injury and Intervention

Since 1987, the Kibale Chimpanzee Project (KCP) team has routinely observed a habituated chimpanzee community of 40–58 individuals in the Kanyawara area of Kibale National Park, Uganda (see Figure 5.2; Emery Thompson *et al.*, 2020). In 1997, in collaboration with the Uganda Wildlife Authority, KCP established its conservation arm, the Kibale Snare Removal Program, in response to the high rate of chimpanzee snare injuries: 45% of the living and deceased individuals had been snared,[5] and the majority of them (88%) had suffered permanent injury (Hartel *et al.*, 2020). While the program has helped to reduce the probability of a chimpanzee being snared, it has not eliminated the threat, such that interventions are still needed. Since 2006, KCP has conducted seven veterinary-supervised snare removal interventions, which have resulted in a reduction in the likelihood of permanent injuries and injury severity (Hartel *et al.*, 2020).

In January 2020, in collaboration with the Jane Goodall Institute and the Uganda Wildlife Authority, KCP established the Chimpanzee Health, Intervention, and Monitoring Program (CHIMP), which takes a One Health approach (ASP, n.d.; see also Chapters 2 and 4). CHIMP is led by a wildlife veterinarian who lives onsite at Kanyawara but is available for interventions throughout Kibale and elsewhere. Since elapsed time is the biggest enemy of a snare injury, CHIMP has been instrumental in reducing the emergency response time, which can increase the chance of a successful intervention and reduce the likelihood of permanent injury.

The discussion below considers three cases of snare injuries with and without intervention prior to CHIMP's establishment, as well as one successful intervention following CHIMP's establishment.

No Intervention

Sometimes interventions are not possible because the snare has already resulted in permanent damage by the time qualified veterinarians become aware of the injury. Such was the case with Max, a male chimpanzee who was snared twice as a juvenile (see photo below). Max's mother, Mususu, was a somewhat peripheral female who would spend weeks in the northern sector of the Kanyawara home range without being seen by researchers.

When Mususu appeared in June 2004, after not having been sighted for 21 weeks, six-year-old Max had a snare around his right ankle. About ten days later, Max was observed carrying his foot, which was still attached to his leg by some connective tissue. As he struggled to travel and climb trees on

his own, Max often screamed or whimpered for help from his mother. He was seen again two days later without his foot; only a bloody stump remained.

Three years later, in March 2007, Max was snared around his left ankle. As was the case with the first snare, this injury led to the loss of his foot before the team had the chance to intervene. In July of the same year, Mususu was seen without Max, and the team assumed he had died. Surprisingly, after

FIGURE 5.2

Kibale National Park and Surrounds

Photo: Max, who was snared twice as a juvenile. © Ronan Donovan

a three-month absence, Max returned. He moved slowly on his stumps, which sometimes still bled.

Even without feet, Max, now an adult, is able to walk and climb trees. He has struggled socially and reproductively, however. He remains the lowest-ranking adult male, is often the target of aggression and has yet to sire any offspring. If not for these permanent snare injuries, his social life would probably have been very different (Cohen, 2010).

Unsuccessful Intervention

When protocols are not well-defined or followed, risk levels increase and interventions can result in adverse outcomes. An example involves Mandela, an eight-year-old orphaned male chimpanzee. In April 2007, KCP field assistants saw Mandela with a snare around the toes of his left foot. An external veterinary team was notified and responded within three days. Unfortunately, the veterinarian darted Mandela although adult male conspecifics were nearby. As a result, Mandela fled to the other males for support and reassurance.

As Mandela's anesthesia took effect, he fell unconscious, and the adult males prevented the veterinarian from approaching by acting aggressively towards him. When the anesthesia wore off, Mandela awoke to find the snare still around his toes. The adult males departed and Mandela, still groggy, followed them (D. Hyeroba, personal communication, 2007; Hartel *et al.*, 2020). The KCP team never saw him again.

Since male chimpanzees remain in their natal group for life, Mandela was presumed dead after his conspecifics were seen without him for several months. While the KCP team cannot be certain of what caused his death—the snare injury, the anesthesia, an infection or something else—his remains the only instance of death following a snare removal intervention at Kanyawara. This incident occurred prior to the 2009 Uganda great ape health workshop, after which standardized protocols became part of every intervention.

Successful Interventions

Under ideal conditions, an intervention alleviates pain and suffering, mitigates an injury's severity and reduces the probability of permanent damage, as illustrated by Special, a 12-year-old orphaned female. On July 28, 2012, the research team encountered Special after not having sighted her in a week. Seeing a snare cutting deeply into her right wrist, they immediately contacted an external veterinary team. As field conditions were favorable, darting was carried out successfully the very next day. The veterinarian removed the snare wire (which had already cut to the bone), cleaned and sutured the wound, and administered an antibiotic to fight infection.

The intervention undoubtedly saved Special's hand, and the intervention was so successful that she used her hand to grasp branches while climbing only three months later (see Figure 5.3). Today, Special has full functionality of her snared hand, and all that remains of her injury is a faint scar (see Figure 5.3). Special remained in her natal community and gave birth to her first offspring when she was 14 years old; she is a caring mother (KCP, n.d.).[6]

Following the establishment of CHIMP and the hiring of a permanent onsite wildlife veterinarian in January 2020, a 12-year-old natal female named Wenka was seen by field assistants with a snare around her left wrist. The veterinarian mobilized an emergency response that same day, immediately began to follow Wenka to assess her snare injury, and formulated an action plan. Five days later, the veterinary team successfully darted Wenka and removed the snare, in an action that ultimately saved her hand. Four years prior, in 2016, Wenka had lost all of her fingers on the same hand to a snare injury, leaving her with only her thumb and palm intact. Despite this permanent injury, she had adapted well. Losing her entire hand would have called for a much tougher adjustment (N. Bukamba, personal communication, 2020).[7]

Having an onsite veterinarian has greatly expedited the team's emergency response. The team expects to improve the probability of success as the chimpanzees become habituated to the veterinarian, who visits the chimpanzees regularly to monitor their health. In Wenka's case, the darting intervention was delayed a few days because the CHIMP veterinarian was still in training; it took time for an experienced veterinarian to arrive with the necessary intervention drugs (N. Bukamba, personal communication, 2020). Nowadays, however, two fully trained CHIMP veterinarians are able to initiate a snare removal intervention independently, which should further reduce the risk of permanent injury.

FIGURE 5.3

Special's Recovery from a Snare Injury Following Human Intervention

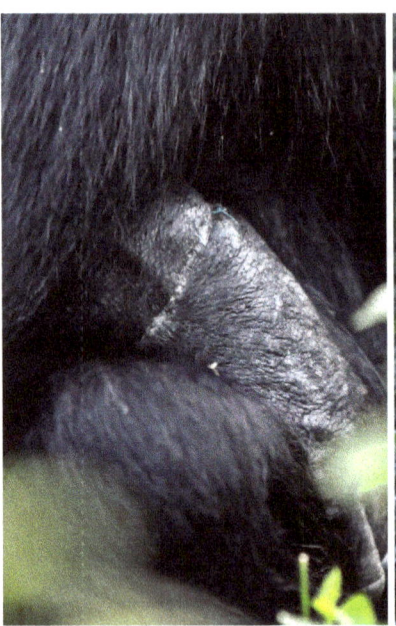

Special two months after intervention (left and middle) and gripping a branch three months after intervention (right).
© Andrew Bernard

Provision of Health Care in Sanctuary Settings[8]

Negra, a chimpanzee who is now living in the Chimpanzee Sanctuary Northwest in the US state of Washington, was captured as an infant in Africa in the early 1970s (Ferdowsian *et al.*, 2011; see photo below). For the next 35 years, she was confined in cages in various biomedical research facilities across the United States for use in experimental surgery and vaccine development. Throughout that time, she was routinely anesthetized by dart, often for the purpose of experimental intervention but also for routine procedures such as physical exams and teeth cleanings. Incomplete records indicate that she experienced at least 60 anesthetic events, although the exact number is probably several times higher.[9]

Today Negra is one of several thousand chimpanzees living in sanctuaries across five continents. Despite their many differences, all sanctuaries share a common goal in that they strive, above all else, to provide for the health and wellbeing of their residents. Beneath this seemingly straightforward objective lies a considerable moral dilemma, however. Owing to their size, strength and volatility, adult chimpanzees like Negra cannot simply be walked into a veterinary clinic for an annual exam, nor can they be restrained like a fractious dog or cat. Instead, most medical interventions for chimpanzees require anesthesia, whose use can potentially lead to significant negative psychological and medical consequences. How then do caregivers weigh the benefits of frequent routine exams against the medical risks and the trauma of anesthesia?

Maintaining quality of life for captive animals requires a broad and balanced approach. Historically, efforts to conceptualize animal welfare have focused on three overlapping categories: maintaining physical health; minimizing negative affective states, such as pain and distress, while allowing for normal pleasures; and allowing captive animals to experience life in as natural a way as possible. When considered together, these three criteria can serve as a roadmap to optimal welfare. When pursued on their own, however, these categories sometimes lead to conflicting outcomes (Fraser, 2009). For example, too much emphasis on promoting pleasure, as in the provision of unhealthy kinds or quantities of food, may result in heart disease or diabetes, while a singular focus on what is "natural" may preclude vaccination and thus lead to preventable illness and suffering. Similarly, the consequences of efforts to maintain the physical health of the chimpanzees in sanctuary care through frequent routine examination may have undesirable consequences when measured against other criteria.

In an ideal world, veterinarians and others responsible for managing captive chimpanzee populations would have sufficient health data on which to base their decisions without the need for anesthesia. With the greater institutional adoption of positive reinforcement training in the United States, this approach is becoming increasingly possible. By employing the principles of operant conditioning, whereby trainers provide food and other rewards for actions that increasingly approximate desired medical behaviors, sanctuary personnel can teach captive chimpanzees to become active participants in their own health care. With relatively little training, chimpanzees learn to sit on a scale to be weighed or to present parts of their body for visual inspection or auscultation. With greater investment of time and resources, urine collection and basic echocardiography become possible. The greater the discomfort or physical restriction required to achieve such behaviors, however, the less likely widespread compliance becomes, for example with respect to blood pressure monitoring or venipuncture.

In practice, thorough medical examinations still require chimpanzees to be fully anesthetized. The benefits of such exams are obvious. Dental prophylaxis, deep palpation, chest X-rays and other procedures that are difficult if not impossible to achieve through positive reinforcement training or across a safety barrier are powerful tools in the prevention and early diagnosis of disease. Still, the drawbacks demand serious consideration.

While recovery can be eased with the use of reversible anesthetics, the effects of which can be rapidly diminished via administration of a reversal agent, such drugs may not be available or affordable for all institutions. Not all chimpanzees are good candidates for these classes of anesthetics, be it due to old age, obesity, cardiac disease or other health conditions. Other anesthetics are safer but more likely to result in difficult recoveries, which may involve stress, confusion and unease, even when combined with adjuncts meant to attenuate such effects.

Most anesthetics must be administered via intramuscular injection. Positive reinforcement training can play a crucial role by giving chimpanzees the choice to willingly present an arm or leg for the needle. In practice, however, not all chimpanzees can be sufficiently trained, either due to their history, behavioral predisposition, or a lack of available time and resources. When transmucosal administration and hand injections are not an option, anesthesia must be administered remotely by dart—a traumatic and potentially dangerous route of administration (Cunningham, Unwin and Setchell, 2015; see Chapter 4).

Beyond the risks and trauma of anesthetic administration, there is the risk of complications from the anesthesia itself. Humans and companion animals set to undergo anesthesia, such as those scheduled for surgery, are often given a preanesthetic blood test to evaluate their ability to tolerate the procedure. Such a test is not possible for chimpanzees if the

Photo: Negra was initially reluctant to go outdoors but now frequently explores her two-acre habitat. © Chimpanzee Sanctuary Northwest

blood draw itself requires anesthesia. In the case of apes who are undergoing their first documented physical exam or for those with undiagnosed, subclinical illness, this lack of information can be dangerous and, in rare cases, deadly.

Perhaps most importantly, anesthesia is a threat to the already diminished autonomy of the individuals in captive care. For chimpanzees like Negra, sanctuary is a place to recover from the trauma of laboratory life. After three and a half decades in research, Negra arrived at the sanctuary exhibiting many of the clinical signs of post-traumatic stress disorder and depression (Ferdowsian *et al.*, 2011). She was socially withdrawn, reluctant to try new things and prone to fearful outbursts when touched. And while she has made great strides in sanctuary, progress has been slow and hard won. It has been measured in small steps ventured farther outdoors, in brief social interactions with her group mates and in tenuous moments of trust in her caregivers. As her caregivers attempt to protect her health, they run the risk of damaging it further.

The issue is not whether anesthesia is ever warranted; anesthesia is commonly indicated for emergent illness and acute trauma on the grounds of limiting short-term pain and distress alone. Similarly, preventive care and the detection of preclinical or subclinical illness each have significant potential to outweigh any short-term negative consequences of anesthesia. Rather, caregivers are forced to wrestle with the question of how often it is warranted, and to what end. Put simply, is the acquisition of health data through routine exams under anesthesia always in the best interest of sanctuary residents?

The sheer range of protocols evident across accredited institutions—from those that perform annual exams on all residents to those that do not conduct exams in the absence of a specific clinical concern—signals a lack of consensus on the issue and a need for further consideration and dialogue.

Choosing when to conduct a routine exam under anesthesia or deciding for whom such exams are not warranted requires caregivers to think beyond their desire to prevent all illness and disease towards a more holistic idea of welfare. It challenges a team to acknowledge the risks of both inaction and action and place them in a context that considers not only physical health, but also broader components of wellbeing, such as agency and autonomy. Most importantly, it demands that caregivers imagine the world from the perspective of those in their care. In human medicine, patients who lack medical competency—that is, those who cannot understand why they require medical intervention—as well as those who are unable to provide informed consent, have designated proxies. These proxies are able to make informed judgments based on shared biological and cultural experience (Ferdowsian *et al.*, 2020). When caregivers serve as proxies for chimpanzee residents, they are tasked with making difficult decisions from the chimpanzees' perspectives but risk unwittingly substituting their own. These decisions, therefore, call for a chimpanzee-centered ethical framework that encourages caregivers to balance the complex and often conflicting values that define wellbeing.

Tertiary Prevention: Respect for Autonomy and Care for Wellbeing in Sanctuaries

Veterinary medicine can be invaluable in the care of apes in natural and captive settings. Nonetheless, veterinarians who specialize in wildlife or work in sanctuaries, other professionals and staff all face vexing dilemmas when it comes to managing ape health, as the very interventions that are designed to improve the health of animals can also effectively restrict or harm them. In reviewing potential courses of action, decision-makers inevitably seek to achieve a balance between the benefits of intervention and an animal's freedom from interference.

One of the key values to consider in determining whether to engage in a medical intervention is wellbeing—a concept that may be open to different interpretations. A good understanding of wellbeing can help in the assessment of whether an intervention appears to be in the interest of an individual ape (see Case Study 5.3).

Moral Agency and Moral Courage in Ape Care and Conservation

Reflections on the ethics of caring for the health of apes can quickly become overly theoretical and detached from the reality of moral dilemmas. Policy-makers, funders and other stakeholders need to be careful not to overlook the professionals who face these dilemmas on a daily basis. Many professionals who seek to provide care for apes are changed positively by their experiences. Nonetheless, threats to their moral resilience and moral courage remain, underlining the need to identify successful preventive techniques, coping mechanisms and approaches to resolve these challenges.

What is required of professionals working in care and conservation for them to navigate the sometimes-vexing moral problems that arise in their fields? And how can institutions and organizations support professionals to realize the principles enshrined in their mission statements? Using the lens of compassion, organizations and supporters can do a lot to help professionals build moral courage to solve complex problems and to help them build resilience to address the next set of challenges (see Box 5.2).

Conclusion

When practitioners, policy-makers, funders and other decision-makers take the health of apes seriously in care and conservation, they are often pulled in different directions. On the one hand, decision-makers are invited to consider threats to the health of, for example, an individual gibbon. On the other hand, decision-makers are also quickly swept the other way, upstream and outwards to untangle the innumerable ways in which the health of an individual ape is inextricably connected to their conspecifics, to animals of other species living in shared habitat and to the many other living beings who populate the earth. Ideally, the various conservation approaches would consider the health of both individuals and groups of apes, within their own ecological context.

A One Health approach provides a way of integrating human and ape health within a shared ecology. Crossing species bounda-

BOX 5.2

Building Resilience and Moral Courage

Professionals who work with apes in risky in-situ settings or in sanctuaries can be vulnerable to burnout, vicarious traumatization and moral distress, particularly in morally complex situations. As a result, caretakers and other professionals need to develop skills to preserve their own health and wellbeing and to maintain an authentic practice of moral decision-making. Development of these competencies can benefit from individual and institutional efforts.

Burnout, Vicarious Traumatization and Moral Distress

Whereas burnout is a cumulative process associated with increased occupational stress, energy depletion, emotional exhaustion, withdrawal from work and reduced professional efficacy, it is typically unrelated to trauma and therefore distinguishable from vicarious traumatization (WHO, 2019).

In 1995, psychologists Laurie Ann Pearlman and Karen Saakvitne first used the term "vicarious traumatization" to document signs of secondary (indirect) traumatic stress in trauma therapists (Pearlman and Saakvitne, 1995). Vicarious traumatization—which is sometimes referred to as compassion fatigue—involves mental and emotional changes in caretakers and other individuals who witness and empathize with others' suffering (Figley, 1995). Changes can include alterations in a professional's world view, sense of self, psychological needs, cognitive processes and sensory experiences (Pearlman and Saakvitne, 1995). Symptoms often mirror post-traumatic stress, which can manifest as reexperienced trauma, avoidant behaviors, negative cognitions or mood, or hyperarousal. Harmful coping mechanisms may include denial, detachment or self-numbing behaviors such as substance abuse (Dunkley and Whelan, 2006). Affected professionals may also experience reductions in their ability to exercise cognitive flexibility, thereby complicating the process of ethical problem solving and sound moral judgment (Bryant, 2006).

Typically, vicarious traumatization occurs insidiously over time. However, it can occur due to a single exposure, such as in response to witnessing a severe injury or being part of a failed intervention. Individual risk factors for vicarious traumatization in professionals include a personal history of neglect or abuse, increased exposure to trauma, too little recovery time, insufficient professional experience and a lack of supervision (Tabor, 2011). Protocols may reduce the risk for vicarious traumatization (see Case Study 5.2).

Vicarious traumatization may also worsen moral distress. Andrew Jameton introduced the concept of moral distress in 1984 to describe the distress nurses face when they identify a morally defensible action but are constrained from carrying it out. Since then, the definition has been expanded to include professionals other than nurses and morally challenging situations that cause distress without feelings of constraint, including moral uncertainty (Fourie, 2017; Jameton, 1984). Moral distress may be aggravated by structural factors such as time constraints and poor teamwork or oversight, and it may also adversely affect worker morale, retention and ethical decision-making (Pauly, Varcoe and Storch, 2012). For instance, when conservationists are confronted with crises—such as political turmoil or natural hazards—the volume of need may increase, especially if support or recovery time is insufficient.

Promoting Resilience and Moral Courage in Institutions

Professionals who work with apes in their natural settings and in sanctuaries can experience vicarious resilience, which has also been called compassion satisfaction. Rather than being traumatized by work, professionals who experience vicarious resilience can become inspired and empowered by the coping capacities and resilience of the individuals and populations they serve (Hernández, Gangsei and Engstrom, 2007). For example, sanctuary care and health interventions performed in situ may invoke a sense of satisfaction among caregivers and other professionals.

Through the development of clear ethical protocols and open ethical discussion of complex dilemmas, professionals can aim to intervene appropriately and raise questions about morally dubious interventions (see Case Studies 5.1, 5.2 and 5.3). They can therefore exercise greater moral courage, which reflects a commitment to stand up for and act on fundamental ethical principles, despite potential adversity such as threats to interpersonal or professional relationships, financial hardship or retaliation.

Establishing a supportive organizational culture is essential to the prevention of vicarious traumatization and moral distress and to the promotion of resilience and moral courage among team members (Bell, Kulkarni and Dalton, 2003). Institutions can help professionals avoid vicarious traumatization by clarifying the roles of various professionals, developing peer networks, encouraging professionals to develop positive coping strategies and offering support.

Finally, organizations can create a compassionate culture that inspires resilience and moral courage by also promoting humility, professionalism, anti-retaliation policies, diversity of thought grounded in evidence and consistent ethical principles, and institutional responsiveness and reform (Aultman, 2008; Murray, 2010; Sekerka and Bagozzi, 2007).

Case studies, role modeling, simulation and practice can be used to help professionals develop ethical reasoning, reflection and communication skills so that they can deepen their understanding and articulation of existing and emerging moral quandaries (Murray, 2010). Promotion of self-awareness and the power to recognize and articulate when ethical principles have been violated can also nurture moral courage (Aultman, 2008). Professionals can be empowered to inquire into related facts and determine potential areas of action. By promoting tools and techniques important to the development of resilience and moral courage, institutions can set the stage for robust moral decision-making within and across organizations.

ries presents a range of scientific and moral challenges, including questions about the ethical distribution of benefits and risks to individuals and populations. Creative and effective solutions require collaboration across disciplines and geographical boundaries, as well as open and respectful discussion about divergent perspectives and values (Nieuwland, 2020; Verweij and Bovenkerk, 2016). Decisions about human and animal health require scientific, moral and political integration, towards a genuine interspecies health policy (Nieuwland, 2020).

In addition to the complexity of viewing human and ape health as interdependent, professionals working to safeguard ape health also contend with diverse emergent medical technologies, including disease monitoring, medication, vaccination and diagnostics. Whether and how they implement such technologies raises questions about scientific uncertainty, potential adverse consequences and the best interests of individual apes, in the light of the vulnerability of apes and ape populations in the 21st century.

The chimpanzees Negra and Special demonstrate how, despite the complexity of moral dilemmas in managing ape health, caregivers and other professionals are still confronted with concrete and context-specific moral problems (see Case Studies 5.2 and 5.3). Professionals are highly motivated to promote the health and wellbeing of apes, and they often know individuals by name, making decisions more difficult and personal. Such a personal, affective bond plays an important role in on-the-ground decision-making (Palmer, 2020).

Moral decisions impact the lives of apes involved, as well as those who make such decisions (see Box 5.2). Fostering moral courage therefore becomes a crucial aspect of managing ape health in conservation and care settings. Supporting moral decision-making, for instance by making use of the framework of compassionate conservation, requires that professionals anticipate moral issues that inevitably arise in the course of caring for apes. Successful efforts require an exchange of moral perspectives among colleagues, as well as the establishment of institutions that can exercise efficient ethical oversight and bolster the preparedness and mobility of organizations and individuals to intervene if such interventions are deemed to be ethically warranted.

Acknowledgments

Principal authors: Joachim Nieuwland[10] and Hope Ferdowsian[11]

Contributors: Nicholas Malone,[12] Emily Otali,[13] Jessica Hartel,[14] J.B. Mulcahy,[15] Diana Goodrich[16] and L. Syd M. Johnson[17]

Box 5.1: Katy Scholfield and Alison White, editors[18]

Box 5.2: Hope Ferdowsian

Case Study 5.1: Derived from Nieuwland (2020)

Case Study 5.2: Emily Otali and Jessica Hartel

Case Study 5.3: J.B. Mulcahy and Diana Goodrich

Endnotes

1 **Primary prevention**: Intervention before a disease, injury or disorder occurs. **Secondary prevention**: Screening to identify and reduce the impact of diseases, injuries or disorders in their early stages. **Tertiary prevention**: Management of the impact of an ongoing illness or injury that has lasting effects.

2 Case Study 5.1 is derived from Nieuwland (2020).

3 Unless otherwise indicated, Case Study 5.2 presents information based on the authors' 30 years of combined experience researching and working on chimpanzee behavior and conservation in Uganda at a long-term field site and with other stakeholders in the region.

4 Cibot *et al.* (2016); Hashimoto (1999); Munn (2006); Newton-Fisher (2003); Stokes and Byrne (2006); Yersin *et al.* (2017).

5 This metric is higher than the one-third stated above because it includes long-term data (not just data on currently living individuals).

6 Internal veterinary reports from the veterinarian, Dr. D. Hyeroba, and the research team, seen by the authors.

7 Internal documents provided by Dr. N. Bukamba and KCP field assistants, seen by the authors.

8 Unless otherwise indicated, the information provided in Case Study 5.3 is based on the authors' experience working at Chimpanzee Sanctuary Northwest, including eight years as co-directors.

9 Chimpanzee Sanctuary Northwest maintains Negra's historical records, but there are no records prior to 1982 and many gaps after that.

10 Wageningen University & Research (www.wur.nl/en.htm) and the Faculty of Veterinary Medicine, Utrecht University.

11 Phoenix Zones Initiative (www.phoenixzonesinitiative.org) and Department of Internal Medicine, University of New Mexico School of Medicine (https://hsc.unm.edu/medicine/departments/internal-medicine).

12 University of Auckland (www.auckland.ac.nz).

13 The Kasiisi Project (www.kasiisiproject.org) and the Kibale Chimpanzee Project (https://kibalechimpanzees.wordpress.com).

14 University of North Georgia (https://ung.edu), Metropolitan Community College – Kansas City (www.mcckc.edu/), Heartland Conservation Alliance (www.heartlandconservationalliance.org/) and the Kibale Chimpanzee Project (https://kibalechimpanzees.wordpress.com).

15 Chimpanzee Sanctuary Northwest (https://chimpsnw.org).

16 Chimpanzee Sanctuary Northwest (https://chimpsnw.org).

17 Upstate Medical University (www.upstate.edu/bioethics).

18 Both: Arcus Foundation (www.arcusfoundation.org).

CHAPTER 6

Disaster Management and the Protection of Apes

Introduction

The dramatic reduction in great ape habitats across Africa and Asia presents an imminent challenge to their survival. Loss of forest cover is primarily due to anthropogenic activities (Estrada *et al.*, 2017; Nellemann and Newton, 2002). The natural habitats of apes have long been exposed to natural hazards, including volcanic eruptions, drought, heat waves, hurricanes and cyclones—which cause flooding, landslides, fires and wind damage to the forest structure. The transition to large-scale farming and industrialization, however, has led to an exponential increase in human population and activities that have steadily eroded ape habitat and thereby increased their vulnerability to natural hazards.

In many ape range states, only isolated habitat fragments remain, surrounded by swathes of cleared land used for agriculture. Logging activities and infrastructure development crisscross landscapes, further subdividing ape habitat and disrupting connectivity, while poaching and hunting have directly reduced local ape populations (Estrada *et al.*, 2017). Consequently, many of these populations are living in small pockets of habitat whose resilience to natural hazard impact has been in decline.

Climate change-induced extreme weather events—which have been growing in intensity and frequency—present a clear threat to apes and their habitats.[1] These hazards and events can lead to potentially detrimental behavioral and physiological adaptations, as recently witnessed in Senegal, where extremely hot daytime temperatures led chimpanzees to become more nocturnal, exhibit changing energy needs and display differences in heat regulation (Pruetz and Bertolani, 2009). Furthermore, apes have increasingly been exposed to environmental threats that are directly associated with human activities, such as deliberate forest fires. In orangutans, fires can cause death or injury (such as through smoke inhalation), change activity patterns and lead to starvation-like physiological responses (Erb *et al.*, 2018; Estrada and Garber, 2022; Estrada *et al.*, 2017; see Box 6.2).

An emerging crisis is the exposure of apes to zoonotic diseases from humans, which can lead to deaths and compromise the viability of populations (Dunay *et al.*, 2018; Negrey *et al.*, 2019; see Case Study 6.3 and Chapter 1). While disease hazards constitute a particular risk to captive apes, wild populations are also at risk from potentially infected hunters, local communities, park staff, tourists and other travelers (Muehlenbein *et al.*, 2010). Moreover, captive and free-ranging apes can be exposed to flooding, chemical poisons and other risks (BBC News, 2002; Kooriyama *et al.*, 2013). For more information on managing ape health, informing interventions, see Chapter 4.

Natural and anthropogenic hazards imperil the survival of apes, especially if multiple threats affect already reduced and fragmented ape populations. Disaster management principles provide a valuable set of tools for mitigating or reducing the impact of natural and anthropogenic hazards on both wild and captive apes. See Box 6.1 for standard definitions of common disaster management terminology and concepts. This chapter presents an overview of these principles and examples of their adoption to mitigate the impact of hazards such as flooding and severe acute respiratory syndrome coronavirus 2 (SARS-CoV-2) on apes (see Case Study 6.1, Case Study 6.2 and Box 6.3).

The chapter highlights consideration of the full range of related risks under the approach known as PEESTOLM, which covers political, environmental, economic, social, technical, operational, legal, and media and communications-related risks (Cooper, 2018; WHA, 2018; see Annex IV). Assessments of each of these risks are used to inform the development of mitigation measures to enhance the survival chances of apes in a disaster context. The spectrum of mitigation measures includes avoiding the risk of hazard impacts, reducing the potential consequences of impact and containment of the risk. Annex V provides an example of an emergency preparedness and response questionnaire and Box 6.4 works through the development of a contingency plan.

Key findings include:

- Although both natural and anthropogenic hazards can strike at any moment, their effects can generally be mitigated through the structured and systematic development of risk mitigation measures in advance of any potential hazard impact. Such measures include preventive action aimed at avoiding any impact; preparedness, including the development of detailed plans; the building of capacity and capability for response and recovery; and exercises and simulations.

■ In the first instance, the delivery of prevention and preparedness goals is associated with local ape populations and any nearby human communities. In addition, national and global support can raise the levels of readiness with respect to ape populations at risk.

■ Targeted research has a part to play in identifying and developing emergency response measures, including by exploring how those measures are likely to benefit at-risk apes.

■ The use of risk mapping for ape populations can inform priorities for developing prevention, preparedness and response.

■ Active reporting and monitoring of preparedness can help to identify gaps and allows for the tracking of progress.

■ Wherever ape-related disaster management structures and arrangements are in place, there are opportunities to adopt a

Photo: When hazards such as forest fires destroy habitat, for instance, apes' access to food and shelter drops significantly faster, causing declines in birth rates and population numbers. Freshly cleared patch of forest that has been burned for agriculture, Gunung Palung National Park, West Kalimantan, Indonesia. © Tim Laman/naturepl.com

comprehensive approach: risk identification, prevention, preparedness, response and recovery.

- At all levels, government and private agencies and organizations that are responsible for at-risk ape populations could potentially make use of established disaster management systems.

Disasters and Apes

To be valuable, assessments of the probability, type and potential severity of different disasters—in relation to apes, their habitats and the people who live alongside them—require a shared understanding of related terminology and concepts (see Box 6.1).

BOX 6.1

Terminology

Crisis: A system-wide disruption that is typically new, unexpected, uncontrollable or abnormal and that requires immediate solutions or interventions involving collaboration among local stakeholders. A crisis typically affects a particular industry, population or community; local stakeholders are able to address the disruption.

Disaster: A serious disruption of the functioning of a community or society due to an interaction of a hazardous event with conditions of exposure, vulnerability and insufficient capacity to cope with the event. The consequences include significant social, built, economic and environmental losses and impacts. Locally impacted communities are unable to cope and require external assistance and coordination.

Emergency: An actual or imminent natural or anthropogenic event that endangers or threatens life, damages infrastructure or destroys the natural environment, thus requiring significant coordinated and time-critical responses, as well as extraordinary measures to save lives, protect vulnerable individuals and limit damage. An emergency tends to be local or regional, so does not result in serious disruption to the broader community or society. Emergencies can be categorized by size of impact area and multiple simultaneous emergencies in one area may be classified as a disaster.

Hazard: A natural, socionatural or anthropogenic process, anomaly or event that is defined by location, magnitude, intensity, frequency and probability, and that has the potential to directly harm life as well as the built and natural environments and ecosystems. A hazard can cause indirect disruptions to an economy.

Sources: AIDR (n.d.); Al-Dahash, Thayaparan and Kulatunga (2016); Staupe-Delgado (2019); UNDRR (n.d.-b); WHO (2020d); WHO/EHA (2002).

Similarly, common terminology can underpin the development of strategies to mitigate and respond to the impacts of such events and associated challenges, whether directly or indirectly. The severity of impacts depends in part on whether affected apes are in their natural habitat or in captive settings, such as sanctuaries.

Natural and anthropogenic disasters can affect apes directly or indirectly. Potential direct impacts on apes include:

- dehydration due to a lack of access to suitable water sources during drought and extended periods of forest fires;
- malnutrition associated with an acute, long-term lack of access to suitable food sources following ecosystem destruction;
- morbidity related to sustained exposure to forest fire smoke, heat exposure and loss of access to nutritious foods;
- poisoning arising from impacts of industrial hazards;
- mortality, from individual flood- or fire-related fatalities to a reduction in the number of apes in a population, to a point at which recovery may be impossible; and
- fragmentation of local populations in response to a hazard-modified natural environment and changes in ape population distribution and structure.

In a crisis, local expertise and resources can be mobilized to cope with and manage a single or small number of affected apes. Monitoring of a crisis includes assessing the capacity of local resources to stop an escalation of the situation, which could otherwise transition to an emergency and require external assistance. An emergency arises when such a direct hazard impact calls for significant coordination and resources to resolve or stabilize the situation. If there is a risk of population collapse, such as when multiple individuals in different ape communities are affected, the situation is considered a disaster.

Ongoing direct threats to apes—such as deforestation, hunting and infectious disease—are compounded by indirect impacts. When hazards such as forest fires destroy habitat, for instance, apes' access to food and shelter drops significantly faster, causing declines in birth rates and population numbers. Such indirect impacts can threaten ape survival, particularly if hazards recur and population decline becomes irreversible (Behie *et al.*, 2019).

Indirect disaster impacts are also possible wherever ape communities and local human populations are interdependent, as loss and damage to human dwellings, food supplies and the economy can translate into reduced support and care for apes or the forest, competition for food, habitat destruction to support rebuilding, and ape hunting.

The scientific literature on disaster management features limited studies of the hazard impacts on apes and responses for impacted apes. Ape exposure to disasters is poorly understood in terms of scale and frequency. Coverage is more likely in the case of high-profile threats, such as the devastating fires that tear through orangutan and gibbon habitat in Indonesia (see Box 6.2).

Natural hazards associated with climate change are expected to continue increasing in frequency, duration and severity. Among them are cyclones, hurricanes, droughts, heat waves, flooding and forest fires caused by lightning (Malhi *et al.*, 2008; Sergio, Blas and Hiraldo, 2018; Wiederholt and Post, 2010).

Disaster preparedness for captive apes focuses on numerous risks, including flooding, and involves preparedness actions to protect resident animals, their keepers and the facilities. This section presents preparedness actions in two captive settings. Case Study 6.1 examines the management of fires and flooding by a chimpanzee sanctuary on an island in Lake Victoria, Uganda; Box 6.3

BOX 6.2

Forest Fires in Indonesia

In 2015, more than 100,000 forest and peat fires burned more than 26,000 km² (2.6 million ha) in Indonesia in 2015 (World Bank, 2016). This burned area included the Sabangau Forest of more than 5,000 km² (500,000 ha), which was home to more than 7,000 orangutans (Vidal, 2015). In Borneo, fires destroyed vast areas of habitat and had harmful impacts on the social, economic and natural environments. Forest fire smoke resulted in 500,000 human respiratory disease cases (Vidal, 2015). There was little evidence of a timely, appropriate or coordinated response to these human-caused forest fires.

The effects of hazardous wildfire smoke on orangutan health included negative impacts of smoke and particle inhalation, which caused apes to increase their rest time and decrease their travel time and distances (Erb *et al.*, 2018). Studies show that ongoing exposure of orangutans to forest fire-scarred landscapes is associated with debilitating changes in their behavior and health. The fires led to the loss of nutritious foods, which caused chronic starvation, poor health, aggression and declines in populations (Jong, 2020; Vogel, 2018).

As orangutans are likely to feel the effects of smoke for a few months, long-term impacts are possible (Erb *et al.*, 2018). The frequent exposure of orangutans to the smoke could have severe implications for the population. Between 1999 and 2015, nearly 100,000 orangutans were lost from intact forests in Kalimantan, indicating that their rapid decline is not due only to habitat loss (Imster, 2018).

Forest fire smoke significantly affects the singing of Bornean white-bearded gibbons (*Hylobates albibarbis*). There is a reduction in both the number of singing days and the duration of singing during the fire-prone dry season (Cheyne, 2008a). Gibbon singing communicates information such as alerts for different predator types, proximity to neighbors, and interaction between males and females (Clarke, Reichard and Zuberbühler, 2006; Coudrat *et al.*, 2015). While the behavioral effects of the smoke are not easy to predict, reduced singing at a time when there are usually peaks in singing could negatively affect territorial spacing and defense, communication and reproduction. Increases in mortality, including of infants and juveniles, have the potential to affect population numbers and, ultimately, survival. While the consequences of the reduced singing and the effects of smoke on health have not been specifically investigated, there is no doubt that smoke negatively impacts forests and wildlife (Cheyne, 2008a; Harrison *et al.*, 2007).

The effects of the smoke on the Indonesian economy were significant, particularly with respect to additional air pollution, poor air quality and excessive carbon emissions (Sumarga, 2017). The World Bank Group estimated the impact on the Indonesian economy at US$16 billion, equivalent to about 2% of the country's gross domestic product and more than twice the recovery costs associated with the 2006 Aceh tsunami in Indonesia (World Bank, 2016).

Although the 2015 fires in Indonesia were smaller than the most devastating 1997–1998 fires, they were disastrous (Cassella, 2019; Dennis, 1999; Jim, 1999; Spessa and Field, 2015). While the estimates of how many hectares of forest were burned in 2015 varied considerably, all accounts described the fires as covering vast areas that had no prior history of burning. The long-term effects of multiple fires over relatively short periods can have irreversible impacts on ecosystems (World Bank, 2016). Having compared government data on economic, human and environmental impacts from the 2015 fires with those of previous environmental hazards, Meijaard (2015) characterizes the more recent fire and haze problems as "the biggest man-made environmental disaster of the 21st century."

Base map data source: OpenStreetMap (n.d.) OpenStreetMap © OpenStreetMap contributors, published under Creative Commons Attribution License CC BY; for more information, see http://creativecommons.org)

CASE STUDY 6.1

Flood Response and Recovery in a Chimpanzee Sanctuary in Uganda[2]

The Sanctuary

The Chimpanzee Sanctuary and Wildlife Conservation Trust manages the 0.4-km² (40-ha) Ngamba Island Chimpanzee Sanctuary, which is home to 52 chimpanzees in Lake Victoria, Uganda (see Figure 6.1).

FIGURE 6.1

Ngamba Island Chimpanzee Sanctuary, Uganda

The chimpanzee sanctuary is a controlled, semi-captive environment. The island's remoteness offers the apes a degree of safety from humans and other wildlife, yet it also presents risks in relation to forest fires, floods and disease. In 2020, the sanctuary team implemented a successful emergency response in the face of rapidly rising coastal waters and flood risk. The sanctuary subsequently developed a response plan for addressing flooding as well as other risks, including disease outbreak and forest fires.

Figure 6.2 presents the layout of infrastructure on the island. A fenced-off area of about 0.03 km² (3 ha) contains two chimpanzee management areas (C and D), staff quarters (B), the veterinary clinic (V) and visitor areas (E). The other 0.37 km² (37 ha) make up the main forest enclosure, where the chimpanzees are free to roam every day. A double electric fence (A) is designed to prevent escapes and to ensure continued functioning in case one fence fails.

FIGURE 6.2

Ngamba Island Sanctuary Infrastructure

Notes: Ngamba Island infrastructure includes a double fence (A) that separates the forest enclosure on the left from the sanctuary structures and facilities on the right, including the staff quarters (B), the sleeping area for chimpanzees requiring additional care (C), the isolation ward (D), the veterinary clinic (V) and the visitor areas (E). The proximity of the staff quarters to the chimpanzee sleeping area is useful for nighttime monitoring and quick response. The visitors are situated as far as possible from the apes within the space available. © Joshua Rukundo

The sanctuary's design takes seasonal fluctuations in the lake water levels into account. All chimpanzee areas are on higher ground, including the overnight housing and the outdoor enclosures. Structures occupied by staff and visitors are all on the lower side of the island, where quick evacuation is possible. While unable to suppress inundation, a retainer wall along the shoreline prevents erosion and protects the housing structures in the area most prone to wave action.

The Floods of 2020

As anthropogenic climate change increases the risk of extreme weather events, the island sanctuary's vulnerability to flooding is growing. Rising lake water levels have already led to the loss of about 30% of the land area (see before and after photos below).

Between October 2019 and April 2020, East Africa—and Uganda in particular—witnessed heavy rains, which led to the saturation of rivers, as well as extensive flooding. By May 2020, the water level in Lake Victoria had increased by nearly 1.5 m—from 12 m to 13.42 m, the highest level ever recorded (Cheptoris, 2020; NBI, 2020).

FIGURE 6.3

Ngamba Island Chimpanzee Sanctuary
Before, During and After the Floods of 2020

A. Before

B. During

C. After—showing new retaining wall

© Joshua Rukundo, Ngamba Island Chimpanzee Sanctuary

While the rising water levels can be attributed to global warming, environmental degradation caused by humans has accelerated the trend. Loss of forest cover, encroachment on wetlands, lakeshores and riverbanks, and poor land use practices have resulted in soil erosion, which leads to siltation and increased water flow into lakes and rivers. Siltation has also reduced the water storage capacities of the bodies of water. At the same time, increased urbanization has introduced highly impermeable surfaces such as roads, roofs and pavements, which reduce water infiltration into the soil, causing declines in water storage. Similarly, the loss of forests and wetlands limits the local environment's interception and evapotranspiration capacity (Aquatic Habitats in Integrated Urban Water Management, n.d.; Cheptoris, 2020; NBI, 2020).

The 2020 flood waters reached an unprecedented level, the highest in living memory. The resulting inundation of the sanctuary put the welfare of the chimpanzees and those who care for them at risk.

The water submerged a 12–20-meter-wide band of land around the entire island, claiming approximately 0.026 km^2 (2.6 ha) of land cover. A further 0.05 km^2 (5 ha) was water-logged and temporarily flooded, reducing the useable land cover by almost 20%. The flooding also affected the electric fence that separates the chimpanzees' outdoor enclosure from the sanctuary buildings for staff and guests. In addition, water stagnation caused by the flooding affected the sanctuary's drainage and sewage systems, as underground septic tanks and drainage soak pits were inundated.

The flood-related damage increased the risk that chimpanzees might escape, enter enclosed areas or be exposed to waterborne diseases. Meanwhile, the water-logged areas provided an ideal environment for fish to breed and attracted schools of tilapia. Fishermen intent on trapping the breeding tilapia subsequently sought to gain illegal access to the island, through local communities.

Flooding is potentially life-threatening to the sanctuary chimpanzees. It reduces their access to food, particularly along the western edges of the island, where the vegetation includes thick, thorny mangrove-like trees and bushes that the chimpanzees like to feed on. These areas become especially treacherous to navigate during flooding, when young chimpanzees are most susceptible to getting stuck and drowning. Some flood-related incidents have required sanctuary team intervention to rescue stranded chimpanzees. These high-risk situations can be dangerous for both the animal and the rescuers, as tranquilizing chimpanzees is rarely an option. Team members are not trained in swamp rescue operations or equipped with the required specialized tools.

Risk Management

In response to the 2020 flooding, the sanctuary management team used a risk management approach to analyze the risks, including the potential impacts. The results informed the development of a response and recovery plan to minimize future exposure to flood hazards and to prioritize areas of intervention.

To safeguard the welfare of the animals and the staff, repairs were first carried out on critical structures such as the electric fence, pier (which provides access to the sanctuary for supplies and possible evacuation) and sleeping quarters. This work involved reinforcing weakened areas of the fencing, building a temporary pier, and damming and reinforcing the shoreline with sandbags and rocks at vulnerable spots around the island. The retainer wall along the shoreline was also bolstered after the 2020 floods.

Non-essential staff members left the island and structures that were flooded or at risk of flooding were evacuated. The caregiving team, led by the head-caregiver, carried out daily surveys of flooded land in the forest enclosure to minimize the risk that chimpanzees or other animals might get stuck in bogged areas. During the monitoring activities, they also cleared areas of brush to reduce the risk of animals getting stuck in this dense vegetation if it flooded. The team located areas of stagnant water, which they filled with sand or drained or, in some cases, applied vegetable oil to minimize mosquito breeding. The sewage systems of the visitor facilities were blocked off as they were not in use. This prevented the backflow of sewage from flooded underground waste tanks and reduced the risk of water contamination and exposure to waterborne diseases. The management team established rapid communication and response procedures to manage potential rapid changes in the scenario, such as increased and more prolonged inundation.

While the flooding crisis was managed effectively, it exposed the need for a rapid response and evacuation plan for the sanctuary. Such a plan, which will allow sanctuary staff to prepare for a catastrophic event, is under development. For a preparedness review of Case Study 6.1, see Annex VI.

considers the impact of the infectious disease caused by the SARS-CoV-2 virus, the COVID-19 pandemic, on rescue and rehabilitation centers and their risk management plans. Such preparedness actions work best when considered alongside well-developed and rehearsed procedures that provide guidance on leadership, risk mitigation measures and resources.

As the frequency and severity of natural hazards increase, so too do the risks of associated impacts on animals and their environments (Zhang *et al.*, 2019). Climate change-induced natural disasters may result in rapid shifts in ape distribution, behavior and diversity (Lehmann, Korstjens and Dunbar, 2010). Such changes demand "tougher choices and more proactive crisis-preparation for conservationists, as well as mentality changes for all" (Sergio, Blas and Hiraldo, 2018, p. 1).

Some apes may have subtle responses to severe weather events. A study of the effects of climate change on bonobos (*Pan paniscus*) in the Democratic Republic of Congo (DRC) shows that as precipitation decreased over a 15-year period, the decay times of the apes' nests increased. The number of storms was the main factor driving decay times. The bonobos also adapted to the changing climatic conditions by strengthening nest structures in response to unpredictable, harsh precipitation (Bessone *et al.*, 2021).

Natural disasters do not always result in long-term negative outcomes. In response to hurricanes and a subsequent forest fire, for example, spider monkeys (*Ateles geoffroyi*) developed successful coping strategies such as changes in diet, activity and fission–fusion dynamics (Champion, 2013; Schaffner *et al.*, 2012). Analysis of data on Cayo Santiago rhesus macaques (*Macaca mulatta*) from 1973 to 2018 shows that hurricanes did not lead to detrimental effects at the population level (Morcillo *et al.*, 2020).

BOX 6.3

COVID-19 and Its Impact on Ape Rescue Centers

Human–wildlife contact can contribute to the global spread of infectious disease.[3] In confronting the infectious disease caused by SARS-CoV-2, the COVID-19 pandemic, the ape rescue and rehabilitation sector has accorded the highest priority to the safety and well-being of staff, their families and the communities around the centers, as well as the wildlife they seek to protect. The centers have long used personal protective equipment (PPE) such as masks and gloves to minimize the transmission of diseases, parasites and bacteria to and among apes. Their approach to food preparation, quarantining and rehabilitation is also aimed at reducing the risk of cross-contamination and transmission, as is their application of best practices and regularly updated policies (Campbell, Cheyne and Rawson, 2015; Cheyne, Campbell and Payne, 2012).

Field monitoring research, especially following the release of rehabilitant apes, remains key to their safety and wellbeing, so long as close contact and unnecessary risks are avoided. Given the high likelihood that COVID-19 or other infections will continue to pose risks to apes, rescue and rehabilitation centers are devising appropriate longer-term primate monitoring strategies that suitably balance the need to conduct post-release monitoring with the potential harm associated with infections. As monitoring activities typically involve small teams working in areas away from human habitation, they have been able to continue relatively unaffected, albeit with some modifications to reduce the COVID-19 risk.

The translocation of endangered apes for conservation purposes has also become more common, especially for species with limited dispersal ability, since they can find themselves confined to shrinking, fragmented habitats where the risk of early extinction is high. Although apes who undergo translocation are tested prior to release, the process of translocation can increase the risk of disease transmission (Campbell, Cheyne and Rawson, 2015). In view of this risk, the International Union for Conservation of Nature recommended that no reintroductions or translocations of great apes take place during the pandemic (IUCN SSC PSG SGA, n.d.-a).

Rescue and rehabilitation centers are continuing to monitor the situation, as well as international guidance from the World Health Organization. They are complying with instructions from local and national governments to minimize contact with apes, while using PPE to reduce the spread of the virus.

Russon, Kuncoro and Ferisa (2015) report that orangutans in forests affected by fires returned to pre-fire food regimes as the forest recovered. A key survival strategy was their ability to vary diet in response to the availability of different food types across an increased foraging area. In setting aside areas for conservation of a species, practitioners can usefully consider the increased area needed for recovery after a major disturbance event, rather than just the area that supports a population in times of stability. Back-to-back events may prevent the recovery of a population. Animals with slow reproductive rates, such as apes, or with very specific dietary requirements can be more negatively impacted by even small drops in numbers due to extreme weather or other destabilizing events (Ameca y Juárez, Ellis and Rodríguez-Luna, 2015; Behie *et al.*, 2019).

Managing Risks to Apes

Historically, risk management has focused primarily on risks directly associated with the impact of one or more hazards. The process can be more effective if it addresses the full range of risks, as encapsulated by the acronym PEESTOLM, which stands for political, environmental, economic, social, technical, operational, legal, and media and communications-related risks (Cooper, 2018; WHA, 2018).

While some disaster and emergency management practitioners, including those in Australia, use the PEESTOLM approach, others apply analogous methods. One example is a recent risk assessment that examines the challenges and opportunities in tropical forest and peatland conservation and restoration in Indonesia, with a particular emphasis on areas affected by fire. It examines political, economic, social, logistical, legal and research challenges, which generally align with PEESTOLM risks (Harrison *et al.*, 2020a). The assessment of risks is fundamentally the same across sectors, including with reference to captive animal and wildlife health issues, plant pests, drought, locust plagues, human pandemics and natural disasters such as flooding and forest fires.

Annex IV presents a PEESTOLM risk register for a zoonosis in apes. It shows that

disease risks involve a much wider range of considerations than risks associated with a single event, such as a fire. The need for a review of risks may be triggered by various changes in context, such as the implementation of mitigation measures, the end of a specific time period or expenditure of a mitigation budget.

Risk management guidance from global organizations can also be instructive. The hazard-based contingency planning flow chart described by the World Health Organization (WHO) starts with risk analysis (European Commission, 2021; UNHCR, 2015; WHO, 2018). Disaster risk management is the basis for the United Nations Sendai Framework for Disaster Risk Reduction 2015–2030, which is a good guide to risk assessment for disasters (UNDRR, 2015, n.d.-c). Integrated risk management, utilizing a One Health and disease risk analysis approach, has been used during the COVID-19 pandemic (see Chapter 2).

Compounding Risks

When multiple risks interact—as compounding or cascading risks—they exacerbate potential impacts and consequences. Annex IV uses PEESTOLM to explore the risks associated with a zoonotic hazard in apes. Within the annex, no single risk can be considered in isolation from the other risks. The compounding relationship between some of the risks translates into a greater risk than for any one risk alone. An inability to mitigate one risk leads to further risks. For example, an inability to address operational risks—such as by providing adequate resources for surveillance—can lead to technical risks, such as a failure to detect disease.

A cost–benefit analysis, undertaken as part of emergency preparedness, can help to identify compounding relationships between risks. As discussed in Case Study 6.2, an analysis that is focused only on the costs of managing technical risks that arise directly

---

COVID-19, Tourism Revenue and Compounding Risks

In general, the exclusive management of technical risk—such as through disease prevention—is likely to lead to compounding risks. A case in point is the management of the technical risks associated with the infectious disease caused by SARS-CoV-2, the COVID-19 pandemic, in the Democratic Republic of Congo (DRC), Rwanda and Uganda, whose economies are tourism-dependent.

In response to the COIVID-19 pandemic, the iconic gorilla conservation parks across these three countries were closed to tourism for protracted periods of 2020 and 2021 (Beament, 2020; Virunga National Park, n.d.-b). This case study reviews the interaction of certain PEESTOLM risks—that is, political, environmental, economic, social, technical, operational, legal and media and communications-related risks—associated with the SARS-CoV-2 hazard/COVID-19 disaster (see Annex IV).

Ramifications of COVID-19 Restrictions on Ape Tourism

In the DRC, Rwanda and Uganda, the negative consequences of halting all ape tourism activities and implementing other measures to prevent the potential infection of apes with

COVID-19 included a loss of employment and livelihoods, reductions in local and national income, and cuts in funding for ape conservation.

In Rwanda, for example, the tourism sector—much of which revolves around gorillas—was the largest foreign exchange earner by 2013 and enabled conservation successes (Maekawa *et al.*, 2013; Nielsen and Spenceley, 2010). Measures designed to address only the technical risks associated with COVID-19 effectively curtailed or severely reduced this source of revenue. As a result, conservation efforts were cut back as local economies suffered from the loss of tourism income (Gilardi *et al.*, 2022; Hockings *et al.*, 2020; Kalema-Zikusoka *et al.*, 2021; Richardson, 2021).

The loss in ape tourism revenue also led to a decline in antipoaching activities and an increase in illegal hunting, including wildlife trapping in Uganda's Bwindi Impenetrable National Park (Guyson, 2021; Ledger, 2020). In the DRC, a baby gorilla was found tangled in a snare in Virunga National Park (Ledger, 2020). The United Nations Educational, Scientific and Cultural Organization (UNESCO) reported that the number of snares increased nearly 40-fold during the year ending in April 2020—from 21 to 822 snares (UNESCO World Heritage Convention, 2020).

Research also indicates that the focus on COVID-19 prevention had a negative impact on protected and conserved

---

areas around the world, as pandemic-related restrictions gave rise to operational risks. Specifically, slashed tourism revenue and tighter budgets for park agencies resulted in lay-offs and related reductions in staff activity, restricted management services and drops in conservation effectiveness. Park staff members were diverted away from conservation work and instead assigned duties associated with the mitigation of wider community risks from COVID-19. Meanwhile, local, tourism-dependent communities suffered a loss of livelihoods, which were compounded by significant reductions in recruitment and employment opportunities for those working in conservation (Corlett et al., 2020; Hockings et al., 2020). Such operational shortfalls can stunt the development of relevant skills and knowledge, further hampering conservation efforts.

Operational deficits caused by an exclusive focus on technical risks can also weaken public support for conservation. By threatening people's livelihoods and thus their wellbeing, a focus on disease prevention measures can place positive attitudes towards wildlife and the forest at risk (Hall et al., 2004). Managing people's views of animals is key to emergency management that supports positive outcomes for both humans and animals affected by a disaster (McCarthy, Bigelow and Taylor, 2018).

These findings underscore the need to address all relevant risks at the same time. Tackling technical risks is a way to manage disease spread, applying measures to address social and economic risks can protect communities dependent on ape tourism, and reducing operational risks can support the conservation of protected natural areas (Hockings et al., 2020). In recognition of the social, economic and environmental risks associated with COVID-19 prevention measures, the International Union for Conservation of Nature (IUCN) Primate Specialist Group and Wildlife Health Specialist Group similarly called for methods to "offset loss of profit and employment from tourism" and support of public health in local human populations (IUCN SSC PSG SGA, n.d.-a; UNEP, 2020).

Precise estimates of the losses due to the suspension of ape tourism and related activities remain elusive. A 2019 estimate puts the global direct value of wildlife tourism at US$120 billion—or US$346 billion when multiplier effects are accounted for—and the number of jobs at 21.8 million (Hockings et al., 2020). In 2016–2017, the travel and tourism industry that focuses on gorilla tourism contributed more than US$400 million to the Rwandan economy, with 10% benefiting local communities (Fitzgerald, 2022). The loss of this contribution to Rwanda would be significant.

Beyond COVID-19 Restrictions

As pandemic restrictions battered local economies, UNESCO provided emergency funding to Bwindi Impenetrable National Park to support the implementation of COVID-19 safe practices by those who interacted with the gorillas (UNESCO

World Heritage Convention, 2020). The funds allowed the staff to continue with the monitoring and surveillance of gorilla health to sustain the early disease detection and response. The funding was also used to extend patrols for the protection of gorillas and to prevent and discourage poaching.

In moves that reflected the economic importance of gorilla tourism, governments decided to loosen travel restrictions despite ongoing risks from COVID-19. Uganda reopened its doors to tourism from October 2020 and Rwanda from April 2021 (ATTA, 2020; Read, 2020; Virunga National Park Congo, n.d.). To encourage the return of tourists, the cost of a gorilla experience was reduced (Bizimungu, 2020). At the same time, requirements and procedures were revised to address the remaining technical risks associated with the gorillas' exposure to humans. The requirements included mask wearing, increased minimum distances between humans and apes, training of ape caregivers and COVID-19 vaccination of humans working with apes (Kalema-Zikusoka et al., 2021; Richardson, 2021).

The DRC, Rwanda and the Uganda Greater Virunga Transboundary Collaboration developed a COVID-19 risk register for mountain gorillas (*Gorilla beringei beringei*), based on prior contingency planning for Ebola virus disease. At this writing, the draft risk assessment appeared focused on technical risks, although other risks could be addressed in future revisions (Gilardi et al., 2022; GVTC, 2020). Separately, the Section on Great Apes of the IUCN Primate Specialist Group produced a list of measures to minimize the risk of SARS-CoV-2 transmission to great apes (IUCN SSC PSG, n.d.).

Risk assessments are more effective if they are directly relevant to a specific context. The COVID-19 risk assessment at Chester Zoo, for example, is specific to that zoo's context (Chester Zoo, 2021). While elements of the assessment may be applicable to another zoo, it cannot be adopted wholesale by a similar facility, as contexts vary across zoos. Moreover, the benefits of developing a shared understanding among key stakeholders and the need to validate the applicability of risk treatments during the process of developing and completing a risk assessment are as important as the final risk assessment itself.

Another consideration relates to the legal risks associated with implementing proposed mitigation measures for COVID-19. Steps can be taken to ensure that measures comply with legislation and policies, while also being acceptable to local and Indigenous communities. The 2016 emergency response plan for Hainan gibbons (*Nomascus hainanus*), for instance, acknowledges the need to have such approvals in place ahead of any response actions (Bryant and Turvey, 2017).

from a hazard impact may exclude costs associated with other risks. Such analysis is more useful if it also factors in the benefits of mitigating social, environmental and economic risks, which are likely to be much greater in both the short and long term than the benefits of managing only the technical risks. Funds allocated for managing social and economic risks usually support the management of technical risks as well.

Risk Treatment

Risk mitigation options may be grouped into five categories: avoiding the risk, reducing the likelihood of a harmful impact, reducing the consequences, transferring the risk and retaining the risk (ENISA, n.d.; see Table 6.1). Successful risk mitigation typically involves more than one of these treatment options.

In the case of disease outbreak, risk mitigation with respect to apes in their natural habitat is generally focused on reducing the technical risk of infection. The key measure is to reduce any contact between humans and apes to the absolute minimum (see Case Study 6.2).[4]

Trevidy (2020) examines the option of retaining the risk of apes becoming infected as a way of balancing the technical risks against economic and environmental risks.

As discussed in Case Study 6.2, a failure to address economic risks resulting from curtailing local tourism is likely to lead to social risks in the local human population. These compounding risks have the potential to increase the exposure of apes to disease. Measures to reduce such consequences are not readily identified in the published literature.

Gorilla tourism has often been interrupted due to emergencies, including insecurity or risk of disease. Contingency plans for such cases could involve alternative sources of funding to communities affected by technical risk mitigation (Litchfield, 2008). In the first instance, support could be drawn from trust funds or emergency assistance that conservation agencies or international donors such as the United Nations Educational, Scientific and Cultural Organization (UNESCO) have specifically earmarked for gorilla conservation (UNESCO World Heritage Convention, 2020; see Case Study 6.2). In addition, sustainable funding mechanisms could receive contributions sourced as premiums or levies from tourism (Litchfield, 2008).

Longer-term risk reduction measures could focus on diversification of the local economy. Local economies that depend on more than one income stream have the potential to reduce their exposure to disaster impacts and provide ongoing support to

TABLE 6.1

Treatment Options for Mitigating Disaster Risks to Apes

Risk treatment option	Description
Avoid the risk	Decide not to proceed with a measure that is likely to generate or involve unacceptable risk.
Reduce the likelihood	Reduce the likelihood of a harmful impact, e.g., through translocation, early warning or vaccination.
Reduce the consequences	Reduce the consequences of a harmful impact, such as by spreading a susceptible population across multiple habitats.
Transfer the risk	Transfer the risk to another party, such as captive care centers or zoos, to share or bear the risk.
Retain the risk	Retain the risk by accepting the level of risk and planning to manage its consequence, for instance using post-impact triaging.

the local ape population. In Uganda, for example, Conservation through Public Health (CTPH) adopted a diversification approach in providing support during the COVID-19 pandemic. In the absence of gorilla tourism revenue, CTPH worked with affected communities to develop a coffee consortium and to distribute rapidly growing seedlings to encourage the cultivation and sale of produce (Guyson, 2021; see Case Study 2.1).

The UN Environment Programme, working with governments and private partners, has supported local communities in efforts to expand their economic base beyond tourism in ways that benefit both the communities and the natural environment (Refisch, 2021). Local communities that receive monetary and non-monetary benefits from the sustainable management of forests and wildlife are more likely to support and enable related conservation efforts (Junker *et al.*, 2017).

The Disaster Management Continuum with a Focus on Apes

The continuum of prevention, preparedness, response and recovery phases can be used to address disaster risks to apes. The Uganda Greater Virunga Transboundary Collaboration COVID-19 risk register for mountain gorillas (*Gorilla beringei beringei*) has made such use of the continuum phases (GVTC, 2020; see Case Study 6.2). This section discusses each phase in turn.

Prevention

Preventive measures are designed to reduce the impact of natural and anthropogenic hazards—or to avert disasters by enhancing the resilience and reducing the vulnerabil-

ity of communities and their environments. Prevention provides the opportunity to apply risk treatments well ahead of any impact (see Table 6.1). While preventive efforts tend to be cost-, time- and labor-intensive, they are generally less expensive than response and recovery (Cusick, 2019; European Commission, n.d.). Moving a village to reduce or remove a flood risk, for instance, is certain to require considerable time, effort and expenditure. Over time, however, the return on investment pays for the outlay many times over, and related benefits exceed those of response and recovery measures (Cusick, 2019).

The scientific literature provides few accounts of significant prevention measures undertaken to protect wild apes from the impacts of natural disasters, although efforts to prevent the spread of diseases such as COVID-19 have been documented (see Case Study 6.2). While the emergency response plan for the Hainan gibbon (*Nomascus hainanus*) is principally a preparedness measure, its desired outcome is prevention. The plan is intended for activation in response to predicted typhoons that could threaten this very small population of critically endangered gibbons (Bryant and Turvey, 2017).

Other prevention plans have been tried and tested. In 2017, when Hurricane Harvey struck the Houston Zoo in Texas, the facility had already activated its emergency plan. For four days, staff kept thousands of animals safe in spacious indoor enclosures, which were equipped with food, medicine and an emergency electricity supply (Airhart, 2018).

The long-standing practice of creating conservation areas is aimed at reducing the likelihood of impact from anthropogenic hazards such as forest fires, industrial incidents, dam failures, landslides associated with construction, and conflict situations. Buffer zones around ape conservation areas further decrease the risks of harmful effects.

These areas also need to be sufficiently large and to harbor enough biodiversity to support the recovery of ape populations after an anthropogenic disaster, as food and shelter may initially be in short supply.

To make up for these shortfalls, apes may need to range over a larger area than they did before a disaster. Conversely, access to a greater area reduces the likelihood that apes will face hazard-induced shortages of food and shelter. A conservation area's size and ability to meet the needs of apes can also influence the impacts of natural hazard, such as hurricanes, typhoons, lightning-caused forest fires, flooding and earthquakes. The larger the conservation area, the lower is the likelihood that a single hazard would be able to impact the entire area and its ape populations. As noted above, larger areas offer more opportunities to find scarce foods and shelter in and around a hazard-affected landscape (Behie *et al.*, 2019).

Photo: The long-standing practice of creating conservation areas is aimed at reducing the likelihood of impact from anthropogenic hazards such as forest fires, industrial incidents, dam failures, landslides associated with construction, and conflict situations. © Jabruson (www.jabruson.photoshelter.com)

One method used in preventing forest fires is risk mapping, which identifies at-risk areas and thus enables the implementation of targeted prevention measures. Risk mapping is used in the forests of Central Kalimantan, Indonesia, where human-made fires have caused widespread damage to the ecosystem and biodiversity, as well as to human health and the livelihoods of local people (Lestari and Puspita Ayu, 2020).

Risk mapping is also a common practice in forest fire-prone areas of Australia, where it is used to inform the development and deployment of prevention measures (NSW Rural Fire Service, n.d.-a). Such measures include developing asset protection zones around high-value areas, thinning understory plants and other fine forest fuels, planting low-fire-risk vegetation, placing buffers around forest fuels that are responsible for expanding high-intensity fire behavior in high-risk areas, developing a network of fire trails to form compartments for fire management and providing community fire safety education (Building Code & Bushfire Solutions, n.d.; NSW Rural Fire Service, n.d.-b; SCS, 2017). Like other risk management tools, risk mapping requires sustained effort to ensure the information it provides is accurate and relevant. The integration of risk mapping with seasonal fire forecasting may help to identify areas of higher risk in any given season (Spessa *et al.*, 2015; Sumarga, 2017).

Another tool used to identify fire-risk distribution and inform fire management is the fire hotspot distribution model, which was described as critical to the success of the peatland restoration program in Indonesia (Sumarga, 2017). Disaster risk reduction efforts also make use of hazard analysis, examples of which can be found on the UN Space-based Information for Disaster Management and Emergency Response (UN-SPIDER) Knowledge Portal (UNOOSA, n.d.). The portal presents information on

> Preparedness enables the identification of measures to avoid, mitigate and respond during a disaster.

locations that are at risk of landslides, for example (Cozannet, 2007).

In the case of disease outbreaks such as COVID-19, which can have direct impacts on ape mortality and morbidity, prevention is the most important strategy that protected area authorities and other stakeholders can take (see Case Study 6.3). Many of the risks associated with a disease outbreak, including reputational risk to a sanctuary or conservation authority, are also critical in managing potential impacts (see Annex V; PCI, 2022).

In some large conservation areas, risk mitigation measures include the controversial use of engineered fences. While such fences are not widely used to protect apes, they have had deleterious impacts on other wildlife and ecosystems, such as by disrupting non-target species' movement patterns, isolating populations and exacerbating mortality linked to entanglement in fences. Many have been taken down or modified because they were acting as barriers to wildlife (McInturff *et al.*, 2020).

In other areas, fences are used to reduce the risk of some anthropogenic hazards, including disease and invasive species that present a threat to target wildlife. In Australia, for instance, some engineered fences are credited with the successful protection of target wildlife species that had been disappearing from their natural habitat due to predation by and competition from introduced feral animals, such as cats, foxes and rabbits. Fences are best used where the fence perimeter is clear of trees that may otherwise fall over it and provide a way for animals to get to the other side. A fence thus needs to be at the edge of the forested area or have a wide buffer free of trees. Ongoing management, including maintenance and patrols by people and technology, such as cameras and other imagery, support the ongoing integrity of a fence (BCT, 2020; Long and Robley, 2004).

COVID-19 and Mountain Gorillas

In efforts to protect apes from the infectious disease caused by SARS-CoV-2, COVID-19, prevention is the key risk mitigation activity. On that front, action by various stakeholders is essential, as highlighted in the stringent measures of the International Union for Conservation of Nature (IUCN) best practice guidelines for health monitoring and disease control in great ape populations.[5] The stakeholders run the gamut from ape researchers, academics, veterinarians and other health personnel to decision-makers and staff in the ape-based tourism and conservation sectors, all of whom can lead in their areas of responsibility.

The development of control and coordination frameworks for this wide variety of stakeholders has enabled the prevention of outbreaks, the effective use of resources and consistency across emergency response and preparedness activities, while also providing a platform for sharing lessons learned. Typically, government takes the lead in initiating and developing such frameworks.

In March 2020, IUCN's Section on Great Apes and Wildlife Health Specialist Group issued a joint statement on great apes and COVID-19, recommending that visits by humans be reduced to the minimum needed to continue the monitoring of ape safety and health (IUCN, 2020b; UNEP, 2020). The development of risk mitigation measures to prevent exposure of apes to COVID-19 began around the same time (Gillespie and Leendertz, 2020; Reid, 2020; Trivedy, 2020).

Initially, all ape-related tourism was halted in the range states of mountain gorillas (*Gorilla beringei beringei*)—the Democratic Republic of Congo, Rwanda and Uganda. Some restrictions were lifted beginning in October 2020, when Uganda reopened its doors to tourists (ATTA, 2020; Guyson, 2021; see Case Study 6.2). In Bwindi Impenetrable National Park, the training of 130 Uganda Wildlife Authority rangers helped to keep the virus away from the gorillas and enabled monitoring for illness. Additional measures required ape researchers to quarantine for up to 14 days prior to any contact with the gorillas, and all visitors and staff to wear masks, keep a safe distance from the apes, and apply best practice guidelines during visits and health monitoring activities (UNESCO, 2020).

Meanwhile, conservationists worked with local people to support livelihood activities that did not require entry into the forest. They provided goats to reduce the need for hunting and support for growing cash crops, while discouraging apes from entering the areas populated by people (Gibbons, 2020). To prevent the exposure of apes to COVID-19, one practitioner proposed postponing all fieldwork until a vaccine could be secured or the pandemic ended (Reid, 2020). Following this advice could potentially have created negative outcomes for ape conservation and poor social and economic outcomes for the local human populations working for the parks, the researchers and the tourism sector (Trivedy, 2020).

Before the pandemic, research found that more than 98% of the tourist groups visiting the mountain gorillas in Bwindi Impenetrable National Park in Uganda got closer than the recommended 7 meters (Weber, Kalema-Zikusoka and Stevens, 2020; see Chapter 3). Compliance checks for adherence to the biosecurity requirements can help indicate where action needs to be taken to avoid risks.

Businesses and groups associated with ape conservation can validate their compliance with the stated biosecurity guidelines by becoming accredited or certified by a third party. The Wildlife Friendly Enterprise Network, in close partnership with the International Gorilla Conservation Programme—a coalition of Conservation International, Fauna and Flora International and the World Wildlife Fund—is piloting species-focused Gorilla Friendly™ tourism and product certification (IGCP, n.d.; WFEN, n.d.). The program uses best practice based on IUCN guidelines and up-to-date expert advice. The model may also be applicable to other ape species and their environments.

Preparedness

Preparedness is defined as the measures and actions taken for and by a community and community partners prior to an impact by a hazard, ensuring a timely and effective response to hazard impacts. Preparedness enables the identification of measures to avoid, mitigate and respond during a disaster. Many key stakeholder organizations with an interest in disaster management have a guide or manual for preparedness (AIDR, 2020; European Commission, 2021; UNHCR, 2015; WHO, 2017a). The actions underpinning effective preparedness include:

- identification of risks and completion of risk assessments for each risk;
- development of treatments to mitigate risks, starting with the ones that have a high risk rating;
- development of warning systems to alert communities and responders;

■ development of command and control arrangements for a response;

■ development and validation of information systems to support timely decision-making, resource deployment and communication to the local community and identified stakeholders;

■ development of documentation, including policies, procedures and contingency plans that identify who is responsible for what, where and when;

■ acquisition and, if necessary, stockpiling of resources, including stores, personnel, equipment and facilities;

■ rehearsal and exercises involving personnel and the community; and

■ updating through monitoring and evaluation of each activity in response to changes in context, including with respect to risks, resources and lessons learned regarding responses and exercises (AIDR, 2020; European Commission, 2021; Nelson *et al.*, 2007; UNHCR, 2015; WHO, 2017a).

These preparedness actions are generally carried out sequentially. Once underway, however, the process becomes iterative and can move seamlessly between the activities, with completion of some actions dependent on progress in others.

Preparedness regarding at-risk apes in captive or natural settings involves clarity on who is responsible for what actions to protect the apes, including any evacuations or translocations, and who has the authority to make decisions on requisite resources and support actions (Beck *et al.*, 2007). Trigger points—either in time or as specified events—contribute to a shared understanding of who is to do what at which moment. Fire drills and other types of training are part of preparedness, as exemplified at the Oakland Zoo in the US state of California (Airhart, 2018).

Preparedness allows for a planned, timely, structured and systematic response to an imminent or actual impact by a natural or anthropogenic hazard, rather than a reactive approach to an unfolding, known or potentially dangerous situation. A key overall outcome is to ensure that any impacted local community is resilient and thus better able to cope with a disaster. Resilient communities are characterized by:

■ their awareness of the hazards and risks that affect their local area, and of actions they can take to prepare for and mitigate these risks;

the actions they have taken to anticipate disasters following a hazard impact and to protect their social, built, economic and natural environments before, during and after a hazard impact; and

■ their understanding of the arrangements for recovery assistance (Royal Commission into National Natural Disaster Arrangements, 2020b).

A natural or anthropogenic hazard that overwhelms a local community's ability to cope can trigger an emergency and escalate to a disaster. In such situations, communities are forced to rely on the provision of planned and coordinated support and resources from outside.

The lead responsibility for preparedness may rest with a particular local authority, such as a fire authority or land manager in the case of forest fires. The lead authority directs preparedness efforts for the community and other stakeholders, including businesses and relevant government agencies that support the lead agency. For all hazards, this iterative process engages the community, harnesses local knowledge and builds commitment to the preparedness outcomes (Dunlop *et al.*, 2016; Nelson *et al.*, 2007; Redshaw *et al.*, 2017).

Under exceptional circumstances, a single lead agency may need to manage the majority of preparedness activities for a specified hazard impact. Preparedness based on the involvement of many government and non-government agencies and organizations is the preferred approach, however. A multiagency preparedness approach brings together all the agencies with a vested interest in mitigating given hazard impacts to access the knowledge and expertise that is needed to address the full range of risks. While a fire authority focuses on operational and technical risks, it is unlikely to be able to address social, economic or environmental risks, such as the management

of apes. Other agencies can take responsibility for mitigating these risks in support of the lead authority.

As discussed in greater detail below, the five key elements underpinning good preparedness are:

■ capacity and capability (all resource categories are fit for purpose and in the right place at the right time, including response-specific training);

■ documentation (plans, agreements, manuals, guides, policies and procedures);

■ governance (oversight, leadership and management arrangements including command control coordination);

■ management systems (such as those for warnings, operations, resource tracking, finance and health and safety); and

■ usage (drills, exercises, rehearsal and practice) (Cooper, 2018).

Collectively, these preparedness elements work to deliver a timely, structured and systematic mobilization of resources, including personnel and stores, as well as information management during a response. Effective information management, for instance, is critical to facilitating decision-making, resource utilization and timely communication. Each of the five elements is applicable to a wide range of disasters. A resource management system that helps to tackle forest fires, for instance, is equally applicable to flooding (Cooper, 2018).

The preparedness elements are comparable to the key elements of public health emergencies and are closely aligned with preparedness activities for animal health emergencies in North America, which include:

■ education and training;

■ exercises or simulations;

■ monitoring and surveillance;

■ networks of key personnel and roles; and

■ response plan development and enhancement (Bowman and Arnoldi, 1999; Nelson *et al.*, 2007).

Preparedness is undertaken at the local, regional, national and international levels. The risks are different at each level and thus necessitate different treatment measures. For a preparedness review of Case Study 6.1, see Annex VI.

Capacity and Capability

The capacity and capability to respond to an emergency in a timely manner depends on available skills, structures and resources—including personnel, equipment, facilities, services and transport (Nelson *et al.*, 2007).

Capacity building involves developing community resources and knowledge with the corresponding skills, enhancing social relations within the community, and forging links between policy and the community (Quijano *et al.*, 2016). These outcomes are more likely to be achieved when the process is led from within a community. The same principle applies to a community's ability to boost resource levels significantly in a short time, a process known as surge capacity, as local ownership and local knowledge contribute to a community's resilience and independence. Surge capacity may include the use of spontaneous volunteers from local and more distant communities to assist. Preparedness needs to consider the significant logistical challenges that are typically

associated with these volunteers (AIDR, 2017; Daddoust *et al.*, 2021; DHS, 2019).

Community leaders typically have a sound understanding of their context, including any weaknesses in relevant expertise. Preparedness begins with a risk assessment that considers local capabilities, such as the availability of expertise needed to address technical, social and economic risks. It also involves developing and continually updating a local contingency plan that documents any need for technical support and options to address shortfalls. Facilitators from external organizations have an important role to play in supporting local leaders ensure preparedness and guiding them and their community.

A key element of preparedness is the development of adequate non-human resources, such as stores and equipment. Stockpiling enables timely availability of appropriate resources during disasters, when a community may be isolated from the usual supply chains or sources.

Documentation

Documentation developed as part of preparedness is used to inform response and recovery actions. The suite of documentation includes risk assessments, policies, procedures, manuals, guidelines and plans at different levels to address various risks, such as hazard-specific plans, business continuity plans and plans designed to address economic risks, which are usually applied at the regional or national level. A contingency plan is part of the documentation and is not a substitute for the full range of preparedness activities.

A contingency plan is effective so long as it is up-to-date, appropriate, and understood by the relevant community and those who are expected to implement it. It identifies courses of action to be taken by individuals in various roles, the allocation of resources and information-handling pro-

cesses (IFRC, 2021; Nadler, 2019; UNDRR, n.d.-a; WHO, 2018; ZAHP, 2017). Warning activation and levels of alert may be included in the plan. Box 6.4 presents a short guide to the development and contents of a contingency plan. Some of the following contingency plans and similar documents explicitly cover apes:

- COVID-19 Pandemic Guidelines (OVAG, 2020a);
- *Emergency Response Plan for the Hainan Gibbon* (Bryant and Turvey, 2017);
- the Greater Virunga Transboundary Collaboration's "Regional EVD and COVID-19 Contingency Plans for Mountain Gorillas" (GVTC, 2020);
- Orangutan Veterinary Advisory Group Contingency Plan (Appendix 2) for facilities and services associated with orangutans in captivity (OVAG, 2020b);
- Rwanda's *Ebola Virus Disease (EVD) Contingency Plan* (The Republic of Rwanda, 2018);
- "Contingency Planning for All Hazards and Foreign Disease" in *Fowler's Zoo and Wild Animal Medicine Current Therapy* (Nadler, 2019);
- the Food and Agriculture Organization and Network of Aquaculture Centres in Asia-Pacific's "Contingency Planning" (FAO and NACA, 2001);
- the Global Federation of Animal Sanctuaries' webinar on "Contingency Planning for Sanctuaries and Rehabilitation Centers" (GFAS, 2017);
- the Zoo and Aquarium All Hazards Partnership's *Contingency Planning for the Exotic Animal Industry: Workshop* (ZAHP, 2017); and
- the Zoological Best Practices Working Group's *Planning Roadmap: A Basic Guide for Emergency Planners for Managed Wildlife Facilities* (ZBPWG, 2011).

Contingency Plan Development and Ape-Related Contents

The process of developing a contingency plan is as important as the final plan. The development process engages community members and other stakeholders who intend to use the plan as a basis for the response. Features of an effective contingency plan include clarity, operational relevance, feasibility and realistic intended use of resources (WHO, 2018). The development process involves:

- carrying out risk analysis at the appropriate level for the plan;
- identifying risk mitigation measures;
- developing preparedness actions and evaluating risk mitigation measures;
- drafting the plan based on preparedness actions, specifying details of warning services and who is to do what, when and where;
- undertaking evaluation, exercises and reviews of the plan; and
- updating the plan through periodic reviews.

Contingency plans typically contain the following elements, usually in this order:

- an overview or context, including of ape populations at risk, disaster history and specific emergency arrangements, such as legislation and policies;
- an assessment of ape populations and potential hazards at the time of plan drafting;
- an up-to-date assessment of potential hazards and risk assessments;
- a forecast of the most likely outcomes for each risk event (consequences);
- risk mitigation measures for reducing both likelihood and consequences; and
- the assignment of responsibilities for each functional area and detailed requirements with respect to expertise and other resources.

A contingency plan does not include information described in procedures, policies or guidelines. These stand-alone elements underpin response actions and tasks (IFRC, 2021; Nadler, 2019; WHO, 2018; ZAHP, 2017).

The above-mentioned contingency plan to protect gorillas from COVID-19 in the Greater Virunga Landscape of the DRC, Rwanda and Uganda was developed by the Greater Virunga Transboundary Collaboration, with support from Gorilla Doctors, the International Gorilla Conservation Programme, UNESCO, the World Wildlife Fund and Partners In Conservation at the Columbus Zoo and Aquarium in the US

state of Ohio (Gilardi *et al.*, 2022; UNESCO, 2020). The plan aims to "protect mountain gorillas, conservation personnel, tourists and park adjacent communities from SARS CoV-2, the emergent coronavirus that causes the human disease COVID-19" (GVTC, 2020, slide 5). The plan is for both COVID-19 and Ebola virus disease, which was the basis for the initial version of the plan (Gilardi *et al.*, 2022).

In captive or semi-captive settings such as sanctuaries, rehabilitation centers and zoos, the framework for a contingency plan includes the resident animals, their keepers and the facilities. Such plans consider animal behavior, veterinary care and temporary shelter away from the hazard impact area. The most effective procedures are broad enough to cover multiple types of disasters but also specific enough to address unique local characteristics (Quijano *et al.*, 2016; ZBPWG, 2011).

Governance

Clear command, control and coordination structures for an emergency response can support critical decision-making and actions by those with the authority to protect apes. The most common governance structures are based on incident command and control systems and are used across a wide range of emergencies and disasters, including natural disasters, medical emergencies and industrial disasters (AFAC, 2017; FEMA, 2017; HHS, 2012). All agencies and organizations within an emergency response—including those involved with the conservation of apes—fall under a command and control structure to deliver the shared objective through the best use of available resources (AFAC, 2017; FEMA, 2017).

An example of good governance is the structure that underpinned the successful transboundary collaboration between the DRC, Rwanda and Uganda across the

Greater Virunga Landscape, including after the outbreak of COVID-19 (Gilardi *et al.*, 2022; Refisch and Jenson, 2016). The collaboration supported management in the interest of gorilla conservation across a conflict-sensitive landscape.

Good leadership skills are necessary to secure optimum outcomes from a governance structure. People tend to be more responsive to leaders who come from and understand their community (Polygeia, 2016; Toppenberg-Pejcic *et al.*, 2019). If local leadership lacks requisite skills and abilities, it may be necessary to overcome those shortcomings during the preparedness process, to avoid having to recruit an external expert during a response. Research indicates that community leadership plays an important part in building trust with a community and, in so doing, community support and commitment (Sakamoto *et al.*, 2020; Waugh and Liu, 2014).

In 2015 the WHO released a statement on the lessons learned from its Ebola response, which had begun in December 2013 (ReliefWeb, 2015). Subsequent analysis of the response identified governance weaknesses, proposed that local authorities be charged with greater accountability and responsibility, and called for enhanced governance structures around information policy and resource management (Moon *et al.*, 2015; Park, 2022).

Management Systems

Management systems are the arrangements, policies, procedures and structures that are required to manage information during a response. They can come in the form of digital software or simple, manual systems, depending on the context. During responses, such systems are commonly expected to:

- inform the selection and management of strategies, tactics and taskings;

- develop and maintain high levels of shared situational awareness across the numerous stakeholder groups;

- inform critical decision processes at all levels;

- manage information collection, processing, analysis, interpretation and visualization; and

- manage the risks associated with human and organizational factors (Royal Commission into National Natural Disaster Arrangements, 2020b; Sakurai and Murayama, 2019; UNDRR DesInventar Sendai, n.d.).

In addition to capturing, holding, analyzing and interpreting information, including through modeling, management systems can contain and serve as a platform for the use of policies, doctrines, procedures and manuals, as well as information technology (Royal Commission into National Natural Disaster Arrangements, 2020b; Sakurai and Murayama, 2019; UNDRR DesInventar Sendai, n.d.).

The past decade has seen a proliferation of management systems. The recent trend has been to develop systems that can integrate the management of information in a single domain that was previously undertaken by multiple stand-alone products. Disaster managers have also benefited from significant advances in the systematic collection, collation, analysis and sharing of information among stakeholders in the field, community members and social media users.[6]

These management systems involve real-time data mining (extraction), including analysis of social media posts and the real-time use of social media to keep communities up to date (Elichai, 2018; Yin *et al.*, 2012; Zheng *et al.*, 2013). When Hurricane Sandy battered New York in 2012, the mining of social media by emergency services to track damage, warn the public and prioritize actions demonstrated the value of data

> The process of developing a contingency plan is as important as the final plan.

mining (Cohen, 2013; Stewart and Wilson, 2016). Data mining of social media messages became essential to disaster managers by 2017, when it was used to inform response and recovery actions during Hurricane Harvey (Ngamassi *et al.*, 2022).

Systems for the early detection of natural hazard impacts such as tsunamis and forest fires are increasingly being integrated into warning systems for at-risk communities and responders (UNISDR, 2010). An early warning system is defined as:

the set of capacities needed to generate and disseminate timely and meaningful warning information to enable individuals, communi-

ties and organizations threatened by a hazard to prepare and to act appropriately and in sufficient time to reduce the possibility of harm or loss (GDPC, n.d.).

The Global Tsunami Warning System, for example, implements tsunami early warning systems that are used to assess tsunami risk and educate communities about preparedness measures (IOC-UNESCO, n.d.). In the case of forest fires, the earlier the detection, the more readily contingency plan actions can be carried out to reduce the exposure of apes to smoke and heat. Uncrewed forest fire detection systems are designed to sense smoke and heat

signatures, even in remote areas (Dampage *et al.*, 2022).

In Australia, the national bushfire monitoring system Digital Earth Australia Hotspots presents information about hotspots across Australia, Indonesia and Papua New Guinea (Australian Government, 2021). During the 2019–2020 Australian bushfires, integrated automatic and manual systems—including the hotspots system, aircraft and satellite-based imagery of fire behavior, weather activity, and forest fuel loads—informed the distribution of targeted early warnings by social media regarding catastrophic fire danger. One of these early warnings allowed for the translocation of some Australian native wildlife from a sanctuary. At-risk wallabies, bettongs and koalas (*Phascolarctos cinereus*) were captured and moved out of harm's way. Months later, the animals were returned to their recovering habitat (Nobel, Rybicki and Martin, 2020).

There is an absence of examples of early warnings for the presence of disease in apes. The early detection of disease in wildlife increases the chances of successful disease management (Mörner *et al.*, 2002). Establishing preventive surveillance that supports early detection of pests and disease, including zoonoses, is crucial to ape conservation, as is instituting control measures such as biosecurity protocols in ape populations (Guimarães *et al.*, 2020). A global, systematic approach to zoonosis surveillance to support an early warning system and decision-making would assist responders in protecting apes and other species.

During the preparedness phase, the following management systems are typically integrated to facilitate collection, collation and integration between two or more of the systems:

- an operations management system for diverse information sources associated with response operations, such as technical investigations and case management; field observations, including local knowledge, areas impacted; imagery and maps; loss and damage reports and casualties (human and animals); and for actions relating to analyses, such as by wildlife specialists and other experts;

- a records management system to provide a single point of reference for all records;

- a resource management system to manage all resources—including people—throughout each phase, not only during response, but also through prevention and to recovery;

- a health, safety and wellbeing system to meet the legislative requirements and ethical obligations for response personnel and the local human population (see Chapter 5); and

- a financial management system to track the costs of personnel and resources in addition to actual expenditure (Myers and Zrinski, 2022).

Each of these systems has a role during preparedness, response and recovery. During preparedness, all the resources that are likely to be activated and deployed in a response can be added to a resource management system, along with the details about each resource—such as contract details for equipment and contact details, next of kin and qualifications for people.

In addition to managing information about individual at-risk apes and all actions relating to apes, the systems can generate related reports and analysis, such as for wildlife specialists and decision-makers.

The activation of each of these systems is usually subject to policy and procedures. Procedures usually include triggers for activation. Activating the systems early and even preemptively is essential if they are part of an integrated early warning system.

Photo: If ape populations are mismanaged during a response, for example, actions such as translocations can have negative consequences that may necessitate additional recovery actions. Similarly, at the group level, the disruption of an ape family during a rescue can complicate the recovery of individuals or the whole family.
© IAR Indonesia (YIARI)/ MoEF of Indonesia

The value of the information in each system extends beyond response and recovery. The information can be analyzed to inform reviews and debriefs, support the development of major reports and provide a critical resource for peer-reviewed research.

Usage and Exercises

Preparedness entails the routine and regular use of governance arrangements, documentation, management systems and resources, including during everyday activities such as drills, exercises, simulations and small-scale responses. These activities provide opportunities to practice, validate and assess the following elements:

- proposed response measures;
- documentation, such as a contingency plan;
- information systems;
- capacity and capability, including training; and
- the relationships between responders and the community (AIDR, 2017).

To maximize preparedness, exercises can be run at the community, regional and national levels, as well as for each of the stakeholder groups, including ape conservationists, first responders, logistics personnel, technical analysts and communication specialists (Bowman and Arnoldi, 1999). Reviews of exercises and simulations help to identify strengths as well as opportunities for improvements. The WHO's guide to contingency planning includes a section on exercises (WHO, 2018). Various manuals describe the development of exercises that are to be conducted in different contexts (AIDR, 2017; WHO, 2017b).

There is a dearth of information on exercises and simulations regarding impacts of hazards on apes in the wild. Formal reporting on exercises and simulations linked to at-risk apes could help to fill this knowledge gap.

Response

Emergency responses to an imminent or actual hazard impact tend to focus on the immediate and short-term needs of ape populations and neighboring human communities. A situation assessment informs actions to limit (further) damage and meet the medical and wellbeing needs of the ape and human populations. As a response progresses, these plans are continually adapted to the evolving situation.

In late 2018, when forest fires raged in California, some zoos were forced to choose between minimizing resident animals' exposure to the widespread smoke and restricting their freedom to roam. Both options can lead to stress. The Los Angeles Zoo, for one, evacuated small primates and birds to prevent their exposure to smoke from a nearby fire (Airhart, 2018). In such situations, preparedness is crucial, as there is too little time to plan evacuations from scratch.

Evacuation management includes decisions that can be made during preparedness, when there is time and space to consider and develop the best options. The optimal resources to support the best evacuation outcomes for animals can be put in place during preparedness. Clarity of who does what, where and when is best determined during preparedness. Further, preparedness provides an opportunity to test a proposed plan and allow the animals to become comfortable with each aspect of the evacuation.

Responses to a hazard impact can include the transfer of apes to previously identified safe havens. The emergency response plan for Hainan gibbons proposes translocation as an option for some emergency situations

(Bryant and Turvey, 2017). It acknowledges that translocations are not without risks, but that the risks have been successfully managed elsewhere in the past (see Case Study 4.1). Wherever such transfers are considered, effective preparedness covers both apes and humans.

Priority actions in response to a hazard's impact on apes usually include ensuring the animals' safety and access to shelter, water and food. Triage and treatment are likely if apes are injured or unwell due to the effects of forest fire smoke, heatwaves or falling debris. Since resource needs are generally significant and often overwhelm local capacity and capability, resourcing is a key risk to be considered during the preparedness phase.

In response to an infectious disease hazard, including zoonosis, the focus is initially on control and containment actions designed to limit any spread to and within ape populations, and from apes to humans. Early detection—including provisional advice and early response—can deliver the best outcomes (Moon *et al.*, 2015; National Research Council (US), 2001; WHO, 2014). Responses to a disease in apes benefit from well-developed and rehearsed command and control, policies and procedures to support robust and timely decision-making on priorities, and resource allocation.

Many emergency responses attract spontaneous volunteers from near and far, including if the aim is to protect wildlife. The successful management of volunteers can be a key success factor in achieving the best outcomes for apes. Planning for volunteers can be part of preparedness (AIDR, 2017; Daddoust *et al.*, 2021; DHS, 2019).

While recovery is often described as the phase that follows a response, it actually commences during the response phase. Actions taken during a response can affect the extent and delivery of recovery. If ape populations are mismanaged during a response, for example, actions such as trans-

locations can have negative consequences that may necessitate additional recovery actions. Similarly, at the group level, the disruption of an ape family during a rescue can complicate the recovery of individuals or the whole family (Bryant and Turvey, 2017; Palmer, 2018; Sherman, Ancrenaz and Meijaard, 2020).

Recovery

For ape and human communities alike, recovery from the impacts of a hazard typically takes place under conditions that constitute a "new normal." Recovery activities may include the restoration of ape habitat, which can involve planting species that meet apes' shelter and food requirements. If a habitat is or has become prone to forest fires, recovery actions can be designed to support the relocation of apes to lower-risk landscapes, for example, through translocations.

A recovery is successful if it delivers resilience to the affected community, such that they would be able to cope with future hazard impacts and any ensuing disaster. The best outcome of recovery is when a community is no longer at risk from a hazard impact. This recovery outcome was achieved by rebuilding Grantham in the Lockyer Valley on higher, non-flood-prone ground following the catastrophic 2011 floods in southeast Queensland, Australia (QRA, 2011).

Delivering resilience is possible if recovery actions simultaneously seek to address needs in the natural, economic, social and built environments, as the revitalization of each contributes to that of a community. Allocating support for the rehabilitation of the natural environment, for example, generates local employment opportunities that give rise to social and economic benefits. Sustained effort and financial support are required throughout the recovery phase. Post-emergency funding earmarked for the recovery of communities can also help to build support for conservation efforts (Dinsi and Eyebe, 2016).

An example of recovery in action is the return of gorilla-based tourism in Rwanda during the post-conflict period of the late 1990s. After gorilla-based tourism was initiated in 1979, the number of tourists per year grew steadily, reaching 6,900 in 1989, only to plummet during the genocide of 1994 (Maekawa et al., 2013). The recovery of tourism was slow, with just 417 tourists in 1999, yet by 2008 that figure had soared to more than 17,000 (Nielsen and Spenceley, 2010). The strategy developed by Rwanda was key to the ongoing development of the ape-based tourism sector, which by 2013 was the largest foreign exchange earner of the national economy (Maekawa et al., 2013).

> From prevention through to recovery, community engagement is essential to successful outcomes of each phase of disaster management.

Community Engagement in Disaster Management

From prevention through to recovery, community engagement is essential to successful outcomes of each phase. During prevention and preparedness, the process may be driven by lead government agencies and organizations, but it also requires engagement and mobilization from the local community, including residents, community organizations, institutions and businesses (Dunlop et al., 2016; Isakov et al., 2014; Nelson et al., 2007; Redshaw et al., 2017). Community engagement is critical to all phases of disaster management—prevention, response and recovery (Sakamoto et al., 2020; Waugh and Liu, 2014).

Key to community engagement are opportunities for local residents and groups to become involved in protecting their built, natural and social environments (Royal Commission into National Natural Disaster Arrangements, 2020a). Technical advisors and external support personnel can facilitate and support local engagement, so long as they resist the temptation to take over. Indeed, concerns regarding the ownership of outcomes—along with differences in organizational cultures and approaches—can act as barriers to community engagement. Residents may need to air grievances, resolve existing conflicts or simply become familiar with government or agency representatives before they can begin to place their trust in an emergency management system (Dunlop et al., 2016).

Lessons from the response to the Ebola epidemic in 2013–2014 point to serious shortfalls in outreach and engagement strategies (Oosterhoff, Mokuwa and Wilkinson, 2015; ReliefWeb, 2015; The Ebola Gbalo Research Group, 2019; Toppenberg-Pejcic et al., 2019). In Guinea, Liberia and Sierra Leone, authorities displayed a lack of understanding of community culture and social norms, which could otherwise have served as a means of harnessing community participation and support. In practice, they dismissed local procedures for activities such as burials and instead implemented their own (Halter, 2018; Mokuwa and Richards, 2020). Nevertheless, local communities demonstrated their capacity to deliver positive outcomes by blending their cultural understanding with the authorities' expectations (Mokuwa and Richards, 2020; Richards, 2016).

In Sierra Leone, where community care centers were established as part of the Ebola response, authorities suffered from insufficient specialized community engagement expertise as well as concerns around a lack of local ownership, poor coordination and the exclusion of certain communities. These weaknesses affected levels of support and participation in response actions. Given the opportunity to deliver on required outcomes, however, communities did demonstrate their abilities (Oosterhoff, Mokuwa and Wilkinson, 2015).

In a statement on the 2013 Ebola experience and subsequent internal reforms, the WHO itself recognized that a significant obstacle to an effective response had been inadequate engagement with affected communities and families (ReliefWeb, 2015). Subsequent research has confirmed the importance of local communities and that engagement approaches are more effective when they are adjusted to the needs of target communities (The Ebola Gbalo Research Group, 2019; Toppenberg-Pejcic *et al.*, 2019).

Monitoring and Review

Ongoing monitoring and review of ape-related disaster management, which remains in its infancy, can enhance outcomes across all phases in ways that meet the needs and expectations of governments at all levels,

international organizations, local communities and key stakeholders, such as those in conservation. The learnings from monitoring and review are the basis for enhancing future responses to hazard impacts on apes.

Monitoring and review are commonly associated with response, although they are equally applicable to the prevention, preparedness and recovery phases. With reference to prevention and preparedness, monitoring and evaluation questions can include the following:

- Are the preparedness measures delivering specified outcomes to deliver a mission readiness response?
- Are the preparedness measures relevant and appropriate?
- Are the preparedness measures still relevant and can they still be implemented?
- Did the risk assessment process identify the full range of hazards and risks?
- Have there been changes to the hazards and risks?
- How effective was prevention and preparedness in addressing the risks?

Monitoring and review activities can produce any of several outputs:

- reviews at specified intervals, conducted by internal or external reviewers;
- formal or informal investigations and research;
- internal and external audits, such as those required for maintenance of accreditation or maintenance of certification;
- lessons identified from exercises; and
- operational debriefs or after-action reviews.

In turn, these outputs—referred to as "lessons identified"—can inform and underpin future actions in each phase. Under

> " The learnings from monitoring and review are the basis for enhancing future responses to hazard impacts on apes. "

ideal circumstances, the lessons identified are transformed into lessons learned. If the identified lesson is that a community was not adequately consulted or acknowledged during the preparedness phase, that lesson can become learned through the establishment of a structured and systematic consultative process that successfully engages the community.

In its 2015 statement, the WHO reported on the lessons that were identified and learned during the 2013 Ebola response. The statement made clear that the world was ill prepared for a large-scale, sustained disease outbreak (ReliefWeb, 2015). Many of the lessons in that statement and related reports are equally applicable to potential hazard impacts on apes and ape conservation (Moon *et al.*, 2015; Park, 2022). The lessons—which align well with the above-mentioned elements of good preparedness, namely capacity and capability, documentation, governance, management systems, and usage and exercises—include the following:

- A timely and rapid response to outbreaks requires allocated contingency funds.

- Ensuring that the development of information systems is fit for purpose and operational use at short notice. The information system data remain up to date through data integration and report production for logistics, resources, laboratory services and coordination.

- The provision of timely, relevant information on health emergencies is critical in meeting the needs and expectations of different stakeholders, including response decision-makers, the various levels of government, communities and non-government actors.

- Developing expertise in community engagement benefits both the preparedness and the response phases.

- Improvements in governance, such as clarified command and control arrangements, can encourage seamless collaboration between all response levels—from the central coordination to regional offices and operational control. Clarity regarding the assignment of roles, accountability and responsibility at the local, regional and national levels can further enable cooperation.

- Developing local, regional, national and international capacity and capabilities can serve to support a timely, effective response.

During international responses, the lead agency guides global prevention, preparedness and response actions and helps to keep related concerns at the top of national and global agendas (Moon *et al.*, 2015; Park, 2022; ReliefWeb, 2015).

Separately, the Lessons Learned Annex provides useful findings from the literature on emergencies and crises affecting zoos (ZAHN, 2011). Compiled by the Zoo Animal Health Network, the annex features topics such as administration, procedures and communication, as well as detailed lessons on issues including:

- approvals and formal permissions for specific roles to access facilities, use information systems and move through roadblocks;

- command and control, including who is in charge;

- updating of documents such as contact lists;

- expertise requirements for managing media to maintain reputation and image;

- personnel availability and training, including specifically for responses and cross-training across roles and tasks;

- use of protocols;

- record-keeping and management;

- relationships with local authorities; and

- suitability of technology (ZAHN, 2011).

Conclusion

There are opportunities for reducing the impact of natural and anthropogenic hazard impacts on ape populations. The structured and systematic disaster management approach that comprises prevention, preparedness, response and recovery phases can serve as a framework for appropriate planning and action.

Best practice applied in a range of sectors—including One Health, public health, biosecurity emergencies and disaster risk reduction—addresses the full range of strategic risks, as well as the compounding effects across risks, rather than focusing solely on the immediate impacts of a single hazard. In efforts to achieve effective risk management, social, economic and environmental risks have parity with technical and operational risks. The following actions can improve the outcomes for hazard-affected free-living and captive apes:

- prioritizing global and national agenda activities, including research, that can inform best practice for ape populations at risk from hazard impacts;

- mapping global hazard risks for ape populations;

- targeting research to develop an understanding of minimum requirements for apes to survive hazard impacts, including successive impacts from different or the same hazard;

- prioritizing the building of capacity and capability across jurisdictions to support disaster management for at-risk ape populations;

- monitoring and evaluating the occurrence and severity of hazards that affect apes;

Photo: Ongoing monitoring and review of ape-related disaster management, which remains in its infancy, can enhance outcomes across all phases in ways that meet the needs and expectations of governments at all levels, international organizations, local communities and key stakeholders, such as those in conservation. © Roland Seitre/Minden/naturepl.com

- encouraging increased reporting of disaster management activities and creating a global library of documentation for managing hazard impacts on apes;

- ensuring clarity of command and control arrangements at all governance levels in and across jurisdictions;

- fostering high levels of community engagement in areas where apes are at risk of being affected by hazard impacts on the social, built, economic and natural environments, and building on existing local knowledge and practices;

- developing information management systems that support local and global shared situational awareness and critical decision-making;

- establishing alternative funding models for local communities that are exposed to social and economic risks in response to hazard impacts on apes and those communities; and

- setting up contingency funds for dealing with disasters that affect ape populations and can subsequently lead to economic, social and other risks to human communities.

Through collaboration, agencies and organizations involved in conservation and development in ape range states can share their knowledge, skills and documentation to maximize the return on efforts without requiring all stakeholders to follow the same script.

Acknowledgments

Principal authors: Kevin Cooper,[7] Fabian Leendertz[8] and George Omondi[9]

Contributors: Susan Cheyne[10] and Joshua Rukundo[11]

Box 6.1: Kevin Cooper

Box 6.2: Susan Cheyne

Box 6.3: George Omondi

Box 6.4: Kevin Cooper

Case Study 6.1: Joshua Rukundo

Case Study 6.2: Kevin Cooper

Case Study 6.3: Kevin Cooper

Endnotes

1   Alvarez-Berríos and Mitchell Aide (2015); C2ES (2022); Estrada *et al.* (2018); Graham, Matthews and Turner (2016); Lehmann, Korstjens and Dunbar (2010); McBean (2004); Meehl *et al.* (2000); Mirza (2003); Seneviratne *et al.* (2012); Wiederholt and Post (2010); Zhang *et al.* (2019).

2   Unless otherwise indicated, the information presented in Case Study 6.1 is based on the author's knowledge and 12 years' experience working at the Chimpanzee Sanctuary and Wildlife Conservation Trust, including as executive director since 2020, and on internal documents and reports to which he had access.

3   Calvignac-Spencer *et al.* (2012); Harrison *et al.* (2020b); Kilbourn *et al.* (2003); Rwego *et al.* (2008); Santos, Guiraldi and Lucheis (2020).

4   Gillespie (2019); Gillespie and Leendertz (2020); de Haas (2020); Lappan *et al.* (2020); Melin *et al.* (2020); Reid (2020); Santos, Guiraldi and Lucheis (2020).

5   Bales (2020); Gilardi *et al.* (2015); Gillespie (2019); Gillespie and Leendertz (2020); de Haas (2020); Lappan *et al.* (2020); Melin *et al.* (2020); Reid (2020); Santos, Guiraldi and Lucheis (2020); IUCN SSC PSG SGA (n.d.-a).

6   Royal Commission into National Natural Disaster Arrangements (2020b); Beydoun (2018); Ogie *et al.* (2018); Sakurai and Murayama (2019); UNDRR DesInventar Sendai (n.d.).

7   All-hazards emergency management consultant specializing in biosecurity emergencies and forest fires.

8   Helmholtz Institute for One Health (www.helmholtz-hzi.de/en) and Robert Koch Institute (https://www.rki.de).

9   University of Minnesota (https://twin-cities.umn.edu).

10 Borneo Nature Foundation (www.borneonaturefoundation.org).

11 Chimpanzee Sanctuary and Wildlife Conservation Trust (https://ngambaisland.org).

INTRODUCTION

Section 2: The Status and Welfare of Great Apes and Gibbons

I n each volume of *State of the Apes*, Section 2 covers issues that affect the broader in-situ and ex-situ populations of apes globally, complementing the theme explored in Section 1. In this volume, Chapter 7 examines the impacts of industrial development projects on apes living in their natural habitat and updates case studies presented in previous volumes. Chapter 8 offers an overview of activities and initiatives developed to improve captive ape welfare and assesses the status of apes in captivity.

The online Abundance Annex—which is available at www.stateoftheapes.com—presents updated population estimates for in-situ great apes across their ranges, as well as a mix of population and density estimates for gibbons across their ranges. When compared to data provided in the previous volumes in this series, the annex allows for the tracking of population trends and patterns over time.

Chapter Highlights

Chapter 7. The Status of Apes: Impacts of Industrial Development Projects on Apes

This chapter looks at industrial development projects in ape range states, summarizes the potential impacts of these projects on the different ape species based on their

Photo:
© Arif Setiawan, SwaraOwa

socioecology and explores species-specific responses to identified impacts. It also identifies the best mitigation strategies currently available to ensure positive conservation outcomes for apes, including national legislation, international lender requirements, certification schemes and the International Union for the Conservation of Nature's Avoid, Reduce, Restore and Conserve (ARRC) Task Force.

Since the first volume of *State of the Apes* was published, two main factors have led to enhanced mitigation efforts: national and lending standards around biodiversity management have improved, while more research has shed light on how industrial development projects affect apes. Taking these factors into consideration, the chapter updates case studies that appeared in previous volumes, including on the Congo Basin's Sangha Trinational region, the Wildlife Wood Project in Cameroon and the Simandou mine in Guinea.

Chapter 8. The Welfare and Status of Captive Apes

The first part of this chapter aims to measure and build a better understanding of the welfare of captive apes. The second part updates statistics on captive ape populations in zoos, rescue and rehabilitation centers, and sanctuaries around the world.

Animal welfare refers to an animal's quality of life. Welfare issues may be linked to a complex set of factors, including cultural, economic, political, religious and social conditions. As discussed in the chapter, one tool for assessing and managing a captive ape's welfare is the Five Domains Model, which reflects how behavioral interaction, environment, nutrition and physical health contribute to an animal's mental state.

In relation to ape welfare, the chapter explores species-specific requirements, trade, reintroduction, standards of practice, assessments and related tools, such as the Enclo-

Photo:
© Center for Great Apes, Wauchula, FL

sure Design Tool. It also considers examples of shared learning and action to support ape welfare, such as the Gorilla Rehabilitation and Conservation Education Center in the Democratic Republic of Congo.

The second part of the chapter updates the population statistics of captive apes in zoos, rescue and rehabilitation centers, and sanctuaries around the globe. These statistics are not exhaustive: not all countries are covered, and a complete picture of the extent of captivity is difficult to obtain. The available data suggest that the number of captive apes is relatively static, although some individuals have been moved from one type of captive facility to another, and there are notable exceptions. Where there are variations, the reasons for them are not always clear or well understood, largely due to a lack of comprehensive data. Another impediment is a lack of information sharing, not only between zoos, but also with captive animal census databases (such as Species360), zoo associations and studbooks. Poor communication may be linked to capacity issues, language barriers, other priorities, technology or low levels of confidence in collaborative initiatives.

200

CHAPTER 7

Status of Apes: Impacts of Industrial Development Projects on Apes

Introduction

Industrial development projects—defined here as agribusiness, hydropower, infrastructure, logging and mining activities—are present in all ape range countries and are among the greatest threats to their survival (Arcus Foundation, 2014). The direct footprint of a single project may be limited and easily quantifiable; cumulatively, however, these projects have significant impacts on apes and their habitats, especially if their indirect impacts and associated infrastructure are taken into consideration (Arcus Foundation, 2018). Indeed, projects that are intertwined, such as dams that are built to provide electricity for mining companies, contribute to cumulative impacts on ape populations over large areas and extended periods of time.

The global human population growth rate and associated demand for land and natural resources indicate a trend that is likely to worsen and become more unsustainable (Arcus Foundation, 2018; Bologna and Aquino, 2020). While populations need places to grow and thrive, the affluent citizens of the world tend to engage in unsustainable consumption patterns and lifestyles that are contributing to most of the deleterious environmental and social impacts noticeable today (Marques *et al.*, 2019; Wiedmann *et al.*, 2020).

Part of the solution is societal change, including a transition to an economy that meets people's needs without exceeding planetary boundaries (Hickel, 2019). The potential benefits of such a change are promising, but they are not likely to materialize unless the conceptual models and frameworks that underpin this transition consider all possible ramifications. In shifting their reliance from fossil fuels to renewable energy, for example, governments may depend on dams and other "green" projects whose effects on biodiversity can also be devastating. In some cases, such projects drive the sourcing of additional minerals from ape habitats to satisfy new demand, such as the nickel required for the construction of wind turbines (World Bank, 2017).

Overall, many projects of concern are being developed and planned in ape ranges, some of them large-scale. The Belt and Road Initiative, for instance, will impact many protected areas that are home to apes, leading to significant fragmentation of their habitat (Arcus Foundation, 2018; Ng *et al.*, 2020). It is difficult to estimate what proportion of apes may be affected, although it is likely to be high, given the significant overlap between industrial development projects and ape ranges (Sloan *et al.*, 2018; Wich *et al.*, 2014a).

Without measures to mitigate the impact of these projects, the future for apes appears bleak (Sloan *et al.*, 2018; Wich *et al.*, 2014a). National laws and regulations are improving within ape range states, however. Many countries are requiring biodiversity offsets, yet their use for great apes remains controversial (Arcus Foundation, 2018; see Box 7.1). Meanwhile, some banks have improved their lending standards and are taking certain biodiversity risks more seriously, which sometimes prevents the funding of projects that could have significant impacts on biodiversity (WWF, 2018). The public is also increasingly putting pressure on companies to provide products that are sourced ethically and sustainably (Viciunaite and Alfnes, 2020; Zhang and Zhu, 2019). Over the past few years, many certification schemes and auditing systems have been created or expanded to include a biodiversity component (IRMA, 2018; ResponsibleSteel, 2022; Tayleur *et al.*, 2017). In addition, an increasing number of large-scale industrial projects, such as mining projects, are developing ape-specific mitigation measures and action plans (IUCN, 2014; IUCN SSC PSG, 2020b).

Nevertheless, much more needs to be accomplished to ensure the long-term survival of all ape species. Stakeholders still lack a basic understanding of the long-term impacts of different types of industry on apes, and of the effectiveness of mitigation measures (Junker *et al.*, 2020). Furthermore, many industrial development projects occur in remote locations, far from external scrutiny, rendering independent monitoring difficult (Arcus Foundation, 2015).

Another factor to consider in developing mitigation strategies is that industrial development projects have different effects across ape taxa, in line with each taxon's distinctive ecological requirements, social systems, demographics and ranging patterns (see the Apes Overview). The cultural behaviors of each species—and the personality of each individual ape—can also influence responses to impacts, making it difficult

to develop uniform mitigation strategies (Morgan *et al.*, 2018; Pederson, King and Landau, 2005).

Ape mitigation measures are more likely to be effective if they are tailored to the specific species, subpopulations and unique context of a project site, and if they take into consideration the interconnectedness of people, wildlife and their shared environments—including how disturbances in one of these can lead to disturbances in the others (see Chapter 2). Examples of such transmissions include the recent spread of zoonotic diseases linked to human encroachment into wild habitat, and the wildlife trade (Jones *et al.*, 2008; Wilkinson *et al.*, 2018).

Since the first volume of *State of the Apes* was published, two main factors have

led to enhanced mitigation efforts: national and lending standards around biodiversity management have improved, while more research has shed light on how industrial development projects affect apes (Arcus Foundation, 2014; Lindshield *et al.*, 2019). This chapter provides summaries of the potential impacts of these projects on apes, species-specific responses to identified impacts, and the best mitigation strategies currently available to ensure positive conservation outcomes for apes. It also presents an update on development projects examined in the different volumes of the *State of the Apes* series. As discussed in Case Study 7.1, 7.2 and 7.3, some of these proceeded as planned, while others have been halted or changed ownership.

Photo: Wherever several industrial development projects occur in a single ape habitat, their cumulative impacts hinder ape population connectivity and represent a serious threat to their long-term viability. © HUTAN-Kinabatangan Orang-utan Conservation Project

The key findings from this chapter are:

- The number of industrial development projects in ape habitat is significant and likely to grow in tandem with the global demand for infrastructure, technology and energy.

- Wherever several industrial development projects occur in a single ape habitat, their cumulative impacts hinder ape population connectivity and represent a serious threat to their long-term viability.

- The impacts of industrial development differ across ape subgroups and species, pointing to a need for mitigation measures that are tailored to individual populations or species.

- Various mitigation strategies aim to minimize the impacts of industrial development projects on biodiversity, but few specifically address impacts on apes.

- By facilitating the exchange of advice, the Avoid, Reduce, Restore and Conserve (ARRC) Task Force of the International Union for the Conservation of Nature (IUCN) Species Survival Commission may help to bridge the gap between stakeholders of industrial development projects, on the one hand, and primatologists and conservationists, on the other.

- More long-term research studies are needed to assess the impacts of industrial development projects on apes; in turn, the findings can be used to improve mitigation efforts.

Ape Behavioral and Ecological Characteristics

The five ape taxa differ in their type of social systems, their diet and other behavioral and ecological characteristics. These traits and features can influence how they might respond to impacts from industrial development projects, and which mitigation approaches are most effective. Table 7.1 summarizes these characteristics for each ape taxon; for more details, see the Apes Overview.

Variations also exist across different species and subspecies, especially when they occur in different habitat types (Furuichi, 2009; Moore *et al.*, 2017). By taking into account both site-specific ecological conditions and apes' behavioral characteristics, survey planning and mitigation measures can better protect and respond to species- and habitat-specific needs. Chimpanzees, for example, are found at greater density in forested areas than in forest–savanna mosaic habitats, where they make use of larger territory to access sufficient resources (Lindshield *et al.*, 2021). A survey area may thus need to be larger in a forest–savanna environment than in forest regions. Regardless of the habitat type, however, a survey that extends beyond a development project's physical boundaries is more likely to identify all the areas of ape territory that overlap with the project area.

Some of the ape taxa occur sympatrically, or in the same geographic area; such is the case for gorillas and chimpanzees, as well as orangutans and gibbons (see the Apes Overview). Wherever these species overlap, they have developed mechanisms to avoid each other and cohabit within the same landscape (Basabose and Yamagiwa, 2002; Marshall, Cannon and Leighton, 2009). With further reduction of their habitat, however, that coexistence might be put under pressure, as feeding competition can increase given that the different taxa depend on some of the same resources. Furthermore, the killing of gorillas by chimpanzees was recently documented for the first time and could increase with further disturbances to their habitat (Southern, Deschner and Pika, 2021).

TABLE 7.1

Behavioral and Ecological Characteristics of Apes, by Taxon

Characteristics	Bonobo	Chimpanzee	Gibbon	Gorilla	Orangutan
Number of species	1	1	20	2	3
Countries within their range	1	21	10	9	2
Habitat types	Forest; forest–savanna mosaic	Forest; forest–savanna mosaic	Forest	Forest; forest–savanna mosaic	Forest
Locomotion	Terrestrial	Terrestrial	Arboreal	Terrestrial	Arboreal
Diet	Generalist; frugivore	Generalist; frugivore	Generalist; frugivore	Generalist; frugivore/herbivore	Generalist; frugivore
Territoriality	No	Yes	Yes	No	No
Social organization	Community; multi-male–multi-female fission–fusion	Community; multi-male–multi-female fission–fusion	Groups; adult pair and offspring	Groups; variable	Most frequently mother–offspring social unit

Source: Apes Overview

Overall, apes are effective flagship species: they are charismatic, live mainly in pristine habitat, range over large areas and are good seed dispersers. Owing to their sensitivity to habitat disturbance, most ape species also serve as a protective umbrella for equally or less sensitive species (see the Apes Overview). The conservation of ape habitat may thus benefit a wider range of biodiversity and restricted-range species.

Industries Prevalent in Ape Ranges

Industrial development projects are present in all ape ranges, where their numbers have grown since the publication of the first volume of *State of the Apes* (Arcus Foundation, 2014; UNGA, 2019). The representation of different industrial sectors varies across ranges, reflecting factors such as the histori-

cal and political context, the local geological formations and the presence of major rivers (Arcus Foundation, 2014, 2015, 2018). A detailed understanding of these projects can help conservationists and other practitioners anticipate where major threats lie for each taxon and genus; it can also assist them in determining where urgent action is required. Hydroelectric power dams, for instance, are absent in the bonobo range, but they are a significant threat to gibbons, as 165 dams are planned or under construction in their range (Arcus Foundation, 2018). This knowledge could help prioritize the development of appropriate mitigation strategies to tackle impacts of dams on gibbons, whereas no such action would be required for bonobos.

To provide an indication of the risks to each of the five ape taxa, Annex VII ranks the anticipated prevalence of agribusiness, dams, infrastructure, logging and mining projects in their ranges over the 2020–25 period. These various industrial projects impinge on apes at different spatial and temporal scales. For example, logging normally occurs temporarily in one area over several months, which can give habitats a chance to recover. In contrast, mining activities tend to be sustained in the same areas for decades, and dam projects flood areas permanently (Arcus Foundation, 2014, 2015, 2018).

Impacts on Apes

Types of Impact

The negative effects of industrial projects on apes can be grouped into three categories: 1) direct; 2) indirect; and 3) cumulative (Arcus Foundation, 2014). Direct impacts are associated with the project's activities, or its ecological "footprint," which may involve removing all the local vegetation to build access roads and camps, for example. These impacts are usually the easiest to anticipate as they are related to project-specific components.

More difficult to predict—but usually more significant—are indirect impacts, which are "by-products" triggered by the presence of a project. These impacts usually extend beyond a project's physical boundary (Arcus Foundation, 2014). Project managers typically attempt to defer responsibility for managing indirect effects to regional government authorities who act at a larger scale. One type of indirect impact is an increase in the population of villages in or close to a project area, the result of in-migration of people seeking work. Such influxes usually lead to higher pressure on natural resources and exacerbate existing threats to apes (Arcus Foundation, 2014). Another indirect impact is easier access to remote areas, typically following project-related road construction or rehabilitation. Consequences include the creation or expansion of hunting

grounds and greater interaction between humans and apes, which increases the risk of disease transmission (see Chapter 1).

Cumulative impacts are the successive, incremental and combined direct and indirect effects of several development projects on the same landscape. These impacts can extend over a large area, potentially straddling borders if the projects are based in more than one country (Arcus Foundation, 2014). The need to operate within different countries' legal and regulatory frameworks can complicate the management of cumulative impacts.

Ape Responses to Impacts

As mentioned above, apes respond to impacts according to their socioecological characteristics (see the Apes Overview). Since chimpanzees are more terrestrial than arboreal gibbons, for example, they can more easily travel on the ground between different forest fragments. When development projects cause disturbances or destruction of habitat that force groups of territorial apes to flee into neighboring territories, intergroup conflicts can arise, sometimes with fatal outcomes (Boesch *et al.*, 2008; Mitani, Watts and Amsler, 2010; Morgan and Sanz, 2007). While several studies have investigated how apes respond to varying impacts from different industries, many knowledge gaps remain, as discussed in the section on long-term research and monitoring (see below).

Mitigation Approaches

This section outlines mitigation strategies that are designed to minimize impacts of industrial development projects on biodiversity, including apes. While international lenders, governments, non-governmental organizations (NGOs) and companies tend to refer to these strategies as "best practice,"

they do attract criticism, especially with respect to their implementation and monitoring (Evans, Wingard and Humle, 2021).

Lender standards, for example, have come under fire owing to a lack of empirical evidence that they are achieving conservation objectives over long time scales. This absence of data is linked to a lack of standardized evaluation criteria, limited project monitoring, under-reporting of failing projects and a general dearth of project information (Kormos *et al.*, 2014). Certification schemes have also attracted criticism (Morgans *et al.*, 2018). Opportunities for strengthening mitigation strategies thus abound, especially with respect to tackling impacts on apes.

The Mitigation Hierarchy

Many industrial project developers are using a framework known as the mitigation hierarchy to guide them in reducing negative impacts on biodiversity (BBOP, 2013; CSBI and TBC, 2015). Implementation of the mitigation hierarchy is increasingly required by lenders and voluntarily adopted by companies (de Silva *et al.*, 2019). It is also being incorporated into national legislation in many ape range states (Evans, Wingard and Humle, 2021; GIBOP, 2019).

The mitigation hierarchy can be applied throughout the lifecycle of a project. It promotes an iterative application of the following four sequential steps (see Figure 7.1 and Annex VIII):

- **Avoidance**. This first step is the most crucial and most effective for minimizing negative impacts on ape populations and their habitats (Phalan *et al.*, 2018). It is applied most rigorously to the highest-priority biodiversity features, such as apes. Effective avoidance measures are based on robust baseline data and implemented from the design stage

FIGURE 7.1

The Mitigation Hierarchy

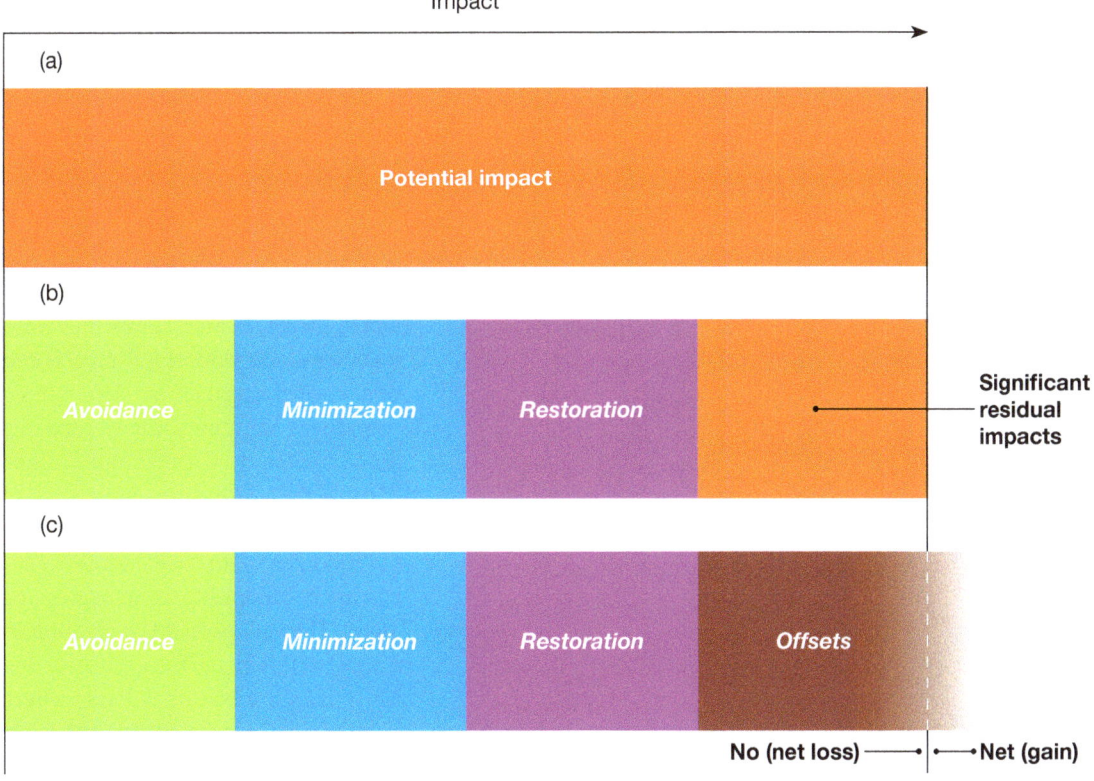

Notes: This diagram shows the application of the four steps of the mitigation hierarchy—avoidance, minimization, restoration and offsets—to reduce a project's impacts. In bar (a), all potential project impacts are identified and estimated. In bar (b), avoidance, minimization and restoration measures have been applied and the project's residual impacts can be quantified. In bar (c), offsets are used to compensate for residual impacts to reach no net loss for natural habitat and to secure a net gain for critical habitat.

Source: CSBI and TBC (2015)

of a project, such as through careful spatial or temporal siting of infrastructure or disturbance away from ape habitat (Arcus Foundation, 2018; CSBI and TBC, 2015). One example of avoidance is the re-rerouting of the Nigerian Cross River superhighway to avoid the Cross River National Park, which harbors important populations of the critically endangered Cross River gorilla (*Gorilla gorilla diehli*) and the endangered Nigeria–Cameroon chimpanzee (*Pan troglodytes ellioti*). The buffer area around the highway was reduced from 19 km to 140 m (Cannon, 2017).

■ **Minimization**. Taken on site, minimization measures aim to reduce the duration, intensity or extent of impacts that cannot be completely avoided (CSBI and TBC, 2015). Effective minimization can eliminate some negative impacts. To reduce the fragmentation effect of roads, for example, canopy bridges can be installed to help maintain a minimum of habitat connectivity for arboreal species (Birot *et al.*, 2020; Linden *et al.*, 2020). Minimization measures require regular updating and adaptation to specific contexts. During the COVID-19 pandemic, for instance, additional guidance was developed to minimize the risk of disease transmission to apes (IUCN SSC PSG, 2020a). Questions remain as to the effectiveness of many minimization measures, including noise reduction efforts, given the poor understanding of what noise levels are tolerable for apes (Arcus Foundation, 2014).

- **Rehabilitation/restoration**. The third step in the hierarchy involves taking on-site measures to improve degraded ecosystems or to reestablish the structure and function of lost ecosystems following exposure to impacts that could not be completely avoided or minimized (CSBI and TBC, 2015). Restoration has the specific goal of returning an area to its prior state or one that is similar; in contrast, rehabilitation aims to restore basic ecological functions or ecosystem services, for example through the planting of exotic trees to stabilize bare soil or the establishment of a lake to provide a recreational facility. The best way to increase the chances of restoration success and decrease costs is to begin restoration trials as early as possible in the first stages of a project. It is often impossible to restore ape habitats to their original state given their ecological complexity and the long time frame over which they were created. As a consequence, the success of rehabilitation and restoration efforts is not ensured, highlighting the need to place more effort on avoiding impacts from the outset (Maron *et al.*, 2012).

- **Offsetting**. This mechanism is designed to compensate for any remaining negative (or residual) impacts following the application of the first three steps of the mitigation hierarchy. The aim is to attain no net loss or to support additional conservation actions to reach a net gain (CSBI and TBC, 2015). Following best practice guidelines, offsetting is required for all projects impacting great apes,

but it is meant to be applied as a last resort, as it triggers moral and ethical considerations (Kormos *et al.*, 2014; see Box 7.1). Offsets are usually off-site measures that fall into two main categories: restoration offsets aim to rehabilitate or restore degraded habitat, and averted loss offsets are intended to reduce or prevent expected biodiversity loss, such as habitat degradation. As offsets are often complex and expensive, the earlier steps in the mitigation hierarchy are preferable. In some cases, companies work together to compensate for impacts and develop aggregated offsets. Two mining companies in Guinea, the Compagnie des Bauxites de Guinée and Guinea Alumina Corporation, have caused residual impacts that affect the western chimpanzee (*Pan troglodytes verus*) and are collaborating on an aggregated offset. They supported the creation of a new national park in Guinea, the Moyen-Bafing National Park, where approximately 5,000 chimpanzees can be safeguarded (A.P.E.S. Wiki Team, 2019a). To be successful, aggregated offsets of this scale and complexity require the formation of multiple partnerships between government entities and NGOs (Maddox *et al.*, 2019).

National Legislation

Levels of national protection for apes vary across range states. A review of the legislation relating to apes in 17 of the 31 range states showed that three countries—Cambodia, the Lao People's Democratic Republic and Viet Nam—do not provide any domestic legal protection to their ape species. This legal gap affects the southern yellow-cheeked crested gibbon (*Nomascus gabriellae*), northern yellow-cheeked crested gibbon (*Nomascus annamensis*) and the southern white-cheeked crested gibbon (*Nomascus siki*) (Rodriguez *et al.*, 2019).

BOX 7.1

Ethical Considerations in the Mitigation of Impacts on Apes

The Ethical Importance of Apes

Apes matter ethically both as individuals and as members of collectives. As individuals, they have capacities such as sentience, self-awareness and sociality that are widely thought to underpin high moral importance (Arcus Foundation, 2020). Their complex cognitive abilities allow them to create unique and valuable local and population-wide cultures (Boesch *et al.*, 2020; Kühl *et al.*, 2019). Ape collectives, such as populations and species, may be seen as having intrinsic value, understood as value that is independent of their usefulness to humans. They also have high ecological value, for instance as seed dispersers; in this sense, they play an important role in maintaining healthy and productive ecosystems that support ecosystem services on which humans and other species depend (Chancellor, Rundus and Nyandwi, 2017; Haurez *et al.*, 2015; McConkey *et al.*, 2018).

Such values may be in tension with each other, however, and they can also be weighed differently, depending on which ethical approaches are taken. First, ethical decisions relating to apes vary based on whether the focus is on individuals or a collective body, such as a population or species; in the latter case, individual interests are given less importance. Second, if assessments prioritize the human-use (extrinsic) value of apes—such as their role in safeguarding ecosystem services—then the development of agricultural or mining projects in ape habitat can be presented as serving human interests better than the conservation of that land. Third, approaches that utilize net value or net harms as tools for making ethical decisions may allow certain harms to be inflicted on individuals, as long as those harms are mitigated or accounted for. In contrast, approaches that grant rights to individuals rule out certain harms, even if such harms could, according to a net-benefit view, be remedied elsewhere (Karlsson and Edvardsson Björnberg, 2021). As discussed below, the distinction between approaches that utilize net benefits versus rights is key in the context of mitigating impacts of industrial development projects on ape habitats.

The net-value approach. The mitigation hierarchy described in this chapter is an influential form of the net-value (or net-benefit) approach. It provides practical guidance designed

Photo: Apes matter ethically both as individuals and as members of collectives. As individuals, they have capacities such as sentience, self-awareness and sociality that are widely thought to underpin high moral importance.
© Martha Robbins/MPI-EVAN

to produce the best possible outcomes wherever development projects affect biodiversity (CSBI and TBC, 2015). While the mitigation hierarchy considers the general value of biodiversity, it fails to incorporate other ethical considerations that are relevant to mitigating impacts on apes, including:

■ any suffering (physical or psychological), other changes in welfare and the death of individual apes;

■ the loss or preservation of ape cultures and their benefit to humans instrumentally (e.g., for research) and/or intrinsically; and

■ the loss or preservation of ape populations or species, in terms of their intrinsic and extrinsic values (that is, their contribution to ecosystem health and services, as well as to biodiversity).

The rights approach. As indicated above, this approach maintains that apes—both as individuals and as collectives— have values that should not be sacrificed, even if doing so could yield a net gain or be carried out without net negative impacts (Karlsson and Edvardsson Björnberg, 2021). Prominent examples of this view defend the rights of individual apes based on their sentience and complex natures. This position is most clearly expressed in the World Declaration on Great Apes, which stipulates that great apes have a right to life, a right to live freely in their habitat and a right not to be subjected to intense physical or psychological pain (Great Ape Project, n.d.). An ape rights approach would parallel a human rights approach, requiring that all development projects respect rights and that industry decision-makers avoid infringing on them.

Since the rights approach requires any proposed development project to respect basic individual ape rights, it is more restrictive than the net-value approach. Nevertheless, industry groups may favor the rights approach, as it creates clear and distinct limits that plainly dictate moral decisions. In contrast, more time, money and resources may be required to make the predictive calculations necessary for the net-value approach.

Ethical Considerations in the Mitigation of Impacts

Regardless of whether a net-value or a rights approach is adopted, avoidance—the first of the four stages in the mitigation hierarchy—is most effective in terms of preventing negative impacts, value loss and rights violations. But what about the other three stages of the mitigation hierarchy? This discussion examines two of these: minimization (stage 2) and offsetting (stage 4). An assessment of rehabilitation/restoration (stage 3) is beyond the scope of this box, as research findings remain inconclusive. Some studies emphasize its ineffectiveness and high costs, while others highlight its successes, suggesting that results may be dependent on local factors (Guy, Curnoe and Banks, 2014; Wilson et al., 2014a).

Minimization (stage 2 of the mitigation strategy). The net-value approach classifies the minimization of impacts as ethically acceptable so long as a development project does not lead to an overall net loss in biodiversity. If a logging project were to be established in ape habitat, for example, it would need to meet two requirements to avoid a net loss. First, apes would need to be protected from any imminent danger, such as impacts from new roads, pollution and noise disturbance, through appropriate mitigation. Second, apes would need sufficient habitat to maintain their social cohesion and cultural behaviors. Project developers would need to undertake dedicated actions to fulfill these requirements, since even if the apes had enough remaining habitat to survive, habitat fragmentation and isolation could threaten their ability to disperse and to find resources, which would reduce genetic flow among populations (Inoue et al., 2013). As apes are key seed dispersers, loss of their habitat may also impact overall ecosystem health. To avoid net biodiversity loss, the available habitat would thus need to be expanded outside the project area or connected to neighboring ape habitat. Either method would be demanding, both technically and financially.

Evaluating the ethical acceptability of this proposed logging development from a rights approach is more difficult. Even if the hypothetical logging project team puts in place mitigation measures to protect apes from serious harm and death, the development nevertheless compromises the apes' right to "live freely in their habitat." Similarly, if the logged forest is within a group's home range and the developers respond by expanding other habitat areas to maintain the amount of available habitat, the group's territory would still be lost. That loss could lead to increased intergroup competition, among other risks (Boesch et al., 2008). From this perspective, removing even a small portion of forest can disproportionally jeopardize and harm the freedom of particular individuals or groups. It is thus very unlikely that such a development would be ethically permissible based on the rights approach.

Another minimization strategy is the translocation of a population from a destroyed habitat to another area. From a net-value approach, translocation is acceptable so long as the number of individuals in a healthy habitat remains unchanged from the start to the completion of a project. In practice, however, the strategy involves a series of risks. The capture, transport and release of primates can have negative impacts on the physical and mental health of individuals; social disruptions within groups and among new neighboring groups can have long-term negative social effects; and new predators and unexpected ecological disruption can come into play (Kavanagh and Caldecott, 2013). A translocation project thus requires extensive planning and stable financial sources to cover high costs (Fischer and Lindenmayer, 2000). Net-value assessments of completed translocations do not consider the social and emotional impacts on individuals, however. From a rights approach, translocation is highly problematic, as individuals lose their right to live freely in their original habitat, lose autonomy in the translocation process, become physically stressed and face significant risks in adapting to a new location.

Offsetting (stage 4 of the mitigation strategy). In view of apes' critical role in maintaining ecological integrity, mitiga-

Photo: The capture, transport and release of primates can have negative impacts on the physical and mental health of individuals, and social disruptions within groups and among new neighboring groups can have long-term negative social effects. © IAR Indonesia (YIARI) / MoEF in Indonesia

tion strategies call for like-for-like replacements as opposed to general biodiversity offsets. In other words, offsets must involve species and habitat types that are identical or equivalent to the ones that are lost (Bull *et al.*, 2013; Ives and Bekessy, 2015). If, for example, a logging project is expected to have residual impacts on an ape population, including significant habitat degradation, an offset strategy might entail improving habitat quality for other ape populations, either locally or farther afield. Legal arrangements could be made to provide their habitat with a higher conservation status or to reduce existing threats, for instance (Bull *et al.*, 2013; Maseyk *et al.*, 2021).

In principle, such offsetting could be acceptable from a net-value approach, but it would be difficult and potentially unachievable. First, it may not be possible to compensate for the suffering inflicted on individual apes in the degraded habitat. If the degradation is so severe that the original population dies out, that population's genetic diversity and cul-

tural uniqueness would be permanently lost and could not be offset. Second, an offset for any individuals lost must result in population growth of the protected offset population that equals or exceeds not just its own projected population levels, but also those of the lost individuals. Additional protections for apes in the offset population would also be needed in relation to any expected anthropogenic threats to that population. Moreover, it would not be ethically acceptable to interpret the prevention of harms in the offset community as a green light for additional harms to apes elsewhere.

From the rights-based perspective, none of these offset options would be acceptable under any circumstances. Negative residual impacts that result in significant habitat degradation risk violating apes' freedom from the infliction of intense physical and psychological pain, their right to continue to live freely in their original habitat and, potentially, their right to life. The benefits to other habitats and individual apes would be irrelevant.

In addition to providing legal protections for individual species, countries are now beginning to enact legislation designed to compensate for adverse impacts of development projects on biodiversity, once the mitigation hierarchy has been fully applied to avoid and minimize potential impacts, and to rehabilitate or restore disturbed ecosystems (Evans, Wingard and Humle, 2021; GIBOP, 2019). The number of countries with government policies on biodiversity offsets has nearly doubled in the past 15 years. More than 100 countries now have, are developing or are starting to discuss national policies that require, encourage, guide or enable the use of offsets. The Global Inventory of Biodiversity Offset Policies compiles information on the status, scope and implementation of biodiversity compensation policies, including offset policies globally. Its database reviews and scores the national environmental legislation and policy developments of 197 countries. Average scores for ape ranges show that the majority of range states have some form of legislation related to offsetting (GIBOP, 2019; see Table 7.2).

In Guinea and Uganda, the Conservation, Mitigation and Biodiversity Offsets in Africa (COMBO) program is working to mainstream biodiversity conservation into national development policy and practice. Launched in 2016, the initiative aims to reconcile African economic development with conservation efforts by collaborating with governments, developers and industry actors to expand and improve the application of the mitigation hierarchy (WCS, n.d.-b).

International Lender Requirements

Many multilateral lenders now require the recipients of their investments to adhere to environmental and social frameworks to ensure the sustainability of projects in their portfolio (Mendez and Houghton, 2020). Nevertheless, the weakening of some lender standards has raised concerns regarding increasing risks to biodiversity in general and to apes in particular, notably with respect to the financing of high-risk infrastructure projects (Arcus Foundation, 2018). This section

TABLE 7.2

Overall Status of National Legislation Surrounding Offset Policies in Ape Range Countries

Family	Genus	Overall status of provisions for biodiversity compensation and/or offsets
Great apes	Bonobos and chimpanzees (*Pan*)	Voluntary
	Gorillas (*Gorilla*)	Voluntary
	Orangutans (*Pongo*)	Required
Gibbons	*Hoolock*	Required
	Hylobates	Voluntary
	Nomascus	Voluntary
	Siamang (*Symphalangus*)	Required

Note: The table indicates the overall status of legislation for the 32 countries that contain ape ranges: Angola, Bangladesh, Brunei, Burundi, Cambodia, Cameroon, Central African Republic, China, Democratic Republic of Congo, Equatorial Guinea, Gabon, Ghana, Guinea, Guinea-Bissau, India, Indonesia, Ivory Coast, Lao People's Democratic Republic, Liberia, Malaysia, Mali, Myanmar, Nigeria, Republic of Congo, Rwanda, Senegal, Sierra Leone, South Sudan, Tanzania, Thailand, Uganda and Viet Nam (see the Apes Overview).

Source: GIBOP (2019)

considers some examples of rigorous frameworks to mitigate risks to biodiversity and offers related resources.

The International Finance Corporation

The International Finance Corporation (IFC) is part of the World Bank Group and a large source of multilateral, private-sector funding. As part of its lender requirements, IFC has eight performance standards that define a client's responsibility for managing social and environmental risks. Performance Standard 6 on Biodiversity Conservation and Sustainable Management of Living Natural Resources represents international best practice for biodiversity management (IFC, 2012, 2019). Its objectives are to protect and conserve biodiversity and habitats, maintain benefits from ecosystem services and promote sustainable management of living natural resources (IFC, 2012). The basic targets are a net gain for critical habitats, which are of highest importance for biodiversity conservation, and no net loss for natural habitats, which contain natural ecosystems. IFC's updated guidance note of June 2019 includes a specific reference to great apes:

> Special consideration should be given to great apes (gorillas, orangutans, chimpanzees and bonobos) due to their anthropological significance. Where great apes may potentially occur, the IUCN/Species Survival Commission (SSC) Primate Specialist Group (PSG) Section on Great Apes (SGA) must be consulted as early as possible to assist in the determination of the occurrence of great apes in the project's area of influence. Any area where there are great apes is likely to be treated as critical habitat. Projects in such areas will be acceptable only in exceptional circumstances, and individuals from the IUCN/SSC PSG SGA must be involved in the development of any mitigation strategy (IFC, 2019, p. 21).

The Equator Principles

The Equator Principles are part of a risk management framework adopted by financial institutions to determine, assess and manage environmental and social risks in projects. The set of ten principles is primarily intended to provide a minimum standard for due diligence and monitoring to support responsible risk decision-making. As of October 2022, 137 financial institutions and 38 countries had adopted the Equator Principles. Under these principles, projects are grouped in different risk categories, with higher-risk projects (including those in ape habitats) required to adhere to strict environmental principles. Principles 2, 3 and 4 relate to the environment and, more specifically, biodiversity (Equator Principles, 2020). Based on a project's risk category, the Equator Principles also require projects to demonstrate their compliance with other applicable standards, such as:

- IFC Performance Standards on Environmental and Social Sustainability;
- World Bank Group Environmental, Health and Safety Guidelines; and
- relevant host country laws, regulations and permits that pertain to environmental and social issues (Equator Principles, 2020).

Certification Schemes

In response to consumer pressure, many industries are developing voluntary certification standards to show compliance with environmental best practice (de Silva *et al.*, 2019). On the whole, these schemes provide their own environmental frameworks with which companies must comply to ensure certification continues. Regular, independent third-party auditing checks compliance and conformity with a standard. As none of the existing standards addresses impacts to apes specifically, there is an opportunity to contribute to the development of existing or new certification schemes. Standards and their frameworks are normally linked to a particular sector, such as mining or forestry. The following certification schemes are relevant to industries operating in ape habitats.

The Roundtable on Sustainable Palm Oil and Other Agribusiness Certification

The Roundtable on Sustainable Palm Oil (RSPO) was established in 2001, after a group of European retailers, processors and consumer goods manufacturers became concerned about their public image in connection with news about deforestation in Southeast Asia (Arcus Foundation, 2014). As of December 2022, the RSPO comprised 5,466 members across seven different sectors: banks and investors, consumer goods manufacturers, environmental and nature conservation NGOs, oil palm growers, palm oil processors, retailers, and social and development NGOs (RSPO, n.d.).

In 2018 the RSPO revised its principles and criteria to ensure greater measurability and relevance. The next review is scheduled to take place in 2023. The seventh principle aims to protect, conserve and enhance ecosystems and the environment. It includes a land-clearing criterion whose guiding philosophy is that high conservation value, high carbon stock and peatlands must be protected and are not to be cleared for planting oil palm. Although there is no specific mention of apes in the guidance document, these protected areas normally coincide with ape habitat as they harbor threatened species (RSPO, 2020).

In addition to the RSPO, several other agribusiness certification schemes that operate in ape habitat—such as Fair Trade and the Rainforest Alliance—feature principles relating to biodiversity (Grunert, Hieke

and Wills, 2014). None of these schemes explicitly mentions apes.

Forestry Stewardship Council

The Forestry Stewardship Council (FSC) provides standard setting, trademark assurance and accreditation to companies, organizations and communities interested in responsible

forestry. Since its founding in 1994, FSC has certified more than 2.2 million km² (220 million ha) of forest in 89 countries. It has certified nearly 57,000 km² (5.7 million ha) of forest in seven African ape range states and more than 54,000 km² (5.4 million ha) in eight Asian ones (FSC, n.d.).

FSC certifies forests and forest products that are managed in line with its principles and criteria. Two of its ten principles relate to biodiversity; although they make no direct mention of apes, some of the references to endemic, threatened and endangered species apply. Principle six states that conservation zones, protection areas and connectivity should be established to protect rare and threatened species and their habitats. It also calls for effective measures to manage and

control hunting, fishing, trapping and collecting. Principle nine commits the FSC to maintaining and enhancing high conservation value at the management unit level. This process requires assessment of the presence of endemic, rare and threatened species whose status is significant at the global, regional or national levels. The next steps involve developing strategies to maintain and enhance these values, and then monitoring their effectiveness (FSC, 2019, 2023).

Mining Certification Schemes

Mineral deposits are known to overlap with areas of high biodiversity (Murguía, Bringezu and Schaldach, 2016). In the past 15 years, mining companies have made increasing use of certification schemes to demonstrate they are operating responsibly. Most mining certification schemes develop their standards around three distinct criteria:

- **environmental**, as related to air and water quality, waste management, rehabilitation and biodiversity conservation;
- **social**, with a focus on human and labor rights, health and safety, and community development; and
- **governance**, concerning legal compliance, policy, transparency and ethics.

An analysis of 15 different mining certification schemes showed that of the three, environmental criteria were most frequently excluded from certification scheme principles. Even among schemes that considered the environment, only 60% included guidelines on managing biodiversity risk (Mori Junior, Franks and Ali, 2015).

The IUCN ARRC Task Force

The ARRC (Avoid, Reduce, Restore and Conserve) Task Force was launched in 2016 to tackle the impacts of industrial develop-

ment projects on apes (ARRC Task Force, n.d.; Campbell, 2021). It is part of the IUCN Species Survival Commission, Primate Specialist Group—which comprises the Section on Great Apes and the Section on Small Apes. The ARRC task force draws on the expertise of more than 150 ape experts who are members of the Section on Great Apes and is also affiliated with other IUCN programs (Campbell, 2021). The ARRC acronym represents the task force goal, which is to ensure that development projects follow international best practice standards, including the application of the mitigation hierarchy, to secure positive outcomes for apes (ARRC Task Force, n.d.).

Until recently, the ARRC Task Force provided companies with ad hoc advice on standard compliance. In January 2019, however, the IFC took the unprecedented step of requiring clients to consult with IUCN's Section on Great Apes[1] regarding any projects that could affect great ape habitat (IFC, 2019). A project that seeks funding from the IFC, or from one of the banks aligned with its standards, is thus effectively required to consult with the ARRC Task Force to seek advice on collecting accurate baseline data, developing appropriate mitigation measures and ensuring impacts on great apes are kept to a minimum.

The IFC requirement presents a tremendous opportunity for the great ape conservation community to engage with governments, industry actors and banks to avoid and reduce adverse impacts on great apes and their habitats. It also offers banks and companies operating in those habitats a chance to decrease their reputational risks, including by protecting great apes from harm and obtaining the best available advice on how to mitigate adverse impacts wherever they do occur. In early 2020, the Task Force responded to these favorable circumstances by establishing a Steering Committee comprised of 20 primatologists specialized in the different ape taxa, as well as internal policies and principles to guide its activities (ARRC Task Force, n.d.).

To date, the ARRC Task Force has engaged with over 20 projects at different stages of development. Its activities so far have ranged from providing advice about survey methodologies to advocacy concerning projects that should be avoiding critical ape habitat. This work has led to more in-depth consideration of apes in project areas, increased survey efforts, improved survey methodologies for gathering accurate baseline data, better long-term monitoring frameworks, more effective mitigation measures and improved offset requirements. The Task Force also aims to conduct independent audits of projects of concern for ape conservation, as well as a long-term assessment of impacts of a few key projects on which data are lacking. To maintain transparency, the ARRC Task Force makes all its advice and results of independent assessments publicly available on its website (ARRC Task Force, n.d.). By monitoring and keeping track of the projects with which it engages over the long term, the Task Force will be able to assess how successful it is at influencing decision-making and improving conservation outcomes for apes.

The ARRC Task Force works with governments and partners in ape range states to build national capacity, including improved knowledge of best practice standards for apes and better technical capacity to respond rapidly to projects impacting apes. By encouraging early and continued involvement of conservationists and ape experts, the Task Force aims to improve mitigation and reduce both indirect and direct negative impacts on apes.

Other Resources

Many resources on how to avoid harming ape habitat from the onset, strengthen baseline surveys and devise effective mitigation strategies are freely available to ape experts, conservationists, NGOs, the private sector and lenders. Here are some examples:

- **The IUCN Ape Populations, Environments and Surveys (A.P.E.S.) Database** (iucngreatapes.org/apes-database). Launched in 2007, this database aims to centralize all great ape survey data (Kühl *et al.*, 2007). It is a useful tool for verifying what surveys have been conducted in an area, obtaining an idea of ape density and assessing trends. Data on gibbons are being added to the database.

- **IUCN Best Practice Guidelines** (iucngreatapes.org/best-practice-guidelines). Seven best practice guidelines are relevant to great apes, including two that are directly related to industrial development projects—on logging and the FSC. The other guidelines focus on human–wildlife conflict, population surveys and monitoring, disease, reintroduction and tourism. A new best practice guideline on

The *State of the Apes* series (stateoftheapes.com). All volumes, policy and investor briefings, background papers and videos related to this series can be accessed freely.

mitigating the impacts of industrial development projects on apes is forthcoming.

- **The *State of the Apes* series** (stateoftheapes.com). All volumes, policy and investor briefings, background papers and videos related to this series can be accessed freely on the Arcus Foundation's website. Intended for a broad audience of policy-makers, academics, researchers, NGOs and experts in the conservation field, the series explores the interrelated factors that affect apes' wellbeing, including industrial development in ape range states, as well as measures for averting and mitigating harm to ape populations.

- **The Conservation Evidence Project** (conservationevidence.com). This searchable website provides evidence for conservation interventions, with specific actions collated for primates (Conservation Evidence, n.d.; Junker *et al.*, 2017; Petrovan *et al.*, 2018). Evidence-based assessment of different actions is particularly useful for selecting effective mitigation measures and identifying research gaps (Junker *et al.*, 2020).

Managing Cumulative Impacts

Cumulative impacts are growing as landscapes increasingly host multiple development projects. At the regional and national scales, tools such as cumulative impact assessments and strategic environmental assessments (SEAs) can help to identify and manage these impacts on apes and other priority biodiversity features (IFC, 2013; Sadler *et al.*, 2010). Such assessments are seldom conducted at these levels, however, as few developers want to take on the considerable cost and responsibility of carrying them out (Arcus Foundation, 2014). Complicating matters is a widespread lack of inclusivity in land use planning, when it is undertaken at all.

In the absence of a global vision, development projects can have avoidable impacts on local biodiversity, for instance when mining companies build adjacent roads through a landscape although they could have shared a single one. Such disconnected planning can threaten the long-term survival of apes by compromising connectivity across a landscape, and by missing the opportunity to identify important areas for conservation or avoidance measures before potential impacts can occur. This problem typically reflects poor communication between government departments and conservation organizations. If, for example, a national mining department is not aware of or has not received relevant data, it may not be able to take ape priority sites or protected areas into consideration when evaluating mining or other permit applications. Long legal battles can ensue, as can the degazettement of protected areas or the retraction of permits. By the time permits are rescinded, however, projects tend to have had some level of impact on biodiversity.

The situation is changing as some countries, including Uganda, are making SEAs mandatory (Government of Uganda, 2019). Legally embedded SEAs remain rare in ape range countries, however, and even if they are mandatory, the relevant laws tend to be weak and their enforcement poorly executed (Tshibangu, 2018). These shortcomings highlight the need for stronger legislation and SEA practice that can help ensure assessments are conducted before impacts can take shape. SEAs can be useful tools for anticipating and averting potential impacts, as well as for assigning responsibility for addressing specific impacts, for example to a government department or a company (Arcus Foundation, 2014). Their utility, however, rests in large part on the strength of a country's SEA-related legislation, its efforts to reinforce their implementation and its commitment to preserving conservation areas into the distant future.

Long-Term Research and Monitoring

The main benefits of conducting research are threefold: 1) it facilitates an understanding and the quantification of impacts of industrial development projects on apes; 2) it documents the effectiveness of different mitigation measures; and 3) its long-term monitoring results can inform adaptive management. The more research is carried out on effective mitigation measures, the more insight is acquired. There is growing evidence, for instance, that gibbons are using artificial canopy bridges and that these measures help decrease fragmentation impacts (Chan *et al.*, 2020; Das *et al.*, 2009). Such research allows conservationists to support proposals for the use of canopy bridges or other measures, for example in advice provided to companies. Knowledge gaps remain, however, including with respect to impacts on some ape taxa, such as bonobos.

In practice, the involvement of researchers and ape experts in a project's development tends to be limited to discrete stages or tasks, such as participating in stakeholder consultations or conducting baseline surveys. In contrast, best practice suggests that experts should be involved as early as possible in a project to ensure avoidance measures are appropriately considered (IFC, 2019). Since apes are long-lived species with slow rates of reproduction and impacts may only become apparent over many years, there is also a need for more independent studies to assess the long-term effects of industrial development projects, support mitigation and inform regional management plans. In the Congo Basin, for example, long-term monitoring and research have helped to guide decision-making processes concerning sustainable logging (see Case Study 7.1).

CASE STUDY 7.1

Using Great Ape Research in Assessing Risks to High-Conservation-Value Forest in the Congo Basin

A key benefit of long-term great ape research sites in high-conservation-value forests is the potential impact their outputs can have on decision-making around industrial development projects. In addition to steering companies towards more environmentally sound land use and management practices, research findings can support calls for the avoidance of impacts from the onset of a project.

Since the early 1990s, applied research has been a cornerstone of the conservation mandate of the Republic of Congo's Nouabalé-Ndoki National Park (NNNP), part of the Congo Basin's Sangha Trinational (TNS) region, a UNESCO World Heritage Site (see Figures 7.2 and 7.3). Research activities at three research sites within and bordering the NNNP—the Mbeli Bai Project, the Goualougo Triangle Ape Project and the Mondika field station—have been ongoing for more than 20 years (See Figure 7.3; Estienne, 2022). While much of what has been

FIGURE 7.2

The Congo Basin's Sangha Trinational Region

Sources: Protected areas—UNEP-WCMC (2021a, 2021b, 2021e); country boundaries —GADM (n.d.); other base map detail—OpenStreetMap (n.d., © OpenStreetMap contributors, published under Creative Commons Attribution License CC BY; for more information see http://creativecommons.org)

FIGURE 7.3

TNS Protected Areas Where Road Expansion and Timber Concessions Overlap with Intact Forest Landscapes, 2020

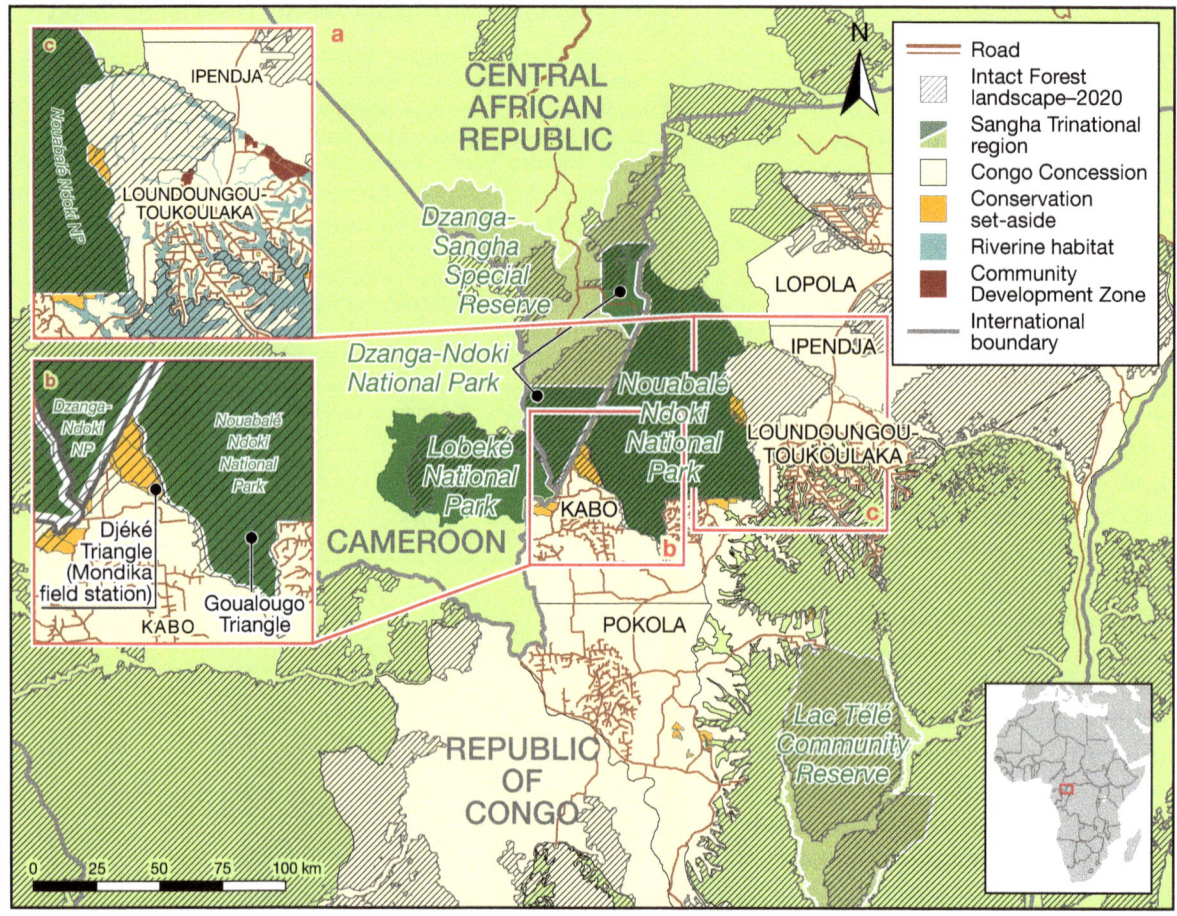

Notes: Located in the Kabo concession, the Djéké Triangle contains intact forest landscapes that are contiguous with the Nouabalé-Ndoki National Park in the Republic of Congo as well as the Ndoki National Park in the Central African Republic. Significant intact forest landscape also remains in the Loundoungou-Toukoulaka concession, including a designated "conservation set-aside" area.

Data sources: Potapov *et al.* (2017); protected areas—UNEP-WCMC (2021a, 2021b, 2021e); country boundaries—GADM (n.d.); other base map detail—OpenStreetMap (n.d., © OpenStreetMap contributors, published under Creative Commons Attribution License CC BY; for more information see http://creativecommons.org)

learned about the behavioral ecology of central chimpanzees (*Pan troglodytes troglodytes*) and western lowland gorillas (*Gorilla gorilla gorilla*) can be attributed to studies at these sites, the research efforts also provide enduring platforms for addressing the most pressing issues facing great apes in the Ndoki landscape and beyond, including threats from logging.

Much of the TNS region still consists of intact forest landscapes (IFLs), defined as large, connected tracts of undisturbed forest. Over time, however, multiple-entry logging has come to typify the landscape. Two main measures can help

to counter the ongoing loss of IFLs and irreplaceable natural attributes, such as large, old-growth trees characteristic of the Congo Basin forests; these measures require urgent application (Bastin *et al.*, 2015; Potapov *et al.*, 2017). One is the creation of permanent conservation set-asides; the other is the adoption of adaptive forest management strategies to reduce forest degradation. With the aim of informing the management strategies in logging concessions neighboring the NNNP, the Goualougo Triangle Ape Project worked with forestry field teams to identify, record and analyze important ape food resources in these areas. In addition to informing assessments of future logging impacts on ape populations, the documentation of high concentrations of food resources in relation to carbon hotspots can feed into climate change scenarios.

Protecting Intact Forest Landscapes

The designation of the NNNP in 1993 was the first step towards ensuring preservation of intact ape habitat in the Ndoki landscape (Estienne, 2022). Shortly thereafter, Wildlife Conservation Society scientists and independent researchers initiated field studies in the Goualougo Triangle and Mondika IFLs, outside of the park's boundaries. The results of their research led to a recognition of the Goualougo Triangle's exceptional conservation value and, in 2003, its annexation to the NNNP. This landmark conservation initiative set a precedent for the use of evidence-based research on apes and forests to inform decision-making on land use and protection (Arcus Foundation, 2014).

Two years later, the Djéké Triangle of the Kabo Forest Management Unit also acquired increased protection, based on the results of long-term gorilla monitoring at the Mondika field station. The Triangle was classified as a "conservation-set-aside" to meet Forest Stewardship Council (FSC) certification criteria, which require forestry companies to spare a percentage of intact forest in leased concessions from timber exploitation for conservation purposes (Morgan et al., 2019; see Figure 7.3).

As beneficial as the set-aside status has been for the Djéké Triangle, it has not entirely eliminated the potential threat of future logging as it lacks the official and long-term protection that a National Park status carries. The recent downgrading of protected areas for resource extraction in other parts of tropical Africa indicates there is no room for complacency with regard to protecting remaining intact forests (Edwards et al., 2014; Qin et al., 2019; Watson et al., 2018).

In addition to presenting opportunities for gorilla research and future tourism, the Djéké Triangle is rich in terms of other wildlife and biodiversity (Harris et al., 2021). In lobbying for the highest level of protected status for the Djéké Triangle, local stakeholders, funding bodies and government agencies are relying on empirical assessments of that biodiversity. One study conducted a standardized comparison of species composition on either side of the NNNP border, in the Triangle (outside the NNNP) and in the more remote forests of the Goualougo Triangle Ape Project (in the NNNP), using video footage from 35 camera traps that served as a "reference of intactness" in the Park. The preliminary results indicate that the two areas harbor a similar number of small to large mammalian species (Morgan and Sanz, 2020).

These findings support arguments for elevating the protected status of the Djéké Triangle. They have been presented to the Nouabalé-Ndoki Foundation, which is responsible for governing the NNNP, as well as the Ministry of Forest Economy of the Republic of Congo.

Transition to Managing Forests and High Conservation Value

Long-term conservation monitoring efforts are urgently needed in the Republic of Congo, not only because roughly 80% of its forests are designated for multi-use resource exploitation, but also because important populations of great apes live outside of the country's protected areas (Strindberg et al., 2018). While outright elevation of unprotected forests to national park status may not be feasible in every context, applied research that assesses IFL loss and identifies high-conservation-value forests can help to underpin ape protection initiatives (Morgan et al., 2020). More specifically, robust scientific data can facilitate the identification of land use options that reduce pressure on IFLs throughout the TNS. This process involves conducting annual updates of the remaining IFLs in the seven logging concessions adjacent to the NNNP and merging these updated data into a large-scale, satellite-based forest integrity index, such as the Forest Intactness Index (Grantham et al., 2020b).

A 2020 concession-scale analysis indicated the loss of 2,014 km², which is equivalent to 19% of remaining IFL, since 2017. This loss of intact forest can largely be attributed to the opening of 2,853 km of logging roads in previously unlogged forests over the past two decades (Morgan et al., 2020). Regular on-the-ground monitoring of the opening and closure of these logging roads can address uncertainties associated with satellite-based accessibility maps, while also verifying that illegal transportation infrastructure is not being built in the concessions. These results support a previous estimate that the remaining IFL in the Republic of Congo will be lost by 2080—unless efforts to halt the decline are implemented immediately and effectively (Potapov et al., 2017).

In this context, the Goualougo Triangle Ape Project is helping to inform the current debate surrounding a FSC Motion, which provides guidance on assessing and protecting IFLs. In particular, the project is linking ape abundance estimates to floral assessments, including those from commercial timber inventories (Morgan et al., 2018). One such assessment concerns the Loundoungou-Toukoulaka concession, which comprises the largest remaining IFL bordering the NNNP (see Figure 7.3). An estimated 1,647 km² (164,700 ha) of intact forest in the concession is contiguous with the national park; however, only 2.2% (128 km²; 12,800 ha) of the IFL is classified as a "conservation set-aside." Meanwhile, 27% of the IFL—core areas of which are floristically unique and merit further study—has been lost since 2017 (Gond et al., 2013; Morgan et al., 2020).

Such research on key features in great ape habitat can inform the high-conservation-value forest approach. Over time, identifying cumulative changes in the quality and quantity of resources can provide a means of establishing thresholds for when a forest can no longer support viable umbrella species such as great apes. This work also provides quantifiable results based on which stakeholders can develop approaches that take ecosystem integrity into account in identifying core IFL (Grantham et al., 2020a; Haurez et al., 2017). It allows for the development of conservation initiatives that explicitly consider the needs of gorillas and chimpanzees in protecting and managing IFLs.

Overall, long-term research projects in the TNS region have collected sufficient data to help stakeholders demonstrate the high conservation value of large areas within the landscape. As a result, protection has been extended to an increasing proportion of ape habitat in the area. In addition, empirical evidence has played a key role in persuading more companies to apply certified sustainable forestry practices to ensure favorable outcomes for apes. Such long-term partnerships among the government, the private sector and researchers can be replicated wherever apes or other umbrella species face a similar demise.

Moving Forward

For conservationists who continue to tackle impacts of industrial development projects on apes, the abovementioned engagements with the private sector point to the following priorities.

- **Bridging the gap between ape conservationists and the private sector**. Poor communication and information sharing between ape conservationists and relevant private sector actors—such as companies and banks—impedes the development and consideration of ape-specific mitigation strategies. The ARRC Task Force can help to bridge the gaps by facilitating and promoting the exchange of information between the sectors.

- **Engaging at the government level**. Ape experts and conservationists usually provide advice on how individual development projects can enhance safeguards for apes. Since such projects are numerous and sited across large areas, however, providing advice at a higher level may be more beneficial. In some cases, site-level mitigation initiatives can be scaled up to inform wider government policies. One example is the Wildlife Wood Project in Cameroon, which the Zoological Society of London launched as a public–private partnership in 2007

(see Case Study 7.2). The project's initial goal was to improve the logging practices of two companies. Today it works with several companies, lobbies the government to standardize logging practices and promotes enhanced sustainability of such projects.

- **Defining ape priority areas**. Identifying and securing priority ape populations is imperative for their long-term protection. Taking this step is among the most effective ways of preventing adverse impacts from the start of a project, so long as priority areas are off limits to development. As indicated above, the avoidance stage of the mitigation hierarchy is key to minimizing adverse impacts on ape populations and their habitats.

- **Improving policy to address impacts during the early stages of a project's operational life**. It is not unusual for companies to change their plans or corporate guidelines, loosen their approach to mitigation once lenders have provided loans, or be sold off to new owners with weaker biodiversity standards (see Case Study 7.3). Following a change of ownership, it can be difficult to require a company to recognize responsibility for impacts attributable to previous management. This problem is common among mining companies which specialize in exploration activities that are poorly regulated. Exploration can last for more than ten years, leading to the destruction and degradation of vast areas and creating roads or access to previously remote areas. After a mine changes ownership, degraded areas tend to serve as a new "baseline," which leads to lower compensation requirements. By linking the avoidance step of the mitigation hierarchy to such a new baseline, new owners can also undermine the potential effectiveness of mitigation measures. These distortions could be prevented if governments and lenders held companies accountable for previous impacts and if policies around exploration were strengthened.

- **Standardizing mitigation metrics**. The most effective mitigation strategies are project- and species-specific; however, consensus among ape conservationists is required to standardize the advice provided, for example with respect to the following questions:

  - What constitutes an adequate baseline?

  - What is the most appropriate way of quantifying residual impacts and determining an appropriate offset to compensate for the loss of apes given a wide range of uncertainties?

  - What is the best way to monitor impacts on apes?

Discussing these questions can help practitioners to arrive at a consensus, which can then be captured in best practice guidelines to inform advice on mitigation strategies.

CASE STUDY 7.2

The Cameroon Wildlife Wood Project: An Update

This case study updates a discussion of the Cameroon-based Wildlife Wood Project (WWP) that was presented in the first volume of *State of the Apes* (Arcus Foundation, 2014, pp. 120–4). Launched in 2007 by the Zoological Society of London (ZSL), the project aims to ensure the persistence of viable wildlife populations through sustainable forest management (ZSL, n.d.).

In the early 2000s, sustainable forestry practices began to spread across Central African countries, and the promotion of environmentally friendly approaches became linked to the prospect of gaining access to the premium market for certified products. Most of the logging concessions in the region—and particularly Cameroon—were European-owned, with progressive, conservation-conscious management.

ZSL began to collaborate with these private-sector entities in Cameroon through its WWP by engaging with key logging partners, undertaking research on best practices and wildlife management methods in logging concessions, agreeing with companies on policies and procedures to put in place, and developing and implementing wildlife management plans alongside staff training. The next stage involved compiling and disseminating lessons learned to all relevant private-sector actors in the country, including those running agroforestry,

FIGURE 7.4

Wildlife Wood Project

Sources: Protected area—UNEP-WCMC (2021a); country boundaries—GADM (n.d.); other base map detail—OpenStreetMap (n.d., © OpenStreetMap contributors, published under Creative Commons Attribution License CC BY; for more information see http://creativecommons.org)

mining and hydroelectric projects. Together with partner logging companies, ZSL also tested a series of technological solutions, such as the surveillance of logging roads with camera traps, as well as real-time acoustic monitoring of gunshots, chainsaw use, and vehicle and motorbike engine noise. In 2016, ZSL released a toolkit featuring guidance and tools to enable the forestry sector to adapt to evolving environmental standards, legal and regulatory frameworks, certification requirements and wildlife protection goals (ZSL, 2016).

Having benefited from more than ten years of direct support from the WWP in implementing wildlife best practices, partner logging companies in Cameroon are taking positive management actions to conserve wildlife in their concessions. Through the implementation of agreed wildlife management plans, they are closing old logging roads; providing workers with competitively priced protein alternatives to wild meat to prevent them from hunting; and adapting their logging activities to minimize impacts on great apes and other local wildlife as much as possible. For Pallisco and Rougier—the two companies with which ZSL was fully involved in wildlife and illegal activity monitoring and management—field-based teams were set up to collect data on a near-permanent basis. Their work was instrumental in the identification of key great ape areas, whose demarcation is used for adaptive management and as a deterrent to illegal activities (Tchakoudeu Kehou, Daïnou and Lagoute, 2021).

After companies in Cameroon, Central African Republic and other neighboring countries expressed interest in the WWP's model of operation, ZSL intended to scale up the project over a wider landscape. Since the 2008 financial crisis, however, the region has favored Asian investors who operate in markets that accept lower-quality wood. These investors have acquired many concessions that were previously owned by European concessionaires, leading to a drop in the number of concessions engaged in certification.

Nevertheless, the WWP has been able to raise awareness, persuade some government representatives to call for the incorporation of wildlife management in the sustainable forest management framework, and ensure that wildlife is recognized as an integral component of forest management standards in Cameroon. Indeed, laws that take wildlife protection in logging concessions into account are being implemented based on the WWP's experience. National forestry and wildlife schools in Mbalmayo and Garoua have included sustainable forest and wildlife management themes as modules for teaching the next generation of conservation leaders.

Based on monitoring data collected in logging concessions in Cameroon, the WWP has shown that illegal exploitation has decreased while the status of great apes and other large and medium-sized mammals seems to have improved. In Pallisco forest concessions, for instance, monitoring teams have documented an increase in the gorilla encounter rate between 2016 and 2019, which indicates that their population is resilient to sustainable logging practices where man-

agement plans are implemented and effective. To date, the WWP's joint work with the Cameroonian government and the logging company Pallisco has been its most successful conservation-focused collaboration, one that has supported economically viable activities.

The WWP model is expanding to two additional logging companies—Alpicam and Cameroon United Forests—both of which own concessions in IUCN great ape priority conservation areas (IUCN, 2014). Alternative schemes, such as ZSL's Sustainability Policy Transparency Toolkit (SPOTT), are also investigating the integration of trackers for sustainable forestry and wildlife management practices (Oppenheimer et al., 2021). Pallisco currently ranks first among all Cameroon-based forest companies on the SPOTT index, which evaluates public disclosure using environmental, social and governance criteria. It is also among the top ten of the 100 timber and pulp producers, processors and traders assessed in the index, demonstrating the positive effects of the WWP model and their commitment to sustainable forestry practices (SPOTT, n.d.).

Forest-dependent communities are sometimes viewed as a hunting threat to local wildlife, including great apes. The WWP sees people as an essential component of the forest ecosystem and recognizes that they need to be part of the solution. The project engages directly with local communities—including those bordering on logging concessions of partner companies—with an eye to empowering them to play a role in managing their natural resources. Through a full process of free, prior and informed consent, the WWP assesses people's willingness to participate in community surveillance networks designed to facilitate information sharing in support of great ape conservation efforts. The project then tracks the networks' progress through regular field visits and community platforms meetings, with the aim of ensuring that the communities develop the capacity required to continue operating over the long term, without (or with minimal) assistance from ZSL. The WWP has also placed emphasis on community behavior change, in particular through educational radio programs on great ape protection and conservation issues.

On the regulatory front, the WWP has provided input into the development of Cameroon's forestry laws and FSC national and regional standards. Next steps include encouraging the Cameroonian government to adopt compulsory best practices in wildlife management and to require their implementation as a prerequisite for the allocation of forest management units, in line with the ZSL toolkit. The WWP also plans to encourage banks and financial institutions that support logging industries to insist that their beneficiaries implement sustainable management practices. ZSL intends to continue working with auditing firms, certification bodies and other platforms to promote best practices in wildlife management.

CASE STUDY 7.3

Change in Mine Ownership: Who Is Responsible for the Long-term Impacts?[2]

Background: The Simandou Mine and Biodiversity

The Simandou massif in the Kankan and Nzérékoré regions of south-eastern Guinea has one of the largest unexploited iron ore deposits in the world. It extends over 110 km from north to south and includes one of the highest peaks in West Africa, the Pic de Fon (1,656 m; see Figure 7.5). This area is critical for biodiversity, harboring highly threatened wildlife species such as the critically endangered western chimpanzee (*Pan troglodytes verus*) and the endangered Diana monkey (*Cercopithecus diana*) (A.P.E.S. Wiki Team, 2019b).

In the late 1990s, Rio Tinto's affiliate Simfer was granted exploration rights at Simandou, initially through four mining exploration licenses covering as many "blocks" (Lewis and Nogueira, 2021). Early on, Simfer developed mitigation plans for Simandou blocks 3 and 4 that included biodiversity targets based on its internal corporate policy and Performance Standard 6 of the IFC, one of the lenders to the project. To guide the development

of mitigation plans, including offsets, Simfer initiated long-term data collection on chimpanzees in the Pic de Fon Classified Forest in the southern blocks starting in 2007. The company also collected chimpanzee survey data along the planned railway between the mine and a port on the Guinean coast, which was to be built in areas of high chimpanzee density. The number of chimpanzees along the planned railway remains unknown, but 2,750 chimpanzee nests were recorded in the rail study area (Kormos *et al.*, 2014; Rio Tinto Simfer S.A., 2012b). Simfer published mitigation plans in its 2012 environmental and social impact assessment (Rio Tinto Simfer S.A., 2012b).

Simfer proposed and committed to mitigation measures for chimpanzees in the southern mining blocks, including controlling hunting, protecting habitat within their range and creating additional habitat for them. Since the mine was expected to impact chimpanzee habitat, Simfer investigated potential options for offset sites to compensate for residual damage (Rio Tinto Simfer S.A., 2012b). The company formed a technical group called the Simandou Biodiversity Offsets Working Group, which brought together representatives from Simfer, Guinea's Ministry of Environment, Water and Forestry, and the non-governmental organization Guinée-Écologie (Kormos *et al.*, 2014).

FIGURE 7.5

Simandou Mine and Transport Routes

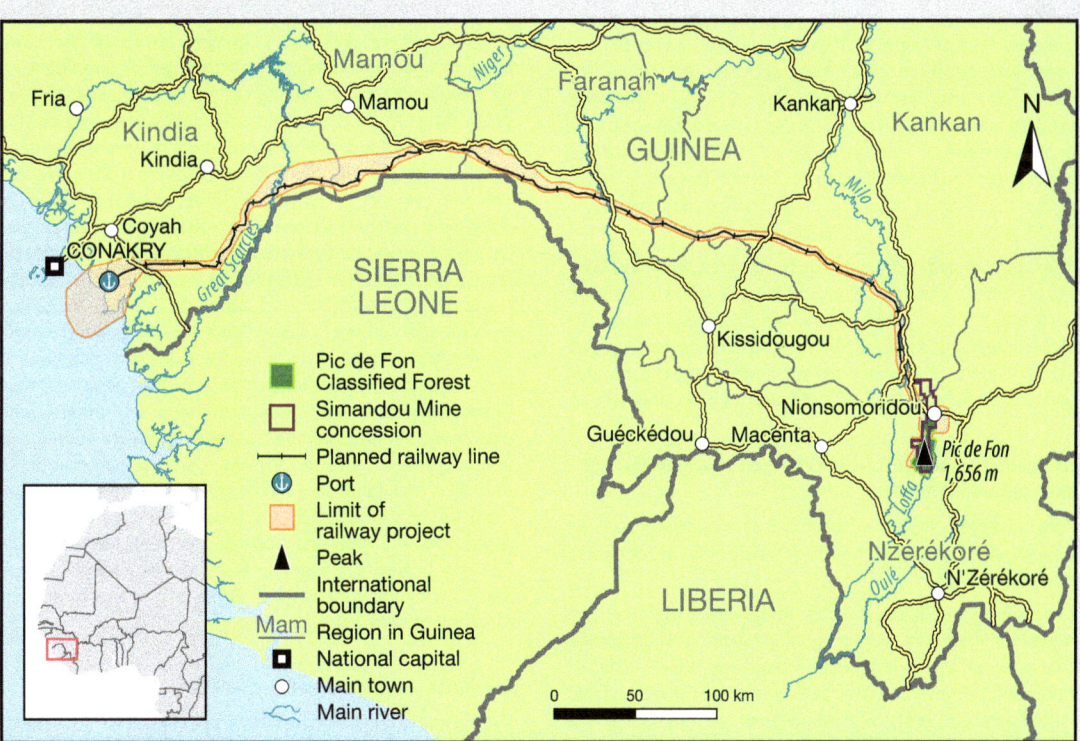

Sources: Rio Tinto (2012a, Figure 1.1); WCS (n.d.-d., Figure 1); country boundaries—GADM (n.d.); other base map detail—OpenStreetMap (n.d., © OpenStreetMap contributors, published under Creative Commons Attribution License CC BY; for more information see http://creativecommons.org)

Divergence of Mitigation Approaches following Change in Mine Ownership

By July 2008, the mining plans were in flux. The government ordered Simfer to retrocede the exploration licenses for Simandou blocks 1 and 2, which lie in the north of the massif. The government awarded the licenses to Beny Steinmetz Group Resources (BSGR) in December 2008. The decision effectively split the Simandou massif into two separate mining projects, both of which would require rail links to the coast to evacuate iron ore. In 2010, BSGR sold 51% of its stake in blocks 1 and 2 to the world's largest iron ore miner, Brazil's Vale (Lewis and Nogueira, 2021). In April 2014, the government of Guinea canceled BSGR and Vale's mining licenses in Simandou. Five years later, the government awarded the northern blocks to the Société Minière de Boké–Winning (SMB–Winning) consortium, which comprised Winning Shipping, a Singaporean maritime firm, United Mining Supply, a Guinean–French logistics company, Shandong Weiqiao, a Chinese aluminum producer, and the government of Guinea.

While iron ore production has not yet started in the Simfer-managed southern blocks (blocks 3 and 4), road building for access and exploration activities has led to some loss of grassland since the early 2000s. Satellite imagery indicates that direct impacts from the footprint of roads and drill pads have been stable in recent years; however, artisanal mining has increased, probably an indirect — or induced — impact that may be linked to an influx of people in search of economic opportunities at the site. This small-scale mining caused some loss of forest cover in the south of the Pic de Fon Classified Forest between 2011 and 2017, as is visible in satellite imagery (see Figure 7.6). No corresponding studies or analyses were conducted for blocks 1 and 2; only recently, in 2020, did the SMB–Winning consortium develop plans to assess the potential impacts of mining in these northern blocks (WCS, n.d.-c).

Management and mitigation of adverse impacts on biodiversity are difficult when a mining project is active. As exemplified here, a hiatus due to a change in ownership can exacerbate these challenges, especially with limited or no presence of personnel to manage the site. If mitigation financing is not provided up front, resources to address impacts may not be available.

Mitigation Issues Linked to Change in Ownership and Development Timeline

This case study highlights a number of mitigation issues for chimpanzees and other wildlife in the context of ownership change.

First, between 2008 and 2019, there was a high risk that impact mitigation would not occur in blocks 1 and 2, as neither national law nor lender or corporate standards required mitigation best practice. The government had not yet updated national policy to align with international standards, such as IFC Performance Standard 6, and companies were not compelled legally to manage their impacts. BSGR and Vale did not produce biodiversity action plans, nor did they establish

mitigation financing for blocks 1 and 2 or the rail line. Neither company published corporate biodiversity no net loss standards between 2008 and 2014.[3] From 2014 to 2019, neither BSGR nor the SMB–Winning consortium had a license for Simandou blocks 1 and 2. During this 11-year period of ownership uncertainty, the risk that biodiversity impacts would not be mitigated was thus high. As it takes time to prepare an environmental and social impact assessment and associated mitigation plans, and because no biodiversity action plan was available for implementation in 2014, biodiversity impacts were likely already occurring and would continue for some time.

Second, even if mitigation plans based on best practices had been in place for the northern blocks and the rail line in 2014, there was an ownership gap of five years between when the government removed the license from BSGR and Vale and when it granted it to the SMB–Winning consortium. Any planned implementation of mitigation measures would have suffered significant delays during that period. In addition to the delays, a change in ownership potentially creates a situation where one company is following one set of standards, while another is operating under a different set. If biodiversity management is not harmonized to reflect best practice across the various blocks and the rail line, the potential exists for indirect and cumulative impacts on critical habitat and chimpanzee populations, despite the company's efforts to adhere to the highest environmental, social and governance (ESG) standards.

Third, current practice — including corporate and lender standards, as well as national policy — does not require financing for mitigation activities during early project phases, such as exploration. In fact, companies often do not even consider the impacts of exploration and only adopt a mitigation hierarchy approach for the design and development of the final project. This gap persists although the critical contribution of early mitigation measures to overall reductions of impacts on biodiversity is well documented and predictable. Examples include the proper siting and mitigation of impacts of new access roads and other linear infrastructures that can facilitate access to previously remote areas. If financing for the mitigation of impacts from exploration had been in place in blocks 1 and 2, then biodiversity risks linked to the ownership gap and delays in the development of mitigation plans could have been addressed. Companies are not likely to secure such funds unless governments include dedicated requirements for financing mechanisms that consider early impacts — especially those that affect highly threatened species such as great apes — in early-stage project licensing agreements. Taking this step is relevant for both mine exploration and for associated infrastructure, such as the proposed rail connecting Simandou to a port.

Solutions: Improved Policy and Green Financing

Typically, the two strongest drivers of enhanced ESG standards are laws and regulations, and, increasingly, lender standards. Guinea and other African countries have integrated relevant improvements in their legislative frameworks

FIGURE 7.6

Satellite imagery showing forest cover change in the Pic de Fon Classified Forest between (a) 2011 and (b) 2017

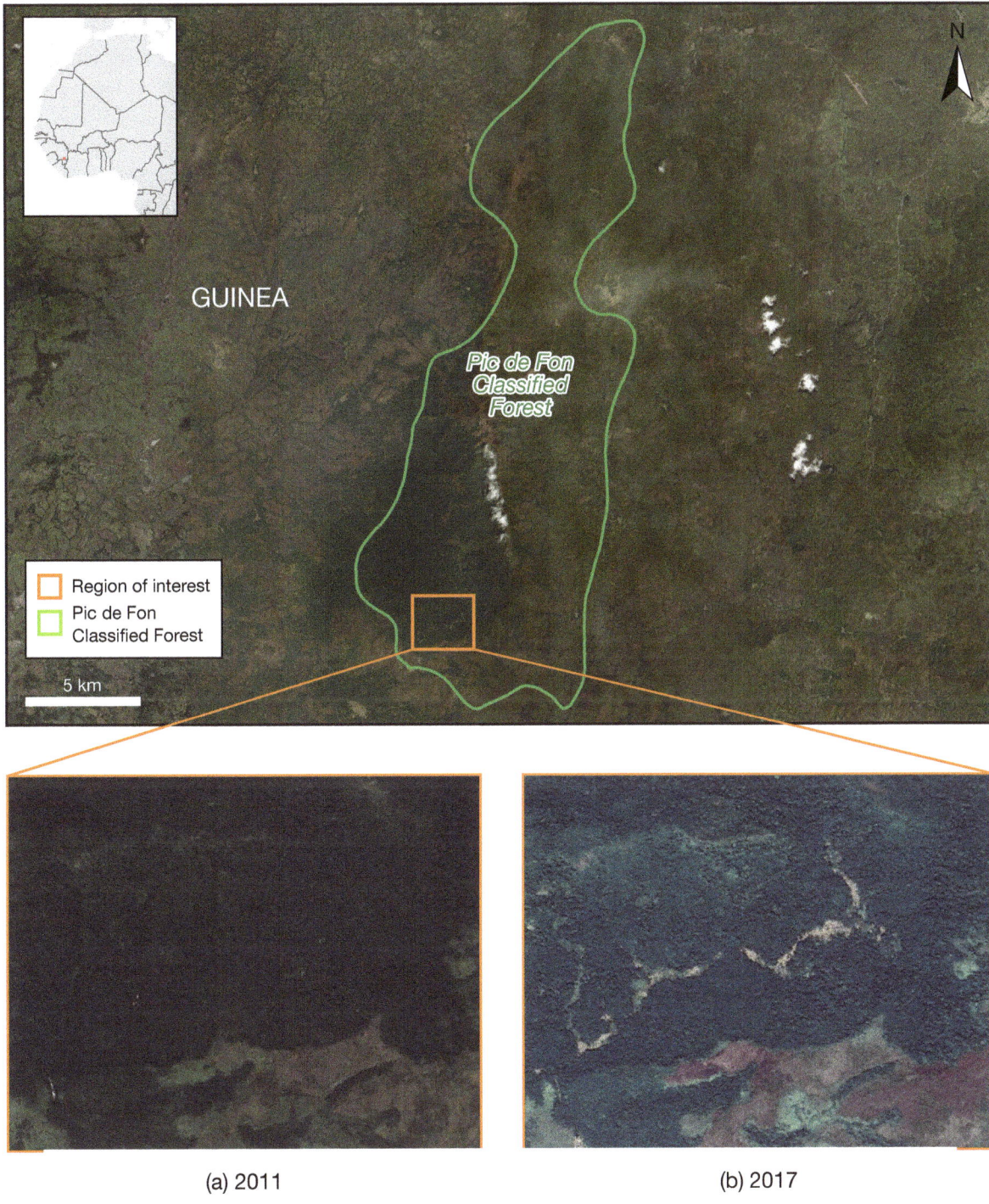

(a) 2011   (b) 2017

Image sources: Top: © 2022 Esri, Maxar, Earthstar Geographics and the GIS User Community. a) © 2022 Maxar Technologies. b) © 2022 CNES / Airbus.
Full figure: © Maegan Fitzgerald

in response to initiatives such as the Conservation, Mitigation and Biodiversity Offsets in Africa (COMBO) project, which is led by the Wildlife Conservation Society, implemented in partnership with Biotope and Guinée Écologie, and funded by the French Development Agency and the French Facility for Global Environment (WCS, n.d.-b). In Guinea, policy on the mitigation hierarchy, including avoidance and offsets, is under development and, once approved, is expected to legally require compliance in places such as Simandou. Such laws, if implemented properly, can ensure that the liabilities for financing mitigation would be transferred to new owners so that long-term benefits to nature and wildlife are not compromised. A third driver of better ESG performance is a government's commitment to granting licenses exclusively to companies with good internal policies and a proven track record of implementing these policies.

New lender standards that focus on reducing biodiversity risk may also form part of a solution. Financial institutions are starting to associate companies that are high carbon emitters or that have weak ESG standards on biodiversity with higher risk. These new standards also consider the extent to which pension funds and re-insurers are divesting from higher-risk companies. As debt financing of mining projects becomes more competitive, companies that do not apply adequate ESG standards may find financing more difficult to secure and more expensive. Chinese institutions are some of the largest lenders for infrastructure in Africa; with evolving national climate policies, they may take on a greater leadership role in addressing climate and biodiversity loss and demanding greater compliance with ESG standards. Greater uptake of these lender and policy requirements can support compliance with conservation commitments, even when project ownership is transferred.

Conclusion

Industrial development projects are likely to increase in number, in line with the demand for resources, infrastructure and energy (Christmann *et al.*, 2022). Fortunately, their impacts on apes are generally taken seriously, as was the case regarding the hydropower dam in the Tapanuli orangutan (*Pongo tapanuliensis*) range in Sumatra (Laurance *et al.*, 2020). Ape-specific mitigation measures continue to be developed, and long-term research provides a better understanding of their effectiveness in terms of minimizing further adverse impacts on apes.

In view of apes' intrinsic and ecological value—as individuals, populations and species—any significant disturbance of their habitat for human development is difficult to justify. The adverse impacts humans have already had on ape populations arguably give rise to an obligation to protect their habitat, both by improving impacted areas and preventing further degradation. Moreover, the 2009 H1N1 (swine flu) and recent COVID-19 pandemics, as well as the risk of zoonotic spillover related to encroachment into ape habitat, call into question whether industry should be allowed to encroach on these ecosystems under any circumstances.

Industrial development is continuing to expand, however, creating a growing need for efforts to reduce its individual and cumulative impacts on apes. The most effective actions are those taken at higher levels, such as through improvements of policy and government regulations. At the local level, the best way to protect the long-term viability of ape populations is through strategic environmental impact assessments and land use planning, which can help identify threats and opportunities for ape conservation before a new development project is launched. In all these approaches, the consideration of apes as an umbrella species can secure better outcomes not only for apes, but also for the ecosystems they help to sustain.

Acknowledgments

Principal author: Genevieve Campbell[4]

Contributors: Dirck Byler,[5] Vanessa Evans,[6] Oliver Fankem,[7] Maegan Fitzgerald,[8] Andrew Fowler,[9] Simon Furnell,[10] Victoria Green,[11] Rebecca Kormos,[12] David Morgan,[13] Clare Palmer,[14] Hugo Rainey,[15] Crickette Sanz,[16] Malcolm Starkey,[17] Ray Victurine[18] and William Winston[19]

Box 7.1: Victoria Green and Clare Palmer

Case Study 7.1: David Morgan, William Winston and Crickette Sanz

Case Study 7.2: Oliver Fankem and Andrew Fowler

Case Study 7.3: Hugo Rainey and Ray Victurine

Endnotes

1 IFC Performance Standard 6, which is quoted above, refers to the IUCN Section on Great Apes and cannot be changed to refer to the ARRC Task Force directly until the next revision of the guidance note. Until then, any request is transferred to the ARRC Task Force.

2 This case study was written by Hugo Rainey and Ray Victurine, based on their experience in Guinea engaging with the mining sector and the government on biodiversity since 2003, including through the COMBO (COnservation, Mitigation and Biodiversity Offsets) program.

3 Vale adopted a corporate policy of no net biodiversity loss in 2019 (Vale, n.d.).

4 Re:wild (www.rewild.org).

5 Re:wild (www.rewild.org).

6 Resolute Mining (www.rml.com.au).

7 Zoological Society of London (www.zsl.org/conservation).

8 Re:wild (www.rewild.org).

9 Zoological Society of London (www.zsl.org/conservation).

10 The Biodiversity Consultancy (www.thebiodiversityconsultancy.com).

11 Texas A&M University (liberalarts.tamu.edu/philosophy).

12 Re:wild (www.rewild.org).

13 Lester E. Fisher Center for the Study and Conservation of Apes, Lincoln Park Zoo (www.lpzoo.org/conservation-science/science-centers/lester-e-fisher-center-for-the-study-and-conservation-of-apes).

14 Texas A&M University (liberalarts.tamu.edu/philosophy).

15 Wildlife Conservation Society (www.wcs.org).

16 Washington University in St Louis (anthropology.wustl.edu).

17 The Biodiversity Consultancy (www.thebiodiversityconsultancy.com).

18 Wildlife Conservation Society (www.wcs.org).

19 Washington University in St. Louis (virtualplanet.wustl.edu).

CHAPTER 8

The Welfare and Status of Captive Apes

Introduction

Apes are kept in various forms of captivity. They can be found in biomedical laboratories; breeder and dealer facilities; entertainment and exhibition contexts such as, circuses, multi-media companies and tourism; private homes; and rescue and rehabilitation centers, sanctuaries and zoos. The movement of apes between these different forms of captivity—for example from a laboratory to a sanctuary—reflects increasing public concern for ape welfare and recognition of ape sentience (Fleury, 2017; Hirata *et al.*, 2020). This chapter focuses on captive apes in rescue and rehabilitation centers, sanctuaries and zoos, hereafter collectively referred to as *captive facilities*, except where further distinction is needed.

This chapter comprises two main sections. Section I focuses on understanding and measuring the welfare of captive apes. Key findings include:

■ Effective ape welfare systems rely on consistently applied governance and operational systems as well as species- and context-specific responses. Partnerships with other organizations can help co-create welfare knowledge, practice and assessment approaches, while also facilitating access to resources.

■ How animal welfare is understood and discussed influences how it is assessed and how the resulting findings are used. Welfare-focused dialogue on neglected topics such as illegal trade and conservation translocation can support both ape welfare and conservation outcomes, particularly if it reflects current legal and scientific thinking—and public opinion—on animal sentience.

■ The welfare of captive apes often falls between the cracks of domestic animal health and wild animal conservation legislation and regulations. Country-level, cross-disciplinary analyses of relevant legal instruments are needed to identify current enforcement gaps and resource requirements. A dedicated convention or agreement could mainstream animal welfare in the global arena.

■ Institutional standards tend to determine what apes experience on a day-to-day basis, influencing their quality of life.

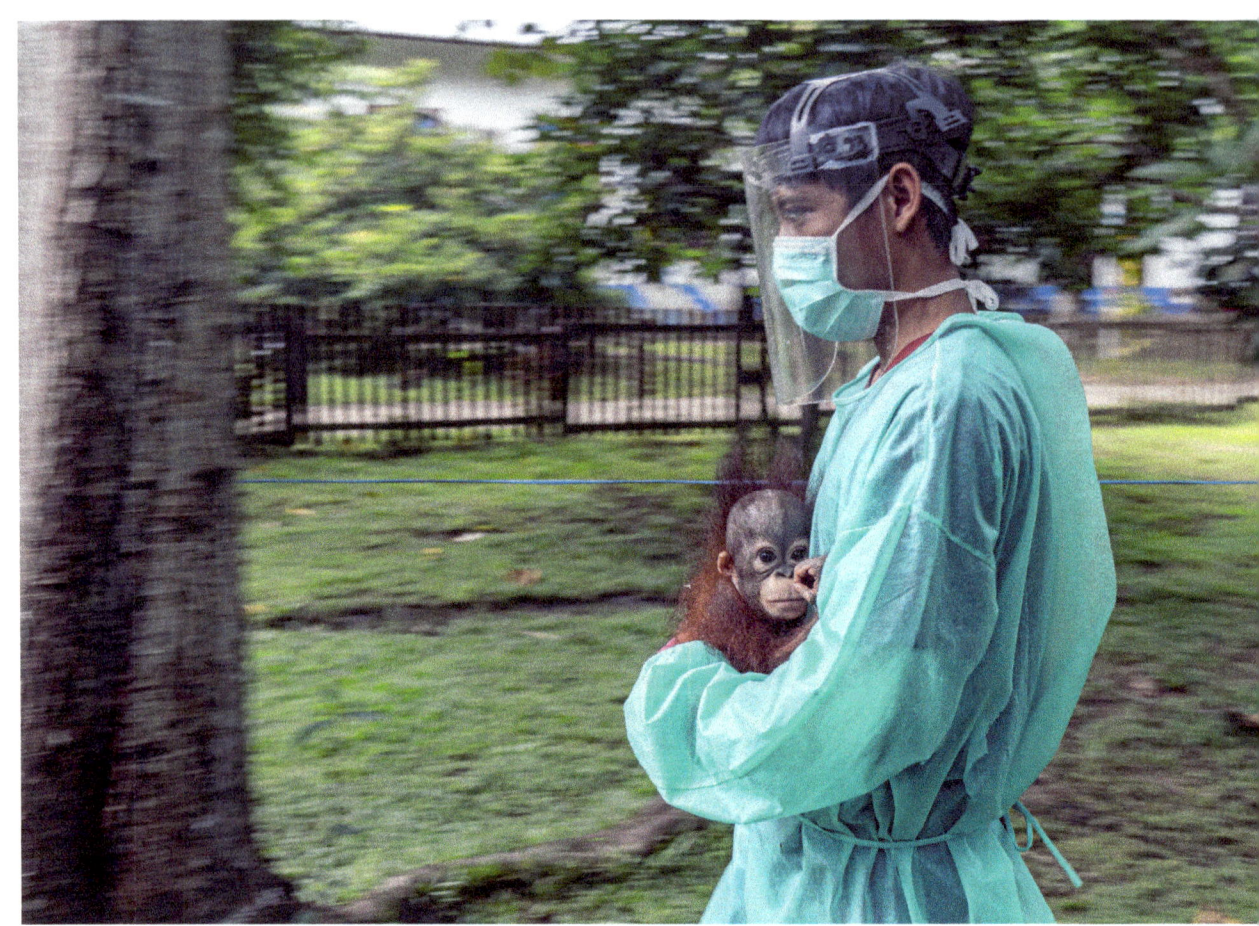

By applying for and attaining professional accreditation, captive facilities can strengthen systems and features that support good animal welfare.

- Appropriate, enforceable standards at multiple levels—legal, professional and institutional—serve as mutually reinforcing insurance policies to support ape welfare.

- A growing body of evidence indicates which welfare features are critical to captive great apes, although less attention has been paid to gibbons. Universal agreement on species-specific ape welfare indicators remains elusive, but ongoing initiatives highlight momentum and synergies.

- The effectiveness of welfare assessments depends on both the utility and immediate applicability of their results. They need to be practical to implement and produce context relevant information, such as to inform management decisions and/or expedite improvements in ape welfare.

Section II updates statistics on captive ape populations in zoos, rescue and rehabilitation centers, and sanctuaries around the globe. The key findings are:

- Available ape census data show little variation overall compared to data presented in previous volumes of *State of the Apes*.

- When they do occur, variations are not well understood, largely due to insufficient data. A lack of information sharing among zoos and with captive animal census databases such as Species360 or studbooks continues to be an impediment. Barriers to information sharing include language, use of different systems and skepticism about the merits of collaboration.

Section I: Improving Captive Ape Welfare

Understanding Animal Welfare

Constructs and Language

Animal welfare refers to how an animal is experiencing life. A balance of emotional, mental and physical components, as well as the attainment of species-specific ethological needs, determines an animal's overall wellbeing and welfare state (Cox and Lennkh, 2016).

As shown in Figure 8.1, the Five Domains Model reflects current thinking on the welfare of animals in human care (Mellor *et al.*, 2020). Four physical domains give rise to negative or positive subjective experiences that contribute to an animal's mental state, known as the fifth domain. Any form of captivity comes with some

FIGURE 8.1

Five Domains Model

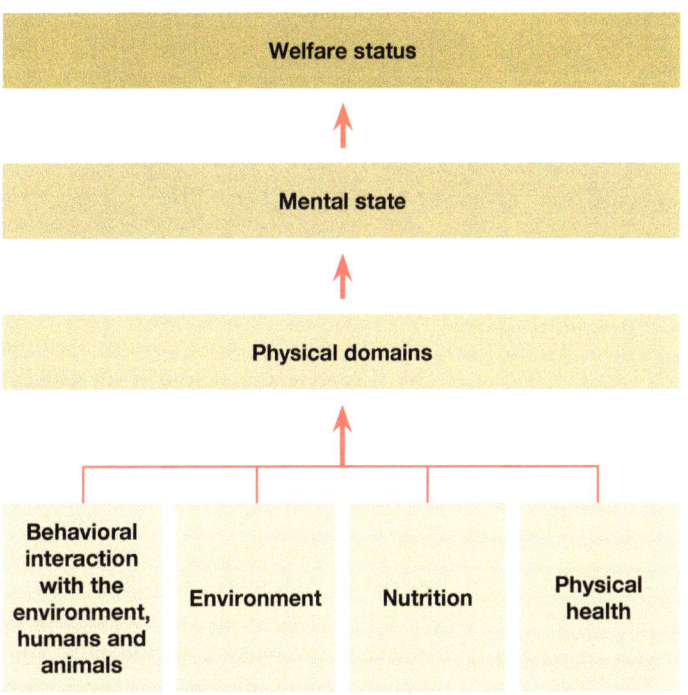

Source: Adapted from Mellor *et al.* (2020)

Photo: The movement of apes between different forms of captivity—for example from a laboratory to a sanctuary—reflects increasing public concern for ape welfare and recognition of ape sentience. Mari was born in a laboratory and one arm was badly broken and the other pulled off when her mother was in an agitated state. She now lives at the Center for Great Apes.
© Jo-Anne McArthur / NEAVS / We Animals Media

poor (Broom, 1999; Spruijt, van den Bos and Pijlman, 2001). Reflecting this continuum, Brando and Buchanan-Smith (2018) propose a 24-hour, seven-day animal welfare framework that includes the animal's life-cycle and natural history.

Issues related to animal welfare are often complex and linked to cultural, economic, political, religious and social factors. Local interventions that recognize the roles of each of these factors are more likely to be effective (Sinclair and Phillips, 2018b). Numerous other considerations also merit inclusion in the welfare dialogue, including age-related changes in captive apes and issues linked to ape captivity in spaces away from public view (Brando and Coe, 2022; Krebs *et al.*, 2018; Ross *et al.*, 2010). This chapter highlights the illegal wildlife trade and translocation, both of which can have profound effects on animal welfare.

Language itself can also influence the welfare conversation. Animal welfare can mean different things in different languages —assuming the term exists in the first place. In Chinese, for instance, there was no conceptual or linguistic equivalent of "animal welfare" until the mid-1990s (Hobson, 2007; Lu, Bayne and Wang, 2013). As discussed in Box 8.1, language can create and empower social change and, conversely, hinder it.

Compassionate conservation sees welfare and conservation outcomes as integrated. The approach, which recognizes animal sentience and personhood, encourages conservation practitioners to consider animals as individuals and not just as members of populations of species (Wallach *et al.*, 2018, 2020). Similarly, the One Welfare approach, which expands on One Health, attempts to bridge the gaps between different disciplines, explicitly recognizing and explaining the interconnections between animal welfare, human wellbeing and the environment (Pinillos *et al.*, 2016; see Chapter 2).

risks, but animals managed in line with the five domains are expected to be healthier physically and psychologically, safer for staff to work with, more likely to qualify for release (if other crucial factors are addressed), and more representative of their species for educational purposes.

It is also generally agreed that animal welfare is not a static construct but a continuum. An individual ape's welfare thus lies somewhere on a scale between good and

Language Matters

Language can devalue the intrinsic worth of animals and disassociate their use or exploitation from their suffering (Kahn, 1992; Stibbe, 2001). Such depreciating usage remains in place in the Convention on International Trade in Endangered Species of Wild Fauna and Flora (CITES) and among zoos. For instance, CITES uses the term "disposal" to refer to the process by which government authorities manage confiscated live animals (CITES, 2016). The Cambridge Dictionary defines "disposal" as an act of getting rid of something or throwing it away. Meanwhile, many zoos describe animals in their care as "stock," which the Cambridge Dictionary equates to a "supply or amount of goods."

The use and impact of language can and does change, however. In the 1960s, Jane Goodall contributed to the current understanding of chimpanzee sentience by naming her study subjects and defying the established scientific approach (Goodall, 1998). More recently, she and other leading advocates requested an update to the Associated Press Stylebook, a style guide used in journalism, to promote the use of gendered or plural (rather than inanimate) personal pronouns for animals (Graef, 2021). By updating their use of language in line with changes in animal science, law and public opinion on animal sentience, entities such as CITES and zoos could encourage greater empathy, more respect and better care for animals worldwide.

Species-Specific Welfare Requirements in a Systems Approach

There is growing evidence that specific, interrelated features of a captive environment underpin and promote good welfare practice. Zoo experts rated great apes' ability to avoid one another, enclosure appearance and furnishings, group size and social structure as the most important indicators

of their welfare. The order of importance of these indicators varies across species: physical attributes of an enclosure are more important for orangutans, while group size and social structure matter more for other species (Fernie *et al.*, 2012). A recent study on chimpanzees underlines the importance of a conspecific for companionship, while other key factors include the caregiver–chimpanzee relationship, opportunities for choice and control, diet, environmental enrichment, and the quality and complexity of space (Ross, 2020; see Figure 8.2).

FIGURE 8.2

Enclosure Design to Facilitate Choice and Control, Center for Great Apes, North America

Note: In North America, the Center for Great Apes provides choice and control for its apes in 19 different habitats with elevated tunnels, which enable apes to move between habitats. Twelve adult male orangutans can move between 11 habitats at different times. If apes want to stay indoors, they can. At night they can choose to sleep inside or outside (P. Ragan, personal communication, 2020).

© Center for Great Apes, Wauchula, FL

A recent revision of the Five Provisions Model also highlights human–animal interaction as a significant welfare indicator (Mellor *et al.*, 2020).

Barber and Mellen (2008) suggest seven programs collectively make up "welfare infrastructure": animal training, environmental enrichment, habitat, husbandry, nutrition, research and veterinary care. The omission or removal of any of these programs destabilizes the welfare infrastructure and may cause it to collapse (Bettinger *et al.*, 2017). Professional accreditation systems that review the whole welfare system —including these seven programs, governance and operational standards, and direct welfare support services—are the most robust and effective.

Animal Welfare and Trade

Cruelty is pervasive throughout the live animal "supply chain." The capture, transport and subsequent "use" of wildlife negatively impact them in all five welfare domains (Baker *et al.*, 2013; Clifford and Steedman, 2021). As a result, captive facilities face the short- and long-term need to manage myriad complex ape health and welfare issues (see Figure 8.3). Case Study 4.3 provides evidence of the many clinical issues and implicit poor welfare affecting privately held gibbons. Even after apes are confiscated, their suffering may be exacerbated, including by transfer processes, which can be complicated by inadequate cooperation among multiple agencies, and by limits on facilities' capacity to provide immediate and appropriate triage and care. Some apes have died before they could arrive at a suitable facility.

In recognition of such risk factors, the Conservation Action Plan for great apes in the eastern Democratic Republic of Congo (DRC) includes an objective to develop a procedure for the safe seizure and transfer of confiscated animals.[1] Such responses

remain limited, however, despite increasing awareness that debates concerning the wildlife trade typically omit the issue of animal welfare. Some observers attribute this inertia to the durability of the view that non-human animals are no more than property, commodities or resources—rather than individuals or sentient beings (D'Cruze *et al.*, 2020; Wyatt *et al.*, 2022).

It is not only the illegal trade that impacts animal welfare, as legal and illegal trading activities are inextricably entwined. In relation to some species, the legal trade is far greater in magnitude than the illegal one (Ban Animal Trading and EMS Foundation, 2020; Nijman, 2021). Based on assessments of the (allegedly) legal and illegal trade in captive chimpanzees from Africa to China, the facilities in which the apes live are not suitably equipped to house or care for them, contravening CITES regulations for an import permit (Ape Alliance, 2018; Ban Animal Trading and EMS Foundation, 2020). More than half of the chimpanzees in these facilities were under ten years of age when they died; about 15% of them had not reached their first birthday (Ape Alliance, 2018). These mortality statistics underscore the potential deleterious impact of trade and transfer on ape welfare in the absence of appropriate oversight and decision-making. They are particularly stark when compared to the high survival rates at Pan African Sanctuary Alliance (PASA) sanctuaries, which had a 2% mortality rate in 2020 (G. Tully, personal communication, 2021).

Overall, there are arguments for and against the legal transfer of animals between zoos. By performing effective due diligence in individual cases, facilities that seek to acquire or transfer an animal have a responsibility to ensure that their behavioral, physiological and psychological requirements can be met (BIAZA, 2019; Pierce and Bekoff, 2018; Rietkerk and Pereboom, 2018).

FIGURE 8.3

Mubaki's Progress from Confiscation through Captive Care in Eastern DRC

Note: About 90% of confiscated individuals who arrive at sanctuaries that are members of the Pan African Sanctuary Alliance are malnourished and sick, suffering from a range of physical and psychological conditions (Farmer, 2002; PASA, 2009). Despite poor arrival conditions, survival rates are similar to those seen in member facilities of the Association of Zoos and Aquariums, although survival itself is not necessarily a sign of good welfare (Faust *et al.*, 2011). Left to right: Mubaki at confiscation and after receiving treatment and care at the Lwiro Primate Rehabilitation Center in the eastern DRC.

Left and center: © Lwiro Primates Rehabilitation Center
Right: © 2Ws Photography

Animal Welfare and Release Programs

Under ideal circumstances, release programs follow international guidance. The reintroduction guidelines of the International Union for Conservation of Nature (IUCN) focus on release for conservation purposes but also define welfare-based release (Beck *et al.*, 2007; Campbell, Cheyne and Rawson, 2015). According to the precautionary principle, however, the conservation of a taxon takes precedence over an individual captive ape's welfare; a welfare-defined release may not meet, or may even contravene, the IUCN principle (Beck *et al.*, 2007; Campbell, Cheyne and Rawson, 2015). Regardless of the goal, designing a rehabilitation and release process that consistently considers animal welfare during each stage is crucial to supporting successful conservation outcomes (see Figure 8.4). Adverse stress may occur at any stage of the release process—from capture, handling, examination and transport to holding, monitoring after release and post-release evaluation (Berg, 2018; Teixeira *et al.*, 2007). While stress responses generally have adaptive value in the short term, they can have devastating impacts on brain function and susceptibility to disease in the long term, depending on the duration, magnitude and nature of the event (McCormick, Shea and Langkilde, 2015; Moberg, 2000).

Just as captivity can compromise welfare, so too can release. In contrast to life in captivity, which is characterized by predictability, life in the wild involves fluctuations and variations in the food supply, predator and competitor types and numbers, and social pressures (Swaisgood, 2010; Teixeira *et al.*, 2007). Nevertheless, some released apes have survived and thrived after release (Goossens *et al.*, 2005; Humle *et al.*, 2011; King, Chamberlan and Courage, 2012; Wedana *et al.*, 2021). In some cases, however, released apes negatively impact the welfare of wild conspecifics.[2]

FIGURE 8.4

Design of the Rehabilitative Environment at Kalaweit Gibbon Project, Indonesia

Note: Wild gibbons are highly territorial, regulating their spacing through direct contact or loud calls. Left: Sufficient space between enclosures is prioritized at Kalaweit. Right: Triangular lozenge-shaped cages are designed to avoid 90-degree angles and prevent intimidation and aggression during integrations. As release becomes possible, gibbons are transferred to a release site where they spend several months in a more natural enclosure to facilitate adaptation (A. Brulé, personal communication, 2020).

© Kalaweit

Standards of Practice

National Legislation and Regulations

By passing appropriate, enforceable animal welfare legislation and regulations, countries can demonstrate a national commitment to captive and wild animal care and management. A Model Animal Welfare Act serves as a basic template and guidance document for governments interested in enacting new or improving existing legislation (Cox and Lennkh, 2016). Two countries—Malawi and Costa Rica—recently saw positive developments in animal welfare legislation, although their impact has yet to be examined (see Annex IX). In countries where some form of animal welfare legislation exists, its configuration, the extent of coverage and enforcement vary greatly (Hassan, 2016). Understanding a country's commitment to animal welfare requires comparing relevant legal requirements, inspection measures and control procedures (Lundmark, Berg and Röcklinsberg, 2018). A recent analysis of the legal protection of apes found that none of the ten African ape range states under review included criminal provisions on the welfare or exhibition of apes (Rodriguez *et al.*, 2019).

Country examples highlight that captive wildlife welfare often falls between the cracks of laws and regulations that govern wildlife conservation and animal health. Conservation law is most often focused on the management and survival of free-roaming wildlife species, while animal welfare law targets domestic animals (Prisner-Levyne, 2020; Whitfort, 2019). Complicating matters is the issue of ape provenance.

Wildlife conservation laws sometimes apply only to apes in their natural habitat or country of origin, or only to wild-born and not to captive-born apes (Beastall, Bouhuys and Ezekiel, 2016; Rodriguez *et al.*, 2019). Indeed, even within the conservation community, a distinction is made between the intrinsic value of apes in their natural habitat versus apes forcibly removed from the same habitat. Country-level cross-disciplinary analyses of relevant legal instruments are urgently required to allow for an assessment of legislative gaps and requisite enforcement resources.

While not a substitute for appropriate, enforceable animal welfare legislation, conservation litigation may help to "remedy" harm caused to affected species and individual animals (Phelps *et al.*, 2021b, 2021c). Remedies are the actions needed to address harm and help an animal recover; an example for an individual orangutan might be rehabilitation and post-release monitoring or long-term care (Phelps *et al.*, 2021b). The provision of good welfare requires money. Captive facilities that are ethically driven yet often short on resources are forced to react as animals are seized or ownership is transferred (Fleury, 2017). Since seizure and post-seizure events involving illegally held and traded live animals are unplanned, however, related costs are not generally included in law enforcement grant budget lines. Nor are they covered in assessments of case severity or decisions about penalties or sentencing, which could potentially help facilities recoup costs associated with seizure and management.

In Indonesia, the environmental group WALHI North Sumatra and the Medan Legal Aid Institute have filed a ground-breaking lawsuit against a company that keeps protected species, including an orangutan, at its zoo without legal permission (Walhisumut, 2021). In court proceedings elsewhere, plaintiffs have focused on the impact of biodiversity loss and the suffering of individual captive animals to strengthen sentencing for wildlife offenses (Knott, 2021; Whitfort, 2019).

International Conventions and Declarations

Convention on International Trade in Endangered Species of Wild Fauna and Flora

CITES signatories have three options: they can maintain confiscated animals in captivity, return them to the wild or perform euthanasia. In most cases, they choose captivity—in its different forms (CITES, 2016; CITES Secretariat, 2017; IUCN, 2019a). Sparse information is provided on the required resources, benefits or challenges associated with each option. Overcrowded Southeast Asian zoos highlight the challenge of having to manage large numbers of confiscated and abandoned animals (Agoramoorthy, 2010; Karokaro, Gokkon and Suriyani, 2017). While CITES is not about welfare, trade and "disposal" decisions can have negative welfare impacts, and good outcomes are not guaranteed with any option (Rivera, Knight and McCulloch, 2021; Ronfot, 2016; Wyatt *et al.*, 2022; see Box 8.1). CITES provides a framework, but each country must adopt its own domestic legislation to ensure national implementation (CITES Secretariat, 2021).

Only ten of the 26 ape range states that are CITES members have laws that satisfy all requirements (Sherman and Greer, 2018). Thailand, for example, has met some CITES requirements, but its wildlife laws have significant shortcomings with respect to native ape species and offer no protection for non-native ape species. Moreover, Thailand exhibits severe welfare problems in government-run captive facilities, where most captive wild animals are held (Beastall, Bouhuys and Ezekiel, 2016; Moore, Prompinchompoo and Beastall, 2016; Ronfot, 2016).

The incoming joint IUCN–CITES task force aims to support each country in developing national action plans to manage seized animals (N. Maddison, personal communication, 2020). Explicitly acknowledging the importance of welfare for positive conservation outcomes within the development of these plans may help to integrate welfare-focused considerations and action.

Universal Declaration on Animal Welfare

Although animal welfare commitments are a prerequisite for greater awareness, they are rarely included at the international policy level. There is no global agreement or treaty to protect the welfare of animals (Bridgers, 2021). The Animal Issues Thematic Cluster and the World Federation for Animals are among the collaborative initiatives that aim to fill this policy gap (AITC, n.d.; WFA, n.d.).

The Universal Declaration on Animal Welfare (UDAW) is a proposed intergovernmental agreement whose aim is to prevent cruelty to animals, reduce their suffering and promote welfare standards (Appleby and Sherwood, 2007). UDAW has received more support than any other international animal welfare initiative: it has been formally endorsed by more than 60 governments and 270 animal welfare organizations from 78 different countries, including the World Organisation for Animal Health (formerly the Office International des Epizooties) and the World Veterinary Association (Gibson, 2011; D.J. Verdonk, personal communication, 2021). UDAW is recognized as an important step towards the proposed UN Convention on Animal Health and Protection, a framework with legally binding provisions (GAL, 2018). A feasibility study would help to identify the challenges and resources needed to adopt and implement such an agreement.

Pledges made at the 4th African Animal Welfare Conference[3] included a call for African governments to support the process of developing and securing the adoption of an animal welfare-focused resolution at the next UN Environment Assembly and to confirm their support for UDAW as an African Union resolution to the UN General Assembly (AAWC, 2020; Chumo, 2021). In 2022, UN Environment Assembly member states adopted the first-ever resolution to make explicit reference to animal welfare, recognizing the link between animal welfare, environment and sustainable development (UNEP, 2022; WFA, 2022).

Professional Associations and Accreditation

Professional systems can be more flexible than legal ones, enabling changes, updates and adaptations to new developments and knowledge (Lundmark, Berg and Röcklinsberg, 2018). In the absence of appropriate, enforceable legislation, the membership of a professional association may facilitate advances in animal welfare through accreditation (Banes *et al.*, 2018). Accreditation processes are usually conducted by a professional body and require assessment against predetermined standards, which are reviewed at designated intervals to ensure they are maintained and adapted as required. The advantage of accreditation, assuming the system is appropriate, is that it can provide a public position on the extent to which captive facilities meet stated standards. As they fall outside of the legal system, accreditation schemes do not normally have the same requirements of transparency and predictability, although the accrediting body may choose to disclose results to increase transparency.

Sanctuary networks and accreditation systems differ in their geographical scope —from international to regional—and in terms of their service provision, which can focus on any combination of accreditation and advocacy, shared learning, and capacity-strengthening programs and events. Their approaches and standards may also vary, as most manage their own accreditation systems internally. Only the North American Primate Sanctuary Alliance requires third-party verification by the Global Federation of Animal Sanctuaries (GFAS) or Association of Zoos and Aquariums (AZA). Each approach has its own merits: the standards developed by internally managed accreditation systems can appeal to captive facilities, while independent third-party verification with species-specific guidance is a more robust benchmark for partners, the public and donors. Separately, Animal Advocacy and Protection has developed a system for assessing the capability of European rescue centers, sanctuaries and zoos to receive its "outplaced" animals (AAP, n.d.). Their system includes assessment of an organization's ability to provide for the welfare of each animal (O. Martin, personal communication, 2020).

The overarching international body for zoos—the World Association of Zoos and Aquariums (WAZA)—has more than 400 members, including regional and national zoo associations. Individual zoos that apply for WAZA membership must be accredited by a recognized association in their own region, if one exists (WAZA, n.d.). WAZA relies on its regional member associations to enforce standards among their member zoos. While standards and compliance control currently vary across regions, WAZA requires all regional associations to have an animal welfare evaluation process in place by 2023 to accredit individual zoos and to ensure compliance among all institutional members (WAZA, 2019). Associations are free to develop their own standards and processes, so long as the standards are based on expert input, best practice and science, and processes that reflect animal-based measures (P. Cerdán Codina, personal communication, 2021).

Internally Developed Standards

The foundations of good animal welfare are standards of practice at the captive facility level; in some cases, these standards may be the only available guidelines for facility staff. While some facilities may have informal standards informed by institutional values, many others have formalized policies, standard operating procedures (SOPs), behavioral management plans and welfare assessment tools. Every institution has its own unique set of approaches to determining what is needed and how best to implement plans on a day-to-day basis. Given variations across facilities, accreditation systems are crucial for ensuring compliance. As part of its new evaluation process, WAZA requires regional associations to verify institutional-level animal welfare policies and evidence of compliance (P. Cerdán Codina, personal communication, 2021).

In North America, Save the Chimps formalized its philosophy of care to ensure consistent and exemplary care for the 230+ chimpanzees in its sanctuary. The philosophy gives rise to a theory of change with which the organization's policies, protocols and methodologies conform and align (see Figure 8.5). The intended outcomes are achieved through standards of care that guide the implementation of SOPs. Care practices are shaped by proven methods and published literature on animal welfare, including more than 100 specific care plans based on chimpanzee past history and current status. Specific care plans are updated based on regular welfare assessments (A. Halloran, personal communication, 2020).

FIGURE 8.5

Save the Chimps Approach to Promoting Chimpanzee Welfare

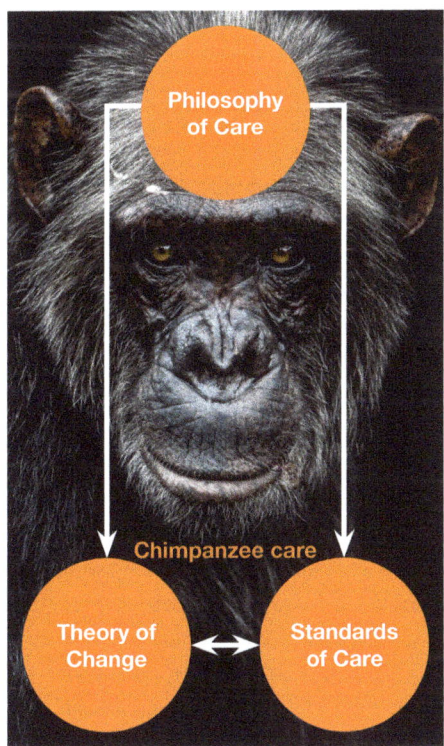

Theory of change	
Ultimate goal	Chimpanzee well-being: provide an evironment in which chimpanzees thrive as a result of the exemplary care they receive
Desired outcomes	■ Safety and well-being of chimpanzees and staff is at forefront of all sanctuary operations ■ Individualized care ■ Enriching environment with freedom of choice ■ Opportunities for resocialization and social integration ■ Exemplary veterinary care
Challenges to achieving goal	■ Difficulty closely monitoring apes in large habitats ■ Need to isolate chimpanzees for veterinary procedures ■ Balancing individual versus group needs
Overcoming barriers	■ Observe apes when indoors and with remote cameras ■ Facilitate cooperation through operant conditioning ■ Accommodate diverse needs whenever possible
Indicators of progress	■ Assurance of safe working environment ■ Alignment of standard operating procedures and Philosophy of Care ■ Fulfillment of individual care plans ■ Welfare, monitoring and intervention assessments ■ Social competence and integration ■ Contribute to improving care practices within the animal care industry

Source: Save the Chimps (n.d.)

The animal welfare organization Four Paws, whose 12 sanctuaries in 11 countries care for more than 250 bears, big cats and orangutans, ensures standardized welfare outcomes, care and management services across its facilities through an articulated quality management system (Four Paws International, 2020a; I. Redtenbacker, personal communication, 2020). As illustrated in Figure 8.6, the system involves four mutually reinforcing, iterative steps:

- **define animal welfare standards** based on sanctuary-specific handbooks, SOPs and guidance on management topics, informed by current scientific knowledge;

- **implement standards** by supporting clear internal communication, a year-long caregiver training program (comprising three in-person workshops complemented by eight e-learning courses) and an annual quality management system workshop to inform staff of standards and expectations;

- **check standards** through internal annual audits and random checks carried out by trained staff to ensure compliance; and

- **improve welfare and standards** based on results from audits and checks as well as recommendations, all of which are shared with staff and management at each facility (I. Redtenbacker, personal communication, 2020).

FIGURE 8.6

Four Paws Quality Management System Promoting Animal Welfare

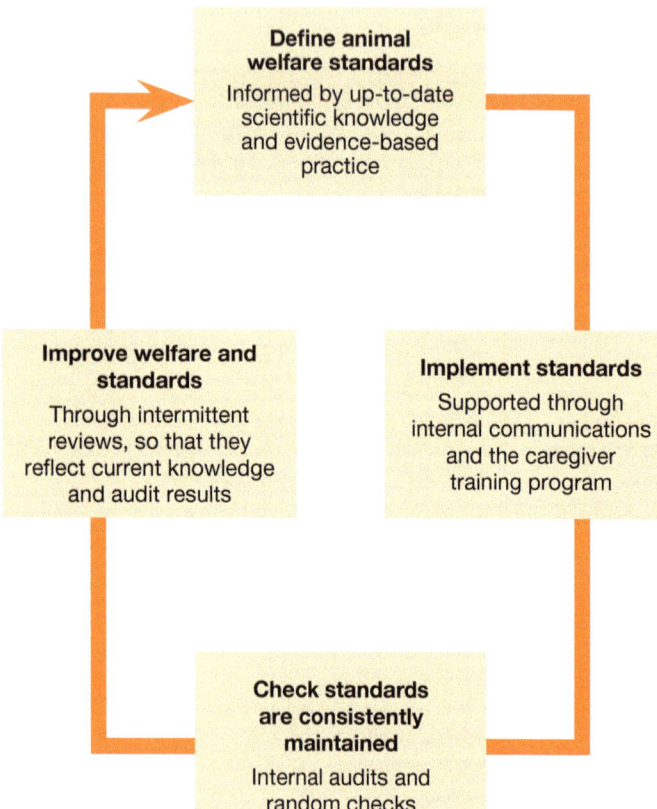

Define animal welfare standards
Informed by up-to-date scientific knowledge and evidence-based practice

Improve welfare and standards
Through intermittent reviews, so that they reflect current knowledge and audit results

Implement standards
Supported through internal communications and the caregiver training program

Check standards are consistently maintained
Internal audits and random checks

Standards for Rehabilitation and Release

For some facilities, rehabilitation for release is a key consideration in determining how an ape is managed. In assessing whether an individual ape has the skills required for life in the wild, including the necessary level of locomotor skill, facilities may use dedicated frameworks such as the Enclosure Design Tool (see Case Study 8.1). As part of the rehabilitation process, they may also aim to minimize human–animal interaction (Russon, Smith and Adams, 2016; see Annex IX). In Indonesia, for instance, subadult rehabilitant female orangutans in the care of International Animal Rescue serve as foster mothers after being paired with rescued infants, which significantly reduces the apes' dependency on and interaction with humans. This approach decreases the duration of rehabilitation from seven years or longer to two–three years until release (K. Sanchez, personal communication, October 2020).

IUCN reintroduction guidelines for great apes and gibbons cover several considerations that influence welfare outcomes, such as behavioral criteria and assessment, phased approaches to release and post-release support to facilitate adaptation (Beck *et al.*, 2007; Campbell, Cheyne and Rawson, 2015). GFAS accreditation standards for sanctuaries include aligned adaptations for rehabilitation centers that are complementary with IUCN guidelines (GFAS, 2022). Capturing lessons learned on the effective features of rehabilitation and release from field practitioners could help underpin the development of agreed welfare indicators for released apes.

Failure to Promote Good Animal Welfare

Failure to provide good animal welfare may be a consequence of simultaneous and interrelated issues. Annex X summarizes common legal, professional and institutional issues as well as barriers to and opportunities for good captive ape welfare, based on shared experience. In short, barriers to good animal welfare include inappropriate accreditation systems, corruption, unfavorable cultural aspects (geographical, sectoral or organizational), inadequate financial or human resources (such as poor leadership or technical deficiencies), pressure from authorities to accept or release animals, and a state's failure to prioritize welfare. A recent analysis focused on Southeast Asia shows a similar level of complexity, highlighting the need for strengthened political will, policy and legislation, as well as enhanced accountability of captive facility management through proper licensing, permitting, regulations and regular inspections (Rivera, Knight and McCulloch, 2021).

Achieving high welfare standards requires an ability to recognize good practice in each part of the system. While high standards in one part can influence outcomes across a system, they can also be specific to a particular area (such as organizational systems or staff competencies) or domain (see Figure 8.1). Under good leadership and with team

buy-in, standards can be learned and become self-sustaining (Sinclair and Phillips, 2018a; Walraven and Duffy, 2017). This process requires ongoing self-reflection and integration of emerging science and practice, as proposed by the Four Paws and Save the Chimps systems (see Figures 8.5 and 8.6).

Facilities that operate outside of professional zoo associations may suffer from weak institutional governance mechanisms, inappropriate leadership, or limited engagement and learning opportunities (Ward *et al.*, 2020). Membership in a zoo association does not necessarily translate into higher animal welfare standards, however, especially if the association lacks an appropriate accreditation system and enforcement mechanism (Draper and Harris, 2012; Rainer *et al.*, 2020). In Indonesia, for example, 50% of zoos are officially accredited, yet a mere 14% are deemed decent and appropriate (Saudale, 2015). Uneven compliance across association members is another complicating factor. Contravening both WAZA and regional association guidelines, zoos in Malaysia, the Philippines and South Africa have permitted direct interaction with chimpanzees, gibbons and orangutans (Corrigan, 2010; WAP, 2019). Poorly run accreditation systems can falsely suggest that facilities are providing an appropriate level of animal care and treatment, which can do more harm than good (Winders, 2017).

The Southeast Asian Zoos and Aquariums Association (SEAZA) acknowledges the challenges inherent in streamlining its members' varying approaches and capacities to achieve and maintain compliance with WAZA's new (2023) requirements (Manansang, 2020). WAZA now requires SEAZA to conduct onsite audits every five years and to seek evidence that facilities are conducting self-assessments of animal welfare (the principles of which are not explicit) at least annually, in between audits. Furthermore, SEAZA is responsible for training inspectors and developing a complaints and disciplinary system (P. Cerdán Codina, personal communication, 2021). By monitoring the impact of permitted variation around the world, WAZA will be able to assess if this flexibility leads to pronounced regional variations in animal welfare outcomes.

The welfare challenge is not restricted to zoos. The terms "rescue center" and "sanctuary" are not legally regulated; facilities that adopt either of them are not necessarily providing good animal welfare (Doyle, 2017; Winders, 2017). Indeed, the quality of care and welfare of apes in sanctuaries has come under scrutiny for not meeting acceptable standards (Grimm, 2020; Sherman and Greer, 2018). The impact of accreditation on the sanctuary sector is likewise limited by the proportion of facilities that are accredited; in Africa, for example, only 19% of great ape sanctuaries have officially met GFAS standards (GFAS, n.d.; see Section II of this chapter). In Southeast Asia, the demand for animal tourism is fueling the rise of "faux rescue centers" (Rivera, Knight and McCulloch, 2021). Unregulated or improperly managed facilities and inappropriate depictions of—and interactions with—animals can have negative welfare impacts that are unrecognized by visitors, enhance demand for exotic pets and undermine conservation goals (Moloney *et al.*, 2021; Moorhouse *et al.*, 2015; Ross *et al.*, 2008).

While no single global body regulates wildlife tourism, an increasing number of tourism-focused initiatives have emerged in response to failures in welfare systems and the lack of communication about appropriate accreditation systems for benchmarking. One such initiative is the Animal Protection Network, designed to help tourism operators and visitors make humane choices (Animondial, n.d.). The IUCN Primate Specialist Group's Section for Human–Primate Interactions' responsible primate watching guidelines include a chapter on primates in

captivity, as well as looking at tourism and primate welfare (Waters *et al.*, 2023).

IUCN reintroduction guidelines state that great apes with significant deficits in knowledge and skills should not be released without sufficient rehabilitation and post-release support (Beck *et al.*, 2007). Nevertheless, limited resources and carrying capacity, along with pressure from authorities, may lead facilities to release animals without proper protocols, with unknown welfare and conservation outcomes (Mitman *et al.*, 2021; Sherman and Greer, 2018). Low long-term survival rates of released orangutans are attributed to a lack of familiarity with current research on orangutan behaviors, the extended periods apes are kept in captivity and the application of a "hard release" strategy, which involves the immediate release of an ape from captivity without post-release support (Sherman, Ancrenaz and Meijaard, 2020). At best, a failure to recognize deteriorating welfare after release can lead to recapture (and re-release), which can cause additional stress; at worst, the result is the animal's death (Sherman, Ancrenaz and Meijaard, 2020; Wilson and McMahon, 2006).

> Identifying welfare indicators is the first step in developing an objective, scientific process by which to assess, monitor and improve welfare.

Assessing Ape Welfare

Assessment is key to the philosophy of continuous welfare improvement. Some zoo associations and sanctuary alliances have developed animal care guidelines and standards for the keeping of ape species.[4] Overall, however, guidance on how to assess species-specific animal welfare is sparse and inconsistent, and the development of assessment processes is often left to individual captive facilities and affiliated academics.[5] This section focuses on key elements of welfare assessments and recent ape-specific developments; it is complemented by Annex XI, which reviews the features of four tools used to assess captive ape welfare.

Understanding Welfare Indicators

Welfare assessments usually consist of a survey with indicators that caregivers score based on whether an animal or group meets given criteria. Identifying welfare indicators is the first step in developing an objective, scientific process by which to assess, monitor and improve welfare. Since welfare is multi-dimensional, assessing it requires a variety of indicators that can generate an overall picture of an animal's physical and psychological wellbeing. The Five Domains Model often underpins the design of assessments, while indicators broadly represent the four physical domains (Sherwen *et al.*, 2018; see Figure 8.1). Ideally, an assessment combines indicators that are resource-based (inputs) and animal-based (outputs):

- **Resource-based indicators** relate to the support and conditions an organization provides to improve animal welfare, including housing, group size and composition, and management (including staff-to-animal ratio and the competency of staff).

- **Animal-based indicators** include direct animal responses to the inputs, such as behavior, body condition and other clinical signs.[6]

These two types of indicators are closely related since an animal's response to resources and consequent welfare status both depend on the quality of resources and how these are applied and managed. For practical reasons, accreditation systems focus on resource-based indicators while institutional welfare assessments primarily use animal-based indicators or a combination of both (see Annex X). Animal-based indicators allow caregivers to assess an animal's welfare more directly, although they could readily check records for basic health conditions or conduct a brief inspection.

As these indicators can be difficult to measure and interpret, however, they require significant time and resources (Brando and Buchanan-Smith, 2018; Crockett and Ha, 2010; Project Chimps, 2020; Truelove *et al.*, 2020).

Specifically, the interpretation of behavioral abnormalities as indicators of welfare remains controversial, mainly because an absence of suffering does not equal good welfare (Bloomsmith *et al.*, 2020; Broom, 1991). Chimpanzees, for example, may not exhibit obvious behavioral abnormalities, even in the most deprived environments. While a facility may not be meeting a chimpanzee's welfare needs appropriately, an absence of such abnormalities may simply demonstrate the species' resilience and adaptability, while stress may be having hidden impacts (S. Ross, personal communication, 2020). Conversely, apes' abnormal behavior does not necessarily signal that they are experiencing poor welfare; any number of factors, past and present, can result in an expression of suffering (Bloomsmith *et al.*, 2020).

Identifying Species-Specific Indicators

The socioecological complexity that characterizes ape species can make measuring welfare a daunting task (Goodall, 1986; Mitra Setia *et al.*, 2009; Ross, 2020). Ape behavior is relatively well understood, as are features of captive (zoo) environments that are important for their welfare, particularly for captive and wild chimpanzees.[7] This body of knowledge enables the development of meaningful welfare indicators for some ape species (see Annex XI).

The preferences of captive facilities—and the available time of their keepers—often determine which indicators and scales they use for scoring. To simplify the assessment process, facilities that care for many species may use more generic measures, such as a mammal template, with adaptations for specific species as required (D. Free and S. Wolfensohn, personal communication, May 2021).[8] A common strategy involves constructing lists of welfare indicators as determined by consensus of expert opinion. An appropriate range of expert input and agreement is crucial for wide uptake and, ultimately, welfare impact. Veasey notes that while there can be a strong agreement between in situ and ex situ expertise, discrepancies can also reflect differing backgrounds (Veasey, 2020a).

The process for soliciting input and agreement ranges from informal, one-off meetings and questionnaire-based surveys to more formalized approaches, such as multiple-round Delphi consultations. Established across many disciplines, the Delphi method enables stakeholders (the expert panel) to address a complex problem collectively and reach consensus, as indicated by agreement across multiple rounds of consultation (Dalkey and Helmer, 1963; Hsu and Sandford, 2007; Millar *et al.*, 2007). Having gained popularity in conservation and animal welfare circles thanks to its adaptability, Delphi was recently used to identify welfare indicators for laboratory-housed macaques.[9] In addition to assisting in the identification and validation of animal-based welfare indicators, Delphi also serves to assess their feasibility and reliability (Truelove *et al.*, 2020). The tools presented in Annex XI reflect some features of the Delphi process, although they lack its rigor.

Engaging an appropriate range of experts from multiple facility types in a systematic process can strengthen the prospects of obtaining universal acceptance of species-specific ape welfare indicators. Work led by the University of Birmingham in the UK, in collaboration with the European Association of Zoos and Aquaria (EAZA) and PASA, includes a consultation process with a range of stakeholders, in particular zoos and sanctuaries, to help identify, gain consensus and validate welfare indicators for

> " The interpretation of behavioral abnormalities as indicators of welfare remains controversial. "

captive great apes (J. Neufuss, personal communication, 2021).

Tools, Technology and Context

Once stakeholders agree on and validate indicators, the development of a welfare assessment tool can normally be explored (Truelove *et al.*, 2020). Many methods and tools to monitor and assess health and welfare, and to fulfill accreditation requirements, are developed in-house. As noted above, Annex XI presents a selection of tools for assessing captive ape welfare.

Behavioral measures are critical to assessing welfare (Wolfensohn *et al.*, 2018). They represent a common and relatively low-cost method, although they can be time-consuming and subjective (Binding *et al.*, 2020; Watters, Margulis and Atsalis, 2009; Whitham and Wielebnowski, 2009, 2015). Behavioral data can be collected by people with varying levels of expertise, resources and objectivity. Digital options for data collection can save time by eliminating the need to enter data manually, reducing the number of errors made during collection or entry, providing immediate feedback through reports and charts, and facilitating opportunities for multi-institutional col-

laboration (Wark *et al.*, 2019). At the same time, facilities that opt for digital data collection may need to purchase commercial software and hardware—or to have the skills and time to design a software solution (McDonald and Johnson, 2014).

One app that is widely used in academic zoo-based behavioral research on apes is ZooMonitor. The app, which is also employed in some sanctuaries, records animal behavior and habitat use using standardized methods, while also logging individual characteristics (Wark *et al.*, 2019). Other software and applications are available (Clegg, 2021; McDonald and Johnson, 2014; Whitham and Miller, 2016). Camera systems, for example, can greatly reduce the amount of time needed to visually access animals in large enclosures. They can also monitor without disturbance, including during times when staff members are absent, and they can store footage for later viewing. Drawbacks include potentially high costs, more limited identification of individual animals, and time-consuming footage viewing (Hansen *et al.*, 2018).

Animal welfare, conservation, farm and laboratory-based researchers are asking different questions, but collectively the answers to these questions could make emerging technology more relevant and adaptable (Buller *et al.*, 2020; Coe and Hoy, 2020; Langford *et al.*, 2010; Wich and Piel, 2021). For example, advances in software are making it possible to measure the body posture of animals over time in a variety of contexts (Graving *et al.*, 2019). Multi-disciplinary collaboration and resource pooling among practitioners, researchers and technical experts can help to develop solutions for specific needs, encourage market developments and strengthen the sustainability of solutions (Allan *et al.*, 2018; Joppa, 2015; Mulero-Pázmány, 2021). A simplified Delphi process may help to bring together key features of an assessment tool fit for purpose across multiple contexts, facilities and species.

Since many captive facilities are strapped for resources and struggle due to limited technical capacity, power and connectivity, technical solutions need to adapt to their needs to be workable. Small, rugged devices, with a relatively long battery life and a simple method for backing up data, can help facilities working in field conditions (McDonald and Johnson, 2014). Software that produces simple graphic outputs can enable easy interpretation and presentation, helping staff to understand the results, which can strengthen ownership of welfare solutions. Until technology becomes affordable, user-friendly enough for widespread use and more inclusive in its taxonomic coverage, it will need to be complemented with more traditional methods. For example, placing photographs of chimpanzees in the order of dominance and sociability, rather than using traditional numerical scales, enables staff with low literacy levels to provide input into integration processes (R. Atencia, personal communication, 2020).

Assessing the Welfare of Released Apes

In the field of conservation, assessments of released animals and reporting about related outcomes tend to focus on their physical state—and their "fitness" in particular (Beausoleil et al., 2018). Post-release assessments that incorporate the contemporary understanding of animal welfare better reflect the resource-intensive nature of rehabilitation. Although released apes may no longer be under human control, they are not necessarily fully competent at the point of release or for some time thereafter. In such cases, releasing individuals may give rise to a "duty of care" not normally provided to free-ranging wild animals (Berg, 2018). Little attention has been paid to the methods for monitoring ape welfare after release, partly because of the delicate balance required between helping and hin-

dering adaptation, and partly because of the challenge of capturing data (Harrington et al., 2013).

The type, frequency and duration of monitoring after release is influenced by several factors, including the goal of the action, site topography, species morphology and the temperament of individual apes. Monitors are advised to locate apes as often as possible—and every day in the beginning of the process—for at least one annual cycle (Campbell, Cheyne and Rawson, 2015; Sherman, Ancrenaz and Meijaard, 2020). While locating released apes to monitor welfare can be challenging, observers can facilitate the task by familiarizing themselves with ape ranging patterns, learning to identify sleeping sites, utilizing bioacoustics technology (to pick up gibbon calls, for example) and employing radio telemetry (Beck et al., 2007; Campbell, Cheyne and Rawson, 2015). Radio telemetry collars have worked well for chimpanzees and there has been some success with implants for orangutans, but animal-borne technology remains elusive for other ape species and its use can present another set of welfare challenges, such as the risk of infection at the surgical site of implantation (Dore et al., 2020; Robins et al., 2019; Trayford and Farmer, 2012).

Whether an ape can be monitored through direct ground observations or remote monitoring determines what type of data and other information can be collected (see Figure 8.7). A single measure may not provide a full picture of ape welfare; a combination of data and other information—such as on behavior, stress hormones, and negative or positive experiences—can offer a more complete indication of an individual's state, especially when collected over time. Unless the monitor is highly familiar with an ape species or individuals, simply collecting data on behavior may miss crucial information on body condition. A monitor who knows

FIGURE 8.7

Monitoring of a Released Orangutan at Jantho, Sumatra

Note: Data collection on released orangutans' locations, positions in trees, behavior, body conditions and health status in Jantho, Sumatra. Tracking movement from nest to nest and acquiring local knowledge can help monitors locate released orangutans.

© PanEco/SOCP

the individual animals may be particularly helpful in the early phases of post-release monitoring. While there are no easy answers to monitoring the welfare of apes after release, explicitly identifying monitoring as a key activity can underpin donor support for the development of appropriate approaches and tools.

Case Study 8.1 specifically considers the rehabilitation of apes for release and introduces the Enclosure Design Tool. To improve adaptation and welfare outcomes, the tool provides a framework to make enclosures less predictable and to ensure they behave like natural habitats, even if they do not look like them.

CASE STUDY 8.1

The Enclosure Design Tool: An Evidence-Based Framework for Improving Captive Ape Wellbeing

Context

All wild apes are highly arboreal, inhabiting forests that are characterized by enormous complexity and spatio-temporal variability (Wessling *et al.*, 2018). In contrast, captive environments are often relatively small, simple and unchanging, offering limited capacity for challenging arboreal activity. Many sanctuaries have large, forested enclosures for their apes during daylight hours, but they house individuals in substantially smaller "night dens" for up to 14–16 hours per day. Captive apes may also spend time (from hours to years) in other kinds of facilities, such as quarantine, clinics or other holding areas for health, welfare or management reasons.

In terms of musculoskeletal health, the repercussions of living in captivity are significant. Captive apes need prolonged natural (arboreal) loading patterns on their muscles and bones, particularly during growth, to gain the strength, skill and stamina needed to express physically demanding natural behaviors and to prevent age-related muscloskeletal degeneration (Chappell and Thorpe, 2022; Sarmiento, 1985). Captive apes also need to experience species-typical wild-type cognitive challenges to interact positively with their environment and to avoid frustration and boredom (Colditz and Hine, 2016). These experiences are particularly critical for apes being rehabilitated for release. While "forest schools" offer valuable exposure to natural habitat, they rarely facilitate the prolonged and diversified exposure to natural challenges required to prepare the apes mentally and physically for life after release (Chappell and Thorpe, 2022).

The Enclosure Design Tool (EDT) is an evidence-led, web-based framework created to address these issues (Chappell and Thorpe, 2022; University of Birmingham, n.d.). Its aim is to allow captive environments in which animals move, sleep, feed and rest to offer species-appropriate challenges, choice and control options that resemble the physical and intellectual challenges they would face in the wild. The first EDT focused on zoo chimpanzees and later versions were adapted to accommodate sanctuary chimpanzees in Cameroon and rehabilitant orangutans in Indonesia.

Photo: In terms of musculoskeletal health, the repercussions of living in captivity are significant. The EDT aims to allow captive environments to offer species-appropriate challenges, choice and control options that resemble the physical and intellectual challenges they would face in the wild. © Ian Bickerstaff

FIGURE 8.8

Summary of Results from the EDT Process

Chimpanzees

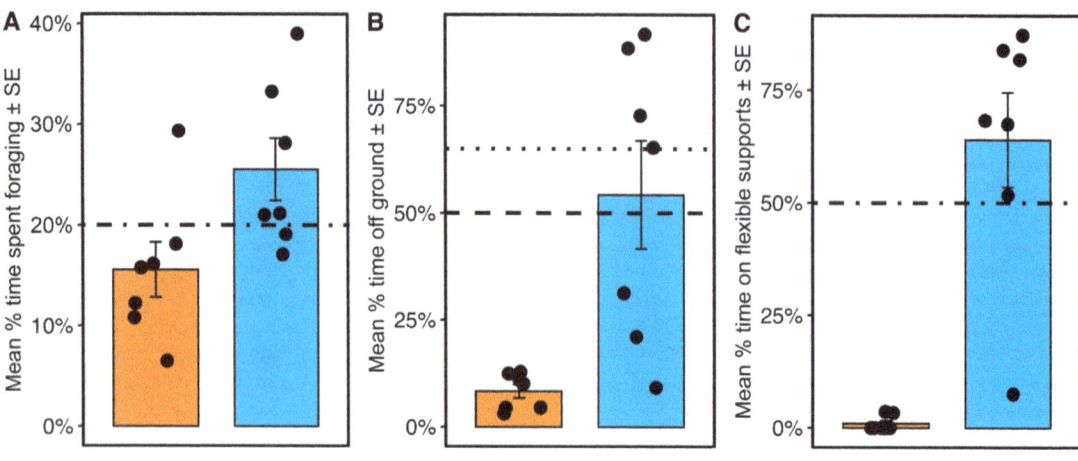

Orangutans

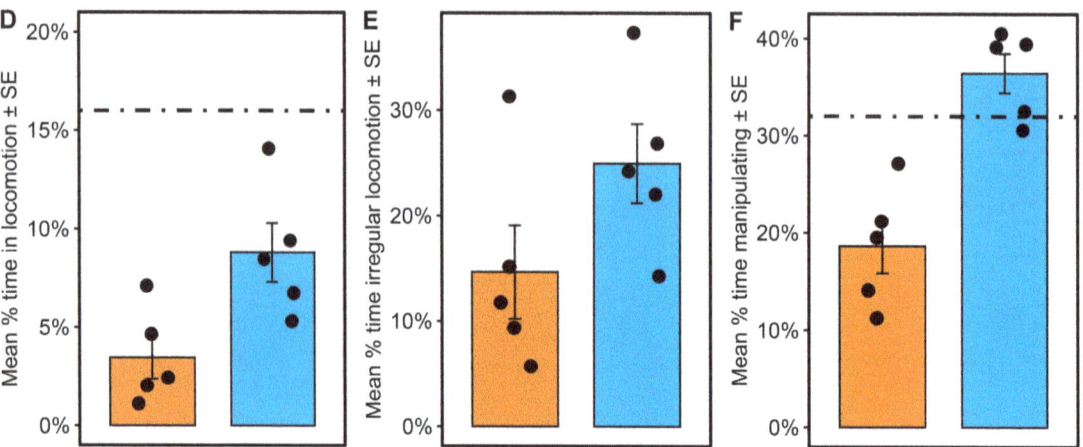

Period of data collection:

 Before modification   After modification

Notes: Panels A–C (chimpanzees) and D–F (orangutans) show means of the variable before (orange bars) and after (blue bars) enclosure modification, with standard error bars. Solid black points show means for each tested individual. Horizontal lines represent the comparison threshold created from data on wild apes: dashed lines show the threshold for adult males, dotted lines for adult females and dotted–dashed lines show the threshold for both groups.

Chimpanzees

A: Percentage of time spent foraging before and after modification; B: Percentage of time spent off the ground; C: Percentage of time spent off the ground on flexible supports.

Orangutans

D: Percentage of time spent in locomotion; E: Percentage of time spent in challenging locomotor modes; F: Percentage of time spent manipulating food items, using tools, exploring the environment or building nests.

Source: Chappell and Thorpe (2021)

The EDT focuses on a series of key locomotor, cognitive and social behaviors, chosen in consultation with end users, to enhance apes' quality of life in captivity and release success. Data collection protocols are suitable for captive facility staff without scientific backgrounds or high levels of literacy. Users can collect data on ape behavior during normal husbandry procedures, upload them to the EDT and then compare them to data on wild individuals. Comparisons are presented graphically, enabling users to explore results. In interpreting the graphs, including data on behaviors in the wild and in captivity, users can ascertain good results and highlight areas that could be improved. Users can also use data analysis to produce bespoke recommendations on how best to modify captive settings to elicit absent or underrepresented wild-type behaviors. Once users have made the modifications, they can collect and upload a post-modification dataset to the EDT, which compares it to wild and baseline data (Chappell and Thorpe, 2022).

Is the EDT Effective?

Results are promising (see Figure 8.8). At Ape Action Africa, for example, implementation of EDT recommendations led to a substantial reduction in the rate of attacks among chimpanzees—from 0.32 to 0.07 per hour of observation—which resulted in a decrease in chimp-induced injuries and the need for veterinary intervention. The results also show a radical increase in activity and engagement thanks to the provision of physical and cognitive stimulation, even within very small cages (5 m × 5 m × 3 m). In particular, there was a significant increase in the percentage of time chimpanzees spent foraging and off the ground, as well as in the percentage of off-the-ground time spent moving on energy-demanding flexible supports (see Figures 8.8A–C).

Similarly, orangutans at the Bornean Orangutan Survival Foundation substantially increased arboreal and complex activity, more than tripling the percentage of time spent in locomotion and doubling the percentage of physically demanding wild-type climbing and clambering behavior (see Figures 8.8D and 8.8E). Taken together, these changes can substantially increase the apes' strength and stamina, which they would need for prolonged bouts of locomotion in the wild, should they be released. Another benefit is the considerable decrease in inactivity, as the percentage of time the orangutans spent manipulating food, nests and other objects doubled (see Figure 8.8F).

One way to ensure uptake of the EDT and keep the process practicable is to balance the amount of data required for meaningful impact with the time required for data collection. The next steps involve creating EDTs for all ape species and rolling out the tool to a wider range of facilities (Chappell and Thorpe, 2022). The results may help to guide the development of environments that are more conducive to rehabilitation for release and that can support—in conjunction with explicit welfare indicators for post-release monitoring—positive welfare outcomes for released apes.

Shared Learning and Action Supporting Animal Welfare

Supporting the many dimensions of animal welfare requires multiple competencies rarely found within a single institution (Kagan, Carter and Allard, 2015; Sinclair and Phillips, 2018a, 2018b). Almost three-quarters of sanctuaries are part of collaborations, with some participating in more than one (Sherman and Greer, 2018). Benefits of collaboration include greater access to expertise and resources; an increase in reach (credibility, visibility and access to relevant connections); opportunities to influence policy through a collective voice; and improved processes, both internal (such as through access to established procedures or adoption of methods and mindsets) and external (such as enhanced approaches to confiscation and ape transfer).

Each collaboration has its own set of goals. For example, the Gorilla Rehabilitation and Conservation Education Center (GRACE) in the eastern DRC was built as a collaborative partnership from the outset, comprised of GRACE staff, based in the DRC and the United States, and advisors with expertise in gorilla management at multiple AZA-accredited US zoos (K. Fawcett, personal communication, 2020). Reciprocal sharing of knowledge and learning has played a critical role in building and strengthening GRACE, enabling the facility to be resilient and maintain welfare during times of crisis and insecurity, including the Ebola epidemic and the COVID-19 pandemic (see Figure 8.9). In 2019, GRACE became the first great ape sanctuary in Africa to receive accreditation from GFAS; in 2020, the facility received the GFAS Annual Outstanding

International Sanctuary Award (GFAS, 2019, 2020).

Research collaboration among facilities, universities and other scientific organizations can serve to bolster welfare knowledge and practice (Ross and Leinwand, 2020; Sherwen *et al.*, 2018). In Spain, for example, collaboration between the MONA Founda-tion sanctuary and the University of Girona has supported hundreds of master's students in primatology. They have conducted non-invasive research at MONA, gaining degrees and developing career pathways while providing the sanctuary with a long-term, continuous monitoring dataset to aid captive management and welfare.

FIGURE 8.9

Strengthening Capacity for Animal Health and Welfare, GRACE, Eastern DRC

Note: Zoo partners have provided GRACE staff members with positive reinforcement training to facilitate examination of gorillas in the eastern DRC. Gorilla training: arm up.

© GRACE Gorillas

In the United States, collaboration between Chimp Haven and the Lincoln Park Zoo Fisher Center represents the first major partnership between an accredited sanctuary and an accredited zoo in North America. Their collaborative projects have focused on outcomes that support the overlapping philosophies of improving chimpanzee care and management. Outputs include Chimp Haven's longitudinal dataset, which is almost exclusively welfare-based (Ross *et al.*, 2019).

Captive Ape Welfare: Conclusion

Previous volumes of *State of the Apes* have reviewed various forms of ape captivity as well as some of the laws that regulate them, observing that what is permitted or prohibited varies, and that current standards do not always meet the needs of apes or promote their wellbeing. The welfare of captive wildlife has yet to be adequately included in relevant dialogue, especially since it often falls between the cracks of domestic animal health and wildlife conservation legislation and regulations. Annex IX considers recent legal "bright spots," including strengthened national standards in Malawi and Costa Rica, whose experiences indicate, for example, that welfare crime is best included in campaigns for legislative reform from the outset.

How animal welfare is understood and discussed influences how it is assessed and how resulting findings are used (Beausoleil *et al.*, 2018). The COVID-19 pandemic has put a spotlight on systems approaches, presenting opportunities to look at the relationship between humans and animals more holistically, and to integrate animal welfare considerations into dialogue, strategy and legal instruments. Efforts to mainstream animal welfare on the global environmental agenda include work on a dedicated UN convention.

The legal systems to combat the illegal wildlife trade, for example, can only be strengthened if each part of the trade chain is targeted, including welfare outcomes for confiscated and surrendered apes. Weak links between the welfare and conservation sectors, the tendency to treat welfare as a low priority and the exclusion of relevant stakeholders from important discourse also need to be addressed if regulatory gaps are to be filled. The inclusion of dialogue on animal welfare in high-level intergovernmental forums on combating the illegal wildlife trade would help to ensure that the full range of costs borne by captive facilities are identified and factored into sentencing and penalty decisions, as well as related decision-making and actions. Both animal welfare and conservation outcomes would benefit as a result. The language used in these discussions, however, needs to reflect current legal and scientific thinking—as well as public opinion—on animal sentience.

While professional body accreditation is not a substitute for national standards, robust systems can provide a benchmark for welfare management. Committing to rigorous standards and a trustworthy compliance system can provide greater credibility and accountability (Lundmark, Berg and Röcklinsberg, 2018; Pierce and Bekoff, 2018). The process of applying for accreditation can be as valuable as the actual stamp of approval, mainly because it demands self-reflection and forces facilities to articulate and formalize crucial internal policies and processes (GRACE, 2019).

Captive facilities' institutional systems serve as bastions of good welfare and tend to determine what apes experience on a day-to-day basis (see Figure 8.10). Key elements that embed animal welfare in a facility's operations include the explicit recognition of a commitment to promote animal wellbeing in its organizational statements, principles and values; an operational framework through which the institutional philosophy is delivered, such as committees, policies and welfare assessment tools; an organizational culture that furthers the approach by investing in staff; and dedicated resources (Farmer, 2012; Kagan, Carter and Allard, 2015; Walraven and Duffy, 2017). While strategy is important, the people part of the system can be an organization's most valuable resource. Initiatives and programs focused on strengthening leadership skills may be increasingly prevalent in the conservation field, but they remain absent for animal welfare (Bruyere *et al.*, 2020).

Appropriate, enforceable standards and control systems at the legislative, professional accreditation and institutional levels serve as mutually reinforcing insurance policies to safeguard welfare and linked conservation outcomes. Understanding why welfare programs go wrong can be complex; poor results may be attributable to shortcomings at different levels, any or all of which could potentially impact features of a welfare system (see Annex X). Society also plays a role in setting welfare standards, in particular by granting or denying industrial outfits, such as captive facilities that are open to the public, a social license to operate. As scientific evidence that animals think and feel grows, so does public concern for animal welfare. As a result,

FIGURE 8.10

Adaptations to the Physical Environment at the Fauna Foundation, Canada

Note: Ropes placed along the ceiling and a smooth floor help Sue Ellen, who has limited use of her legs, move through the tunnel and enclosure at Fauna Foundation, Canada.

© Justin Taus / Fauna Foundation

animal welfare is increasingly becoming a crucial consideration in people's willingness to give "wildlife use industries" a social license to operate (Hampton, Jones and McGreevy, 2020).

Collaboration within and across disciplines—as well as among animal welfare and conservation practitioners, scientists, lawyers and technologists—can harness knowledge and resources. A strong ethos of shared learning and collaboration across the sanctuary and zoo sector has led to a better understanding of the most important features of a captive environment for great apes, although less attention has been paid to gibbons (Fernie *et al.*, 2012; Ross and Leinwand, 2020). The importance of specific features and services may vary according to the species and context, for example with respect to lifetime care and rehabilitation for release.

Universally agreed species-specific welfare indicators would not only allow for a more comprehensive way to assess the captive world for apes, but also to help shape it. They would aid monitoring within and across facilities, help set professional standards and make it easier for authorities to determine whether a welfare crime has been committed (Whitfort, 2019). While such indicators remain elusive, this section and Annex XI present examples of promising tools and initiatives.

Welfare assessment tools must walk the line between being practical for the context, but specific enough to produce useful outcomes (Wark *et al.*, 2019). Relatively few sanctuaries have the resources to employ scientists, and many studies are initiated by external academics and students (Ross and Leinwand, 2020). Such partnerships can help bolster welfare knowledge and practice. Just like conservation, however, welfare needs to be championed locally to shape narratives and gain prominence (Sayektiningsih *et al.*, 2020; Sinclair and Phillips, 2018b).

Section II: The Status and Number of Captive Apes

Overview

This update provides country-level animal welfare scores for 2020, drawn from the Animal Protection Index (API) produced by World Animal Protection (Nizamuddin and Rahman, 2019; WAP, n.d.-a). It covers regions and countries whose scores were not reported in previous volumes of *State of the Apes*.

The scores range from A (highest) to G (lowest). A country's "overall" API score is the average of its scores in ten categories, which cover recognition of animal sentience, the presence of animal welfare legislation, the establishment of supportive government bodies and support for international animal welfare standards. This section presents overall scores as well as scores for two indicators that pertain to apes in captivity and in the wild: "legislation protecting animals in captivity" and "legislation protecting the welfare of wild animals" (WAP, n.d.-c).

This section also provides the best available data on the number of apes in captive facilities in 2020, gathered from reliable and transparent sources such as databases, published annual reports and personal communications whenever possible. In the absence of such sources, data were drawn from facility webpages, facility social media accounts and news articles. As noted in previous volumes of *State of the Apes*, data on captive apes are sometimes incomplete or inconsistent (Durham, 2020).

Africa

Overall, API scores for African countries range from D in Kenya and Tanzania to F in Algeria, Egypt, Ethiopia and Morocco (WAP, n.d.-a; see Table 8.1). The scores indicate that animal welfare conditions across

Africa range from somewhat deficient to very poor. Scores for legislation protecting animals in captivity are equivalent to or weaker than the overall API average scores. Nigeria, the only African ape range state to be scored, was given an E for its "legislation protecting the welfare of wild animals," as the country's anti-cruelty legislation does not extend to wild animals and there is no ban on non-subsistence hunting (WAP, n.d.-a).

Zoos

For the year 2020, African zoos reported holding 53 chimpanzees, 26 gibbons, 5 gorillas and 1 orangutan, accounting for 6% of apes in captivity on the continent (Species360, n.d.). While gorilla and orangutan numbers were similar to those reported in previous volumes of *State of the Apes*, chimpanzee and gibbon numbers increased from 46 and 22 in 2018, respectively. Data on African zoos are limited in scope and may be underestimated, due in part to the voluntary nature and cost of reporting (Durham, 2020). The Species360 database indicates that eight African institutions accounted for 53 chimpanzees in 2020; by contrast, one rescue center's records for the same year show that 27 chimpanzees were in public and private zoos in Ivory Coast alone (E. Raballand, personal communication, 2020).

TABLE 8.1

API Scores for African Countries, 2020

Country	Overall API score	Animals in captivity	Welfare of wild animals
Algeria	F	F	E
Egypt	F	F	E
Ethiopia	F	F	E
Kenya	D	F	A
Morocco	F	F	D
Niger	E	E	D
Nigeria	E	E	E
South Africa	E	F	E
Tanzania	D	G	C

Notes: The only range state on the list, Nigeria, is shaded in gray. Scores in the last two columns relate to legislation that protects animals in captivity and legislation that protects the welfare of wild animals, respectively.

Data source: WAP (n.d.-a)

TABLE 8.2

Number of Apes in African Sanctuaries, 2011–2020

Year	Bonobos	Chimpanzees	Gorillas	Total
2011	55	1,071	83	1,209
2015	72	1,072	127	1,271
2018	70	1,136	118	1,324
2020	70	1,261	75*	1,406

Notes: Figures may include apes in pre-release or other semi-wild release sites. *Indirect accounts suggest there are between one and a few additional gorillas at Projet Protection des Gorilles sites in Gabon and the Republic of Congo (G. Tully, personal communication, 2020).

Data sources: Ambassade de France (2019); Ape Action Africa (n.d.); Chimfunshi Wildlife Orphanage (n.d.); Chimp Eden (n.d.); Chimpanzee Conservation Center (2020); Durham (2018, 2020); Friends of Animals (n.d.); GRACE (2020); HELP Congo (n.d.); J.A.C.K. Sanctuary (n.d.); Jane Goodall Institute (n.d.); Limbe Wildlife Centre (2020); Ngamba Island Chimpanzee Sanctuary (2020); P-WAC (2020); Parc National des Virungas (n.d.); PASA (n.d.a); Prak (2020); Projet Gorille Fernan-Vaz (n.d.); Second Chance Chimpanzee Refuge Liberia (2020); Tacugama Chimpanzee Sanctuary (n.d.); personal communication in 2020 with N. Bachand, K. Cereghino, N. Colwill, J. Desmond, K. Farmer, D. Morel, S. Ngulu, E. Raballand and G. Tully

Sanctuaries

African sanctuaries housed 1,406 apes in 2020 (see Table 8.2). The number of bonobos (*Pan paniscus*) in sanctuaries has remained largely unchanged in recent years. While 14 bonobos from Lola Ya Bonobo sanctuary awaited their release to a dedicated reserve, 15 bonobos were rescued in 2019 and 2020. Sanctuary staff suggest that the increase may be due to greater engagement of local conservation actors, as well as a rise in poaching for wild meat and wildlife trafficking (D. Morel, personal communication, 2020).

The number of gorillas reported in sanctuaries in 2020 is significantly lower than in previous years. As nearly all facilities with gorillas supplied data for 2020 for this update, the level of confidence in the figures is high. Reasons for the drop since 2015 and 2018 are unknown. African sanctuaries, including the Chimp Rehabilitation Project in the Gambia, often include apes released into semi-wild or pre-release settings in reported captive populations (J. Sherman, personal communication, 2020). Hence, one possibility is that released gorillas were counted in previous sanctuary population estimates.

Non-range state sanctuaries account for 23% of African sanctuary chimpanzees reported for 2020 (see Table 8.3). Their numbers are similar to those reported in previous years (Durham, 2020). Captive chimpanzees in range states, however, have increased by 15% since 2018 and 21% since 2011 (Durham, 2015, 2020; see Table 8.4). The increase in intake of chimpanzees at range-state sanctuaries suggests that wild meat hunting and subsequent trafficking of orphans may be increasing (GRASP and IUCN, 2018; Ondoua *et al.*, 2017; J. Desmond, personal communication, 2020). Intake rates could also be influenced by other factors, such as increased rescue efforts, which would be expected given the growth of the Liberia Chimpanzee Rescue and Protection sanctuary

TABLE 8.3

Number of Great Apes in African Sanctuaries, by Country, 2020

Country	Number of sanctuaries	Bonobos	Chimpanzees	Gorillas
Cameroon	4	0	271	40
DRC	6	70	134	18
Gabon	3	0	30	17*
Gambia	1	0	>100	0
Guinea	1	0	64	0
Ivory Coast	1	0	3	0
Kenya	1	0	36	0
Liberia	2	0	127	0
Nigeria	1	0	28	0
Republic of Congo	3	0	172	0*
Sierra Leone	1	0	92	0
South Africa	1	0	33	0
Uganda	1	0	50	0
Zambia	1	0	120	0
Total	**27**	**70**	**>1,260**	**75**

Notes: Figures account for sanctuary populations and may include apes in pre-release or other semi-wild release sites. Range states are shaded in gray.

* Indirect accounts suggest there are between one and a few additional gorillas at Projet Protection des Gorilles sites in Gabon and the Republic of Congo (G. Tully, personal communication, 2020).

Data sources: Ambassade de France (2019); Ape Action Africa (n.d.); Chimfunshi Wildlife Orphanage (n.d.); Chimp Eden (n.d.); Chimpanzee Conservation Center (2020); Friends of Animals (n.d.); GRACE (2020); HELP Congo (n.d.); J.A.C.K. Sanctuary (n.d.); Jane Goodall Institute (n.d.); Limbe Wildlife Centre (2020); Ngamba Island Chimpanzee Sanctuary (2020); PASA (n.d.a); Prak (2020); Projet Gorilles Fernan-Vaz (n.d.); Second Chance Chimpanzee Refuge Liberia (2020); Tacugama Chimpanzee Sanctuary (n.d.); personal communication in 2020 with N. Bachand, K. Cereghino, N. Colwill, J. Desmond, K. Farmer, D. Morel, S. Ngulu, E. Raballand and G. Tully

TABLE 8.4

Number of Chimpanzees in Range-State Sanctuaries, 2011, 2015, 2018 and 2020

Country	2011	2015	2018	2020
Cameroon	244	246	247	271
DRC	85	109	117	134
Gabon	20	20	20	30
Guinea	38	50	46	64
Ivory Coast	4	1	2	3
Liberia	76	63	99	127
Nigeria	28	30	28	28
Republic of Congo	156	145	161	172
Sierra Leone	101	75	74	92
Uganda	45	49	49	50
Total	**797**	**788**	**843**	**971**

Data sources: Ambassade de France (2019); Ape Action Africa (n.d.); Chimfunshi Wildlife Orphanage (n.d.); Chimp Eden (n.d.); Chimpanzee Conservation Center (2020); Durham (2018, 2020); Friends of Animals (n.d.); HELP Congo (n.d.); J.A.C.K. Sanctuary (n.d.); Jane Goodall Institute (n.d.); Limbe Wildlife Centre (2020); Ngamba Island Chimpanzee Sanctuary (2020); P-WAC (2020); PASA (n.d.a); Prak (2020); Second Chance Chimpanzee Refuge Liberia (2020); Tacugama Chimpanzee Sanctuary (n.d.); personal communication in 2020 with K. Cereghino, J. Desmond, K. Farmer, E. Raballand and G. Tully

in Liberia and the expansion of the non-governmental organization Eco-Activists for Government and Law Enforcement (EAGLE) across African ape range states (EAGLE, 2019; Liberia Chimpanzee Rescue & Protection, n.d.).

Increased chimpanzee intake is of concern for three main reasons. First, the slow maturation and reproduction rates of chimpanzees make their populations especially vulnerable to poaching pressure (Ondoua *et al.*, 2017). Second, the number of chimpanzees in sanctuaries is increasing in three of the range states of the critically endangered western chimpanzee (*Pan troglodytes verus*)—Guinea, Liberia and Sierra Leone (Durham, 2018, 2020). Third, the rise in intake may be linked to the increase in commercial and private poaching, due in part to the financial impact of the COVID-19 pandemic on the tourism industry in areas surrounding wild ape populations (Dalton, 2020; Somerville, 2020; Zenda, 2020).

Asia

API scores for Asian states range from C in India and Malaysia to E in China, Indonesia and Japan (see Table 8.5). These scores demonstrate that animal welfare is relatively poor in Asian countries. Most range states have a better score for legislative protection of animals in captivity than they do for protecting the welfare of wild animals. This discrepancy suggests a need for improved legislation to protect the welfare of wild animals, which, if enforced, could decrease the numbers of wild apes entering captivity.

China is the exception to the trend: its score for legislation protecting the welfare of wild animals (D) is higher than its score for laws that protect the welfare of animals in captivity (E). The higher score reflects China's introduction of bans on advertisements concerning the illegal wildlife trade and the ivory trade. Wildlife trafficking and the treatment of wildlife as a resource continue

TABLE 8.5

API Scores for Asian Countries, 2020

Country	Overall API score	Animals in captivity	Welfare of wild animals
Azerbaijan	G	E	E
China	E	E	D
India	C	C	E
Indonesia	E	D	D
Iran	G	G	E
Japan	E	D	E
Korea	D	D	D
Malaysia	C	C	D
Myanmar	F	D	E
Pakistan	E	E	D
Philippines	D	D	E
Thailand	D	D	E
Vietnam	F	E	E

Notes: Range states are shaded in gray. Scores in the last two columns relate to legislation that protects animals in captivity and legislation that protects the welfare of wild animals, respectively.

Data source: WAP (n.d.-a)

to have a deleterious impact on animal welfare in China (WAP, n.d.-a, n.d.-b).

Zoos

The number of apes reported by Asian zoos for the year 2020—excluding those in Japan, which is discussed below—is substantially lower than the figure for 2018 (Durham, 2020; Species360, n.d.; see Table 8.6). Roughly one-third of the drop is due to a change in methodology: Israel, Turkey and the United Arab Emirates were included in Asian regional totals for 2018, but for 2020 these countries are listed among the European totals because their reporting zoos are EAZA members (D. Durham, personal communication, 2020; see Figure 8.11).

Concerning gibbons in particular, experts indicate that the discrepancy between the 2018 and 2020 zoo population numbers is probably due to data and reporting problems

TABLE 8.6

Apes in Asian Zoos Reporting to Species360, excluding Japan, 2018 and 2020

Year	Chimpanzees	Orangutans	Gorillas	Gibbons	Total
2018	220	170	25	436	**851**
2020	137	144	8	280	**569**

Notes: Figures from 2020 represent aggregate zoo data and could include previous years' holdings.[10] No bonobos were reported. As not all zoos report to Species360, this table does not cover all apes in Asian zoos.

Data source: Species360 (n.d.)

(S. Cheyne and B. Lefaux, personal communication, 2020). Zoo numbers can be affected by illegal trade in gibbons, which is not reported to CITES. Since many Chinese and other Asian zoos did not report to Species360, however, no firm conclusions can be drawn (B. Lefaux, personal communication, 2020).

Variations in reported zoo holdings demonstrate the limitations of data from voluntary databases such as Species360. Participation in Species360 and similar databases is limited in regions around the world, including Asia (Banes *et al.*, 2018; Durham, 2020). Two recent studies of Asian zoos underscore this point. The first presents data from 58 Asian, non-Japanese institutions. In comparison, just 30 institutions outside of Japan reported to Species360 in 2020 (Banes *et al.*, 2018; Durham, 2020; Species360, n.d.). The second report counts

213 gibbons, 85 orangutans, 50 chimpanzees and 1 gorilla in zoos in Thailand and Malaysia alone. Of the 42 institutions identified in this report, only seven reported to Species360 in 2020 (Beastall, Bouhuys and Ezekiel, 2016; Species360, n.d.).

Rescue and Rehabilitation Centers and Sanctuaries

Figures provided by Indonesian facilities for 2020 indicate that the number of captive orangutans fell by 12% since 2016 (see Table 8.7). Two key factors help to explain this decrease. First, massive forest fires in 2015 led to an unusually high number of orangutan rescues in 2016 (Sherman, Ancrenaz and Meijaard, 2020). Second, orangutan populations fluctuate year to year, depending on releases of rehabilitated orangutans back into the wild. More than

TABLE 8.7

Number of Orangutans and Gibbons in Asian Rescue and Rehabilitation Centers and Sanctuaries, by Country, 2016 and 2020

Country	Orangutans		Gibbons	
	2016	2020	2016	2020
Cambodia	n/a	n/a	77	93
India	n/a	n/a	–	15
Indonesia	1,147	1,006	293	439
Lao People's Democratic Republic	n/a	n/a	–	5
Malaysia	98	87	–	12
Taiwan	n/a	n/a	–	15
Thailand	2	2	229	163
Viet Nam	n/a	n/a	45	39
Total	**1,247**	**1,095**	**644**	**781**

Notes: Some figures are drawn from aggregated data submitted to Species360 and may thus reflect holdings from previous years. A median was used in a single case for which a range of values had been provided. "–" no data available. "n/a" not applicable, as there are no orangutans in rescue and rehabilitation centers in these countries.

Data sources: BOSF (2020); Durham (2018); Endangered Asian Species Trust (2020a, 2020b, 2020c); Gibbon Rehabilitation Project (n.d.); Highland Farm (n.d.); Lee, Leong and Dzar (2020); Orangutan Appeal UK (n.d.); Orangutan Foundation International (n.d.); SOC (n.d.); Species360 (n.d.); Wildlife Rescue Center Jogja (n.d.); personal communication in 2020 with L. Biddle, B. Chan, A. Brulé, R. Durgut, D. Hendarto, F. Magne, N. Marx, P. Nurantika, K. Pei, A. Pipe, S. Preuschoft, K. Sánchez, J. Sherman, T. Tran and M. Wedana

600 rehabilitants were released between 2007 and 2017, while more than 100 were released in 2019 alone (BOSF, 2020; International Animal Rescue, 2020; PanEco, 2020; Sherman, Ancrenaz and Meijaard, 2020). Intake of newly rescued wild orangutans also continues at high levels, with more than 50 rescued in 2019 alone (BOSF, 2020; International Animal Rescue, 2020; PanEco, 2020).[11]

There are also notable variations across Indonesian reporting facilities. One rescue and rehabilitation center, International Animal Rescue, reported a markedly lower intake of orangutans in 2018 and 2019, which, coupled with releases of rehabilitated orangutans, has resulted in a lower population overall (K. Sánchez, personal communication, 2020).

Continued analysis is necessary to confirm whether the overall orangutan population decline indicated by the data constitutes a trend, especially given the continued pressures of poaching, possession and trafficking facing wild populations (Freund, Rahman and Knott, 2017; Nijman, 2017; Sherman, Ancrenaz and Meijaard, 2020). Recent research suggests captive populations in Indonesia will not decline to zero at current rates of rescue, release and captive birth (Sherman, Ancrenaz and Meijaard, 2020).

Gibbon populations may appear to have dropped by 29% in Thailand and increased by 50% in Indonesia between 2016 and 2020, but inconsistencies in gibbon facility numbers make it difficult to draw conclusions with any degree of accuracy (Ancrenaz et al., 2020, table 1.1; Durham, 2018; see Table 8.7). Moreover, the illegal pet trade, which is increasingly supported by social media, continues to push gibbons into captivity (Gill, 2017; Rainer et al., 2020; Yu and Jia, 2015). Laws protecting apes are often disregarded, as few offenders are prosecuted (Nijman, 2017).

Japanese Captive Facilities

In Japan, figures for apes in captive facilities for 2020 are 3% lower than for 2018, down to 6 bonobos, 303 chimpanzees, 20 gorillas, 46 orangutans and 170 gibbons (Durham, 2020; GAIN, n.d.). Figures for Japan are more reliable than those for some other regions, as the GAIN studbook in which they are reported is complete and consistent (Banes et al., 2018; Durham, 2018). Japan has an overall API score of E (WAP, n.d.-a; see Table 8.5).

Europe

In 2020, European zoos held a total of 2,467 captive apes, one-third of whom were gibbons (Species360, n.d.; see Figure 8.11). Compared to data for 2018, the 2020 figures show that the total ape population increased 3%, with the number of chimpanzees rising by 5% and gibbons by 4% (Durham, 2020). While in keeping with rates for stable captive population growth, these increases could reflect the addition of data from zoos in Israel, Turkey and the United Arab Emirates. As noted above, these countries are included in the European 2020 dataset because their institutions are EAZA members (Species360, n.d.).

The most recent figures for apes in European sanctuaries suggest there could be many as 186 chimpanzees, 15 orangutans and 17 gibbons.[12]

Fewer lone apes in captivity were reported in 2020 (13) than in 2018 (23) (Durham, 2020; Species360, n.d.). On the legislative front, France and Lithuania joined 28 other European countries in enacting bans on the use of wild animals in circuses and more generally (AFP, 2020; Four Paws International, 2020b; GATO, 2020). In Spain, the Castilla–La Mancha region—which is home to more than three-quarters of Spain's population—has banned wild animals in circus performances (AAP, 2020b;

FIGURE 8.11

Number of Apes in European Zoos Reporting to Species360, 2012, 2016, 2018 and 2020

Key: ■ Gibbons ■ Orangutans ■ Gorillas ■ Chimpanzees ■ Bonobos

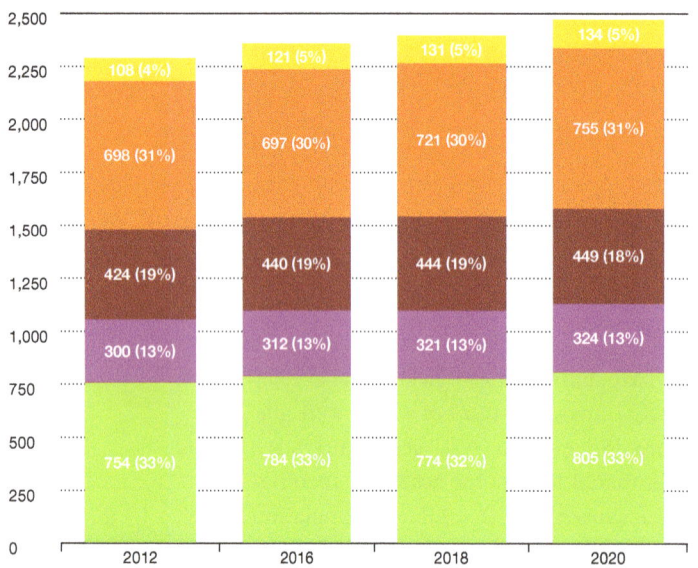

Notes: Figures for 2020 are drawn from aggregated data submitted to Species360 and may thus reflect holdings from previous years. Institutions reporting to Species360 from Israel, Turkey and the United Arab Emirates are included here. As not all zoos report to Species360, this table does not cover all apes in European zoos.

Data sources: Durham (2015, 2018, 2020); Species360 (n.d.)

Pozo, 2020). It is worth noting that trafficking remains a concern in Europe. Research conducted over a six-week period in 2017 shows that 152 live primates were sold via online sales, mainly on Russian platforms (IFAW, 2018).

Latin America

A recent survey of Latin American zoos found that while facilities and training for veterinarians need to be improved, most survey respondents had access to tools to manage large animals, including apes. In addition, the Association of Latin American Zoos (ALPZA) started implementing accreditation standards in 2017 (Riva, Zordan and Sánchez, 2020). All ALPZA members must be accredited under ALPZA standards by 2025 (ALPZA, n.d.).

Latin American facilities—zoos and sanctuaries—reported 183 apes in 2020 (see Table 8.8). These include 79 chimpanzees and one orangutan in four sanctuaries associated with the Great Ape Project (J. Ramos, personal communication, 2020).

The number of apes reported to Species360 for 2020 is lower than the 2018 figure (Durham, 2020; see Table 8.8). As mentioned in Volume IV of *State of the Apes*, official reporting by Latin American zoos, including to databases such as Species360, is limited (Durham, 2020). In 2020, just 16 institutions reported housing apes to Species360.[13] As in 2018, additional information for the 2020 data was collected via personal communication or online searches (Durham, 2020). Variations in figures provided for 2018 and 2020 in Table 8.8 may be partly due to the use of unofficial results, the lack of responses from certain sources and inconsistent reports obtained though indirect inquires.

TABLE 8.8

Number of Apes in Selected Latin America Facilities, 2020

	Chimpanzees	Gorillas	Orangutans	Gibbons	Total
2018	170	12	13	19	**208**
2020	151	9	10	13	**183**

Notes: Zoo data are aggregated and could include previous years' holdings. No bonobos were reported. As not all zoos report to Species360 or share their ape figures on their webpages or social media sites, this table does not cover all apes in Latin American zoos.

Data sources: adnCUBA (2020);[14] Durham (2020); Fauna Silvestre de Nicaragua (2020); G1 (2020); Listín Diario (2019); Matos Mendes (2020); Olhar Animal (2020); Sierra Maestra (2020); Species360 (n.d.); ZooLeón (n.d.);[15] Zoológico de Culiacán (2020); I. Ho and J. Ramos, personal communication, 2020

Oceania

Figures for apes in Australian zoos are consistent with those presented in the previous volume of *State of the Apes*, except figures for chimpanzees (Durham, 2020). While the population size is small, the number of chimpanzees in zoos increased by roughly 35% from 2018 (see Table 8.9). This increase could be the result of a transfer of 12 chimpanzees to Australia from Germany (CITES, n.d.). Facing possible closures due to the budgetary ramifications of the COVID-19 pandemic, the Australian government granted nearly AUS$100 million (US$65 million) to help zoos with animal welfare costs for up to six months (Brown, 2020).

New Zealand zoos reported similar figures for 2020 as they did for 2018, except for a decrease in the number of gibbons (Durham, 2020; see Table 8.9). Reasons for the decrease are not clear. No transfers are indicated in the CITES trade database (CITES, n.d.). The New Zealand government gave out nearly NZ$9.5 million (US$6.2 million) to support zoos in response to COVID-related reductions in revenue (RNZ, 2020).

United States

US facilities reported a total of 2,576 captive apes for 2020 (see Figure 8.12). Compared to 2018, the captive populations of the following taxonomic groups increased: bonobos (up by 8%), orangutans (10%) and gibbons (21%) (Durham, 2020). These increases could be the result of animals reported here that are not typically reported in Species360. The Fort Worth Zoo, for example, does not report to Species360, but the bonobo data it reported on social media are included in Figure 8.12. Similarly, the number for captive orangutans includes individuals outside of AZA-accredited zoos and GFAS-accredited sanctuaries, for which data were obtained by an expert in the field. In the case of gibbons, the increase is probably due to higher num-

TABLE 8.9

Number of Apes in Zoos in Australia and New Zealand Reporting to Species360, 2018 and 2020

	Australia		New Zealand	
	2018	**2020**	**2018**	**2020**
Chimpanzees	37	50	16	16
Gorillas	17	19	3	3
Orangutans	18	18	3	3
Gibbons	54	51	15	9
Total	**126**	**138**	**37**	**31**

Notes: Figures for 2020 are drawn from aggregated data submitted to Species360 and may thus reflect holdings from previous years. As not all zoos in Australia and New Zealand report to Species360, this table does not cover all apes in zoos in these two countries.

Data sources: Durham (2020); Species360 (n.d.)

bers of gibbons reported in zoos outside of Species Survival Plan populations, which are relatively stable (B. Richards, personal communication, 2020).

Durham (2018) reported that changes to the US Department of Agriculture's Animal Welfare Act database dramatically impacted the accessibility of data on apes held in private collections. As of September 2020, the

FIGURE 8.12

Number of Apes in Captivity in the United States, 2020

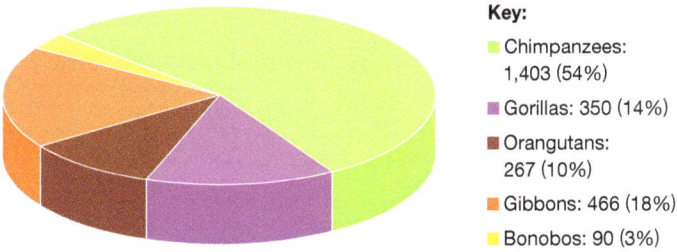

Key:
- Chimpanzees: 1,403 (54%)
- Gorillas: 350 (14%)
- Orangutans: 267 (10%)
- Gibbons: 466 (18%)
- Bonobos: 90 (3%)

Notes: Some figures are drawn from aggregated data submitted to Species360 for 2020 and may thus include previous years' holdings. Bonobo numbers were obtained through Species360, the Bonobo Species Survival Plan website hosted by the Zoological Society of Milwaukie, and the Fort Worth Zoo. Orangutan numbers are from Megan Elder of the Orangutan Species Survival Plan, Patti Ragan at the Center for Great Apes and from Species360; they include animals in non-AZA zoos, entertainment facilities and a private breeder. Gorilla numbers are from Roby Elsner of the Gorilla Species Survival Plan and include non-AZA zoos.

Data sources: A.P.E.S. (n.d.); ChimpCare (n.d.-a); Fort Worth Zoo (2020); GCC (n.d.); IPPL (n.d.); Species360 (n.d.); Zoological Society of Milwaukie (n.d.); personal communication in 2020 with M. Elder, R. Elsner, K. Lukas, and P. Ragan, S. Ross, J. Tagliatatela and L. Wathne

database was again accessible online (PETA, 2020; USDA, 2020). Nonetheless, there is no reliable way to obtain information about the number of gibbons held by private parties or institutions outside of AZA-accredited zoos and GFAS-accredited sanctuaries (Durham, 2018; B. Richards, personal communication, 2020).

While captive gorilla populations have remained stable, captive chimpanzee numbers continue to decline, dropping 27% since 2011 (Durham, 2020; see Figure 8.12 and Table 8.10). Very few chimpanzees are born in private settings and sanctuaries do not have any breeding programs (Ross and Leinwand, 2020; S. Ross, personal communication, 2020). Sanctuary populations grew by 13% between 2018 and 2020, largely as a result of the transfer of chimpanzees from biomedical facilities; this trend will probably continue in the short term, before declining as the captive chimpanzee population ages and eventually dies out (Ross and Leinwand, 2020).

Statistical Update: Conclusion

Poaching, trafficking and increasing human–ape conflict due to pressures associated with the loss of habitat and political instability continue to be a problem for wild ape populations. As reflected in this section, the results include the ongoing intake of apes into sanctuaries and continued demand for rescue–release operations. Sanctuaries, in particular, often operate at, or over, capacity (Karokaro, Gokkon and Suriyani, 2017 ; G. Tully, personal communication, 2020). The pressures apes face in the wild may have been compounded in 2020 due to the knock-on effects of the COVID-19 pandemic on tourism-dependent local economies.

The lack or inadequacies of law enforcement may also have played a role in the 2020 captive ape data. Trafficking of apes and other wildlife online, especially through social media, continues to be a concern. More studies are needed to assess the impact of anti-trafficking measures on social media, law enforcement and offenders.

Based on reliable data that are available, captive ape populations are largely stable. This conclusion can be drawn for locations where data were reported to captive ape databases—such as ChimpCare, GAIN and Species360—and where sanctuary data were shared publicly.

On the whole, however, data continue to be limited in breadth because relatively few

TABLE 8.10

Number of Chimpanzees in Different Forms of Captivity in the United States, 2011–2020

Captivity Type	2011	2014	2016	2018	2020	% change from 2018	% change 2011–2020
Biomedical labs	962	794	658	464	310	-33%	-68%
GFAS* sanctuaries	522	525	556	585	659	13%	26%
AZA zoos**	261	258	259	236	244	3%	-7%
Unaccredited ***	106	196	111	192	154	-20%	45%
Dealer or pet owner	60	52	37	61	25	-59%	-58%
Trainer or media	20	18	13	10	11	10%	-45%
Total	**1,931**	**1,843**	**1,634**	**1,548**	**1,403**	**-9%**	**-27%**

Notes: * Global Federation of Animal Sanctuaries. ** Association of Zoos and Aquariums. *** Unaccredited facilities include zoos that are not AZA members, as well as sanctuaries and institutions that are not accredited by the North American Primate Sanctuary Alliance; these sites may share characteristics with accredited zoos and sanctuaries, as well as dealers.

Data sources: 2011: Durham and Phillipson (2014, fig. 10.2); 2014: Durham (2015, table 8.4); 2016: Durham (2018, table 8.1); 2018: Durham (2020); 2020: ChimpCare (n.d.-a)

institutions report to Species360 or other transparent zoo population databases. The reliability of data also remains of concern, as many sanctuaries do not report any data publicly. This lack of information sharing hinders the accuracy of captive population assessments, while also precluding the establishment of a clear baseline number against which to compare changes over time.

There may be an opportunity for funders, accreditors and sanctuary coalition organizations to assist in the facilitation of data sharing by requiring transparency from their grantees, accredited facilities or members. Greater transparency can contribute to the development of a clearer picture of how the number of apes in captivity changes over time, which, in turn, can help to determine to what extent illegal hunting and trade activities continue to affect these species.

Acknowledgments

Principal authors

Section I: Supporting and Improving Ape Welfare: Kay Farmer[16] and Steve Unwin[17]

Section II: Number and Status of Captive Apes: Christine Caurant[18]

Contributors

Case Study 8.1:

Susannah Thorpe,[19] Johanna Neufuss,[20] Julia Myatt,[21] Emily Tarrega,[22] Gery Wamba,[23] Fransiska Sulistyo,[24] Alejandro Benítez López[25] and Jackie Chappell[26]

The following people generously shared their knowledge and learning for Section I of this chapter: Rebeca Atencia (Jane Goodall Institute), Nicholas Bachand (Projet Gorille Fernan-Vaz), Kari Bagnall (Jungle Friends Primate Sanctuary), Tuan Bendixsen (Animals Asia), Jackie Bennett (Global Federation of Animal Sanctuaries), Tammie Bettinger (Gorilla Rehabilitation and Conservation Education Center), Sabrina Brando (Animal Concepts), Sabine Brels (World Federation for Animals), Jessica Bridgers (World Federation for Animals), Aurélien Brulé (Kalaweit Project), Paula Cerdán Codina (World Association of Zoos and Aquariums), Bosco Chan (Kadoorie Farm and Botanic Garden), Susan Cheyne (Borneo Nature Foundation), Zanna Clay (Durham University), Jon Coe (Jon Coe Design), Christelle Collins (Centre de Conservation pour Chimpanzés), Kathleen Conlee (Humane Society International), Dietmar Crailsheim (MONA Foundation), Ali Crumpacker (Project Chimps), Grettel Delgadillo (Humane Society International), Cindy Dent (Humane Society International), Ioana Dungler (Four Paws), Katie Fawcett (Gorilla Rehabilitation and Conservation Education Center), Olga Feliu (MONA Foundation), Erika Fleury (North American Primate Sanctuary Alliance), Luis Flores (Centre de Réhabilitation des Primates de Lwiro), Danielle Free (Marwell Wildlife), Malene Friis Hansen (IUCN Species Survival Commission Primate Specialist Group (SSC PSG) Section for Human–Primate Interactions), Amy Fultz (Chimp Haven), Andrew Halloran (Save the Chimps), Mark Jones (Born Free Foundation), Miguel Llorente (University of Girona), Chris Lloyd (independent veterinary expert), Roger Lohanan (Thai Animal Guardians Association), Mary Lee Jensvold (Fauna Foundation), Godelieve Kranendonk (Animal Advocacy and Protection), Thirza Loffeld (University of Kent), Neil Maddison (Landscape Conservation Ltd.), Olga Martin (Animal Advocacy and Protection), Tetsuro Matsuzawa (Kyoto University), Linda May (Arcus Foundation), Dominique Morel (Lola ya Bonobo), Tilo Nadler (Endangered Primate Rescue Center), Citrakasih Nente (Sumatran Orangutan Conservation Programme), Johanna Neufuss (University of Birmingham), Signe Preuschoft (Four Paws), Heidi Quine (Animals Asia), Patti Ragan (Center for Great Apes), Shirley Ramírez (Ministerio del Ambiente y Energía, Costa Rica), Irene Redtenbacher (Four Paws), Patti Regan (Center for Great Apes), Lisa Ries (Fours Paws), Steve Ross (Lincoln Park Zoo), Karmele Llano Sánchez (International Animal Rescue), Julie Sherman (Wildlife Impact), Ian Singleton (Sumatran Orangutan Conservation Programme), Sheri Speede (Sanaga-Yong Chimpanzee Rescue), Sumita Sugnaseelan (Universiti Putra Malaysia), Fransiska Sulistyo (Orangutan Veterinary Advisory Group), Susannah Thorpe (University of Birmingham), Gregg Tully (PASA), Daniel Turner (Animondial), Jonny Vaughan (Lilongwe Wildlife Trust), Itsaso Vélez del Burgo (Centre de Réhabilitation des Primates de Lwiro), Dirk-Jan Verdonk (World Animal Protection), Sian Waters (IUCN SSC PSG Section for Human–Primate Interactions), Vernon Weir (American Sanctuary Association), Jessica Whitham (Chicago Zoological Society/Brookfield Zoo), Serge Wich (Liverpool John Moores University) and Sarah Wolfensohn (University of Surrey).

For the status update (Section II of this chapter), the AZA studbook keepers, ChimpCare, GAIN, participating sanctuaries and rescue centers, PASA, Patti Ragan at the Center for Great Apes and Species360 generously shared their data; Susan Cheyne, Debra

Durham, Iris Ho, Brice Lefaux and Beth Richards shared their expertise; and Julie Sherman supported the author throughout the process.

Endnotes

1   Unpublished 2015 revision of Maldonado *et al.* (2012), read by authors.

2   Beck (2017); Berg (2018); Browning and Veit (2021); Guy *et al.* (2014); Sherman, Ancrenaz and Meijaard (2020).

3   The annual African Animal Welfare Conference is co-hosted by the Africa Network for Animal Welfare in collaboration with the United Nations Environment Programme and the African Union Inter-African Bureau for Animal Resources (AAWC, n.d.).

4   AZA has issued care manuals for chimpanzees (AZA Ape TAG, 2010); gorillas (AZA Gorilla Species Survival Plan Program, 2017); and orangutans (AZA Ape Taxon Advisory Group, 2017). The European Association of Zoos and Aquaria has produced best practice guidelines for bonobos (Stevens, 2020); chimpanzees (Carlsen, de Jongh and Pluháčková, 2022); gorillas (Abelló, Rietkerk and Bemment, 2017); and orangutans (Bemment, 2018). GFAS released standards for great apes and gibbons (GFAS, 2022). PASA has guidelines for African apes (PASA, 2016).

5   See, for example, Clegg, Borger-Turner and Eskelinen (2015); Kagan, Carter and Allard (2015); Mellor (2017); Ross (2020b); Sherwen *et al.* (2018); Whitham and Wielebnowski (2015); Wolfensohn *et al.* (2018); and Yon *et al.* (2019). For reviews of zoo-based animal welfare assessments, see Hill and Broom (2009) and Wolfensohn *et al.* (2018).

6   This section focuses on behavior as health indicators are covered in other chapters.

7   See, for example, Boesch, Hohmann and Marchant (2002); Brent (2001); Fernie *et al.* (2012); Goodall (1986); Hopper and Ross (2020); Lonsdorf *et al.* (2010); Nakamara *et al.* (2015); and Ross (2020a).

8   See Brouwers and Duchateau (2021) for adaptations made to the Animal Welfare Assessment Grid for zoo-housed gorillas.

9   Greggor *et al.* (2016); Mukherjee *et al.* (2015); Rioja-Lang *et al.* (2020a; 2020b); Truelove *et al.* (2020); Veasey (2020a; 2020b).

10 Table 8.6 shows the number of apes in Asian zoos reporting to Species360. Reports came from Hong Kong (Hong Kong Zoological and Botanical Gardens); India (Arignar Anna Zoo Park, Lucknow Zoological Park, Mysore Zoo, Nandankanan Biological Park, Nehru Zoological Park, Sanjay Gandhi Biological Park); Indonesia (Bali Safari and Marine Park, Batu Secret Zoo, Gembira Loka Zoo, Maharani Zoo and Caves, Prigen Safari Park, Taman Safari Bogor); Kazakhstan (Almaty State Zoo); Malaysia (Zoo Taiping); Nepal (Central Zoo); Singapore (Singapore Zoological Gardens); South Korea (Everland Zoological Gardens, Seoul Zoo); Taiwan (Taipei Zoo); Thailand (Chiangmai Night Safari, Chiangmai Zoological Garden, Khao Kheow Open Zoo, Khao Suan Kwang Zoo, Nakhon Ratchasema Zoological Park, Songkhla Zoo, Ubon Ratchathani Zoo); Uzbekistan (Tashkent Zoo); and Viet Nam (Saigon Zoo).

11 Unpublished data gathered by J. Sherman covering changes in intake levels of newly rescued wild orangutans between 2016 and 2019, reviewed by the author, 2020.

12 AAP (2020a); Ape Monkey Rescue (n.d.); Gut Aiderbichl (n.d.); Monkey World (n.d.); Monte Adone (n.d.); Rainfer (n.d.); Sh Barcelona (2016).

13 In 2020, 16 Latin American zoos reported housing apes to Species360: Argentina (Ecoparque de Buenos Aires); Brazil (Bioparque de Rio, Fundação de Parques Municipais e Zoobotânica/Belo Horizonte, Fundação Parque Zoológico de São Paulo, Zoológico de Pomerode); Chile (Buin Zoo, Zoológico Nacional/Parquemet); Colombia (Bioparque Ukumari); Dominican Republic (Parque Zoológico Nacional/ZOODOM); Guatemala (Zoológico Nacional La Aurora); Mexico (Africam Safari, Guadalajara Zoo, Zacango Ecological Park, Zoofari, Zoológico de Chapultepec, Zoológico de San Juan de Aragón) (Species360, n.d.).

14 The visual count from film may underestimate the actual number.

15 Data extracted from map and species list with one individual allocated to each, which may underestimate the actual number.

16 Wild Ally Consulting (www.linkedin.com/in/kayfarmer/).

17 University of Birmingham (https://www.birmingham.ac.uk/schools/biosciences/index.aspx) then Wildlife Health Australia (https://wildlifehealth australia.com.au).

18 Independent consultant (www.linkedin.com/in/christine-caurant).

19 University of Birmingham (www.birmingham.ac.uk/schools/biosciences/index.aspx).

20 At time of writing: University of Birmingham (www.birmingham.ac.uk/schools/biosciences/index.aspx).

21 University of Birmingham (www.birmingham.
    ac.uk/schools/biosciences/index.aspx).

22 University of Birmingham (www.birmingham.
    ac.uk/schools/biosciences/index.aspx).

23 University of Birmingham (www.birmingham.
    ac.uk/schools/biosciences/index.aspx).

24 Orangutan Veterinary Advisory Group
    (www.ovag.org).

25 University of Birmingham (www.birmingham.
    ac.uk/schools/biosciences/index.aspx).

26 University of Birmingham (www.birmingham.
    ac.uk/schools/biosciences/index.aspx).

Annex I

Summary of the Five Criteria (A–E) Used to Evaluate if a Taxon Belongs in an IUCN Red List Threatened Category (Critically Endangered, Endangered or Vulnerable)*

A. POPULATION SIZE REDUCTION. POPULATION REDUCTION (MEASURED OVER THE LONGER OF 10 YEARS OR 3 GENERATIONS) BASED ON ANY OF A1 TO A4		Critically Endangered	Endangered	Vulnerable
A1		≥90%	≥70%	≥50%
A2, A3 & A4		≥80%	≥50%	≥30%
A1	Population reduction observed, estimated, inferred, or suspected in the past where the causes of the reduction are clearly reversible AND understood AND have ceased.	based on any of the following:	(a) direct observation [except A3]	
A2	Population reduction observed, estimated, inferred, or suspected in the past where the causes of reduction may not have ceased OR may not be understood OR may not be reversible.		(b) an index of abundance appropriate to the taxon	
A3	Population reduction projected, inferred or suspected to be met in the future (up to a maximum of 100 years). [(a) cannot be used for A3]		(c) a decline in area of occupancy (AOO), extent of occurrence (EOO) and/or habitat quality	
A4	An observed, estimated, inferred, projected or suspected population reduction where the time period must include both the past and the future (up to a max. of 100 years in future), and where the causes of reduction may not have ceased OR may not be understood OR may not be reversible.		(d) actual or potential levels of exploitation (e) effects of introduced taxa, hybridization, pathogens, pollutants, competitors or parasites	

B. GEOGRAPHIC RANGE IN THE FORM OF EITHER B1 (EXTENT OF OCCURRENCE) AND/OR B2 (AREA OF OCCUPANCY)		Critically Endangered	Endangered	Vulnerable
B1	Extent of occurrence (EOO)	<100 km²	<5,000 km²	<20,000 km²
B2	Area of occupancy (AOO)	<10 km²	<500 km²	<2,000 km²
AND at least 2 of the following 3 conditions:				
(a)	Severely fragmented OR Number of locations	=1	≤5	≤10
(b)	Continuing decline observed, estimated, inferred or projected in any of: (i) extent of occurrence; (ii) area of occupancy; (iii) area, extent and/or quality of habitat; (iv) number of locations or subpopulations; (v) number of mature individuals			
(c)	Extreme fluctuations in any of: (i) extent of occurrence; (ii) area of occupancy; (iii) number of locations or subpopulations; (iv) number of mature individuals			

C. SMALL POPULATION SIZE AND DECLINE

		Critically Endangered	Endangered	Vulnerable
Number of mature individuals		**<250**	**<2,500**	**<10,000**
AND at least one of C1 or C2:				
C1	An observed, estimated or projected continuing decline of at least (up to a max. of 100 years in future):	25% in 3 years or 1 generation (whichever is longer)	20% in 5 years or 2 generations (whichever is longer)	10% in 10 years or 3 generations (whichever is longer)
C2	An observed, estimated, projected or inferred continuing decline AND at least 1 of the following 3 conditions:			
(a)	(i) Number of mature individuals in each subpopulation:	≤50	≤250	≤1,000
	(ii) % of mature individuals in one subpopulation =	90–100%	95–100%	100%
(b)	Extreme fluctuations in the number of mature individuals			

D. VERY SMALL OR RESTRICTED POPULATION

		Critically Endangered	Endangered	Vulnerable
Number of mature individuals		**<50**	**<250**	**<1,000**
D1	*Only applies to the VU category* Restricted area of occupancy or number of locations with a plausible future threat that could drive the taxon to CR or EX in a very short time.	–	–	D2. typically: AOO <20 km² or number of locations ≤5

E. QUANTITATIVE ANALYSIS

	Critically Endangered	Endangered	Vulnerable
Indicating the probability of extinction in the wild to be:	**≥50% in 10 years or 3 generations, whichever is longer (100 years max.)**	**≥20% in 20 years or 5 generations, whichever is longer (100 years max.)**	**≥10% in 100 years**

Note: * Use of this summary sheet requires full understanding of the IUCN Red List Categories and Criteria and Guidelines for Using the IUCN Red List Categories and Criteria. Please refer to both documents for explanations of terms and concepts used here.

Source: IUCN (2012, pp. 28–9)

Annex II

Summary Table of Ape Diseases, Infections and Other Health Issues

The list of diseases presented in this annex is not exhaustive; rather, it is designed to illustrate the developing knowledge in this area and to provide a quick and accessible overview of key ape diseases, infections and other health issues. The information is drawn from chapters in this volume, primarily Chapter 1, and from personal experience of the veterinary contributors, unless otherwise indicated.[1]

While all apes, including humans, may be susceptible to the diseases described in this annex, a few caveats apply:

■ Exposure depends on the geographical range of pathogens, which constantly shifts in response to climate change, the wildlife trade and other anthropogenic factors.

■ Most disease data for non-infectious conditions come from captive situations.

■ Significant data gaps preclude verification in many identified disease situations and, consequently, raise uncertainty in disease risk management decisions.

■ Parasite–host balance is an ecological process that is necessary for life. The presence of parasites does not always indicate disease, nor is it always a cause for concern. Conversely, a lack of parasites is unnatural and could actually increase the risk of infection by pathogenic parasites.

■ The table does not consider effects of trauma or accidental injuries, common occurrences in all ape species, both in captivity (ex situ) and in their natural habitat (in situ).

■ The relative risk of each disease occurring in an ape population depends on the interaction between host, parasite (potential pathogen) and the environment. The diseases of concern in a sanctuary in Sierra Leone differ from those in a national park in Uganda, a rehabilitation centre in Kalimantan, an entertainment facility in Thailand or a zoo in Australia, for example.

The reader is recommended to review the *Manual of Procedures for Wildlife Disease Risk Analysis* when making actual risk-based decisions for ape disease (Jakob-Hoff *et al.*, 2014).

---

1   For an extensive disease list for apes, see Volume 8 of *Fowler's Zoo and Wild Animal Medicine*; see also the ape chapters in the subsequent volumes (Miller, Calle and Lamberski, 2023; Miller and Fowler, 2015; Miller, Lamberski and Calle, 2019). Specific aspects of ape parasitology and gorilla pathology are covered in Cooper and Hull (2017) and Modry et al. (2018).

Health issue	Description	In situ	Ex situ	Location	Consequences	Notes
AGE-RELATED ISSUES						
Age-related cardiovascular disease	Degenerative conditions that affect the cardiovascular system (heart and blood vessels)	Suspected	Confirmed	Global	Progressive and fatal. Signs are mild to severe, including aortic dissection, congestive heart failure, malignant arrhythmia, myocardial fibrosis, strokes.	
Age-related dental disease	Degenerative diseases that affect the teeth and jaw	Suspected	Confirmed	Global	Dental attrition, enamel hypoplasia (thin or missing tooth enamel) of deciduous and permanent teeth, tooth loss.	
Age-related liver disease	Degenerative diseases that affect the liver	Suspected	Confirmed	Global	Can be progressive and fatal. Mild to severe signs, including cirrhosis, decreased activity, lethargy, hepatic fibrosis, hepatitis, weight loss.	
Age-related ocular conditions	Degenerative diseases and conditions that affect the eyes	Confirmed	Confirmed	Global	Cataracts and retinal disease, potentially leading to blindness.	
Osteoarthritis	Degenerative condition that results in stiff, painful joints	Confirmed	Confirmed	Global	Commonly affects knees, hips, elbows and lower spine, thereby affecting mobility, which may result in injury and/or malnutrition.	
Age-related renal disease	Degenerative conditions that affect the renal system (kidneys, ureters, bladder and urethra)	Suspected	Confirmed	Global	Progressive and fatal. Mild to severe signs, including chronic interstitial nephritis and glomerular lesions.	
INFECTIOUS DISEASES						
Anthrax	Bacterial infection (*Bacillus anthracis*). Skin, lung and bowel disease	Confirmed	Unknown	Central and West Africa	Fatal. Rapid onset, fever, septicemia (blood poisoning) and a high fatality rate.	
Air sacculitis	Resulting from bacterial infection of the respiratory system	Probable	Confirmed	Global	Can be fatal. Purulent material accumulates within the tiny sacs off the laryngeal tubes,	Part of a syndrome that often includes sinusitis (which often goes undetected) and can also lead to pneumonia. ▶

Health issue	Description	In situ	Ex situ	Location	Consequences	Notes
					with the potential for serious complications, including fatal bronchopneumonia and sepsis.	
Candidatus Sarcina troglodytae	Bacterial infection of the neural and gastrointestinal systems	Unknown	Confirmed	Sierra Leone	Can be fatal. Neurologic and gastrointestinal signs.	A new, highly virulent bacterial *Sarcina* strain has been linked to disease in captive, rehabilitant chimpanzees, termed "epizootic neurologic and gastroenteric syndrome (ENGS)." Potentially emerging.
Clostridium tetani	Neuro-muscular bacterial disease	Unlikely	Confirmed	Global	Death, jaw cramping, muscle spasms and hypertonia, seizures, trouble swallowing.	
"Common cold"	Viral infection of the respiratory system (human rhinovirus C)	Confirmed	Confirmed	Global	Dyspnea (labored breathing), wheezing, mild to heavy cough, lethargy, nasal discharge.	Can make the body susceptible to bacterial infections.
COVID-19	Viral infection of the respiratory and gastrointestinal systems (SARS-CoV-2)	Unknown	Confirmed (gorillas and chimpanzees)	Global	Dyspnea, wheezing, mild to heavy cough, lethargy, nasal discharge.	Identified in gorillas in the zoos of San Diego and Prague. Identified in captive chimpanzees in the Democratic Republic of Congo (L. Flores, personal communication, 2023).
Ebola virus disease, formerly known as Ebola hemorrhagic fever	Viral (ebolaviruses)	Confirmed	Unknown	Central, East and West Africa	Fatal. Bleeding (internal and sometimes external), diarrhea, emaciation, fever, lethargy and vomiting.	Of the six ebolaviruses, only four cause disease in humans (Bundibugyo, Sudan, Taï Forest and Zaire ebolaviruses). No human-pathogenic ebolaviruses are known from Asia; however, Reston ebolavirus, which circulates in bats in the Philippines, can cause disease in apes.

Health issue	Description	In situ	Ex situ	Location	Consequences	Notes
Encephalomyo-carditis	Viral disease that tends to affect the central nervous and cardiovascular systems	Unknown	Confirmed	Several zoos around the world	Sudden death is the most common consequence. Clinical signs may include fever, anorexia, listlessness, trembling, staggering, dyspnea and paralysis.	Gaskin (2022)
Hepatitis A virus	Viral infection of the liver and gastro-intestinal system	Confirmed	Confirmed	Global	May be asymptomatic but has caused fulminant hepatitis in chimpanzees and has been a likely cause of death in gibbons (awaiting confirmation).	Can be from zoonotic transmission, but chimpanzees and other apes have been shown to have their own strains.
Hepatitis B virus	Viral infection of the liver	Confirmed	Confirmed	Global	Often asymptomatic. Can potentially lead to increased liver enzymes and hepatic neoplasia, typically in aged animals.	Chimpanzees, gorillas, orangutans and gibbons all have their own strains, which are distinct from human strains.
Herpes simplex virus	Viral infection of the skin and nervous system	Confirmed	Confirmed	Global	Can be fatal, with mortality reported in captive gibbon, gorilla and orangutan populations. Systemic infections with encephalitis; signs include blisters and sores.	
Human coronavirus OC43	Viral infection of the respiratory and gastrointestinal systems	Suspected	Confirmed	Global	Dyspnea, wheezing, mild to heavy cough, lethargy, nasal discharge.	
Human orthopneumovirus	Viral infection of the respiratory system	Confirmed	Confirmed	Central, East and West Africa	Dyspnea, wheezing, mild to heavy cough, lethargy, nasal discharge.	
Human respirovirus 3	Viral infection of the respiratory system	Confirmed	Confirmed	Global	Can be fatal, especially with secondary bacterial infection. Dyspnea, wheezing, mild to heavy cough, lethargy, nasal discharge.	

Health issue	Description	In situ	Ex situ	Location	Consequences	Notes
Influenza (flu)	Viral infection of the respiratory system	Unknown	Confirmed	Global	Unknown	There is no confirmation of apes being infected with human influenza strains, but the chimpanzee adenovirus shell is used in influenza vaccines for humans. Data are limited on confirmed infection with influenza A, B, C and D. See Annex III for confirmed infections in apes.
Klebsiella pneumonia	Bacterial disease (*Klebsiella pneumoniae*)	Unknown	Confirmed	Global	Can be fatal. Signs depend on which organ is affected, they include air sacculitis, gastrointestinal inflammation, pneumonia and septicemia (blood poisoning).	It is unclear whether the disease can be a primary infection, but it appears to be related to immunosuppression. In humans it is typically a secondary infection related to health or medical care.
Leprosy	Bacterial infection of the nerves, skin, eyes and lining of the nose (*Mycobacterium leprae*)	Confirmed	Confirmed	Sub-Saharan Africa	Lesions, including nodules on the face; hair loss and skin depigmentation; abnormal nail growth and hand deformity; disfigured faces and crippled limbs.	
Meliodosis/ Whitmore's disease	Bacterial infection (*Burkholderia pseudomallei*)	Confirmed	Confirmed	Southeast Asia and northern Australia	Can be fatal, with a wide range of signs of varying severity, from subclinical to subacute. Signs include wasting with subcutaneous and soft-tissue abscesses.	Can be challenging to diagnose and treat because the organism can remain latent for years; it can be mistaken for other infections, such as tuberculosis, and is resistant to many antibiotics.
Monkeypox	Viral infection (*Orthopoxvirus*)	Confirmed	Confirmed	Central and West Africa	Can be fatal. Diverse clinical manifestations, such as maculopapular rash; mild to severe respiratory signs with absent or limited (1–2) skin lesions; or no signs.	

Health issue	Description	In situ	Ex situ	Location	Consequences	Notes
Pasteurella multocida	Bacterial infection of the respiratory system	Confirmed	Confirmed	Global	Can be fatal. Infections include air sacculitis and pneumonia.	
Poliomyelitis	Musculoskeletal viral disease	Confirmed	Suspected	East Africa	Infection of bones and muscles, resulting in paralysis and influencing survival and reproductive success.	Although not definitively diagnosed, a disease with clinical signs similar to polio in humans was seen in 1966 in Gombe chimpanzees (Morbeck *et al.*, 1991). Williams *et al.* (2008) highlight the issue in diagnosing infectious disease in apes, that continues today, the lack of disease surveillance.
Pseudomonas spp. infection	Bacterial disease	Confirmed	Confirmed	Global	Responsible for air sacculitis infection in orangutans and even death after a wound infection (Kanamori *et al.*, 2012; Lawson, Garriga and Galdikas, 2006).	These bacteria do not appear to cause disease in healthy animals or humans.
Retrovirus	Viral infection: simian immuno-deficiency virus in chimpanzees (SIVcpz)	Confirmed	Confirmed	Central, East and West Africa	Fatal; carrier state possible. Disease is usually not seen until long past infection. AIDS-like illness similar to human immuno-deficiency virus (HIV) in humans. The latter stages of infection develop into simian acquired immunodeficiency syndrome (SAIDS).	
Salmonella/ Shigella infection	Bacterial disease of the gastrointestinal system	Confirmed	Confirmed	Global	Can be fatal. Most common signs are abdominal pain and watery diarrhea. May also cause dehydration, fever and vomiting.	
Streptococcus pneumonia	Bacterial disease (*Streptococcus pneumoniae*)	Confirmed	Confirmed	Global	Can be fatal. Dyspnea (labored breathing), wheezing, mild to heavy cough, lethargy,	This is a secondary infection that occurs after an individual has been weakened by a

Health issue	Description	In situ	Ex situ	Location	Consequences	Notes
					nasal discharge. Can result in pneumonia.	respiratory viral infection.
Tuberculosis	Bacterial infection of the respiratory system, but granulomas can appear elsewhere, including in the gastrointestinal system (*Mycobacterium tuberculosis* complex)	Confirmed (chimpanzees)	Confirmed	Global	Can be fatal. May be asymptomatic in early stages; signs are progressive. First signs may include lethargy, decreased activity, wasting, weight loss. Advanced cases can present with respiratory signs (coughing, dyspnea). Gastrointestinal cases can present with diarrhea.	Warning: Extremely complicated to diagnose and confirm. Impacts on apes differ from those on humans. This infection should be considered in all cases of respiratory or gastrointestinal infection and weight loss. Securing an expert opinion is recommended. Tuberculosis can spread from humans to animals and vice versa.
Typhoid fever	Bacterial disease (*Salmonella typhi/ paratyphi*)	Unlikely	Confirmed	Global (more common in developing countries)	High fever, headaches, gastrointestinal signs (diarrhea or constipation) and lethargy.	
Yaws	Musculoskeletal bacterial disease (*Treponema pallidum* subspecies *pertenue*)	Confirmed	Suspected	Sub-Saharan Africa	Infection of the skin, bones and joints, resulting in non-cancerous lumps and ulcers.	
PARASITES						
Filariasis	Nematodes that affect the heart and lungs	Unknown	Confirmed	Global	Mild to fatal signs, including loss of appetite, weight loss, lethargy and difficulty breathing.	*Dirofilaria immitis* has been described in orangutans (Sandosham, 1951). A recent case in an orangutan was detected during a necropsy.
Gastrointestinal protozoa	Protozoa that affect the gastrointestinal and other systems, including *Entamoeba histolytica* (amebic dysentery); *Giardia duodenalis* (giardiasis); *Balantidium coli* and *Dientamoeba fragilis*	Confirmed	Confirmed	Global	Some can be fatal. Mixture of signs, from mild to severe. Acute to subacute necrotizing or granulomatous meningoencephalitis, bloating, cramping, diarrhea, lung or liver abscesses, ulcerative colitis and vomiting.	Consequences are much more severe in captivity. Untreated *Entamoeba histolytica* and *Giardia duodenalis* cause diseases in captive apes, mostly in infants. *Ballantidium coli* is commensal in captive apes and rarely causes diseases.

Health issue	Description	In situ	Ex situ	Location	Consequences	Notes
Internal helminths	Worms—round-worms (nema-todes), tapeworms (cestodes), flukes (trematodes)—that tend to affect the gastrointestinal system, with occa-sional respiratory phases of the life cycle that lead to disease in multiple organs	Confirmed	Confirmed	Global	May be asympto-matic. Heavy burdens may be associated with weight loss, weak-ness, failure to thrive, diarrhea and, occasionally, blood in feces (hematochezia).	These parasites are most common in captive apes and include *Ankylostoma, Ascaris, Capillaria, Enterobius, Oesophagostomum, Strongyloides* and *Trichuris.* The consequences of an infection by gastrointestinal parasites depend on the parasitic load and the animal's immune status. Parasites are com-mensal agents in ape intestines; their presence is not necessarily a risk to health. Problems arise when there is a lack of control of the parasitic load, such as when an animal is in captiv-ity, when natural habitats are over-populated or when an animal's immune status is deficient.
Malaria	Protozoa (single-celled organisms) that affect various organs, with liver and brain infection leading to the most serious consequences (*Plasmodium* spp.)	Confirmed	Confirmed	Tropics	Potentially fatal. Mostly causing asymptomatic infec-tions in apes, but documented signs of malaria range from moderate to severe. Conse-quences depend on the species of *Plasmodium*, the protozoal load of *Plasmodium*, the species of ape and which organ system is affected (Sanchez *et al.*, 2022).	The disease is caused by para-sites transmitted through the bites of infected female *Anopheles* mosquitoes.
Mange/scabies	Ectoparasites (mites: *Sarcoptes scabiei*) that affect the skin	Confirmed	Confirmed	Global	Rarely fatal. Flaky, sore and itchy skin. Can make young apes more suscep-tible to other diseases.	

Health issue	Description	In situ	Ex situ	Location	Consequences	Notes
PSYCHOLOGICAL DISORDERS						
Behavioral disturbances similar to post-traumatic stress disorder (PTSD) following traumatic experiences	Mental/emotional disorder affecting the nervous system	Unknown	Confirmed	Global	Potential for long-term behavioral and physiological issues if not identified. Could manifest many months or years after inciting incident.	To be taken into consideration in the context of orphan ape rescues, translocations of "displaced" apes and confinement of apes in captivity (see Chapter 8).
Chronic stress	Mental/emotional issues affecting the nervous system	Confirmed	Confirmed	Global	May create lethargy, stereotypical behaviors (such as pacing) and other psycho-pathologies, which also require managing, as well as impairment of the immune system in its ability to fight off certain infections or regulate the microbiome. The combination of these factors usually results in a higher disease prevalence under captive conditions.	Limited opportunity or ability to engage in natural behavior, physical exercise and, most importantly, mental exercise increases the chances of the development of psychological disorders, including stereotypical behaviors, accompanied by increased levels of stress hormones such as cortisol.
OTHER ISSUES						
Alcohol, drug and tobacco dependency	Due to abusive captivity	Unknown	Confirmed	Global	Like humans, apes can show behavioral changes and neuro-cognitive deficits, such as memory loss and cognitive impairments.	Such dependency has been documented in illegally kept apes used as photo props and tourist attractions, such as young gibbons at Thai beaches, bars and restaurants and smoking chimpanzees in zoos (Guarino, 2016). They are given alcohol, cigarettes and drugs, such as amphetamines, to keep them awake and ensure they "perform."
Burns	Due to forest fires or contact with uninsulated power lines (see electrocution)	Probable	Confirmed	Africa and Asia	Can be fatal. Depending on their severity, burns can lead to disfigurement; leave individuals susceptible to infection at burn sites; impact mobility, potentially	Most fires in ape ranges are intentionally set by humans or due to human error (Kimbrough, 2020).

Health issue	Description	In situ	Ex situ	Location	Consequences	Notes
					resulting in starvation or exposure to predation; impair the immune system, leaving individuals open to other infections.	
Dental issues	Resulting from a poorly balanced diet	Confirmed	Confirmed	Global	The high-energy content in a poorly balanced diet served in captivity or based on "crop raiding" in situ can lead to dental problems such as cavities, associated toothache and tooth loss.	Anthropogenic disturbances in ape habitat can lead to a decrease in food supply, forcing apes to "crop raid."
Drowning	Related to crossing drainage channels in search of food, clean water or other apes, including potential mates	Confirmed	Confirmed	Global	If the drowning is not fatal, the damage to the respiratory system can make an individual more susceptible to other respiratory infections.	Drainage channels are used in commercial plantations and may run through ape ranges. If they divide populations and cut them off from food, clean water and other apes of the same species, apes may be forced to cross channels, which can result in drowning, even though some apes can swim.
Electrocution	Due to contact with uninsulated electricity pylons or cables	Confirmed	Probable	Global	Can be fatal. Can result in burns, shock, damage to the heart, and falls that cause physical injury, all of which can be immediately fatal or can result in secondary infections, which can then be fatal.	Apes may use electricity pylons and cables to get around in the same way that they use trees, which can result in electrocutions if the pylons and cables are not insulated.
Heart (cardiovascular) disease (non-age-related)	Possibly related to poorly balanced diet and reduced activity levels	Probable	Confirmed	Global	Can be fatal. Fibrosing (replacement of heart muscle by fibrous tissue) or idiopathic cardiomyopathy (reduction in heart's ability to pump blood around the body due to abnormalities in the ventricular wall and/or cavity).	Heart disease is among the leading causes of death in captive great apes, yet the causes are not fully understood. In the long run, the disease could affect the genetic viability needed to sustain a healthy captive population.

Health issue	Description	In situ	Ex situ	Location	Consequences	Notes
Malnutrition: obesity	Related to a poorly balanced diet	Confirmed	Confirmed (more probable in captivity)	Global	Overconsumption leading to obesity predisposes an individual to diseases such as diabetes and heart disease due to high blood pressure.	In zoos, obesity is the most common form of nutritional disorder in apes, due to a high intake of simple carbohydrates, combined with limited physical exercise. In the wild, anthropogenic disturbances can lead to a decrease in food supply, forcing apes to "crop raid" and thus rely on a poorly balanced diet.
Malnutrition: undernutrition	Related to a poorly balanced diet	Confirmed	Confirmed	Global	Can be fatal. Undernutrition leads to emaciation and starvation.	Applicable to apes stranded in very small forest fragments, or even individual trees, within a clear-felled area for plantation agriculture, as well as to captive apes suffering from neglect.
Physical injury: competition and territoriality	Due to intra- or intergroup aggression	Confirmed	Confirmed	Africa	Can lead to physical injuries and subsequent infections, which may be fatal.	Intragroup aggression can involve fighting to contest the position of alpha male or in response to attempts by underlings to procreate. In such cases, apes can be expulsed from a group and thus be left vulnerable, without support.
Physical injury: human–wildlife conflict	Related to confrontations between farm-owning humans and "crop-raiding" apes, or communities and apes	Confirmed	Unknown	Africa and Asia	Can lead to physical injuries, which may be fatal.	The likelihood of conflict between humans and apes is exacerbated by habitat destruction and degradation, which brings them into closer contact.
Physical injury and loss of limbs: hunting snares	Due to snares set by hunters	Confirmed	Unknown	Africa and Asia	Can be fatal. Can result in injury or loss of limbs.	Apes sometimes fall victim to snares that are set by hunters to legally catch other species.

Health issue	Description	In situ	Ex situ	Location	Consequences	Notes
Physical injury and loss of limbs: road and rail accidents	Resulting from the need to cross roads or railway tracks to access food, water and other apes of the same species	Confirmed	Unknown	Africa and Asia	Often fatal. Can lead to physical injuries and loss of limbs.	Roads and railway tracks that run through habitat directly affect apes by dividing populations, cutting them off from food, water supplies and other apes of the same species, including potential mates, and forcing them to cross roads and tracks, which can result in traffic or train accidents.
Poisoning: agriculture	Related to pesticide use in agriculture	Confirmed	Unknown	Africa and Asia	Signs attributed to pesticides include facial dysplasia (abnormal growth) in chimpanzees in Uganda.	In Uganda, DDT/pp-DDE, chlorpyrifos and imidacloprid levels in maize have exceeded recommended limits (Krief et al., 2017).
Poisoning: mining	Related to mining and ore processing that poison soil and water	Confirmed	Unknown	Africa and Asia	Can be fatal. Poisoning can lead to neurological or renal malfunctions.	Mining and ore processing can poison soil and water supplies. Gold ore processing often involves the uncontrolled use of mercury, for example.
Smoke inhalation	Due to forest fires	Confirmed	Confirmed	Africa and Asia	Can be fatal. Smoke inhalation can impair the ability to breathe and make an individual more susceptible to other issues, such as respiratory infections.	Most fires in ape ranges are intentionally set by humans or due to human error (Kimbrough, 2020). Smoke can carry a very long way, impacting apes across wide geographies, both in situ and in captivity. An example is the smog in Singapore that resulted from fires in Indonesian Borneo.

Acknowledgments

Veterinary contributors: Marc Ancrenaz, Luis Flores, Karen Payne, Karmele Llano Sánchez and Steve Unwin

Annex III

Confirmed Transmissions of Viral Pathogens from Humans to Apes in Their Natural Habitats

Host genus	Host species	Virus family	Virus name	Location	Consequences of infection*	References
Pan	Bonobo (Pan paniscus)	Pneumoviridae	Human orthopneumovirus A and B	Malebo Community Reserve, DRC	Severe clinical signs; up to 40% morbidity; 8 fatalities recorded over 2 outbreaks; secondary bacterial infection with Streptococcus pneumoniae	Grützmacher et al. (2018b)
	Eastern chimpanzee (Pan troglodytes schweinfurthii)	Paramyxoviridae	Human respirovirus 3	Kibale National Park, Uganda	Severe clinical signs; 69% morbidity; 1 fatality attributed to weakness and conspecific aggression	Negrey et al. (2019)
		Picornaviridae	Human rhinovirus C	Kibale National Park, Uganda	Severe clinical signs; up to 71% morbidity; 5 fatalities over 3 epidemic phases	Scully et al. (2018)
		Pneumoviridae	Human metapneumovirus	Mahale Mountains National Park, Tanzania	Severe clinical signs; 34% morbidity; 3 fatalities	Kaur et al. (2008)
				Kibale National Park, Uganda	Severe clinical signs; 44% morbidity; 25 fatalities	Negrey et al. (2019)
	Western chimpanzee (Pan t. verus)	Coronaviridae	Human coronavirus OC43	Taï National Park, Ivory Coast	Mild clinical signs; 27% morbidity; 0 fatalities	Patrono et al. (2018)
		Pneumoviridae	Human metapneumovirus	Taï National Park, Ivory Coast	Severe clinical signs; up to 100% morbidity; 8 fatalities recorded over 2 outbreaks; secondary bacterial infection with S. pneumoniae	Köndgen et al. (2008)
			Human orthopneumovirus A and B	Taï National Park, Ivory Coast	Severe clinical signs; up to 100% morbidity; 9 fatalities recorded over 4 outbreaks (more suspected); secondary bacterial infection with S. pneumoniae (of human origin in one instance)	Köndgen et al. (2008, 2010, 2017)

▶

Host genus	Host species	Virus family	Virus name	Location	Consequences of infection*	References
Gorilla	Mountain gorilla (*Gorilla beringei beringei*)	*Pneumoviridae*	Human orthopneumovirus A	Volcanoes National Park, Rwanda	Severe clinical signs; up to 87% morbidity; 0 fatalities over 2 outbreaks	Mazet *et al.* (2020)
		Pneumoviridae	Human metapneumovirus	Volcanoes National Park, Rwanda	Severe clinical signs; 92% morbidity; 2 fatalities; secondary bacterial infection with *S. pneumoniae* and *Klebsiella pneumoniae*	Palacios *et al.* (2011)
	Western lowland gorilla (*Gorilla gorilla gorilla*)	*Pneumoviridae*	Human orthopneumovirus A	Dzanga Sangha Protected Areas, CAR	Severe clinical signs; 88% morbidity; 0 fatalities	Grützmacher *et al.* (2016)

Notes: Ex-situ examples are discussed in Chapter 1; for a more comprehensive list, see Miller and Fowler (2015).

* Severe clinical signs include frequent coughing, sneezing, shortness of breath, oculo-nasal discharge, lethargy and loss of appetite. Mild clinical signs include sporadic coughing and sneezing.

Annex IV

Example of a PEESTOLM Risk Register for Zoonosis in Apes[2]

PEESTOLM covers political, environmental, economic, social, technical, operational, legal and media and communications-related risks.

Hazard: Zoonotic disease.

Context: Zoonosis is known or highly likely to cause mortality and significant disease in apes and humans. Historically, zoonotic diseases such as Ebola in gorillas have caused mortality and debilitating disease.

Objective: Conservation of apes and their habitat using PEESTOLM to assess the full range of zoonotic risks to apes.

Risk type	Risks related to zoonotic disease	Consequences
Political: Risks arising from each level of elected officials and chief executives of large corporations and agencies.	1. In response to zoonosis-related economic fallout, policy or legislative changes are introduced to sustain or enhance the economy, but these changes effectively reduce protective measures for apes or ape habitat. 2. In response to zoonosis-related economic fallout, reduced funding results in cutbacks in resources, such as rangers to manage conservation areas.	▢ Ape health declines; apes die due to habitat loss or degradation; reduced availability of food and shelter ▢ Decline in ape numbers due to increased exposure to humans with zoonoses ▢ Increase in hunting and poaching of apes ▢ Greater competition for food and habitat between individual apes and between ape populations ▢ Increased potential for illegal resource extraction from ape habitat
Environmental: Risks to the natural environment arising from zoonosis and from response measures that address the consequences of zoonosis.	1. A reduction in or loss of ape tourism revenue and a reduction in related conservation activities lead to environmental degradation of ape habitat.	▢ Conservation outcomes are set back years or declines are irreversible
	2. A reduction in or loss of ape tourism revenue and cuts in related conservation activities undermine local communities' environmental stewardship in protected areas.	▢ Increased degradation of habitat leads to decline of biodiversity, including apes ▢ Increased exposure of apes to infection ▢ Increase in hunting and poaching of apes ▢ Greater competition for food and habitat between individual apes and between ape populations ▢ Decline in ape numbers
Economic: Risks to the local, regional and national economies as a result of revenue shortfalls and zoonosis response costs.	1. A reduction in ape tourism results in a drop in economic benefits to local communities. 2. The full recovery of ape tourism takes an extended period of time.	▢ Reduction in funds for ape conservation ▢ Local communities place less value on the apes and habitat ▢ Increase in mental health issues among people and communities that usually benefit from ape tourism ▢ Increase in poverty due to the absence of ape tourism ▢ Communities that usually support ape tourism are fragmented or move away ▢ Reduction in protections for apes and their habitat (see the consequences of environmental risk 2, above)
	3. A drop in overall funding outside the local communities—such as discretionary spending and donations for ape conservation—leads to a significant decline in ape habitat.	▢ Ape conservation programs are set back or curtailed ▢ Ape populations decline ▢ Loss of expertise for ape conservation

---

2   Created by Kevin Cooper.

Risk type	Risks related to zoonotic disease	Consequences
Social: Risks arising from impacts on health, safety, wellbeing and social fabric of local and regional ape and human populations.	1. The prevalence of mental health and wellbeing issues increases among people who usually benefit from ape tourism or have a connection to apes.	Reduction in care and responsibility for apes and their habitat
	2. Local communities move into or return to ape habitats to secure food and shelter materials or to sustain their livelihoods.	Apes and humans compete for resources such as food and shelter materials Increase in hunting and poaching Dislocation of local ape populations Apes exposed to humans and infected
Technical: Risks linked directly to zoonosis, and the risk arising from the control and containment measures used to manage the zoonosis.	1. Local communities that support ape preservation or ape habitat are not protected from zoonotic infection.	Local community is infected with zoonosis Increases the likelihood that local apes will be infected
	2. Apes are infected with zoonosis and mortality is significant.	Ape population declines to the point at which natural population recovery is impossible Ape absence has a ripple effect on biodiversity in the habitat
	3. Potential control measures for zoonosis-infected apes are limited.	Ape population declines to the point at which natural population recovery is impossible Apes lost from local, regional and national habitat: extinction
	4. Zoonosis in apes is poorly understood or unknown.	Likelihood and consequences of zoonosis in apes increase
Operational: Risks around the timeliness and adequacy of resource capability and capacity; health, wellbeing and safety; and structures and management systems of the response.	1. Infection or a high potential for infection of compliance personnel such as rangers causes absences.	Reduction of protections for apes Increased poaching and hunting results in a decline in ape numbers Increased potential for exposure of apes to zoonosis from people due to human encroachment into ape habitat Increased potential for illegal resource extraction from ape habitat
	2. Infection or a high potential of infection among veterinarians and animal keepers causes absences.	Apes exposed to increased likelihood of infection Direct spread of infection in apes results in decline in ape numbers
	3. Isolation (self- or imposed) of veterinarians and animal keepers reduces the ability to undertake surveillance and monitoring of apes.	Reduced capacity for early detection of disease in individual apes and timely implementation of mitigation treatment measures for apes Ape infections lead to mortality and decline in ape populations
	4. Inadequate supplies or a lack of personal protective equipment and related stores for veterinarians and animal keepers limits their ability to manage health and safety risks and biosecurity risks when working with apes.	Increased exposure of veterinarians and animal keepers to zoonosis results in the same consequences as operational risk 3 (above) Increased possibility of exposure of apes to zoonosis and infected apes Increased possibility of spreading zoonosis between and within ape populations
	5. Management absences (due to reduced funding, and actual and potential infection) lead to a reduction in the management of ape populations, including normal animal health management, disaster preparedness and population management.	Increased exposure of apes and ape habitat to pressure (such as fires and human–wildlife conflict) that would usually be controlled through everyday management Decline in ape habitat and ape numbers

Risk type	Risks related to zoonotic disease	Consequences
Legal: Risks relating to the legal authority to complete the mitigation activities and to the alignment of legal obligations.	1. Mitigation activities do not comply with legislative requirements or the equivalent for the conservation of apes and ape habitat.	Decline in ape numbers and their habitat translates into decline in overall biodiversity Ape population unable to recover naturally
Media and communications: Risks that arise from the need to provide stakeholders with timely and accurate information.	1. The local human population does not support the management or control measures because they do not receive timely and appropriate information.	Increased likelihood of the local community becoming infected with zoonosis The management or control measures fail in the absence of local community support Increased likelihood of apes being infected with zoonosis Failure of community support leads to inaction on conservation measures and an increased likelihood of degradation of ape habitat, which can lead to a reduction in ape numbers

Annex V

Reputational Risk Assessment for Animal Sanctuaries and Crisis Communications: A Planning Workbook

Emergency Preparedness and Response Questionnaire

Question	Yes (check)	No*
Do you have an emergency response plan that has been updated in the past three years?		
Does your emergency response plan include a crisis communications plan for internal and external audiences?		
Do you, at least twice per year, conduct emergency response drills that cover a range of potential emergency response scenarios?		
Are staff members routinely cross-trained to work in multiple areas to ensure all animals receive appropriate care and management in an emergency?		
Are apes routinely desensitized to crates or trained to shift in and out of enclosures when asked, so that they can be moved with minimal stress?		
Do you have an adequate number of transfer crates or temporary housing options for all sanctuary residents?		
Do you have secure shelter for all apes, including access to supplemental off-grid heating or cooling and water options, if needed?		
Do you have an established evacuation site and coordinated transportation options for moving apes if needed?		
Do you have provisions available for staff members if they must stay overnight? Do you have round-the-clock veterinary care available during an emergency?		
Have you developed a trusted working relationship with local fire and police agencies to ensure effective collaboration in the event of an emergency at the sanctuary?		
Do you have access to reliable legal counsel?		
Are managers trained to handle emotional situations with their teams?		
Can the sanctuary easily access counseling resources if needed?		
Score*		

Notes: * Guide for "No" column: low risk=0–3; medium risk=4–7; high risk=8 or more.

Source: PCI (2022, p. 10)

Annex VI

Risk Assessment and the Disaster Management Continuum in Relation to Case Study 6.1

Case Study 6.1 outlines reactive response and recovery actions aimed at overcoming the 2020 flood crisis that affected the apes and staff of the Ngamba Island Chimpanzee Sanctuary in Uganda. Risk reduction through prevention and preparedness is essential in the face of potential hazards, such as the one described in the case study. Typically, however, response measures and, subsequently, recovery measures are given priority over prevention and preparedness.

The following risk assessment is based on the risks described in Case Study 6.1. The disaster management continuum presented thereafter may help to mitigate these—as well as other—risks.

a. Risk Assessment

Table A6.1 ranks and describes the risks identified in Case Study 6.1. This type of risk assessment is best informed by ongoing risk mapping, informed by such tools as modeling, as well as regular reviews including consideration of projected climate change-induced extreme weather events.

TABLE A6.1

Risk Assessment on the Basis of Case Study 6.1*

Risk	Likelihood (LH)	Consequence/ Impact (CS)	Risk Rating (RR)	Description
Flooding	High	Medium–high	Medium–high	Ngamba Island has a history of flooding, with the 2020 event characterized by the highest water levels on record. Extreme weather events, including increased rainfall linked to climate change, result in higher water levels and ongoing flooding events, which are expected to last longer as the planet warms. Potential consequences beyond those described in the case study include an inability to evacuate animals or people, as the external support may not be available due to widespread local and regional flooding, an inability to obtain adequate provisions, more permanent damage to infrastructure, and more sustained and prolonged submersion of habitat. In turn, inundated habitat may exacerbate the deterioration or loss of natural food sources for the animals. Anticipated major floods present an increased threat to the chimpanzees and the staff.
Significant disease among the chimpanzees and humans	Low	Medium	Medium	Significant disease was not reported in the case study. The likelihood of disease affecting the chimpanzees and staff is currently low, but emerging diseases and longer-term flooding may increase disease risk. Based on other situations in similar settings, the consequences, once a significant disease is present, are at least medium.

Risk	Likelihood (LH)	Consequence/ Impact (CS)	Risk Rating (RR)	Description
Inadequate food and poor conditions for the chimpanzees and humans	Low	Low	Low	This risk reflects an inability to secure the delivery of food and other provisions to the island at times of crisis, as well as flood-related damage to infrastructure housing chimpanzees and staff. Future extreme weather events may affect the availability of transport to and from the island and thus the delivery of supplies and opportunities for evacuation. Extreme weather may also result in more substantial damage to infrastructure.
Inadequate care for the chimpanzees	Low	Low–medium	Low–medium	Under what were previously considered normal conditions, this risk was not reported, yet future extreme weather events that are landscape-wide may reduce the staff's ability to care for the chimpanzees.
Injury to staff or chimpanzees during tasks associated with rescuing or evacuating chimpanzees	Low	Low–medium	Low–medium	Increases in the frequency and duration of flooding have the potential to increase the risk of injuries sustained during rescue operations and evacuations.

* This risk assessment is based on the information presented in the case study and should therefore not be taken as a full risk assessment because a more in-depth review would likely identify other risks. The authors aimed to build the information included here around the "experiences" in the case study rather than attempting to be comprehensive. Risk assessment steps consider LH independent of CS and then CS independent of LH. The RR = LH × CS. The final RR typically reflects the higher value for either LH or CS (always go up with RR). Treatments aim to reduce LH or CS—most commonly reducing LH in the first instance. However, where the LH of the risk materializing is reduced, if it still occurs the CS will remain the same unless some treatment is also aimed at reducing them.

b. Disaster Management Continuum

The disaster management continuum can be employed to address the abovementioned risks. Some of the following risk mitigation measures can also reduce risks associated with other hazard impacts, such as fire. Prevention and preparedness offer the greatest opportunities to mitigate the risks.

Prevention

Elimination is the preferred option for risk mitigation as it covers both likelihood and consequences. In Case Study 6.1, the relocation and re-establishment of island infrastructure at an elevation above the predicted worst-case-scenario water levels eliminate flood risks for infrastructure. This long-term objective removes the need for future response and recovery resources and actions linked to flooding of the infrastructure.

After elimination, substitution is the next best risk mitigation. It primarily reduces the likelihood of risks, although it can also minimize consequences. The substitution of the existing fixed-level pier with a floating pier that rises and falls with water levels ensures that boats can dock to deliver supplies or evacuate residents. Resupply by air, such as by helicopter, is also an example of substitution.

If neither elimination nor substitution can be applied to reduce risk, the next best option is isolation, such as via a barrier. On Ngamba Island, an example of isolation is the retainer wall between the water and the infrastructure, which reduces erosion caused by breaking waves. This approach can involve modifying the existing wall or constructing a new purpose-built wall (a levee) to hold the lake water back. Any new wall would be farther up the shoreline, closer to the infrastructure. The case study mentions temporary measures such as the use of sandbags and rocks to reduce the area of inundation; more permanent structures would be required for an isolation approach.

Given that future extreme weather events are likely to further increase and intensify flooding, new walls and levees will not be fit for purpose unless they exceed the height of historical flood levels.

Engineering risk mitigation measures constitute a fourth option. They operate automatically to address an impending risk. On Ngamba Island, such a measure might involve installing an automatic pump-out system to remove water from designated areas that are at risk of being inundated. Pumps can work with a retaining wall around high-value infrastructure. In some areas, automated sanitary treatment is an engineering control that—when used every day as well as during floods—reduces the disease risks associated with waste.

Safe places can be identified and prepared in advance of any flooding. On Ngamba Island, designated safe places would need to be clearly identified, known to all those on the island, readily accessible and designed to accommodate all staff and visitors—and potentially the chimpanzees. One or more safe places may be needed.

In the absence of a safe place, a "lifeboat" can provide support if flood waters are life-threatening. The boat could also be used for other purposes, as long as it is well maintained and ready for use during flooding, with trained personnel to crew it.

Preparedness

Capacity and Capability

Case Study 6.1 alludes to the need to train staff to ensure the safe rescue of chimpanzees and the safety of personnel. Suitable resources and upkeep are similarly required for higher-risk tasks. Sanctuary staff and regular visitors who are suitably trained are able to minimize the risks associated with rescues and other high-risk activities.

The stockpiling of provisions—including reserves or supplies for use during an emergency—reduces risks that can arise from an inability to get timely deliveries to the sanctuary. Having arrangements in place for alternative delivery methods, such as by air, if needed, can help to ensure the delivery of provisions if the usual arrangements fail.

Documentation

The case study mentions an evacuation plan. Triggers for activation of the plan are essential, such as the prediction of an extreme weather event, to ensure all stakeholders know when and why the plan will be activated. Triggers ensure activation of the plan before and in response to a hazard impact. The timing of the activation trigger must ensure there is adequate time between the activation and the predicted weather event to complete a full evacuation.

With respect to high-consequence mitigation, which may be applied in life-threatening situations, it is good practice to identify more than one possible response measure. The evacuation plan can suggest alternative evacuation pathways, for instance.

Effective contingency plans are developed in consultation with all stakeholders, including local communities; once validated by exercise and finalized for use, their currency can be maintained by review at specified intervals or following actual events or exercises. The responsibility for a contingency plan is best assigned to a person who has the authority to ensure its ongoing relevance and currency.

Documentation of procedures and policies for high-risk tasks associated with the Ngamba Island chimpanzees—especially during flooding, but also for less frequent routine tasks—is best developed in consultation with the individuals who are to implement the procedures and policies. The clarification of procedures in different formats, such as pictures or cartoons, can enhance the effectiveness of training and exercises. This documentation becomes the basis for continuous improvement, training and the sharing of knowledge (one generation to next) – a "how to." Users will seldom refer to the documentation during an event – it is all about preparedness.

A condition of entry for all those who come to the island is an induction, which should include the procedures to be followed in case of an imminent or actual emergency situation, health and safety requirements and logging the relevant emergency/crisis skills and abilities of those arriving on the island.

Management Systems

An automated warning system can be installed to support early intervention actions in the case of flooding. Such a system detects the rising water levels and automatically issues an alarm. Cameras may be part of the system.

A resource management system is essential for the tracking of resources including personnel, stores and equipment before and during a flood. This system supports the maintenance of the stockpile of provisions for use during flooding when the usual deliveries may not be possible. It should be integrated with the standard resource management system used every day at the sanctuary.

Exercises and Drills

Regular exercises and drills are routine for staff and anyone who is part of any plan or procedure for the sanctuary. The evacuation plan, for example, is only effective if it is routinely and regularly practiced. High-risk procedures can be exercised to inform review at specified intervals, to ensure their ongoing relevance and the ability of staff to complete the procedures.

There may be some merit in exercising the animals as part of the preparedness for actions to be undertaken during a flood. High-risk tasks are likely to challenge most animals and their staff if they are carried out for the first time during a flood emergency. Exercising is an opportunity to explore all options, learn the lessons and provide opportunities for animals and their carers to become familiar and comfortable with actions in a controlled environment compared with the challenges of undertaking these actions for the first time in a response context.

Annex VII

Expected Prevalence of Industrial Development Projects for the Period 2020–2025 and Corresponding Risks to Apes

Ape taxon	Industry sector				
	Agribusiness	Hydroelectric dams	Infrastructure	Logging	Mining
Bonobo	**++** Of all of great ape ranges, the bonobo range has the highest overlap with land suitable for oil palm development (99.2%), suggesting that such development could become prevalent in the future (Wich *et al.*, 2014a).	**Not available** No known project is active or planned within the bonobo range.	**++** As most of the bonobo range is fairly remote, any improvement of roads or the creation of new access roads associated with logging, agribusiness or other projects poses a threat to bonobos, mainly by facilitating access for poaching (Arcus Foundation, 2018).	**+** About 10% of forests in the Democratic Republic of Congo (DRC) are under logging concessions, but a moratorium on the attribution of any new industrial logging titles has been in place since 2002, theoretically limiting this threat. Despite the moratorium, however, the government granted two new logging concessions to Chinese companies in 2018—in areas that overlap with bonobo habitat (Belmaker, 2018).	**+** Commercial mining is not currently a prominent threat, but since the bonobo range is rich in mineral reserves, the situation could change. Any construction of infrastructure to facilitate the export of such commodities would pose a risk to bonobo populations (Arcus Foundation, 2014).
Chimpanzee	**++** Countries in the chimpanzee range are suitable for the cultivation of industrial-scale crops, such as coffee, cocoa, rubber and oil palm (Wich *et al.*, 2014a). Most of this development is concentrated in West African countries, where the expansion of oil palm and cocoa plantations has already had severe impacts on chimpanzees (Bitty *et al.*, 2015).	**+++** Many dams are planned throughout the chimpanzee range. Some of them may have significant impacts on chimpanzee populations. One example is the Koukoutamba dam in Guinea, which could lead to the deaths of up to 1,500 western chimpanzees (*Pan troglodytes verus*) (Watts, 2019).	**+++** "Development corridors" are planned across Africa and the chimpanzee range, mainly in the form of new roads and highways (Laurance *et al.*, 2015). The construction of other linear infrastructure, such as power lines, also occurs alongside dams and other development projects.	**++** Threats from logging are most prevalent in Central Africa, where 47% of the central chimpanzee (*Pan troglodytes troglodytes*) range falls within timber concessions (Arcus Foundation, 2014).	**++** The mining threat is most significant for the western chimpanzee, whose range overlaps with high-grade mineral deposits (such as gold, bauxite and iron ore) and many active and planned mines (Arcus Foundation, 2014).

Ape taxon	Industry sector				
	Agribusiness	Hydroelectric dams	Infrastructure	Logging	Mining
Gibbon	++ Agricultural concessions overlap with most gibbon ranges. They pose particularly significant threats to species found in Indonesia and Cambodia (Arcus Foundation, 2014).	+++ Fifty-five hydroelectric dams have been installed in gibbon ranges. A further 165 dams are either planned or under construction (Arcus Foundation, 2018).	++ Two of the six planned corridors under the Belt and Road Initiative are to cut large swaths through gibbon habitat: the Bangladesh–China–India–Myanmar corridor and the China–Indochina corridor (Hughes, 2019). The relocation of the Indonesian capital within gibbon habitat may also pose direct and indirect threats to several species (Teo *et al.*, 2020).	++ Exact data surrounding the size and location of timber concessions, and their overlap with gibbon ranges, is scarce.	++ Only two species of gibbon have no industrial mining projects within their range: the Hainan gibbon (*Nomascus hainanus*) and the Cao Vit gibbon (*Nomascus nasutus*) (Arcus Foundation, 2014).
Gorilla	++ Many commercial crops threaten gorilla habitat. Oil palm production in Africa is expected to intensify and could become an increasing threat, mainly for the western lowland gorilla (*Gorilla gorilla gorilla*) (Wich *et al.*, 2014a).	++ Several dams have already impacted gorillas throughout their range, including in Cameroon. Many more dams are planned, increasing the risks to the gorilla population (Arcus Foundation, 2018).	+++ The development of roads, railways and power lines is expected to fragment gorilla habitat, while also facilitating access for hunters and farmers to some of the more remote areas of their range (Arcus Foundation, 2018).	++ A large number of timber concessions are within the range of the western lowland gorilla (Morgan and Sanz, 2007).	++ Since mineral deposit extraction is not as formalized in East and Central Africa as it is in West Africa, range overlap with commercial mining activities is limited. In the eastern DRC and other areas where such activities do occur, they tend to be poorly regulated. The impacts from artisanal mining are more significant for this genus (Arcus Foundation, 2014).
Orangutan	+++ Industrial agriculture (mainly oil palm and paper pulp) overlaps with a large part of the orangutan range (Arcus Foundation, 2015).	++ Many dams are already operational in orangutan habitat. Significant impacts may result from several others that are being planned, including the well-publicized Batang Toru hydropower project in the range of the Tapanuli orangutan (*Pongo tapanuliensis*) (Wich *et al.*, 2019).	++ Several linear infrastructure development projects are planned in orangutan habitat. Among them is the Trans-Sumatra Highway, which is to pass through the north-eastern area of the Leuser ecosystem (Sloan *et al.*, 2019).	++ Logging concessions overlap with 29% of the range of the Bornean orangutan (*Pongo pygmaeus* spp.) and 4% of the range of the Sumatran orangutan (*Pongo abelii*) (Wich *et al.*, 2012b).	++ Mining activity overlaps with 9% of the range of the Sumatran orangutan (Meijaard, 2014). Only one project—the Martabe gold mine—is present in the Tapanuli orangutan range. Its expansion could have a significant impact on the species (Wich *et al.*, 2019).

Notes: Agribusiness includes large-scale oil palm, cocoa and rubber plantations; infrastructure includes roads, railways and ports. Prevalence and associated risks are scored[3] as follows:

+++ **High prevalence/risk:** could lead to a significant decrease in ape populations that is difficult to mitigate.

++ **Intermediate prevalence/risk:** could lead to a decrease in ape populations.

+ **Low prevalence/risk:** could lead to a decrease in ape populations, some of which can be mitigated.

3   The scoring system is based on a Google search using a combination of keywords to assess the approximate number of each type of project within each taxon's range.

Annex VIII

Application of the Mitigation Hierarchy in Practice: The Mako Gold Project in Senegal[4]

The Mako Gold Project in southeastern Senegal is owned and operated by the Petowal Mining Company, a subsidiary of Resolute Mining Ltd, in which the Government of Senegal has a 10% interest (Figure A1). The western chimpanzee (*Pan troglodytes verus*) is among the priority species in the project area. To accomplish its corporate goal of achieving a "net gain" for chimpanzees, the project implemented the mitigation hierarchy. In particular, it used the following measures to avoid and minimize impacts related to the construction of the mine and associated infrastructure, rehabilitate or restore damaged habitat, and offset residual impacts (Earth Systems, 2015).

FIGURE A1

Mako Gold Project, Senegal

Sources: Protected area—UNEP-WCMC (2021d); country boundaries—GADM (n.d.); other base map detail—OpenStreetMap (n.d., © OpenStreetMap contributors, published under Creative Commons Attribution License CC BY; for more information see http://creativecommons.org)

Avoidance

To **reduce the size of the mine footprint**, the Mako Gold Project made significant changes to the mine design and layout in the feasibility study. The changes resulted in the consolidation and containment of all major mine infrastructure—the open pit, waste rock, tailings and the processing plant—within one catchment area measuring about 3 km² (300 ha), or about half the size of the original design footprint. As a result, the project avoided some

---

4   Annex VII is written by Vanessa Evans, general manager of environment and community at Resolute Mining Limited, based on her experience leading and implementing biodiversity aspects of the Mako Gold Project.

direct loss of chimpanzee habitat and prevented land disturbance within adjacent catchments that drain into core nesting habitat.

The project **re-routed the main access road** to the mine as it would otherwise have impacted chimpanzees by fragmenting their habitat and impeding access to an important dry-season water source, a gallery forest and foraging habitat at the eastern extent of their range. Re-routing involved co-aligning the road with existing local community infrastructure to avoid these impacts to chimpanzees.

Minimization

The Mako Gold Project minimized disturbance to chimpanzees from **noise, vibration and air blasts** by restricting the use of particular machinery and vehicles at dusk, dawn and during the night. Wherever possible, staff retained natural barriers—such as stands of tree and mounds—during land clearing to buffer noise and vibration, especially near sensitive chimpanzee habitats.

The project also introduced reduced **speed limits** for its vehicles and developed an **injured wildlife protocol** to be followed in the event of an incident. The protocol included a mandatory reporting system to prompt further actions or mitigation measures, if required.

Project staff and contractors were **banned from hunting, buying and trading** chimpanzees, and **environmental education** and awareness programs were conducted for project staff and contractors.

Rehabilitation/restoration

To mitigate impacts on chimpanzees and other fauna during its decommissioning and closure, the project intends to implement rehabilitation measures. The aim is for rehabilitation and closure to reestablish an ecosystem that functions much like it did before mining-related disturbance. Wherever feasible, revegetation efforts are to include the establishment of self-sustaining tree savannah, wooded savannah or shrub savannah vegetative communities to promote connectivity between areas of natural habitat, benefiting foraging, nesting and commuting chimpanzees and other wildlife. The rehabilitation is to be "like for like" to minimize the loss of high-value habitat from the project footprint, and revegetation is to utilize native species of local provenance. Species selection is to include vegetation known to provide nesting or foraging value for chimpanzees, and riparian corridors are to be planted to provide cover for migration, to the extent feasible. The project does not expect to achieve like-for-like habitat restoration in the medium term, however.

Offsetting

To mitigate the residual impacts of the Mako mine on biodiversity, Resolute Mining is implementing the Petowal Biodiversity Offset Program (PBOP). The PBOP encompasses areas within and adjacent to Niokolo-Koba National Park, with the goal of achieving a net gain in biodiversity, including species protection and improved habitat connectivity. The PBOP is being implemented through an innovative partnership that comprises protected area authorities, communities and non-governmental organizations, based on integrated and participatory approaches to land use planning. The design and implementation of the PBOP is guided by an independent advisory panel comprising national and international conservation and resource management experts (Resolute, 2019).

Annex IX

Positive Developments in Wildlife Welfare Legislation

This annex discusses recent developments in Malawi and Costa Rica, both of which recently passed legislation and regulations designed to meet and surpass best practice standards for captive wildlife welfare.

Malawi

Under the revised version of Malawi's National Parks and Wildlife Act 2017, it is an offense to cause unnecessary or undue suffering to any wild animal, whether that animal lives in the wild or is kept in captivity (Ministry of Natural Resources Energy and Mining, 2017, s. 83). The Wild Animal Captivity Licensing Regulations that bring the legislation into effect are two-fold. First, wild animals may not be kept in captivity without a license. The license application process involves an inspection and regular spot checks are to be conducted once a license has been granted. Second, new captive care standards define the requirements and conditions for obtaining an animal captivity license and for the keeping of captive wildlife. The standards classify species according to their requirements and whether they are suited to being kept in captivity. They cover enclosure, health and safety, husbandry, management, and nutritional and veterinary requirements. Minimum care standards are provided for each taxonomic family of mammal species, as well as for individual species with specific requirements. The standards can also be used in assessing potential offenses associated with unnecessary or undue suffering (Lempena and Sal, 2018).

The Wild Animal Captivity Licensing Regulations target individual persons, small non-commercial operations and facilities that could potentially operate as sanctuaries. Commercial breeding facilities are to be covered under ranching guidelines, which remain to be developed (J. Vaughan, personal communication, 2020).

The Lilongwe Wildlife Trust (LWT) established an enforcement unit to support the government with implementation of the revised legislation and regulations. The regulatory system issues warnings to first-time offenders, most of whom do not reoffend; those who go on to commit a crime may be charged. Since the Wild Animal Captivity Licensing Regulations were passed in 2018, six inspections have been conducted and only two licenses have been granted. Two cases of wildlife trafficking, involving a baboon and pangolin, have gone to court on charges of welfare crime (J. Vaughan, personal communication, 2020).

LWT worked closely with the government to support the amendment of the law and the development of the regulations to strengthen conservation and curb wildlife trafficking. The issue of welfare crime had not featured prominently in LWT's campaigning as it was not expected to resonate with policy-makers. While welfare offenses are still treated as lesser crimes in Malawi, more focused lobbying for the inclusion of welfare crime may be able to attract media attention and raise public awareness of the issue (J. Vaughan, personal communication, 2020).

Costa Rica

Driven by mass tourism, about 250 captive wildlife facilities (known as wildlife management sites) operate across Costa Rica (S. Ramirez, personal communication, 2020). In 2017, the country was ranked as the seventh worst country for selfies with wild animals. Efforts by the government and non-governmental organizations are underway to stop direct and inappropriate contact between visitors and wildlife, including through the #StopAnimalSelfies campaign launched in 2019 (Stop Animal Selfies, n.d.; WAP, 2017; 2019; C. Dent, personal communication, 2020).

Animal welfare laws in Costa Rica only cover companion and farm animals, yet the updated Wildlife Conservation Law No. 7317 helps to bridge the legislative gap for captive wildlife (MINAE, 2017b; Silva, 2018). In view of the large number of wildlife management sites, the initial challenge faced by the authorities was to categorize them according to purpose and develop associated requirements and standards (G. Delgadillo, personal communication, 2020). The regulations that accompany the law separate the facilities into four categories: rescue centers, which focus on rehabilitation and release; zoos—both commercial, which can take in animals from other countries, and non-commercial, which can serve as a sanctuaries; breeding programs—be they commercial, conservation-driven, or consumption- or subsistence-based; and aquariums. The regulations outline requirements for each set of facilities (MINAE, 2017a).

If a facility operates both as a rescue and rehabilitation center and as a sanctuary, it must have two separate permits and fulfil distinct requirements (G. Delgadillo, personal communication, 2020). This double requirement reflects an understanding that the rehabilitation of animals for release is fundamentally different from lifetime care. Among the regulations designed to safeguard the rehabilitation process is a ban on public visits to rescue centers that are focused on the rehabilitation and release of wildlife. All wildlife management sites must develop a management plan that includes animal care and operational considerations, such as animal diet and health, contingency plans for emergencies, contraception, enclosure design and size, an organizational chart, species carrying capacity, staff training and a contingency plan in case of facility closure. Regardless of registration status, a facility must also have a conservation focus and an education program for species conservation (MINAE, 2017a). At this writing, 30 facilities had been closed down due to poor welfare standards (S. Ramirez, personal communication, 2020).

While all wildlife management facilities require permits to operate legally, rescue centers that wish to be officially recognized by the government as priority sites that receive confiscated wildlife must be accredited by the Global Federation of Animal Sanctuaries. The special status and external validation enable facilities to play a greater conservation role in rehabilitating native wildlife and potentially allow them to attract funding. Humane Society International is collaborating with GFAS to enable effective execution of the accreditation process; together with the government, it is also developing a range of accompanying protocols on animal intake, biosecurity (animal and human), emergency and evacuation, euthanasia, quarantine, rehabilitation and release (G. Delgadillo, personal communication, 2020).

Annex X

Understanding Barriers to and Opportunities for Good Captive Ape Welfare

Level	Issue	Barriers	Opportunities and action
Legislation and supporting regulations, articles (national, regional)	Inadequate legal mechanisms due to omission of captive wildlife welfare in animal welfare legislation (covering domestic and farm animals only) or conservation legislation (covering wild animals in the wild or in trade).	Animal welfare is not seen as a priority; lack of political will. Funding bias towards wildlife conservation.	Raise awareness of the links between animal, human and environmental health and wellbeing (One Welfare–One Health). Adopt appropriate language that reflects current knowledge on animal sentience among conservation, environmental, human and animal health, sustainable development and trade bodies. Acknowledge the role of confiscation and captive facilities in the law enforcement chain; add associated activities to conservation or illegal wildlife trade grant budget lines. Include welfare components in lobbying campaigns when seeking to amend legislation to ensure welfare is included. Undertake gap analysis of policy, legislation, regulations, and control and enforcement capacity underpinning captive wildlife welfare and management options.
	Limited understanding of legislation and the role of different agencies. Excessively demanding regulations (including the detention of animals in short-term stay transit centers as evidence for court cases).	Animal welfare is not seen as a priority; lack of political will. Poor understanding of how welfare is negatively impacted by excessive regulations. Lack of resources (human and financial).	Communicate clearly and appropriately with stakeholders (from the government to the public) about legislation and the role of different agencies. Raise awareness of any negative welfare impacts of regulations. Add to appropriate curriculums for sustainability of learning outcomes.
	Lack of formally articulated CITES National Action Plans (NAPs) for the seizure and management of live animals.		For CITES parties: develop NAPs, including clear messaging on the importance of welfare to conservation. Ensure the NAP is developed in a way that supports national ownership to aid implementation and that is appropriate for the context. Connect the relevant agency to collaborating experts.
	Lack of resources among national enforcement agencies for effective seizure and management of live animals, and for oversight of wildlife in captivity generally.		Conduct a needs assessment; an articulation of resources (financial, human, infrastructure) is required to implement the NAP and to provide oversight of relevant legislation and regulations. Develop appropriate management protocols, guidelines, standards and a welfare assessment system with species-specific requirements.

Level	Issue	Barriers	Opportunities and action
			Create a learning environment in which approaches and standards are reviewed and updated in line with emerging research and practice.
			Provide training on animal handling and care during seizures for government personnel, as appropriate.
			Facilitate animal transfer by using clemency periods during which owners can surrender animals without penalty to minimize the animals killed or hidden.
			Minimize mass confiscations by using a clause for owners to keep animals (uniquely identified) who are already in their possession for a fixed time period.
			Strengthen capacity for agency and partner staff: embed knowledge and skills required into existing professional training programs for the broad range of agencies involved (including the police, customs, rangers and the judiciary) and into other relevant government, academic and professional training or courses on sustainability. Include a blended approach to embedding capacity—such as through training, secondment and mentoring—and evaluate impact.
			Collaborate with relevant experts and partner organizations.
			Lobby for inclusion of resources into national budgets.
	Inadequate deterrents to conservation and welfare crimes.	Wildlife conservation and animal welfare are not seen as a priority.	Include dialogue on the full range of costs (environmental, social, conservation and animal welfare), action and resources required in high-level intergovernmental forums on combatting the illegal wildlife trade.
		Provision of inadequate welfare services not seen as a crime.	Support the development of a legal system that allows for criminal actors to bear the financial costs for the seizure and management of live animals.
		Lack of resources (human, financial) and technical expertise.	Capture and communicate the full cost of seizure and management options for each animal (in terms of financial, human and infrastructure costs).
		Substandard investigations into criminal activity.	Strengthen capacity (in operational procedures, skills and competencies, and financial resources) to conduct investigations.
		Corruption.	Develop appropriate guidelines, standards, indicators and a welfare assessment system with species-specific requirements to facilitate identification of a welfare crime.

Level	Issue	Barriers	Opportunities and action
Professional accreditation systems (international, regional)	Limited understanding of what is required to create and manage an active governance structure and organization.	Lack of resources (human, financial). Competing and conflicting interests. Unfavorable cultural context. Fear of losing face and control. Difficulty in finding board members and keeping them active and engaged.	Seek expert input into good and appropriate governance systems; include a focus on how to engage and manage a board, working groups or steering committees. Ensure governance structure members are independent and without competing interests. Reach out and collaborate with other accreditation systems elsewhere to learn what works and why; adapt practice to the context.
	Inadequate systems and standards supporting the accrediting system.	Lack of resources (human, financial). Lack of technical expertise. Animal welfare not seen as a priority. Need to accommodate the diversity of constituents.	Reach out and collaborate with other accreditation systems to learn what works and why; adapt best practice to the context. Establish technical committees or working groups. For zoo systems: explicitly acknowledge that any role in conservation is underpinned by good animal welfare. Develop appropriate guidelines, standards, indicators and a welfare assessment system with species-specific requirements to ensure the overall system is outcomes-based and permits adaptation to the context. Develop specific guidance, standards and indicators for rehabilitation and post-release support and monitoring. Create a learning environment in which approaches and standards are reviewed and updated in line with emerging research and practice.
	Inadequate capacity to support and enforce the system and standards.	Lack of resources (human, financial). Fear of reprisal.	Develop detailed guidelines and standards with accompanying programs to strengthen capacity (blended approach) to ensure the system is transparent and consistently applied. Take a phased approach to implementation. Partner with relevant organizations.
	Poor uptake by the community (captive facilities).	Inappropriate system. Value not seen by the community. Community unwilling to feel judged or afraid to fail. Lack of resources (human and financial) to go through the process.	Demonstrate the value of accreditation. Ensure the system is supportive and appropriate for the context; have regional representation if the overarching body is based outside the region. Solicit input into standard development from targeted practitioners to facilitate buy-in. Create a system of peer-to-peer learning and co-support.
	Lack of awareness among relevant partner organizations (such as tourism providers).	Lack of understanding and trust in the system of accreditation. Lack of resources (human, financial).	Demonstrate the value of accreditation. Ensure the system is appropriate for the context. Solicit input into the development of the standards and accreditation process from targeted partners to facilitate buy-in. Ensure targeted and transparent communication on the system and results.

Level	Issue	Barriers	Opportunities and action
Individual facility (governmental and non-governmental)	Need for better understanding of what is required to create and maintain an active, strong governance structure, management team and effective and workable policies.	Lack of resources (human, financial). Competing and conflicting interests. Fear of losing control. Unfavorable cultural context. Difficulty in finding board members and keeping them active and engaged.	Seek expert input into developing and maintaining an effective and appropriate governance system. Reach out to other facilities and accreditation systems to learn what works and why; adapt practice to the context. Ensure that the difference between governance and management is understood; transition to a system in which the director is a non-voting member of the board. Prepare and distribute a board information pack to ensure that board members know the organization and requirements of the role. Develop a board self-assessment process to guide understanding of gaps in knowledge, skills and performance. Seek input from other facilities and accreditation systems on effective management systems and essential policies. Develop a system to keep the board, working groups and steering committees active and engaged. Ensure the management team shares responsibility, potentially by creating departments to spread responsibility.
	Lack of planning.	Planning (strategic, succession, action) not seen as valuable or a priority. Lack of management expertise. Lack of resources (human, financial). Fear of losing control.	Seek expert input from other facilities and accreditation systems on how to develop a strategic plan and succession plan; reach out to learn what works and why; and adapt practice to the context. Ensure that plans are used and updated. Develop operational processes, systems and policies for organizational sustainability; engage staff in their development and keep them informed. Ensure the board or a coach supports the director to facilitate a vision of sustainable services.
	Lack of an employee-oriented human resources approach to support staff retention.	Limited expertise in and poor appreciation of the importance of looking after and investing in staff. Lack of management expertise. Limited resources (human, financial).	Seek expert input from other facilities and accreditation systems to understand appropriate ways to look after and invest in staff; consider economical ways to demonstrate appreciation and value. Ensure each position has a job description, including board members and the director. Ensure that all key tasks and roles can be carried out by more than one person for succession and sustainability. Assess different ways to strengthen capacity (through a blended approach) and ways to benefit more than one staff person.
	Reliance on a single donor or funding mechanism.	Complacency. Lack of resources (human, financial). Limited professional fundraising expertise. Competitive market.	Seek expert input from other facilities and accreditation systems to understand appropriate ways to diversify fundraising sources. Develop a fundraising plan. Create a financial reserve and add to it as possible.

Level	Issue	Barriers	Opportunities and action
	Lack of technical expertise.	Poor awareness and skill gaps. Lack of planning to fully understand and predict knowledge and skills required—currently and in the future. Lack of resources (human, financial).	Explicitly acknowledge the importance of animal welfare. Seek partnerships with organizations with the required expertise. Understand the potential complex needs of the animals in the facility and the required knowledge and skills to properly manage them—currently and in the future. Develop appropriate standards and a welfare assessment system with species-specific requirements for captive and released apes. Create a learning environment in which approaches and standards are reviewed and updated in line with emerging research and practice. Recognize the funding required to employ people with the necessary expertise; consider what benefits can be offered to people in lieu of higher salary scales. Plan ahead for knowledge and skills required, keeping in mind the geriatric requirements of aging apes, the complex medical needs of chimpanzees from laboratories, and the need for post-release monitors or trackers with the knowledge and skills to assess welfare.
	Carrying capacity exceeded.	Policy on carrying capacity absent or not followed. Planning not conducted or not followed. Government pressure to accept more animals.	Create a policy on carrying capacity (per enclosure, species, and for the facility generally); review the policy if situations arise that would lead the facility to exceed its carrying capacity and determine how to acquire the resources required for intake. Communicate with the relevant government agencies about the facility's carrying capacity and what it means to exceed it—for the facility's reputation, animal welfare, and financial and other resources.
	Inadequate government understanding of the requirements that underpin good welfare and good outcomes for release.	Government pressure to accept more animals Government pressure to release animals in contravention of IUCN guidelines, in ways that undermine welfare and conservation outcomes.	Raise awareness of the links between animal, human and environmental health and wellbeing (One Welfare–One Health). Work with and get support from the relevant accrediting body to communicate what good practice looks like for care, welfare, rehabilitation and release. Communicate with the relevant government agencies about the facility's carrying capacity and what it means to exceed it—for the facility's reputation, animal welfare, and financial and other resources.

Sources: Based on author observations, supplemented by Baker *et al.* (2013); D'Cruze and McDonald (2016); Farmer (2012, 2018); IUCN (2019a); Mitman *et al.* (2021); Phelps *et al.* (2021a); Pinillos (2016); Rivera, Knight and McCulloch (2021); Rodriguez *et al.* (2019); Ronfot (2016); Sherman and Greer (2018); Sinclair and Phillips (2018b); Sollund (2022); Wyatt *et al.* (2022); personal communication in 2020 with N. Maddison, O. Martin and J. Vaughan

Annex XI

Selected Tools for Assessing Captive Ape Welfare: Key Features

Name of tool	Details

Animal Welfare Assessment Grid (AWAG)®
 	
University of Surrey and Reuben Digital
	Focus
	◦ Originally designed to monitor laboratory primate welfare; adapted for individual and groups of primates (and other species) in zoos.
	◦ Used for daily welfare monitoring of Siamang gibbons (*Symphalangus syndactylus*) and several non-ape species at Marwell Zoo, UK.
	◦ Trialled on gorillas at Safaripark Beekse Bergen, the Netherlands.
	◦ Currently being adapted for farm and pet animals.
	◦ Intended for use by captive facility staff.
	
	Indicators and parameters
	◦ Input and output indicators. Factors can be adapted for the species based on adaptations made for gorillas.
	◦ Four parameters, each with several factors:
	  ▪ environmental: access/events, enclosure furnishings, group size, housing, nutrition;
	  ▪ physical: activity level, clinical assessment, food/water intake, general condition;
	  ▪ procedural: change in daily routine, restraint, sedation/anesthesia, vet procedures; and
	  ▪ psychological: abnormal behaviors, aversion to routine events/animal training, enrichment provision/use, response to catching events, social disruption within groups.
	
	Tool development and application
	◦ Development: Factors are scored 1–10 (good to poor) and chosen/adapted by zoo staff (animal welfare advisors, keepers, veterinarians, zoologists). Research into known abnormal behaviors for each species is conducted to facilitate recognition and scoring. Each scoring sheet is independently scored by three people.
	◦ Application: Previously the score was calculated retrospectively from the daily reports generated by animal staff members. Cloud-based software enables staff to score in real time, with the option to add comments.
	◦ The software analyses the data and presents it in graphical form.
	
	Outputs
	◦ Software outputs are numerical scores and a visual polygon. The averages of the four parameters can be plotted as a radar chart to form a two-dimensional polygon, representing the impact of each category on an animal's welfare. The cumulative welfare assessment score (CWAS) is equal to the surface area of this polygon (not just the average) and increases when parameter classes are compromised, indicating a potential welfare issue.
	◦ The radar chart can be used to capture long-term trends, whereas the CWAS can be plotted over time to identify short-term events that impact welfare.
	
	Additional information
	◦ By monitoring the changes in the aggregate scores over time, users can determine the factors affecting the welfare of an animal or group. AWAG can also be used to assess the potential welfare impact of planned interventions. The software is best used to highlight perceived positive and negative welfare impacts, complemented by more traditional auditing methods.
	◦ AWAG does not allow comparison between species or between individuals held in different institutions, but it could be used for individuals within institutions (for example, to monitor moves to different enclosures).
	◦ An adapted version for gorillas was tested for usability and reliability. AWAG provided a good indication of individual and group welfare, and potential welfare issues. Daily audits may not be required as welfare appeared to be stable in the long term. Inter-rater agreement (between keepers and researcher) was good. More frequent and longer observations, reduced scoring options, regular staff meetings and staff training to make scoring unambiguous could improve usability and increase accuracy.
	◦ The aim is to make the tool available for other facilities and species, and to integrate it with the Zoological Information Management System (ZIMS).

Name of tool	Details
Great Ape Welfare Index (GAWI)	**Focus** The GAWI focuses on the assessment of chimpanzee welfare at the group level, with ongoing work to include individual welfare. This description is focused on input measures as the validation of output measures is ongoing. Derived from expert opinion and validated by behavioral observations, the GAWI identified the most important attributes of a great ape captive management system. **Indicators and parameters** Current input indicators: dietary provision, the physical and social environment, and management aspects (including indoor/outdoor enclosure availability and staff qualifications). The following output measures are being created and validated: body condition score* (correlated with hematology and other physiological values), bilateral alopecia and fecal cortisol as indicators of stress, and wounds (frequency, location, presence), in conjunction with behavior. **Tool development and application** Development: JGI Tchimpounga Chimpanzee Rehabilitation Center (Republic of Congo) refined and tested the GAWI. It was further tested at three other range state chimpanzee sanctuaries but found to be open to subjective interpretation. To combat this issue, a working group of experts from African sanctuaries, European and North American zoos and an Australasian university was established to review and adjust the welfare index. Input on welfare indicators was further validated by a range of persons including caregivers, managers and veterinarians. Application: Each chimpanzee group is scored on a scale of 1 to 5 (poor to good) on each input indicator. The average score represents a welfare index for each group. Data are manually collected and entered onto an Excel spreadsheet. **Outputs** Input indicators are scores for each indicator and group, and an average index. Graphical representation of results would need to be manually generated. **Additional information** The index was found to be helpful for assessing the welfare of chimpanzee groups but not individuals, hence the ongoing work to develop output indicators. The GAWI places emphasis on useability for African facilities, by caregivers with non-academic backgrounds and in the relevant resource context. The tool also aims to ensure that caregivers can see the results of their efforts.
Project ChimpCARE Chimpanzee Assessment Lincoln Park Zoo Lester E. Fisher Center for the Conservation of Apes	**Focus** Designed to provide a practical yet empirical assessment tool that can add an extra species-specific (chimpanzee) layer to a full organizational assessment and facilitate comparisons across facilities. Data are predominately collected by an external person or assessor. **Indicators and parameters** The tool is predominantly input indicator-focused, with a smaller proportion of output measures designed to present a comparative assessment of how chimpanzees utilize the resources. Three areas of assessment, each comprising several variables: programs: daily management practices, diet, staff experience and veterinary care; social: composition, size and stability; and space: complexity and size. **Tool development and application** Development: Specific metrics underlying the design of the assessment are derived from a process that gathered 20 experts in captive chimpanzee care, working in research centers, sanctuaries and zoos around the world. Application: Each of the three areas has several variables that are scored and weighted, with a resulting score between 0 and 100 (poor to good). The assessment has been trialled at Project Chimps and scores were compared with spontaneously chosen observations of practice from AZA zoos and GFAS-accredited sanctuaries. Data are collected during scheduled and unscheduled site visits, as well as by the organization under the direction of the assessor. In the latter case, measures are verified by the assessor during unscheduled visits.

Name of tool	Details
	Outputs The result is a score for each area and an overall score that is the average of the three, representing the overall capacity of the organization to address the welfare of the chimpanzees. The scores reflect what was seen and assessed during scheduled and unscheduled site visits only. **Additional information** During the trial, two key challenges were experienced: ■ The first was difficulty in objectively measuring what represents sufficient complexity of space, relative to space availability. This step involves assessing which resources make the space complex enough and functionally relevant for chimpanzees. Based on scientific literature and expert opinion, key elements include elevated resting areas, substrate coverage, vertical climbing opportunities and visual barriers. ■ The second challenge relates to characterizing and assessing spaces according to the standard binomial "indoor" and "outdoor." In response, a "mixed/hybrid" category was added, defined by the proportion of the perimeter that is open-air. It allows for better assessments of spaces that provide some, but not all, of the benefits of outdoor access. The assessment will be refined as it is tested at other facilities, with the aim of expanding it to other types of chimpanzee facilities and helping to create similar tools for other species.

WelfareTrak®
 	
Created and managed by the Chicago Zoological Society's Center for the Science of Animal Care and Welfare
 	
As of November 2022, the program had begun to close down. New customers were no longer being accepted.
	Focus
	□ This web-based application provided a mechanism for tracking zookeepers' assessments of individual animal welfare over time, across multiple species (20 species-specific surveys).
	□ It was designed to monitor individuals over time (weekly), not to compare animals within or between facilities.
	
	Indicators and parameters
	□ Species-specific, animal-based output (positive and negative) indicators. Each species-specific survey was composed of 10–15 indicators rated on a five-point Likert-type scale.
	□ WelfareTrak® was used more than 60 times to run species-level surveys, for animals ranging from geckos to gorillas.
	
	Tool development and application
	□ Development: A panel of experts (comprising zookeepers, animal managers, veterinarians and wildlife biologists) helped to develop each species-specific tool; the chimpanzee version, for example, solicited input from 17 experts. Questionnaires were used to establish a consensus of opinion on the most useful welfare indicators and definitions (including emotional, mental and physical states).
	□ During the initial trial period, the application was tested with nearly 50 animal care specialists representing five AZA-accredited facilities.
	□ Application: Staff members provided input into a species-specific welfare survey that could only be used online. Observers rated indicators on a five-point Likert-type scale (1–5 for poor to excellent *or* never to always). It was also possible to document special events that may have impacted welfare scores.
	□ The tool permitted entry of score ratings from multiple raters for comparison.
	□ A nominal fee was collected for each species that was monitored and this fee was used to cover server and site maintenance costs. Data were stored on the Center's server but users' data were confidential and not viewed unless requested (for example, if a user had a question about interpretation).
	
	Outputs
	□ Two types of reports could be viewed:
	  ■ Trend reports generated separate graphs for each welfare indicator over time and gave users the ability to view scores of individual raters and mean scores.
	  ■ Reports on individual wellbeing generated separate tables for each welfare indicator and "flagged" potential changes in scores using symbols and banners.
	□ Both options permitted viewing of special events.

Notes: This table includes selected examples of tools used to assess captive ape welfare; it does not provide an exhaustive list.

* A study in zoo gorillas found that environmental variables (visitor density and noise levels) and modifications showed significant effects on behavior but not on fecal glucocorticoid measurements, demonstrating the importance of aligning fecal hormone studies with behavioral monitoring (Clark *et al.*, 2012). The body condition score has been used to assess body weight of orangutans without touching or weighing to avoid intervention and to reduce stress (C. Nente, personal communication, 2020; see Chapter 4). Although the body condition score represents a reliable system for comparing scores over time and can provide for some objectivity in multiple situations, challenges remain, such as effectively assessing body condition in large, long-haired males.

Sources: Based on author knowledge and experience, supplemented by the following: AWAG®: Brouwers and Duchateau (2021); Justice *et al.* (2017); Wolfensohn *et al.* (2018); D. Free and S. Wolfensohn, personal communication, 2021; GAWI: Fernie (2008); Fernie *et al.* (2012); R. Atencia, personal communication, 2020; Project ChimpCARE Chimpanzee Assessment: ChimpCARE (n.d.-b); Project Chimps (2020); Ross (2020); S. Ross, personal communication, 2020; WelfareTrak®: CZS (n.d.); Whitham and Wielebnowski (2015); J. Whitham and L. Miller, personal communication, 2021 and 2022

ACRONYMS AND ABBREVIATIONS

AIDS	Acquired immune deficiency syndrome
ALI	Allostatic load index
A.P.E.S.	Ape Population, Environments and Surveys
ARRC	Avoidance, Reduction, Restoration, and Conservation
asl	Above sea level
AZA	Association of Zoos and Aquariums
BAL	Bronchoalveolar lavage
Bcbva	*Bacillus cereus* biovar *anthracis*
BMP	Best management practice
BNF	Borneo Nature Foundation
BOSF	Borneo Orangutan Survival Foundation
BSGR	Beny Steinmetz Group Resources
CHIMP	Chimpanzee Health, Intervention, and Monitoring Program
CITES	Convention on International Trade in Endangered Species of Wild Fauna and Flora
COVID-19	Coronavirus disease 2019 (infectious disease caused by the SARS-CoV-2 virus)
CTPH	Conservation through Public Health
CWAS	Cumulative welfare assessment score
DNA	Deoxyribonucleic acid
DRC	Democratic Republic of Congo
EAZA	European Association of Zoos and Aquaria
EDT	Enclosure Design Tool
ESG	Environmental, social and governance
EVD	Ebola virus disease
FAO	Food and Agriculture Organization of the United Nations
FSC	Forest Stewardship Council
GC	Glucocorticoid
GDP	Gross domestic product
GFAS	Global Federation of Animal Sanctuaries
HDI	Human Development Index
HIV	Human immunodeficiency virus
HMPV	Human metapneumovirus
HRSV	Human respiratory syncytial virus
IFC	International Financial Corporation
IFL	Intact forest landscape
IGCP	International Gorilla Conservation Programme
IPHP	International Primate Heart Project
IPLCs	Indigenous Peoples and local communities
IUCN	International Union for Conservation of Nature

KCP	Kibale Chimpanzee Project
Lao PDR	Lao People's Democratic Republic
LWT	Lilongwe Wildlife Trust
MPXV	Monkeypox virus
MYR	Malaysian Ringgit
NAP	National Action Plan
NGO	Non-governmental organization
NGS	Next-generation sequencing
NNNP	Nouabalé-Ndoki National Park
OIE	World Organisation for Animal Health (originally founded as the Office International des Epizooties)
OVAG	Orangutan Veterinary Advisory Group
PASA	Pan African Sanctuary Alliance
PBOP	Petowal Biodiversity Offset Programme
PCR	Polymerase chain reaction
PEESTOLM	Political, environmental, economic, social, technical, operational, legal, media and communications
PPE	Personal protective equipment
PTSD	Post-traumatic stress disorder
RNA	Ribonucleic acid
RSPO	Roundtable on Sustainable Palm Oil
SAIDS	Simian acquired immunodeficiency syndrome
SARS-CoV-2	Severe acute respiratory syndrome coronavirus 2
SDG	Sustainable Development Goal
SEA	Strategic environmental assessment
SEAZA	Southeast Asian Zoos and Aquariums Association
SGA	Section on Great Apes
SIVcpz	Simian immunodeficiency virus in chimpanzees
SIVgor	Simian immunodeficiency virus in gorillas
SMB	Société Minière de Boké
SOCP	Sumatran Orangutan Conservation Programme
SOP	Standard operating procedure
sp.	Species (plural spp.)
SPOTT	Sustainability Policy Transparency Toolkit
TB	Tuberculosis
TNS	Sangha Trinational
TPE	*Treponema pallidum pertenue*
UAE	United Arab Emirates
UDAW	Universal Declaration on Animal Welfare
UNESCO	United Nations Educational, Scientific and Cultural Organization
WAZA	World Association of Zoos and Aquariums
WCS	Wildlife Conservation Society
WHO	World Health Organization
WWP	Wildlife Wood Project

YEL	Yayasan Ekosistem Lestari
ZEBOV	Zaire Ebola Virus
ZIMS	Zoological Information Management Software
ZSL	Zoological Society of London

GLOSSARY

Abiotic: Not derived from living organisms.

Abscess: An enclosed, pus-filled cavity in any tissue, usually caused by bacteria.

Accidental host: An organism that does not usually allow for the transmission of a pathogen to the definitive or typical target species; also referred to as an incidental or dead-end host.

Adaptive management: An iterative process of improving the management of natural resources by incorporating monitoring results in the decision-making process.

Agonist: A substance that activates a receptor inside a cell or on its surface.

Air blast: A sudden rush of air through openings, such as tunnels, caused when unsupported rock spanning a void in an underground mining system collapses.

Air sacculitis: A common inflammatory condition of air sacs, the tiny sacs off the laryngeal tubes of apes (and many other animals) that act as resonating chambers, amplifying vocalizations.

Allelic dropout: The failure of an allele (one of two or more versions of DNA sequence) to increase the number of copies of a gene in a polymerase chain reaction (PCR).

Amplifying host: An organism in which a pathogen can multiply to high levels, which facilitates the pathogen's spread.

Anemia: A lack of red blood cells (or hemoglobin) that leads to decreased oxygen transport.

Animal reservoir: see **reservoir**.

Animal welfare: Physical and mental wellbeing of animals.

Anogenital: The area around the anus and genitalia.

Anthelmintic: Any drug that acts against infections caused by parasitic worms (helminths).

Anthropocene: Unofficially, the current geologic epoch, during which human activity started to have a significant impact on Earth's climate and ecosystems.

Anthropogenic: Caused by humans or human activity.

Anthroponosis: Infectious disease that is transmitted from humans to other animals. Related terms: anthroponoses, anthroponotic. See also: **zoonosis**.

Arboreal: Living in trees.

Asset protection zone: An area surrounding a built asset or structure, such as a residence, farm, commercial building or landmark, where the level of forest fuel has been reduced to a level that does not support intensive fire behavior. The width of the area between the at-risk asset and the forest is determined by the predicted fire behavior. Commonly referred to as APZ.

Asymptomatic: Infected with a pathogen but not showing signs of disease.

Autonomy: Self-determination of individuals.

Bacteria: Single-celled, microscopic organisms that are ubiquitous in humans, other animals and the environment, and that can be beneficial (for example by promoting digestion) or cause illness (singular: bacterium).

Best management practice (BMP): Methods or means that have been determined to be the most effective and practical means of preventing or reducing disease risks from humans to apes and vice versa.

Bimaturism: Development characterized by differing stages or timings within a species or within a sex; among orangutans, mature males are flanged or unflanged (see **flanged**).

Bioacoustics: Digital technology, including equipment and software, used to record and analyze animal sounds.

Biopsy: The removal of a small piece of tissue from a living animal for diagnostic purposes.

Biosafety: A set of measures designed to prevent the accidental spread of pathogens to humans, animals or the environment, including safe handling measures of potentially infectious substances, for example through the use of personal protective equipment (such as gloves and masks), disinfection and garbage disposal protocols. *Biosafety* aims to protect public health and environment from accidental exposure to biological agents. See also: **Biosecurity**.

Biosecurity: *A combination of management practices and protocols designed to prevent the transmission of diseases and disease-causing agents.* Biosecurity deals with the prevention of misuse through loss, theft, diversion or intentional release of pathogens, toxins and any other biological materials. See also: **Biosafety**.

Biotic: Relating to living organisms.

Bloating: The accumulation of gas in the intestines, often uncomfortable or painful.

Brachiation: Arboreal locomotion that relies exclusively on the arms to propel the body forward.

Bronchoalveolar lavage (BAL): A safe, simple and inexpensive diagnostic test of the lower respiratory system, which can be performed in the field without sophisticated equipment. BAL involves instilling sterile saline via a tube into an anesthetized animal's airways. The method provides an excellent sample of epithelial lining fluid for characterization of diffuse lung diseases (such as tuberculosis on polymerase chain reaction) and airway inflammation.

Burnout: A syndrome that results from chronic, unsuccessfully managed work-related stress.

Captive facilities: Sites of wildlife captivity, such as rescue and rehabilitation centers, sanctuaries and zoos.

Cardiologist: A doctor who specializes in treating diseases of the cardiovascular system—mainly the heart and blood vessels.

Cardiomyopathy: Diseases of the heart muscle involving stretching, thickening or stiffening of the walls of the heart chambers, which affect the heart's ability to pump blood around the body.

Cardiovascular: Related to the circulatory system—the heart and blood vessels.

Cardiovascular disease: A group of disorders of the heart and blood vessels.

Carry-over: Released apes carrying human pathogens to wild apes.

Case fatality rate: In epidemiology, the proportion of individuals who die from a specified disease among all those who have been diagnosed with it during a certain time period.

Catheter: A flexible tube inserted into a vein to administer or extract fluids intravenously.

Causative agent: A (micro)organism that causes a disease.

Chlorpyrifos: An insecticide, acaricide or miticide used to protect plants against damage.

Colitis: Inflammation of the bowels or large intestine (colon).

Command and control system: A structured system used to organize functional management and leadership, as well as to provide authorities with clear lines for planning, organizing and directing operations during response and recovery at the strategic and tactical levels.

Commensal: Relating to a relationship in which one organism obtains food or other benefits from another without damaging or benefiting the host organism; an organism in such a relationship.

Compassionate conservation: A discipline that combines the fields of conservation and animal welfare.

Compounding (or cascading) risks: Interacting risks whose collective effect exceeds those associated with the individual risks.

Conservation litigation: The use of liability lawsuits to ensure that companies, organizations and people who damage the environment or biodiversity are held responsible and have to take action to remedy the damage caused.

Conservation medicine: A field that combines veterinary science, conservation biology and public health to address animal, human and ecological health holistically, rather than as siloed branches of knowledge.

Conspecific: An individual of the same species.

Core area: The portion of the home range of a group or individual that is most frequently used.

Costs: In disaster management, the estimated or known monetary value of facilities, goods, services and provision of personnel provided by an agency or organization in support of activities during preparedness, response and recovery for which the agency or organization receives no payment; may be referred to as in-kind contributions.

Cramping: The sudden, extreme, involuntary tensing of muscle that results in pain.

Crisis: A system-wide disruption that is typically new, unexpected, uncontrollable or abnormal and that requires immediate solutions or interventions involving collaboration among local stakeholders. A crisis typically affects a particular industry, population or community; local stakeholders are able to address the disruption.

Cross-reactivity: The ability of some antibodies to target or "react to" parts of different pathogens other than the one with which the antibodies typically bind. If a test is not very specific, cross-reactivity can result in a false positive. See also: **reactivity**.

Cytomegalovirus (CMV): A common virus that can cause herpes infections that may become dormant and later reactivate. In individuals whose immune systems are compromised, symptoms may include fever, jaundice and lesions, as well as neurological and respiratory issues. Humans and other primates serve as natural hosts.

Data mining: The sorting and analyzing of very large datasets to find patterns and relationships that can serve as intelligence for facilitating planning and decision-making.

Deciduous: Pertaining to trees that lose their leaves for part of the year.

Deciduous teeth: Baby teeth, also known as milk or primary teeth, which are later replaced by permanent, or adult, teeth.

Degenerative: Related to the deterioration of organs, often caused by age.

Depigmentation: Loss of color, such as of hair or skin.

Dermatophyte: A common label for a group of fungi of the *Arthrodermataceae*, which commonly cause skin disease.

Design thinking: An iterative process used to understand and redefine problems with the aim of creating innovative solutions.

Diagnostic protocol: A practical text that can guide clinical approaches and facilitate diagnosis, for example by recommending questions and examinations.

Dichromatic: Exhibiting two color variations independent of sex and age.

Dimorphic: Having two distinct forms.

Dipterocarp: A tall hardwood tree of the family *Dipterocarpaceae* that grows primarily in Asian rainforests and that is the source of valuable timber, aromatic oils and resins.

Disaster: A serious disruption of the functioning of a community or society due to an interaction of a hazardous event with conditions of exposure, vulnerability and insufficient capacity to cope with the event. The consequences include significant social, built, economic and environmental losses and impacts. Locally impacted communities are unable to cope and require external assistance and coordination.

Disease emergence: A first-time outbreak of illness in a species or area, or a rapidly increasing incidence of outbreaks, typically with reference to zoonotic infectious disease. In contrast, re-emerging (infectious) diseases are ones that appear in species or areas in which they have not been seen for a long time.

Dispersing sex: Either male or female apes who, upon reaching sexual maturity, depart from their birth area to establish their own range, leaving the opposite sex behind.

Diurnal: Daily or active during the day.

DNA: Deoxyribonucleic acid, the hereditary material in almost all organisms.

Dysbiosis: An imbalance in the gut microbial community that is associated with disease and may be due to the gain or loss of microbial community members or changes in the relative abundance of microbes.

Dyspnea: Labored breathing.

Echocardiographer: A health care professional trained to use imaging technology to help physicians diagnose heart problems in patients, specifically by operating ultrasound equipment that provides moving 2-D or 3-D images of the heart and its chambers.

Ecosystem health: A paradigm or model that integrates environmental conditions with the effects of anthropogenic activities to yield information for the sustainable use and management of natural resources. An ecosystem in good health is one whose dynamic attributes are expressed within the normal ranges of activity relative to its ecological state of development.

Ecotourism: A sustainable version of nature-based tourism that contributes to biodiversity conservation and the wellbeing of local communities.

Ecotoxicology: The study of the toxic effects of chemicals on organisms and their ecosystems.

Ectoparasite: An organism that lives off or feeds on the surface of the body, such as skin and hair, including ticks and lice.

Ectopy: A mislocation of a body part. In the heart, ectopic contractions start in an abnormal location, are irregular and are associated with decreased function.

Emergency: An actual or imminent natural or anthropogenic event that endangers or threatens life, damages infrastructure or destroys the natural environment, thus requiring significant coordinated and time-critical responses, as well as extraordinary measures to save lives, protect vulnerable individuals and limit damage. An emergency tends to be local or regional, so does not result in serious disruption to the broader community or society. Emergencies can be categorized by size of impact area, and multiple simultaneous emergencies in one area may be classified as a disaster.

Encephalomyocarditis: An acute febrile disease, especially of swine and some primates, caused by a picornavirus and marked by degeneration and inflammation of skeletal and cardiac muscle and lesions of the central nervous system.

Endangered: Threatened with extinction.

Endemic: Native to or only found in a certain place; indigenous.

Endoparasite: A parasite that lives inside its host.

Epidemic: Rapid spread of a disease that causes a high number of cases in a limited space and time, such as a sudden outbreak.

Epithelium: The outer cell layer of skin or mucosae.

Epizootic: Relating to a disease that is temporarily prevalent and widespread in an animal population; such a disease.

Ethical: Pertaining to standards of "right and wrong" imposed by an external source, e.g., a community, profession.

Ethnoprimatology: A practice that combines primatology and anthropology to view humans and other primates as living in shared, integrated ecological and social spaces.

Ethology: The scientific study of animal behavior under natural conditions.

Evapotranspiration capacity: Water demand.

Ex-situ: Outside of a natural environment; in captivity.

Facial dysplasia: A disease in which bone in the skull is replaced by softer tissue, causing abnormal growth and an altered facial appearance.

Fauna: Animals.

Fission–fusion: Pertaining to communities whose size and composition are dynamic due to the coming together (fusion) and moving away (fission) of individuals.

Flagship species: A charismatic species that is selected to serve as ambassador for the protection of an ecosystem or an area inhabited by many species that are less well known.

Flanged: Pertaining to one of two morphs of adult male orangutan, the other being "unflanged"; characterized by large cheek pads, greater size, a long coat of dark hair on the back and a throat sac used for "long calls."

Flora: Plant life.

Folivore: Any chiefly leaf-eating animal. Related terms: folivorous, folivory.

Fomite: An object or material likely to carry infection, such as clothes.

Food security: Constant physical, social and economic access to sufficient, safe and nutritious food that meets food preferences and dietary needs for an active and healthy life.

Footprint: In relation to industrial development, the direct deforestation and disturbance area related to a project and its infrastructure.

Forest fuel: In a wildland forest fire context, the combustible materials such as fine live and dead leaves, twigs, and branches from the ground level to the treetops. Fine fuels are the primary fuel source for intense forest fires. Forest fire behavior is proportional to the level of fine fuels within the overall combustible biomass. High levels of fine fuels, which may be expressed as fuel load (weight per area, such as tons per hectare), are associated with more intensive fire behavior.

Formulation: The process of selecting the types and amounts of ingredients in an animal diet that is to contain planned concentrations of nutrients.

Frugivore: Any chiefly fruit-eating animal. Related terms: frugivorous, frugivory.

Fulminant: In medicine, rapidly progressing and severe.

Fungi: Spore-producing organisms that feed on organic matter—moulds, mushrooms, toadstools and yeast.

Gastrointestinal: Referring to the digestive system—from the esophagus, via the stomach and intestines, to the anus.

Genome: The complete genetic material of an organism.

Genotypic: Related to genetic information (see also: **phenotypic**).

Glomerular: Related to the kidneys' glomeruli, bundles of capillaries that transfer waste products from the blood into the urine.

Glucocorticoids: Anti-inflammatory steroid hormones that are involved in the metabolism of carbohydrates, proteins and fats.

Granulomatous: Forming nodes of immune cells in chronic inflammation.

Guarded prognosis: A prediction about an individual's health outcome based on insufficient information, such that the outcome is in doubt.

Habituated: Accustomed to the presence of humans, as achieved by frequent or prolonged exposure to people. Related term: habituation.

Habituation: A process by which animals are repeatedly exposed to the same stimuli, such as the presence of humans, until they no longer respond to those stimuli.

Hazard: A natural, socionatural or anthropogenic process, anomaly or event that is defined by location, magnitude, intensity, frequency and probability, and that has the potential to directly harm life as well as the built and natural environments and ecosystems. A hazard can cause indirect disruptions to an economy.

Helminth: Flat or round-bodied worm.

Hepadnavirus: A group of DNA viruses, such as the hepatitis B virus, that can cause liver damage.

High-value areas: In the forest fire context, the areas around valuable built assets including public infrastructure for transport, health and communications; private industry, such as agriculture, tourism and mining; environmental areas of significant biodiversity, and those important for endangered species, or of significant cultural value; and some managed water catchment areas.

Histopathology: A branch of pathology evidencing tissue changes characteristic of disease; microscopic evidence. Related term: histopathological.

Holism or **collectivism:** Approaches that value wholes, such as species and ecosystems, over individuals.

Homeostasis: In medicine, the state of balance among all body systems that is required for the body to function correctly.

Home range: An area in which individuals or groups regularly spend time and which territorial animals may defend from others.

Hominines: The evolutionary subfamily of great apes, including African great apes and humans. A similar Asian great ape subgroup is referred to as Ponginae.

Hormone: A biological messenger substance that is produced in the body and transported via the blood stream to different organs and tissues, where it influences the metabolism.

Hybrid: The offspring of two different species or varieties of plant or animal; something that is formed by combining different elements.

Hybrid zone: Area where closely related but genetically distinct populations meet, mate and give birth to cross-fertilized offspring.

Hyperendemic: Regarding a pathogen, persistently present at high levels in a region or population.

Hypertension: A condition in which the blood vessels have persistently raised pressure; also known as high or raised blood pressure.

Hypertrophy: Growth of muscle cells.

Hypometabolism: A condition marked by an abnormally low metabolic rate and decreased glucose consumption.

Hypoplasia: Genetic disorder that leads to underdevelopment of tissue, organs or organisms based on an insufficient number of cells.

Hypotension: Low blood pressure, as is commonly caused by anesthesia (because it expands the volume of the cardiovascular system).

Hypothermia: A drop in the core body temperature below the level at which the body can self-regulate.

Hypothesis: A proposed, testable explanation for an observation.

Idiopathic myocardial fibrosis: A significant increase in the collagen volume of muscular tissue of the heart due to an unknown cause.

Imidacloprid: An insecticide widely used to control pests in agriculture.

Immunocompetence: A body's ability to produce a normal immune response following exposure to an antigen.

Infanticide: The act of killing an infant.

Influenza: A contagious upper respiratory viral infection, commonly referred to as "the flu".

In-situ: In a natural environment; in the wild.

Instrumentalism: A view that ascribes value to an individual or a collective for its extrinsic worth, disregarding or denying any intrinsic value.

Intact forest landscape: Large, connected tracts of undisturbed woodland.

Inter-: Between.

Interbirth interval: The biologically determined period of time between consecutive births.

Interception capacity: The amount of rainwater that can be retained by the leaves and branches of plants and trees, as well as the debris on the forest floor, and so does not reach the soil.

Interferon-gamma assay: Medical test used in the diagnosis of some infectious diseases, especially tuberculosis.

Interstitial: Relating to spaces between cells, tissues or organs in the body.

Intra-: Within.

Intravenous: Into or within a vein.

Intrinsic value: Worth of an individual irrespective of extrinsic valuing, such as in instrumentalism.

Isotopes: Atoms of the same element that contain the same number of protons but a different number of neutrons.

Laryngeal: Referring to the larynx (voice box).

Lesion: Abnormal tissue growth resulting from injury or disease.

Macroparasite: Parasites that are large enough to be visible to the naked eye, such as worms and ticks.

Maculo-papular: Related to skin lesions typically characterized by red, flat and raised bumps.

Mast fruiting: The simultaneous production of fruit by a large number of trees every 2–10 years, without any seasonal change in temperature or rainfall.

Metabarcoding: A method of species identification that uses a section of DNA or RNA in a way that allows for the simultaneous identification of many species within the same sample.

Metabolic syndrome: A combination of conditions that increase the risk of heart disease, stroke and diabetes.

Metazoan: A multicellular organism with differentiated cells. Parasite examples include helminths and arthropods.

Microbiome: All microorganisms in a certain habitat, such as the gut (gastrointestinal system) or skin.

Microcosm: A small subset of a whole that is generally considered representative of the whole.

Mitigation hierarchy: A tool used to limit the negative impacts of development projects on biodiversity.

Mixed methods: The combination of quantitative and qualitative data collection and analysis within the same study.

Molecular: In this volume, related to DNA- or RNA-based methods, such as polymerase chain reaction (PCR) tests.

Monogamy: The practice of having a single mate over a period of time.

Moral: Pertaining to what a person feels they ought to do—distinguishing between "good and bad" and "right and wrong"—based on guiding principles shaped by their environment and sometimes their belief system.

Moral courage: The ability to take action in line with ethical values despite the risk of adverse consequences.

Moral distress: Unease that results from discrepancies between what an individual perceives as an ethically correct action and what the individual is tasked with doing.

Moral resilience: The ability to deal with an ethically adverse situation without experiencing lasting effects of moral distress.

Moral standing or **status:** In ethics, an ascribed quality based on which an individual is worthy of moral consideration and moral significance.

Morbidity: Sickness. In epidemiology, the number of individuals in a population who suffer from a disease compared to the total number of individuals in the population.

Morph: A distinct form of an organism or species.

Mortality: Deaths. In epidemiology, the number of individuals in a population who die from a disease compared to the total number of individuals in the population.

Multi-agency response: In disaster management, collective work undertaken by a number of agencies, each with its own command and communications system, to deliver shared aims and objectives; as part of the response, they also pursue their own, separate priorities, which reflect individual agency strengths, including resource type and expertise.

Multimodal approach: Characterized by multiple means of communication, such as a combination of text, video, photos and audio. With respect to ape conservation, its relevance is consistency of scientific communication between various communities.

Multiple-use zones: Land characterized by a range of integrated uses, such as residential, commercial, industrial, agricultural and conservation-related.

Mycobacterium tuberculosis **complex:** A genetically related group of bacteria that cause tuberculosis.

Nasal flora: Microorganisms in the nose.

Nasopharyngeal: Related to the area covering the back of the nose and the throat.

Natal philopatry: The tendency of some animals to remain in or return to the area of their birth.

Nature-based tourism: Travel-related experiences that are centered on wild and natural environments.

Necropsy: Examination and dissection of an animal corpse to assess health prior to death and cause of death.

Necrotizing: Related to an inflammation type defined by dead cells or tissue.

Necrotizing dermatitis: An inflammation of the skin defined by dead cells or tissue.

Nematode: A worm of the phylum *Nematoda*, commonly referred to as a roundworm.

Nephritis: An infection or inflammation of the kidney.

Nervous system: The brain and network of nerves that transmit information from the brain to the rest of the body and vice versa, controlling the entire organism, including breathing, moving, thinking and feeling.

Net gain: In an ecological context, a positive outcome for biodiversity following a development project and the application of targeted conservation measures.

Neurological: Related to the nervous system.

Next-generation sequencing (NGS): A technology that is able to run multiple parallel RNA/DNA sequences at high speed and throughput to reveal entire genomes or larger RNA/DNA sequences than is possible using standard sequencing.

Nodules: Little nodes.

No net loss: In an ecological context, an outcome that avoids an overall loss of biodiversity and ecosystem services following a development project and targeted conservation activities. This term is often used in association with the **mitigation hierarchy**.

Nucleic acid: A molecule that carries genetic information, such as DNA and RNA.

Nulliparous: Never having given birth to a live baby.

Nutritional security: See **food security**.

Omnivorous: Related to animals who eat a large variety of foods, both animal- and plant-based.

One Health: A transdisciplinary approach that addresses the interdependence of human, animal and environmental health, with the aim of producing shared optimal health outcomes through international collaboration.

One Welfare: A framework designed to help improve animal welfare, human wellbeing and the integrity of the environment by emphasizing their interconnections.

Operant conditioning: A method of learning using reward and punishment to impact behavior, i.e., behavior that is rewarded will likely be repeated and behavior that is punished is less likely to be repeated. Also known as instrumental conditioning.

Osteoarthritis: Inflammation of a bone that has spread to a joint.

Pandemic: A global epidemic.

Parasite: An organism that lives off or in another organism, to the detriment of the host organism.

Parasitemia: Parasites in the blood stream.

Pathogen: A microorganism that causes sickness or disease.

Pathogenicity: An organism's ability to cause disease.

Pelage: Fur; coat.

Personhood: The legal status of a person. Under the law, a "person" has one or more rights while a "thing" has none. Historically, some humans were classified as "things" under the law; today, personhood is not restricted to human beings.

Phage: A virus that infects bacteria.

Phenotypic: Related to the appearance or observable characteristics of an organism (related words: phenotype).

Phylogenetic: Relating to the evolutionary development, diversification and relatedness of a species or group of organisms, or to a particular feature of an organism.

Physiology: The study of how a living system works.

Pith: The spongy tissue in the stems and branches of many plants.

Planetary health: A paradigm that focuses on the health of humans and the state of the natural systems on which they depend.

Plasma: The cell-free part of blood.

Plasmodium **sp.:** Single-celled parasites that cause malaria.

Pneumoviruses: Viruses of the family *Pneumoviridae*, which cause respiratory diseases, including common colds in humans.

Poaching: Illegal hunting, killing, capturing or taking of wildlife in violation of local or international wildlife conservation laws.

Polyandrous: Pertaining to a mating system that involves one female and two or more males.

Polygynandrous: Pertaining to an exclusive mating system that involves two or more males and two or more females. The numbers of males and females are not necessarily equal.

Polygynous: Pertaining to a mating system that involves one male and two or more females.

Polymerase chain reaction (PCR): A laboratory technique for the rapid production (amplification) of millions to billions of copies of a specific segment of DNA, which can then be studied in greater detail.

Positive reinforcement training: Use of a reward (praise, treat, prize, . . .) to encourage the repetition of desired activities/behavior.

Post-mortem: After death.

Prediabetes: A condition characterized by blood sugar levels that are higher than normal, but not high enough for a type 2 diabetes diagnosis.

Preparedness: Measures and actions taken for and by a community and community partners prior to an impact by a hazard, ensuring a timely and effective response to hazard impacts.

Preputial: Relating to the foreskin or clitoral hood.

Primary prevention: Intervention before a disease, injury or disorder occurs.

Priority biodiversity feature: Irreplaceable or highly vulnerable habitats, species, structures, functions or other natural elements needed to maintain ecological viability.

Pro-poor tourism: Travel-related experiences that generate net economic, social, environmental or cultural benefits for low-income communities.

Proprietary mixture: A combination of ingredients, often unique to a particular product, such as medication.

Proteomics: The study of proteomes, which are protein-based systems within the body.

Protozoa: Single-celled organisms.

Psychopathology: The scientific study of mental health issues and disorders.

Purulent: Containing or producing pus.

Quarantine: A state, place or period of isolation, the length of which is dictated by the time needed to detect diseases of concern. The most common period is 30 days, although a minimum of 90 days is common if tuberculosis is of concern, or 6 months for rabies.

Radiology: In this volume, X-ray and other medical imaging techniques that use radiation.

Radio telemetry: A technique employed to track the movement and behavior of animals, using the transmission of radio signals to locate a transmitter attached to the animal of interest.

Range state: A country that exercises jurisdiction over any part of the native distribution or range of a particular species or group of species.

Reactivity: In serological tests, a reaction/change in blood that has been mixed with an antigen that may indicate the presence of antibodies.

Reintroduction: The planned release of an organism into its natural habitat after life in captivity.

Renal: Concerning the kidneys.

Rescue and rehabilitation center: A facility that treats and cares for injured, orphaned or sick wild animals with the aim of restoring essential capabilities and releasing them back to into their natural habitat.

Reservoir: Any living being or substance in which an infectious agent normally lives, grows and multiplies, and from where it can spill over to (other) species, where it may cause disease.

Retrovirus: A microbe that inserts DNA copies of its own RNA genome into the host cells that it invades; an enzyme called reverse transcriptase transcribes RNA into DNA.

RNA: Ribonucleic acid; genetic material that is structurally similar to DNA, although it is typically single-stranded rather than double-stranded.

Roving strategy: Actively searching for, or roaming between, females.

Sanctuary: A non-profit facility dedicated to providing care (often lifelong) for orphaned, confiscated or injured wildlife.

Secondary prevention: Screening to identify and reduce the impact of diseases, injuries or disorders in their early stages.

Semi-captive: Individuals ranging, foraging and socializing in their natural environment, with some level of human intervention such as enclosure management, shelter, supplementary feeding and veterinary care.

Sentience: The capability of animals to experience and feel different emotions, such as joy, pleasure, pain and fear.

Septicemia: Blood poisoning, especially that caused by bacteria or their toxins.

Seroconversion: A process by which the body's immune system produces specific antibodies in the blood to combat infection. These antibodies can be produced by infection or vaccination, which itself is aimed at enhancing the body's immune response to a particular infection.

Serology: The study of blood serum and other bodily fluids, with a focus on their immunological properties; testing of blood serum for the presence of antibodies.

Shared situational awareness: A common understanding among two or more people or agencies, based on three elements: the perception in time and space of the specified observations and outcomes for an area or activity (what has happened), the comprehension of what the observations and outcomes mean and a projection of that meaning into the future. Commonly abbreviated to SSA.

Silverback: An adult male gorilla who has reached maturity and developed silver hairs on the saddle of his back.

Sociality: The level to which individuals in a species or population associate with one another and form social groups.

Socionatural: Featuring both natural and anthropogenic characteristics.

Spillback: Disease transmission from any species to the species in which the disease originated. Tuberculosis, for example, was initially transmitted from humans to other species, but the disease is now spilling back into humans from these species.

Spillover event: A pathogen's jump from one species to another, where it can cause disease. A spillover can be self-limiting, not resulting in further transmission, or can result in spread of the pathogen in the new host population, which can lead to an epidemic (or pandemic) and potentially become endemic.

Spontaneous volunteers: Members of a community or the public who are not affiliated with a response or recovery agency or organization and volunteer their time and services.

Standard operating procedures (SOP): A set of written instructions describing a step-by-step process that must be taken to perform a routine activity properly.

Stochastic: Having a random probability of occurrence or pattern that cannot be predicted precisely.

Strongyloides: A genus of parasitic roundworm, known as threadworm in the United States. The main species that infects humans is *Strongyloides stercoralis*, but it can affect other primates as well. The main species in primates is *Strongyloides fuelleborni*.

Subacute: Concerning the duration of a disease (between 14 and 30 days), between acute and chronic.

Subadult: Not yet having attained all adult characteristics; an animal in the subadult stage.

Substrate: Any organic substance that may be used as bedding or nesting material, insulation, food or as an absorptive layer for animal excreta.

Superinfection: A complication following or adding to a prior infection, whereby a pathogen can overgrow and overwhelm a host. Superinfections are due to antimicrobial resistance or immunosuppression.

Surge capacity: The ability to increase resources (stores, equipment, facilities, services, personnel) significantly in a short period of time to deliver a timely and effective response.

Sympatric: Pertaining to species or populations that occupy the same geographic ranges.

Syndromic surveillance: Collection and analysis of health data for the rapid detection of health threats.

Taxon (plural: **taxa**): Any unit used in the science of biological classification or taxonomy.

Terrestriality: Adaptation to living on the ground.

Tertiary education: University or other post-school studies.

Tertiary prevention: Management of the impact of an ongoing illness or injury that has lasting effects.

Tracker: An observer employed to follow the movements of a specific group of apes or other animals.

Transit facility: A site that temporarily accommodates seized and rescued wild animals, either prior to their transfer to a rehabilitation center or sanctuary, or prior to their release into natural habitat.

Translational medicine: A field that links medical research, medical practice and community to produce a holistic approach to medicine, connecting the patient to the environment.

Translocation: In conservation, the process of moving organisms (animals or plants) from one area to another, in captive or wild settings. Conservation translocations can reinforce an existing population, reintroduce one that has disappeared, or introduce a species outside its indigenous range.

Ulcerative: Related to inflammation that is defined by ulcer formation, a lesion of skin or mucosa.

Ultrasound: An imaging method and diagnostic tool that uses high-energy sound waves to produce images of tissues and organs inside the body. Unlike X-rays, ultrasound can be used to portray soft tissues in detail, such as the uterus. Also known as a sonogram.

Umbrella species: A species whose conservation leads to the protection of many other species in the same ecosystem or landscape.

Understory: The shrubby plant layer below the forest canopy.

Utilitarianism: An ethical theory that includes all sentient beings in moral consideration so as to maximize overall wellbeing.

Vascular changes: Alterations in blood vessels that could be associated with, or could lead to, disease.

Vector: An organism that carries a pathogen (on the inside or outside) and spreads it. Examples include mosquitoes carrying blood parasites from one person to another when feeding, as well as vectors that transfer pathogens mechanically, such as rodents who move from stable to stable carrying dirt under their feet.

Venous: Relating to veins.

Vicarious resilience: An experience of personal growth based on witnessing growth in others, such as among professionals.

Vicarious traumatization or **compassion fatigue:** An indirect traumatic reaction to the traumatic experiences of others.

Virus: An infectious organic structure whose replication depends on a living organism. Some can cause disease, many are commensals.

Wadge: A lump of fibrous fruit matter, which apes spit out after chewing the fruit to extract the juices.

Wean: To accustom a young animal to nourishment other than the mother's milk.

Wildlife tourism: Travel-related experiences that provide close contact with wild animals and nature.

Zoo: A captive facility that puts animals on display for public viewing. Zoos typically have integrated reproduction strategies, whereas sanctuaries generally prevent captive breeding.

Zoonoses: Diseases that can be spread from animals to humans and vice versa (see also: **zoonosis, zoonotic**).

Zoonosis: Infectious disease that is transmitted from animals to humans and vice versa. Related terms: zoonoses, zoonotic. See also: **anthroponosis**.

Zoonotic: Related to diseases that can be spread from animals to humans and vice versa (see also: **zoonosis, zoonoses**).

REFERENCES

AAP (2020a). *Jaarverslag 2019*. Almere, the Netherlands: Animal Advisory and Protection (AAP). Available at: https://www.aap.nl/wp-content/uploads/2021/11/Jaarverslag-2019-DEF-gecomprimeerd.pdf.

AAP (2020b). *Ook Castilla-La Mancha verbiedt circussen met wilde dieren*. Almere, the Netherlands: Animal Advisory and Protection (AAP). Available at: https://www.aap.nl/nieuwsbericht/ook-castilla-la-mancha-verbiedt-circussen-met-wilde-dieren/.

AAP (n.d.). *Outplacement*. Almere, the Netherlands: Animal Advisory and Protection (AAP). Available at: https://en.aap.eu/outplacement/. Accessed: May, 2022.

AAWC (2020). *4th Africa Animal Welfare Conference: Action 2020 Resolutions*. Nairobi, Kenya: Africa Animal Welfare Conference (AAWC). Available at: https://www.aawconference.org/2020/4th_Africa_Animal_Welfare_Conference_Action_2020_Resolutions.pdf.

AAWC (n.d.). *About*. Nairobi, Kenya: Africa Animal Welfare Conference (AAWC). Available at: https://www.aawconference.org/index.php/about-us. Accessed: May, 2022.

Abbott, R.C. (2020). Wildlife vaccination – growing in feasibility? *Cornell Wildlife Health Laboratory*, February 17, 2020. Available at: https://cwhl.vet.cornell.edu/article/wildlife-vaccination-growing-feasibility.

Abelló, M.T., Rietkerk, F. and Bemment, N. (2017). *EAZA Great Ape TAG: Best Practice Guidelines Gorilla (Gorilla gorilla gorilla)*. Barcelona, Spain: Barcelona Zoo. Available at: https://www.eaza.net/conservation/programmes/#BPG.

Acevedo-Whitehouse, K. and Duffus, A.L. (2009). Effects of environmental change on wildlife health. *Philosophical Transactions of the Royal Society B: Biological Sciences*, **364**(1534), 3429–38.

Adamo, S.A. (2012). The effects of the stress response on immune function in invertebrates: an evolutionary perspective on an ancient connection. *Hormones and Behavior*, **62**(3), 324–30. DOI: 10.1016/j.yhbeh.2012.02.012.

Adams, W.M. and Infield, M. (2003). Who is on the gorilla's payroll? Claims on tourist revenue from a Ugandan National Park. *World Development*, **31**(1), 177–90.

Addison, C. and Malone, N. (2018). An experimental ethics, but an ethical experiment? Anthropological perspectives on using unproven vaccines on endangered primates. *American Journal of Bioethics*, **18**(10), 53–5. DOI: 10.1080/15265161.2018.1513592.

Adefuye, M.A., Manjunatha, N., Ganduri, V., *et al.* (2022). Tuberculosis and cardiovascular complications: an overview. *Cureus*, **14**(8), e28268. DOI: 10.7759/cureus.28268.

adnCUBA (2020). Monos del Zoológico de 26 en La Habana atacan a un periodista oficialista mientras trabajaba. *adnCUBA*, July 8, 2020. Available at: https://adncuba.com/noticias-de-cuba/entretenimiento/monos-del-zoologico-de-26-en-la-habana-atacan-un-periodista.

AFAC (2017). *AIIMS: Australasian Inter-Service Incident Management System*. Melbourne, Australia: Australasian Fire and Emergency Service Authorities Council (AFAC). Available at: https://www.afac.com.au/initiative/aiims.

AFP (2020). "Fin progressive" des animaux sauvages dans les cirques itinérants. *Le Point*, September 29, 2020. Available at: https://www.lepoint.fr/societe/fin-progressive-des-animaux-sauvages-dans-les-cirques-itinerants-29-09-2020-2394102_23.php.

Agoramoorthy, G. (2010). Setting standards for evaluation of captive facilities, Southeast Asia. In *Wild Mammals in Captivity: Principles and Techniques*, ed. D. G. Kleiman, K. V. Thompson and C. K. Baer. Chicago, IL: University of Chicago Press, pp. 28–31.

Aguilera, R., Corringham, T., Gershunov, A. and Benmarhnia, T. (2021). Wildfire smoke impacts respiratory health more than fine particles from other sources: observational evidence from Southern California. *Nature Communications*, **12**(1), 1493. DOI: 10.1038/s41467-021-21708-0.

Ahebwa, W.M., van der Duim, R. and Sandbrook, C. (2012). Tourism revenue sharing policy at Bwindi Impenetrable National Park, Uganda: a policy arrangements approach. *Journal of Sustainable Tourism*, **20**(3), 377–94. DOI: 10.1080/09669582.2011.622768.

AIDR (2017). *Managing Exercises Handbook 3*. Melbourne, Australia: Australian Institute for Disaster Resilience (AIDR). Available at: https://knowledge.aidr.org.au/media/3547/handbook-3-managing-exercises.pdf.

AIDR (2020). *Emergency Planning*. Melbourne, Australia: Australian Institute for Disaster Resilience (AIDR). Available at: https://www.aidr.org.au/media/8313/aidr_handbookcollection_emergencyplanning_2020.pdf.

AIDR (n.d.). *Australian Disaster Resilience Glossary*. Melbourne, Australia: Australian Institute for Disaster Resilience (AIDR). Available at: https://knowledge.aidr.org.au/resources/adr-glossary/. Accessed: July, 2022.

Ainerukundo, E., Gaffikin, L. and Kalema-Zikusoka, G. (2019). Evaluation of a community-based health and conservation model at Bwindi Impenetrable National Park. In *2nd African Primatological Society Conference. Primate Conservation in Africa: Challenges and Opportunities*, ed. African Primatological Society (APS). Entebbe, Uganda: APS, p. 87. Available at: https://apsuganda.africanprimatologicalsociety.org/book-of-abstracts/.

Airhart, E. (2018). Even zoos are learning the art of doomsday prepping. *Wired*, December 13, 2018. Available at: https://www.wired.com/story/even-zoos-are-learning-the-art-of-doomsday-prepping/.

AITC [The Animal Issues Thematic Cluster] (n.d.). *The Animal Issues Thematic Cluster*. Available at: https://animalissuesun.org/. Accessed: May, 2022.

Akinyi, M.Y., Tung, J., Jeneby, M., *et al.* (2013). Role of grooming in reducing tick load in wild baboons (*Papio cynocephalus*). *Animal Behaviour*, **85**(3), 559–68. DOI: 10.1016/j.anbehav.2012.12.012.

Al-Dahash, H., Thayaparan, M. and Kulatunga, U. (2016). Understanding the terminologies: disaster, crisis and emergency. In *Proceedings of the 3rd Annual ARCOM Conference, 5–7 September 2016, Vol. 2*, ed. P. Chan and C. Nielson. Manchester, UK: Association of Researchers in Construction Management (ARCOM), pp. 1191–200.

Allan, B.M., Nimmo, D.G., Ierodiaconou, D., *et al.* (2018). Futurecasting ecological research: the rise of technoecology. *Ecosphere*, **9**(5), e02163. DOI: 10.1002/ecs2.2163.

Allela, L., Boury, O., Pouillot, R., *et al.* (2005). Ebola virus antibody prevalence in dogs and human risk. *Emerging Infectious Diseases*, **11**(3), 385–90. DOI: 10.3201/eid1103.040981.

ALPZA (n.d.). *Acreditación*. Santiago, Chile: La Asociación Latinoamericana de Parques Zoológicos y Acuarios (ALPZA). Available at: https://www.alpza.com/acreditacion. Accessed: October, 2020.

Altizer, S., Nunn, C.L., Thrall, P.H., *et al.* (2003). Social organization and parasite risk in mammals: integrating theory and empirical studies. *Annual Review of Ecology, Evolution, and Systematics*, **34**(1), 517–47. DOI: 10.1146/annurev.ecolsys.34.030102.151725.

Alvarez-Berríos, N.L. and Mitchell Aide, T. (2015). Global demand for gold is another threat for tropical forests. *Environmental Research Letters*, **10**(1), 014006. DOI: 10.1088/1748-9326/10/1/014006.

Ambassade de France (2019). *Rencontre avec Pauline Grentzinger Docteur vétérinaire au parc de la Lékédi*. Libreville, Gabon: Ambassade de France. Available at: https://ga.ambafrance.org/Rencontre-avec-Pauline-Grentzinger-Docteur-veterinaire-au-parc-de-la-Lekedi.

Ameca y Juárez, E.I., Ellis, E.A. and Rodríguez-Luna, E. (2015). Quantifying the severity of hurricanes on extinction probabilities of a primate population: insights into "island" extirpations. *American Journal of Primatology*, **77**(7), 786–800. DOI: 10.1002/ajp.22402.

Ampumuza, C. and Driessen, C. (2021). Gorilla habituation and the role of animal agency in conservation and tourism development at Bwindi, South Western Uganda. *Environment and Planning E: Nature and Space*, **4**(4), 1601–21. DOI: 10.1177/2514848620966502.

Ancrenaz, M. (2015). *The conservation management and conservation medicine of orang-utan* (Pongo pygmaeus morio) *in Sabah, Malaysia*. PhD thesis. Berlin, Germany: Freien Universität.

Ancrenaz, M. (2018). *Arcus Disease Strategy*. Unpublished work. New York, NY: Arcus Foundation.

Ancrenaz, M., Ambu, L., Sunjoto, I., *et al.* (2010). Recent surveys in the forests of Ulu Segama Malua, Sabah, Malaysia, show that orang-utans (*P. p. morio*) can be maintained in slightly logged forests. *PLoS ONE*, **5**(7), e11510. DOI: 10.1371/journal.pone.0011510.

Ancrenaz, M., Calaque, R. and Lackman-Ancrenaz, I. (2004). Orangutan nesting behavior in disturbed forest of Sabah, Malaysia: implications for nest census. *International Journal of Primatology*, **25**(5), 983–1000.

Ancrenaz, M., Cheyne, S.M., Humle, T. and Robbins, M.M. (2020). The impact of killing, capture and trade on apes and their habitat. In *State of the Apes: Killing, Capture, Trade and Conservation*, ed. Arcus Foundation. Cambridge, UK: Cambridge University Press, pp. 25–47. Available at: https://www.stateoftheapes.com/volume-4-killing-capture-trade/.

Ancrenaz, M., Dabek, L. and O'Neil, S. (2007). The costs of exclusion: recognizing a role for local communities in biodiversity conservation. *PLoS Biology*, **5**(11), e289. DOI: 10.1371/journal.pbio.0050289.

Ancrenaz, M., Gumal, M., Marshall, A.J., *et al.* (2016). Pongo pygmaeus *(errata version published in 2018). The IUCN Red List of Threatened Species 2016: e.T17975A123809220*. Gland, Switzerland: International Union for Conservation of Nature (IUCN). DOI: 10.2305/IUCN.UK.2016-1.RLTS.T17975A17966347.en.

Ancrenaz, M., Oram, F., Ambu, L., *et al.* (2015). Of *Pongo*, palms and perceptions: a multidisciplinary assessment of Bornean orang-utans *Pongo pygmaeus* in an oil palm context. *Oryx*, **49**(3), 465–72. DOI: 10.1017/S0030605313001270.

Ancrenaz, M., Oram, F., Nardiyono, N., *et al.* (2021). Importance of small forest fragments in agricultural landscapes for maintaining orangutan metapopulations. *Frontiers in Forests and Global Change*, **4**, 560944. DOI: 10.3389/ffgc.2021.560944.

Ancrenaz, M., Sollmann, R., Meijaard, E., *et al.* (2014). Coming down from the trees: is terrestrial activity in Bornean orangutans natural or disturbance driven? *Scientific Reports*, **4**, 4024. DOI: 10.1038/srep04024.

Anderson, D.P., Nordheim, E.V. and Boesch, C. (2006). Environmental factors influencing the seasonality of estrus in chimpanzees. *Primates*, **47**(1), 43–50. DOI: 10.1007/s10329-005-0143-y.

Ando, C., Iwata, Y. and Yamagiwa, J. (2008). Progress of habituation of western lowland gorillas and their reaction to observers in Moukalaba-Doudou National Park, Gabon. *African Study Monographs*, **39**, 55–69.

Andrews, K., Comstock, G., Crozier, G.K.D., *et al.* (2018). *Chimpanzee Rights: The Philosophers' Brief*. London, UK: Routledge. DOI: 10.4324/9780429461071.

Animondial (n.d.). *Animal Protection Network*. Hove, UK: Animondial. Available at: https://animondial.com/animal-protection-network. Accessed: May, 2022.

Anthes, E. (2022). When people take pandemic precautions, gorillas breathe easier. *The New York Times*, February 21, 2022. Available at: https://www.nytimes.com/2022/02/21/health/gorillas-respiratory-illness-colds.html.

Antonation, K.S., Grützmacher, K., Dupke, S., *et al.* (2016). *Bacillus cereus* biovar *anthracis* causing anthrax in sub-Saharan Africa – chromosomal monophyly and broad geographic distribution. *PLoS Neglected Tropical Diseases*, **10**(9), e0004923. DOI: 10.1371/journal.pntd.0004923.

Ape Action Africa (n.d.). *Mefou Primate Sanctuary*. Bristol, UK: Ape Action Africa. Available at: https://www.apeactionafrica.org/mefou-primate-sanctuary. Accessed: October, 2020.

Ape Alliance (2018). Chimpanzees in Chinese captive wild animal facilities. *Ape Alliance News*, August 17, 2018. Available at: https://4apes.com/news/ape-alliance/item/1614-new-ape-alliance-report-on-chimpanzees-in-chinese-captive-wild-animal-facilities/.

Ape Monkey Rescue (n.d.). *Chimpanzees*. Abercrave, UK: Wales Ape and Monkey Sanctuary. Available at: http://www.ape-monkey-rescue.org.uk/chimpanzees.html. Accessed: December, 2020.

A.P.E.S. (n.d.). *Meet the Primates*. Blacklick, OH: American Primate Educational Sanctuary (A.P.E.S.). Available at: https://apesohio.weebly.com/meet-the-primates.html. Accessed: October, 2020.

A.P.E.S. Wiki Team (2019a). *Moyen-Bafing National Park. A.P.E.S. Wiki*. Munich, Germany: Max Planck Society for the Advancement of Science e.V. Available at: https://wiki.iucnapesportal.org/index.php/Moyen-Bafing_National_Park.

A.P.E.S. Wiki Team (2019b). *Pic de Fon Classified Forest. A.P.E.S. Wiki*. Munich, Germany: Max Planck Society for the Advancement of Science e.V. Available at: https://wiki.iucnapesportal.org/index.php/Pic_de_Fon_Classified_Forest.

Appleby, M.C. and Sherwood, L. (2007). *Animal Welfare Matters to Animals, People and the Environment: the Case for a Universal Declaration on Animal Welfare*. London, UK: World Society for the Protection of Animals (WSPA). Available at: https://www.worldanimalprotection.ca/sites/default/files/media/ca_-_en_files/case_for_a_udaw_tcm22-8305.pdf.

Aquatic Habitats in Integrated Urban Water Management (n.d.). *Water Cycle in Urban Areas*. Paris, France: United Nations Educational, Scientific and Cultural Organization (UNESCO). Available at: http://www.aquatic.unesco.lodz.pl/index.php?p=water_cycle. Accessed: October, 2022.

Arandjelovic, M., Head, J., Kühl, H., *et al.* (2010). Effective non-invasive genetic monitoring of multiple wild western gorilla groups. *Biological Conservation*, **143**(7), 1780–91. DOI: 10.1016/j.biocon.2010.04.030.

Arandjelovic, M., Head, J., Rabanal, L.I., *et al.* (2011). Non-invasive genetic monitoring of wild central chimpanzees. *PLoS ONE*, **6**(3), e14761. DOI: 10.1371/journal.pone.0014761.

Archabald, K. and Naughton-Treves, L. (2001). Tourism revenue-sharing around national parks in western Uganda: early efforts to identify and reward local communities. *Environmental Conservation*, **28**(2), 135–49. DOI: 10.1017/S0376892901000145.

Arcus Foundation (2014). *State of the Apes: Extractive Industries and Ape Conservation*. Cambridge, UK: Cambridge University Press. Available at: https://www.stateoftheapes.com/volume-1-extractive-industries/.

Arcus Foundation (2015). *State of the Apes: Industrial Agriculture and Ape Conservation*. Cambridge, UK: Cambridge University Press. Available at: https://www.stateoftheapes.com/volume-2-industrial-agriculture/.

Arcus Foundation (2018). *State of the Apes: Infrastructure Development and Ape Conservation*. Cambridge, UK: Cambridge University Press. Available at: https://www.stateoftheapes.com/volume-3-infrastructure-development/.

Arcus Foundation (2020). *State of the Apes: Killing, Capture, Trade and Conservation*. Cambridge, UK: Cambridge University Press Available at: https://www.stateoftheapes.com/volume-4-killing-capture-trade/.

Arlian, L.G., Vyszenski-Moher, D.L. and Pole, M. (1989). Survival of adults and developmental stages of *Sarcoptes scabiei* var. *canis* when off the host. *Experimental and Applied Acarology*, **6**, 181–7. DOI: 10.1007/BF01193978.

Armstrong-Mensah, E.A. and Ndiaye, S.M. (2018). Global health security agenda implementation: a case for community engagement. *Health Security*, **16**(4), 217–23. DOI: 10.1089/hs.2017.0097.

Arora, N., van Noordwijk, M.A., Ackermann, C., *et al.* (2012). Parentage-based pedigree reconstruction reveals female matrilineal clusters and male-biased dispersal in nongregarious Asian great apes, the Bornean orang-utans (*Pongo pygmaeus*). *Molecular Ecology*, **21**(13), 3352–62. DOI: 10.1111/j.1365-294X.2012.05608.x.

ARRC Task Force (n.d.). *IUCN SSC Primate Specialist Group Section on Great Apes & Section on Small Apes ARRC Task Force. Avoid, Reduce, Restore Negative Impacts from Energy, Extractive and Associated Infrastructure Projects on Apes and Contribute Positively to their Conservation*. ARRC Task Force. Available at: https://www.arrctaskforce.org/. Accessed: December, 2022.

Ashbury, A.M., Willems, E.P., Utami-Atmoko, S.S., *et al.* (2020). Home range establishment and the mechanisms of philopatry among female Bornean orangutans (*Pongo pygmaeus wurmbii*) at Tuanan. *Behavioral Ecology and Sociobiology*, **74**(4), 42. DOI: 10.1007/s00265-020-2818-1.

ASP (n.d.). *Kibale Snare Removal Program*. American Society of Primatologists (ASP). Available at: https://www.asp.org/2020/08/20/kibale-snare-removal-program/. Accessed: October, 2022.

ATTA [Adventure Travel Trade Association] (2020). Classic Africa Safaris reports gorilla "baby boom" in Uganda; tourism reopens. *Adventure Travel News*, September 24, 2020. Available at: https://www.adventuretravelnews.com/classic-africa-safaris-reports-gorilla-baby-boom-in-uganda-tourism-reopens.

Aultman, J. (2008). Moral courage through a collective voice. *American Journal of Bioethics*, **8**(4), 67–9. DOI: 10.1080/15265160802147140.

Aung, P., Lwin, N., Aung, T.H., *et al.* (2023). Confirmation of skywalker hoolock gibbon (*Hoolock tianxing*) in Myanmar extends known geographic range of an endangered primate. *International Journal of Primatology*, in press.

Australian Government (2021). *Digital Earth Australia Hotspots*. Canberra, Australia: Commonwealth of Australia (Geoscience Australia). Available at: https://hotspots.dea.ga.gov.au/#/.

Avanzi, C., del-Pozo, J., Benjak, A., *et al.* (2016). Red squirrels in the British Isles are infected with leprosy bacilli. *Science*, **354**(6313), 744–7. DOI: 10.1126/science.aah3783.

Avoi, R. and Liaw, Y.C. (2021). Tuberculosis death epidemiology and its associated risk factors in Sabah, Malaysia. *International Journal of Environmental Research and Public Health*, **18**(18), 9740. DOI: 10.3390/ijerph18189740.

AZA Ape TAG [Taxon Advisory Group] (2010). *Chimpanzee (*Pan troglodytes*) Care Manual*. Silver Spring, MD: Association of Zoos and Aquariums (AZA). Available at: https://nagonline.net/wp-content/uploads/2014/05/ChimpanzeeCareManual2010-NAG-EDIT.pdf.

AZA Ape TAG [Taxon Advisory Group] (2017). *Orangutan (*Pongo*) Care Manual*. Silver Spring, MD: Association of Zoos and Aquariums (AZA). Available at: https://ams.aza.org/iweb/upload/Orangutan%20Care%20Manual%202017-de54741f.pdf.

AZA Gorilla Species Survival Plan Program (2017). *Western Lowland Gorilla (*Gorilla gorilla gorilla*) Care Manual*. Silver Spring, MD: Association of Zoos and Aquariums (AZA). Available at: https://assets.speakcdn.com/assets/2332/gorilla_care_manual_2018.pdf.

Baker, J., Milner-Gulland, E.J. and Leader-Williams, N. (2012). Park gazettement and integrated conservation and development as factors in community conflict at Bwindi Impenetrable Forest, Uganda: drivers of community conflict at Bwindi. *Conservation Biology*, **26**(1), 160–70. DOI: 10.1111/j.1523-1739.2011.01777.x.

Baker, L. (2017). Translocation biology and the clear case for compassionate conservation. *Israel Journal of Ecology and Evolution*, **63**(3–4), 52–60. DOI: 10.1163/22244662-20181026.

Baker, L. and Winkler, R. (2020). Asian elephant rescue, rehabilitation and rewilding. *Animal Sentience*, **28**(1). DOI: 10.51291/2377-7478.1506.

Baker, S.E., Cain, R., van Kesteren, F., *et al.* (2013). Rough trade: animal welfare in the global wildlife trade. *BioScience*, **63**(12), 928–38. DOI: 10.1525/bio.2013.63.12.6.

Balasubramaniam, K.N., Aiempichitkijkarn, N., Kaburu, S.S.K., *et al.* (2022). Impact of joint interactions with humans and social interactions with conspecifics on the risk of zooanthroponotic outbreaks among wildlife populations. *Scientific Reports*, **12**, 11600. DOI: 10.1038/s41598-022-15713-6.

Bales, K.L. (2020). Introduction to special section on COVID-19 in primatology. *American Journal of Primatology*, **82**(8), e23174. DOI: 10.1002/ajp.23174.

Ban Animal Trading and EMS Foundation (2020). *Breaking Point: Uncovering South Africa's Shameful Live Wildlife Trade with China*. Johannesburg, South Africa: EMS Foundation. Available at: https://emsfoundation.org.za/the-breaking-point-uncovering-south-africas-shameful-live-wildlife-trade-with-china/.

Banes, G.L., Chua, W., Elder, M. and Kao, J. (2018). Orang-utans *Pongo* spp in Asian zoos: current status, challenges and progress towards long-term population sustainability. *International Zoo Yearbook*, **52**(1), 150–63. DOI: 10.1111/izy.12178.

Barber, J.C.E. and Mellen, J. (2008). Assessing animal welfare in zoos and aquariums: is it possible? In *The Well-Being of Animals in Zoo and Aquarium Sponsored Research: Putting Best Practices Forward*, ed. T. L. Bettinger and J. T. Bielitzki. Greenbelt, MD: Scientists Center for Animal Welfare, pp. 39–52.

Barnhill, A., Joffe, S. and Miller, F.G. (2016). The ethics of infection challenges in primates. *Hastings Center Report*, **46**(4), 20–6. DOI: 10.1002/hast.580.

Barone, J. (2015). Gorilla doctors: these veterinarians are saving Africa's gorillas, one patient at a time. *Science World/Current Science*, **71**(7), 8–12.

Bartlett, T.Q. (2011). The Hylobatidae: small apes of Asia. In *Primates in Perspective*, ed. C. Campbell, A. Fuentes, K. C. Mackinnon, S. K. Bearder and R. M. Stumpf. New York, NY: Oxford University Press, pp. 300–12.

Basabose, A.K. and Yamagiwa, J. (2002). Factors affecting nesting site choice in chimpanzees at Tshibati, Kahuzi-Biega National Park: influence of sympatric gorillas. *International Journal of Primatology*, **23**(2), 263–82. DOI: 10.1023/A:1013879427335.

Bastin, J.F., Barbier, N., Réjou-Méchain, M., *et al.* (2015). Seeing Central African forests through their largest trees. *Scientific Reports*, **5**, 13156. DOI: 10.1038/srep13156.

Batavia, C., Nelson, M.P., Bruskotter, J.T., *et al.* (2021). Emotion as a source of moral understanding in conservation. *Conservation Biology*, **35**(5), 1380–7. DOI: 10.1111/cobi.13689.

Batavia, C., Nelson, M.P. and Wallach, A.D. (2020). The moral residue of conservation. *Conservation Biology*, **34**(5), 1114–21. DOI: 10.1111/cobi.13463.

Baum, S.E., Machalaba, C., Daszak, P., Salerno, R.H. and Karesh, W.B. (2017). Evaluating One Health: are we demonstrating effectiveness? *One Health*, **3**, 5–10. DOI: 10.1016/j.onehlt.2016.10.004.

Baylet, R., Thivolet, J., Sepetjian, M., Nouhouay, Y. and Baylet, M. (1971). La tréponématose naturelle ouverte du singe *Papio papio* en Casamance [Natural open treponematosis in the *Papio papio* baboon in Casamance] [in French]. *Bulletin de la Société de Pathologie Exotique et de ses Filiales*, **64**(6), 842–6.

BBC (2016). Twycross Zoo begins great ape heart disease study. *BBC News*, July 21, 2016. Available at: https://www.bbc.com/news/uk-england-leicestershire-36847743.

BBC (2020). Rafiki, Uganda's rare silverback mountain gorilla, killed by hunters. *BBC News*, June 12, 2020. Available at: https://www.bbc.com/news/world-africa-53024073.

BBC News (2002). Zoo animals killed in Prague floods. *BBC News World: Europe*, August 14, 2002. Available at: http://news.bbc.co.uk/2/hi/europe/2193483.stm.

BBOP (2013). *To No Net Loss and Beyond: An Overview of the Business and Biodiversity Offsets Programme (BBOP)*. Washington DC: Business and Biodiversity Offsets Programme (BBOP). Available at: https://www.forest-trends.org/publications/to-no-net-loss-and-beyond/.

BCT (2020). *Essential Conservation Fencing Infrastructure. Guidelines, Standards and Cost Benchmarks*. Lismore, Australia: NSW Government Biodiversity Conservation Trust (BCT). Available at: https://www.bct.nsw.gov.au/sites/default/files/2020-11/BCT%20Essential%20Conservation%20Fencing%20guide%20Nov%202020.pdf.

Beament, E. (2020). "Ecotourism" shut down to protect mountain gorillas. *The Ecologist*, March 25, 2020. Available at: https://theecologist.org/2020/mar/25/ecotourism-shut-down-protect-mountain-gorillas.

Bearder, S.K. and Martin, R.D. (1980). The social organization of a nocturnal primate revealed by radio tracking. In *A Handbook on Biotelemetry and Radio Tracking*, ed. C. J. Amlaner and D. W. Macdonald. Oxford, UK: Pergamon, pp. 633–48. DOI: 10.1016/B978-0-08-024928-5.50082-8.

Beastall, C.A., Bouhuys, J. and Ezekiel, A. (2016). *Apes in Demand: For Zoo and Wildlife Attractions in Peninsular Malaysia and Thailand*. Selangor, Malaysia: TRAFFIC. Available at: http://www.trafficj.org/publication/16_Apes_in_Demand.pdf.

Beausoleil, N.J., Mellor, D.J., Baker, L., *et al.* (2018). "Feelings and fitness" not "feelings or fitness" – the raison d'être of conservation welfare, which aligns conservation and animal welfare objectives. *Frontiers in Veterinary Science*, **5**, November 27 2018. DOI: 10.3389/fvets.2018.00296.

Beck, B. (2017). *Unwitting Travelers: A History of Primate Reintroduction*. Berlin, MD: Salt Water Media.

Beck, B., Walkup, K., Rodrigues, M., *et al.* (2007). *Best Practice Guidelines for the Re-introduction of Great Apes*. Gland, Switzerland: International Union for Conservation of Nature (IUCN) Species Survival Commission (SSC) Primate Specialist Group (PSG). Available at: https://portals.iucn.org/library/sites/library/files/documents/SSC-OP-035.pdf.

Becker, D.J., Albery, G.F., Kessler, M.K., *et al.* (2020). Macroimmunology: the drivers and consequences of spatial patterns in wildlife immune defence. *Journal of Animal Ecology*, **89**(4), 972–95. DOI: 10.1111/1365-2656.13166.

Behie, A.M., Pavelka, M.S.M., Hartwell, K., Champion, J. and Notman, H. (2019). Alas the storm has come again! The impact of frequent natural disasters on primate conservation. In *Primate Research and Conservation in the Anthropocene*, ed. A. M. Behie, J. A. Teichroeb and N. Malone. Cambridge, UK: Cambridge University Press, pp. 237–56. DOI: 10.1017/9781316662021.014.

Behringer, V. and Deschner, T. (2017). Non-invasive monitoring of physiological markers in primates. *Hormones and Behavior*, **91**, 3–18. DOI: 10.1016/j.yhbeh.2017.02.001.

Behringer, V., Stevens, J.M.G., Hohmann, G., *et al.* (2014). Testing the effect of medical positive reinforcement training on salivary cortisol levels in bonobos and orangutans. *PLoS ONE*, **9**(9), e108664. DOI: 10.1371/journal.pone.0108664.

Bell, H., Kulkarni, S. and Dalton, L. (2003). Organizational prevention of vicarious trauma. *Families in Society: The Journal of Contemporary Human Services*, **84**(4), 463–70. DOI: 10.1606/1044-3894.131.

Belmaker, G. (2018). DRC breaches logging moratorium for Chinese-owned companies. *Mongabay*, 28 February 2018. Available at: https://news.mongabay.com/2018/02/drc-breaches-logging-moratorium-for-chinese-owned-companies/#.

Bemment, N., ed. (2018). *Orangutan EEP Best Practice Guidelines*, 1st edn. Amsterdam, the Netherlands: European Association of Zoos and Aquaria (EAZA) Great Ape Taxon Advisory Group (TAG). Available at: https://www.eaza.net/assets/Uploads/CCC/BPG-new-version/2018-OU-EEP-Best-Practice-Guidelines-final-NV.pdf.

Bennett, N.J., Roth, R., Klain, S.C., *et al.* (2017). Conservation social science: understanding and integrating human dimensions to improve conservation. *Biological Conservation*, **205**, 93–108. DOI: 10.1016/j.biocon.2016.10.006.

Berg, C. (2018). Restoring what we have destroyed: animal welfare aspects of wildlife conservation, reintroduction and rewilding programmes. In *Animal Welfare in a Changing World*, ed. A. Butterworth. Wallingford, UK: CABI International, pp. 68–79.

Berga, S.L. (2008). Stress and reproduction: a tale of false dichotomy? *Endocrinology*, **149**(3), 867–8. DOI: 10.1210/en.2008-0004.

Bergl, R.A., Dunn, A., Fowler, A., *et al.* (2016). Gorilla gorilla *ssp.* diehli *(errata version published in 2016)*. *The IUCN Red List of Threatened Species 2016: e.T39998A102326240*. Gland, Switzerland: International Union for Conservation of Nature (IUCN). DOI: 10.2305/IUCN.UK.2016-2.RLTS.T39998A17989492.en.

Bermejo, M., Rodríguez-Teijeiro, J.D., Illera, G., *et al.* (2006). Ebola outbreak killed 5000 gorillas. *Science*, **314**(5805), 1564. DOI: 10.1126/science.1133105.

Bertolani, P. and Boesch, C. (2008). Habituation of wild chimpanzees (*Pan troglodytes*) of the South Group at Taï Forest, Côte d'Ivoire: empirical measure of progress. *Folia Primatologica*, **79**(3), 162–71. DOI: 10.1159/000111720.

BES [British Ecological Society] Press Office (2022). Scientists study tourists to protect great apes from disease transmission. *British Ecological Society News and Opinion*, September 5, 2022. Available at: https://www.britishecologicalsociety.org/scientists-study-tourists-to-protect-great-apes-from-disease-transmission.

Bessone, M., Booto, L., Santos, A.R., Kühl, H.S. and Fruth, B. (2021). No time to rest: how the effects of climate change on nest decay threaten the conservation of apes in the wild. *PLoS ONE*, **16**(6), e0252527. DOI: 10.1371/journal.pone.0252527.

Bettinger, T., Cox, D., Kuhar, C. and Leighty, K. (2021). Human engagement and great ape conservation in Africa. *American Journal of Primatology*, **83**(4), e23216. DOI: 10.1002/ajp.23216.

Bettinger, T.L., Leighty, K.A., Daneault, R.B., Richards, E.A. and Bielitzki, J.T. (2017). Behavioral management: the environment and animal welfare. In *Handbook of Primate Behavioral Management*, ed. S. J. Schapiro. Boca Raton, FL: CRC Press, pp. 37–51.

Beydoun, G., Dascalu, S., Dominey-Howes, D. and Sheehan, A. (2018). Disaster management and information systems: insights to emerging challenges. *Information Systems Frontiers*, **20**(4), 649–52. DOI: 10.1007/s10796-018-9871-6.

Beyer, W. and Turnbull, P.C.B. (2009). Anthrax in animals. *Molecular Aspects of Medicine*, **30**(6), 481–9.

Bhat, S.A., Mounsey, K.E., Liu, X. and Walton, S.F. (2017). Host immune responses to the itch mite, *Sarcoptes scabiei*, in humans. *Parasites & Vectors*, **10**(1), 385. DOI: 10.1186/s13071-017-2320-4.

BIAZA (2019). *BIAZA Animal Transfer Policy (ATP)*. London, UK: British and Irish Association of Zoos and Aquariums (BIAZA). Available at: https://biaza.org.uk/downloader/41.

Binding, S., Farmer, H., Krusin, L. and Cronin, K. (2020). Status of animal welfare research in zoos and aquariums: where are we, where are we to next? *Journal of Zoo and Aquarium Research*, **8**(3), 166–74. DOI: 10.19227/jzar.v8i3.505.

Birke, L. (2002). Effects of browse, human visitors and noise on the behaviour of captive orang utans. *Animal Welfare*, **11**(2), 189–202. DOI: 10.1017/S0962728600028141.

Birot, H., Campera, M., Imron, M.A. and Nekaris, K.A.I. (2020). Artificial canopy bridges improve connectivity in fragmented landscapes: the case of Javan slow lorises in an agroforest environment. *American Journal of Primatology*, **82**(4), e23076. DOI: 10.1002/ajp.23076.

Bitariho, R., Akampurira, E. and Mugerwa, B. (2020). Regulated access to wild climbers has enhanced food security and minimized use of plastics by frontline households at a premier African protected area. *Conservation Science and Practice*, **2**(10), e275. DOI: 10.1111/csp2.275.

Bitty, E.A., Bi, S.G., Bene, J.-C.K., Kouassi, P.K. and McGraw, W.S. (2015). Cocoa farming and primate extirpation inside Cote D'ivoire's protected areas. *Tropical Conservation Science*, **8**(1), 95–113. DOI: 10.1177/194008291500800110.

Bizimungu, J. (2020). Rwanda announces promotional prices of gorilla-trekking permits. *The New Times*, June 18, 2020. Available at: https://www.newtimes.co.rw/news/rwanda-announces-promotional-prices-gorilla-trekking-permits.

Björk, J.R., Dasari, M., Grieneisen, L. and Archie, E.A. (2019). Primate microbiomes over time: longitudinal answers to standing questions in microbiome research. *American Journal of Primatology*, **81**(10–11), e22970. DOI: 10.1002/ajp.22970.

Blackett, T.A., McKenna, C., Kavanagh, L. and Morgan, D.R. (2017). The welfare of wild animals in zoological institutions: are we meeting our duty of care? *International Zoo Yearbook*, **51**(1), 187–202. DOI: 10.1111/izy.12143.

Blom, A. (2001a). *Ecological and economic impacts of gorilla-based tourism in Dzanga-Sangha, Central African Republic*. PhD thesis. Wageningen, the Netherlands: Wageningen University.

Blom, A. (2001b). Potentials and pitfalls of tourism in Dzanga-Sangha. *Gorilla Journal*, **22**, 40–1.

Blom, A., Cipolletta, C., Brunsting, A.M.H. and Prins, H.H.T. (2004). Behavioral responses of gorillas to habituation in the Dzanga-Ndoki National Park, Central African Republic. *International Journal of Primatology*, **25**(1), 179–96. DOI: 10.1023/B:IJOP.0000014649.15973.3a.

Bloom, P. (2017). *Against Empathy: The Case for Rational Compassion*. New York, NY: HarperCollins Publishers.

Bloomsmith, M.A., Clay, A.W., Ross, S.R., *et al.* (2020). Chimpanzees in US zoos, sanctuaries, and research facilities: a survey-based comparison of atypical behaviors. In *Chimpanzees in Context: A Comparative Perspective on Chimpanzee Behavior, Cognition, Conservation, and Welfare*, ed. L. M. Hopper and S. R. Ross. Chicago, IL: University of Chicago Press, pp. 481–508. DOI: 10.7208/chicago/9780226728032.003.0021.

Bloomsmith, M.A., Laule, G.E., Alford, P.L. and Thurston, R.H. (1994). Using training to moderate chimpanzee aggression during feeding. *Zoo Biology*, **13**(6), 557–66. DOI: 10.1002/zoo.1430130605.

Bloomsmith, M.A., Neu, K., Franklin, A., Griffis, C. and McMillan, J. (2015). Positive reinforcement methods to train chimpanzees to cooperate with urine collection. *Journal of the American Association for Laboratory Animal Science*, **54**(1), 66–9.

BNF (n.d.-a). *About Us*. Kalimantan, Indonesia: Borneo Nature Foundation (BNF). Available at: https://www.borneonaturefoundation.org/about/. Accessed: September, 2022.

BNF (n.d.-b). *Drones for Conservation*. Kalimantan, Indonesia: Borneo Nature Foundation (BNF). Available at: https://www.borneonaturefoundation.org/project/drones-for-conservation/. Accessed: September, 2022.

BNF (n.d.-c). *Fire-Fighting & Prevention*. Kalimantan, Indonesia: Borneo Nature Foundation (BNF). Available at: https://www.borneonaturefoundation.org/project/firefighting-prevention/. Accessed: September, 2022.

BNF (n.d.-d). *Understanding the Natural World is at the Heart of Effective, Sustainable Conservation Strategies*. Kalimantan, Indonesia: Borneo Nature Foundation (BNF). Available at: https://www.borneonaturefoundation.org/scientific-research/. Accessed: September, 2022.

BNF (n.d.-e). *Youth Education and Empowerment for Nature Conservation on Borneo*. Kalimantan, Indonesia: Borneo Nature Foundation (BNF). Available at: https://www.borneonaturefoundation.org/environmental-education/. Accessed: September, 2022.

Boesch, C., Crockford, C., Herbinger, I., *et al.* (2008). Intergroup conflicts among chimpanzees in Taï National Park: lethal violence and the female perspective. *American Journal of Primatology*, **70**(6), 519–32. DOI: 10.1002/ajp.20524.

Boesch, C., Hohmann, G. and Marchant, L., ed. (2002). *Behavioural Diversity in Chimpanzees and Bonobos*. Cambridge, UK: Cambridge University Press. DOI: 10.1017/CBO9780511606397.

Boesch, C., Kalan, A.K., Mundry, R., *et al.* (2020). Chimpanzee ethnography reveals unexpected cultural diversity. *Nature Human Behaviour*, **4**(9), 910–16. DOI: 10.1038/s41562-020-0890-1.

Bologna, M. and Aquino, G. (2020). Deforestation and world population sustainability: a quantitative analysis. *Scientific Reports*, **10**, 7631. DOI: 10.1038/s41598-020-63657-6.

BOSF (2020). *Annual Report 2019*. Bogor, Indonesia: Borneo Orangutan Survival Foundation (BOSF). Available at: https://www.orangutan.or.id/cfind/source/files/annual-reports/bosf-annual-report-2019.pdf.

BOSF (n.d.). *Our Story*. Bogor, Indonesia: Borneo Orangutan Survival Foundation (BOSF). Available at: https://www.orangutan.or.id/our-story. Accessed: September, 2022.

Botha, C.J., Coetser, H., Labuschagne, L. and Basson, A.T. (2015). Confirmed organophosphorus and carbamate pesticide poisonings in South African wildlife (2009–2014). *Journal of the South African Veterinary Association*, **86**(1), 1–4.

Bowman, Q.P. and Arnoldi, J.M. (1999). Management of animal health emergencies in North America: prevention, preparedness, response and recovery. *Revue Scientifique et Technique de l'Office International des Épizooties*, **18**(1), 76–103. DOI: 10.20506/rst.18.1.1149.

Boyer-Ontl, K.M. and Pruetz, J.D. (2014). Giving the forest eyes: the benefits of using camera traps to study unhabituated chimpanzees (*Pan troglodytes verus*) in southeastern Senegal. *International Journal of Primatology*, **35**(5), 881–94. DOI: 10.1007/s10764-014-9783-3.

Brando, S. and Buchanan-Smith, H.M. (2018). The 24/7 approach to promoting optimal welfare for captive wild animals. *Behavioural Processes*, **156**, 83–95. DOI: 10.1016/j.beproc.2017.09.010.

Brando, S. and Coe, J. (2022). Confronting back-of-house traditions: primates as a case study. *Journal of Zoological and Botanical Gardens*, **3**(3), 366–97. DOI: 10.3390/jzbg3030029.

Brant, H.L., Ewers, R.M., Vythilingam, I., *et al.* (2016). Vertical stratification of adult mosquitoes (Diptera: Culicidae) within a tropical rainforest in Sabah, Malaysia. *Malaria Journal*, **15**(1), 370. DOI: 10.1186/s12936-016-1416-1.

Brent, L. (2001). *The Care and Management of Captive Chimpanzees*. San Antonio, TX: The American Society of Primatologists.

Bridgers, J. (2021). How has COVID19 shifted the global dialogue on animal welfare? And how to move forward to make a global convention for animals a reality. Presented at: *Expert Panel Discussion "UNCAHP a Better World for All Animals", January 15, 2021*. Global Animal Law (GAL) Association and Global Research Network (GRN) Animals and Biodiversity Think Tank Programme. Available at: https://www.youtube.com/watch?v=xZIxZPB2uXo.

Brncic, T.M., Amarasekaran, B. and McKenna, A. (2010). *Sierra Leone National Chimpanzee Census Project August 2010*. Freetown, Sierra Leone: Tacugama Chimpanzee Sanctuary. Available at: http://www.tacugama.com/wp-content/uploads/2017/12/2010_Brncic_SLNCCP_Final_Report.pdf.

Brockelman, W. and Geissmann, T. (2019). Hoolock leuconedys. *The IUCN Red List of Threatened Species 2019: e.T118355453A17968300*. Gland, Switzerland: International Union for Conservation of Nature (IUCN). DOI: 10.2305/IUCN.UK.2019-1.RLTS.T118355453A17968300.en.

Brockelman, W., Molur, S. and Geissmann, T. (2019). Hoolock hoolock. *The IUCN Red List of Threatened Species 2019: e.T39876A17968083*. Gland, Switzerland: International Union for Conservation of Nature (IUCN). DOI: 10.2305/IUCN.UK.2019-3.RLTS.T39876A17968083.en.

Broom, D.M. (1991). Animal welfare: concepts and measurement. *Journal of Animal Science*, **69**(10), 4167–75. DOI: 10.2527/1991.69104167x.

Broom, D.M. (1999). Animal welfare: the concept of the issues. In *Attitudes to Animals: Views in Animal Welfare*, ed. F. L. Dolins. Cambridge, UK: Cambridge University Press, pp. 129–42. DOI: 10.1017/CBO9780511608476.009.

Brouwers, S. and Duchateau, M.J. (2021). Feasibility and validity of the animal welfare assessment grid to monitor the welfare of zoo-housed gorillas *Gorilla gorilla gorilla*. *Journal of Zoo and Aquarium Research*, **9**(4), 208–17. DOI: 10.19227/jzar.v9i4.607.

Brown, G.C. (2019). The endotoxin hypothesis of neurodegeneration. *Journal of Neuroinflammation*, **16**(1), 180. DOI: 10.1186/s12974-019-1564-7.

Brown, K.M. and Leggat, P.A. (2016). Human monkeypox: current state of knowledge and implications for the future. *Tropical Medicine and Infectious Disease*, **1**(1), 8.

Brown, S.L., Anderson, D.C., Dick Jr, E.J., *et al.* (2009). Neoplasia in the chimpanzee (*Pan* spp.). *Journal of Medical Primatology*, **38**(2), 137–44. DOI: 10.1111/j.1600-0684.2008.00321.x.

Brown, V. (2020). Federal government gives zoos $95 million coronavirus lifeline. *News.Com.AU*, April 28, 2020. Available at: https://www.news.com.au/travel/australian-holidays/federal-government-gives-zoos-95-million-coronavirus-lifeline/news-story/085e372f08a1c921b55e1c170f4ef8d1.

Browne, E., Driessen, M.M., Ross, R., Roach, M. and Carver, S. (2021). Environmental suitability of bare-nosed wombat burrows for *Sarcoptes scabiei*. *International Journal for Parasitology: Parasites and Wildlife*, **16**, 37–47. DOI: 10.1016/j.ijppaw.2021.08.003.

Browning, H. and Veit, W. (2021). Freedom and animal welfare. *Animals*, **11**(4), 1148. DOI: 10.3390/ani11041148.

Bruskotter, J.T., Vucetich, J.A., Dietsch, A., *et al.* (2019). Conservationists' moral obligations toward wildlife: values and identity promote conservation conflict. *Biological Conservation*, **240**, 108296. DOI: 10.1016/j.biocon.2019.108296.

Bruyere, B., Bynum, N., Copsey, J., Porzecanski, A. and Sterling, E. (2020). *Conservation Leadership Capacity Building: A Landscape Study*. New York, NY: American Museum of Natural History. Available at: https://www.amnh.org/research/center-for-biodiversity-conservation/resources-and-publications/conservation-action-and-planning/conservation-leadership-capacity-building-a-landscape-study.

Bryant, J.V., Olson, V.A., Chatterjee, H.J. and Turvey, S.T. (2015). Identifying environmental versus phylogenetic correlates of behavioural ecology in gibbons: implications for conservation management of the world's rarest ape. *BMC Evolutionary Biology*, **15**(1), 171. DOI: 10.1186/s12862-015-0430-1.

Bryant, J.V. and Turvey, S.T. (2017). *Emergency Response Plan for the Hainan Gibbon: Report and Recommendations of the Emergency Response Plan Advisory Meeting, Haikou, Hainan, China, 8–9 September, 2016*. London, UK: Zoological Society of London (ZSL). Available at: https://gibbons.asia/wp-content/uploads/2018/08/Hainan_Gibbon_Emergency_Response_Planning_Meeting_2016_Report-2.pdf.

Bryant, T.L. (2006). Trauma, law and advocacy for animals. *Journal of Animal Law and Ethics*, **1**, 63–138.

Buckley, R.C., Morrison, C. and Castley, J.G. (2016). Net effects of ecotourism on threatened species survival. *PLoS ONE*, **11**(2), e0147988. DOI: 10.1371/journal.pone.0147988.

Buddle, B.M., Vordermeier, H.M., Chambers, M.A. and de Klerk-Lorist, L.-M. (2018). Efficacy and safety of BCG vaccine for control of tuberculosis in domestic livestock and wildlife. *Frontiers in Veterinary Science*, **5**, 259. DOI: 10.3389/fvets.2018.00259.

Bueno de Mesquita, C.P., Nichols, L.M., Gebert, M.J., *et al.* (2021). Structure of chimpanzee gut microbiomes across tropical Africa. *mSystems*, **6**(3), e01269-20. DOI: 10.1128/mSystems.01269-20.

Building Code & Bushfire Solutions (n.d.). *Asset Protection Zone Maintenance*. Mount Kuringai, Australia: Building Code & Bushfire Solutions. Available at: https://www.bushfirehazardsolutions.com.au/services/asset-protection-zone-maintenance/. Accessed: July, 2022.

Buitendijk, H., Fagrouch, Z.C., Niphuis, H., *et al.* (2014). Retrospective serology study of respiratory virus infections in captive great apes. *Viruses*, **6**(3), 1442–53.

Bull, J.W., Suttle, K.B., Gordon, A., Singh, N.J. and Milner-Gulland, E.J. (2013). Biodiversity offsets in theory and practice. *Oryx*, **47**(3), 369–80. DOI: 10.1017/S003060531200172X.

Buller, H., Blokhuis, H., Lokhorst, K., Silberberg, M. and Veissier, I. (2020). Animal welfare management in a digital world. *Animals*, **10**(10), 1779. DOI: 10.3390/ani10101779.

Bunge, E.M., Hoet, B., Chen, L., *et al.* (2022). The changing epidemiology of human monkeypox – a potential threat? A systematic review. *PLoS Neglected Tropical Diseases*, **16**(2), e0010141. DOI: 10.1371/journal.pntd.0010141.

Burt, E., Quinn, E., Quinn, R., Cranfield, M. and Sibbald, S.L. (2017). Case 3: providing continuing professional development in a developing country – the One Health initiative. In *Western Public Health Casebook 2017*, ed. A. John-Baptiste and G. McKinley. London, Canada: Public Health Casebook Publishing, pp. 43–54.

Buttke, D.E., Decker, D.J. and Wild, M.A. (2015). The role of one health in wildlife conservation: a challenge and opportunity. *Journal of Wildlife Diseases*, **51**(1), 1–8. DOI: 10.7589/2014-01-004.

Butynski, T.M. and Kalina, J. (1998). Gorilla tourism: a critical look. In *Conservation of Biological Resources*, ed. E. J. Milner-Gulland and R. Mace. Oxford, UK: Blackwell Science, pp. 294–313. DOI: 10.1002/9781444313598.ch12.

C2ES (2022). *Extreme Weather and Climate Change. Centre for Climate and Energy Solutions*. Arlington, VA: Centre for Climate and Energy Solutions (C2ES). Available at: https://www.c2es.org/content/extreme-weather-and-climate-change/.

Cabana, F., Jasmi, R.A. and Maguire, R. (2018). Great ape nutrition: low-sugar and high-fibre diets can lead to increased natural behaviours, decreased regurgitation and reingestion, and reversal of prediabetes. *International Zoo Yearbook*, **52**, 48–61.

Cabezas, S., Calvete, C. and Moreno, S. (2006). Vaccination success and body condition in the European wild rabbit: applications for conservation strategies. *Journal of Wildlife Management*, **70**(4), 1125–31, 7. DOI: 10.2193/0022-541X(2006)70[1125:VSABCI]2.0.CO;2.

Caillaud, D., Eckardt, W., Vecellio, V., *et al.* (2020). Violent encounters between social units hinder the growth of a high-density mountain gorilla population. *Science Advances*, **6**(45), eaba0724. DOI: 10.1126/sciadv.aba0724.

Caillaud, D., Levréro, F., Cristescu, R.H., *et al.* (2006). Gorilla susceptibility to Ebola virus: the cost of sociality. *Current Biology*, **16**, R489–91.

Caillaud, D., Ndagijimana, F., Giarrusso, A.J., Vecellio, V. and Stoinski, T.S. (2014). Mountain gorilla ranging patterns: influence of group size and group dynamics. *American Journal of Primatology*, **76**(8), 730–46. DOI: 10.1002/ajp.22265.

Calvignac-Spencer, S., Düx, A., Gogarten, J.F., Leendertz, F.H. and Patrono, L.V. (2021). A great ape perspective on the origins and evolution of human viruses. *Advances in Virus Research*, **110**, 1–26. DOI: 10.1016/bs.aivir.2021.06.001.

Calvignac-Spencer, S., Leendertz, S.A.J., Gillespie, T.R. and Leendertz, F.H. (2012). Wild great apes as sentinels and sources of infectious disease. *Clinical Microbiology and Infection*, **18**(6), 521–7. DOI: 10.1111/j.1469-0691.2012.03816.x.

Cambre, R.C., Wilson, H.L., Spraker, T.R. and Favara, B.E. (1980). Fatal airsacculitis and pneumonia, with abortion, in an orangutan. *Journal of the American Veterinary Medical Association*, **177**(9), 822–4.

Cameron, K. and Reed, P. (2019). Ebola virus disease in great apes. In *Fowler's Zoo and Wild Animal Medicine Current Therapy, Volume 9*, ed. R. E. Miller, N. Lamberski and P. P. Calle. St Louis, MO: W.B. Saunders, pp. 233–8. DOI: 10.1016/B978-0-323-55228-8.00034-5.

Campbell, A.F. and Sussman, R.W. (1994). The value of radio tracking in the study of neotropical rain forest monkeys. *American Journal of Primatology*, **32**(4), 291–301. DOI: 10.1002/ajp.1350320406.

Campbell, C.O., Cheyne, S.M. and Rawson, B.M. (2015). *Best Practice Guidelines for the Rehabilitation and Translocation of Gibbons*. Gland, Switzerland: International Union for Conservation of Nature (IUCN) Species Survival Commission (SSC) Primate Specialist Group (PSG). Available at: https://portals.iucn.org/library/sites/library/files/documents/SSC-OP-051.pdf.

Campbell, G. (2021). Primate Specialist Group ARRC Task Force. *Oryx*, **55**(4), 495–6. DOI: 10.1017/S0030605321000533.

Campbell, T.P., Sun, X., Patel, V.H., *et al.* (2020). The microbiome and resistome of chimpanzees, gorillas, and humans across host lifestyle and geography. *The ISME Journal*, **14**(6), 1584–99. DOI: 10.1038/s41396-020-0634-2.

Campbell-Smith, G., Campbell-Smith, M., Singleton, I. and Linkie, M. (2011a). Apes in space: saving an imperilled orangutan population in Sumatra. *PLoS ONE*, **6**(2), e17210. DOI: 10.1371/journal.pone.0017210.

Campbell-Smith, G., Campbell-Smith, M., Singleton, I. and Linkie, M. (2011b). Raiders of the lost bark: orangutan foraging strategies in a degraded landscape. *PLoS ONE*, **6**(6), e20962. DOI: 10.1371/journal.pone.0020962.

Canfield, P.J., Vogelnest, L.J., Cunningham, M.L. and Visvesvara, G.S. (1997). Amoebic meningoencephalitis caused by *Balamuthia mandrillaris* in an orang utan. *Australian Veterinary Journal*, **75**(2), 97–100. DOI: 10.1111/j.1751-0813.1997.tb14165.x.

Cannon, J.C. (2017). Cross River superhighway changes course in Nigeria. *Mongabay*, April 28, 2017. Available at: https://news.mongabay.com/2017/04/cross-river-superhighway-changes-course-in-nigeria.

Capps, B. and Lederman, Z. (2015). One Health, vaccines and Ebola: the opportunities for shared benefits. *Journal of Agricultural and Environmental Ethics*, **28**(6), 1011–32. DOI: 10.1007/s10806-015-9574-7.

Capps, B. and Lederman, Z. (2016). Responding to a public health objection to vaccinating the great apes. *Journal of Agricultural and Environmental Ethics*, **29**(5), 883–95. DOI: 10.1007/s10806-016-9633-8.

Capua, I. and Cattoli, G. (2018). One Health (r)evolution: learning from the past to build a new future. *Viruses*, **10**(12), 725. DOI: 10.3390/v10120725.

Cardiff Metropolitan University (n.d.). *International Primate Heart Project*. Cardiff, UK: Cardiff Metropolitan University. Available at: http://primateheartproject.co.uk/. Accessed: January, 2022.

Carlitz, E.H.D., Miller, R., Kirschbaum, C., *et al.* (2016). Measuring hair cortisol concentrations to assess the effect of anthropogenic impacts on wild chimpanzees (*Pan troglodytes*). *PLoS ONE*, **11**(4), e0151870. DOI: 10.1371/journal.pone.0151870.

Carlsen, F., de Jongh, T. and Pluháčková, J. (2022). *EAZA Best Practice Guidelines Great Ape Taxon Advisory Group Chimpanzees (*Pan troglodytes*)*, 1st edn. Amsterdam, the Netherlands: European Association of Zoos and Aquaria (EAZA). Available at: https://www.eaza.net/assets/Uploads/CCC/BPG-2022/Chimpanzee-BPG22.pdf.

Carne, C., Semple, S., Morrogh-Bernard, H., Zuberbühler, K. and Lehmann, J. (2013). Predicting the vulnerability of great apes to disease: the role of superspreaders and their potential vaccination. *PLoS ONE*, **8**(12), e84642. DOI: 10.1371/journal.pone.0084642.

Carne, C., Semple, S., Morrogh-Bernard, H., Zuberbühler, K. and Lehmann, J. (2014). The risk of disease to great apes: simulating disease spread in orang-utan (*Pongo pygmaeus wurmbii*) and chimpanzee (*Pan troglodytes schweinfurthii*) association networks. *PLoS ONE*, **9**(4), e95039. DOI: 10.1371/journal.pone.0095039.

Carr, N. (2016). An analysis of zoo visitors' favourite and least favourite animals. *Tourism Management Perspectives*, **20**, 70–6. DOI: 10.1016/j.tmp.2016.07.006.

Carver, S., Peters, A. and Richards, S.A. (2022). Model integrated disease management to facilitate effective translatable solutions for wildlife disease issues. *Journal of Applied Ecology*, **59**(12), 2902–10. DOI: 10.1111/1365-2664.14298.

Cassella, C. (2019). Forest fires in Indonesia a decade ago may have stunted the growth of children today. *Science Alert*, February 22, 2019. Available at: https://www.sciencealert.com/past-forest-fires-in-indonesia-may-have-stunted-the-growth-of-children.

Cavalieri, P. and Singer, P., ed. (1996). *The Great Ape Project: Equality Beyond Humanity*. New York, NY: Macmillan.

CBD (2020). *Ecosystem Approach*. Montreal, Canada: Convention on Biological Diversity (CBD). Available at: https://www.cbd.int/ecosystem/.

CDC (2017). *Hansen's Disease (Leprosy): Diagnosis and Treatment*. Atlanta, GA: Centers for Disease Control and Prevention (CDC). Available at: https://www.cdc.gov/leprosy/treatment/index.html#:~:text=Hansen's%20disease%20is%20treated%20with,This%20is%20called%20multidrug%20therapy.

CDC (2020a). *Anthrax: Treatment of Anthrax Infection*. Atlanta, GA: Centers for Disease Control and Prevention (CDC). Available at: https://www.cdc.gov/anthrax/treatment/index.html.

CDC (2020b). *Anthrax: Types of Anthrax*. Atlanta, GA: Centers for Disease Control and Prevention (CDC). Available at: https://www.cdc.gov/anthrax/basics/types/index.html.

CDC (2022). *Ebola (Ebola Virus Disease): History of Ebola Virus Disease (EVD) Outbreaks*. Atlanta, GA: Centers for Disease Control and Prevention (CDC). Available at: https://www.cdc.gov/vhf/ebola/history/chronology.html.

CDC (n.d.-a). *Key Achievements of the GHSA*. Atlanta, GA: Centers for Disease Control and Prevention (CDC). Available at: https://www.cdc.gov/globalhealth/resources/factsheets/5-years-of-ghsa.html. Accessed: October, 2022.

CDC (n.d.-b). *One Health Basics*. Atlanta, GA: Centers for Disease Control and Prevention (CDC). Available at: https://www.cdc.gov/onehealth/basics/index.html. Accessed: June, 2021.

Čejková, D., Zobaníková, M., Chen, L., *et al.* (2012). Whole genome sequences of three *Treponema pallidum* ssp. *pertenue* strains: yaws and syphilis treponemes differ in less than 0.2% of the genome sequence. *PLoS Neglected Tropical Diseases*, **6**(1), e1471. DOI: 10.1371/journal.pntd.0001471.

Celestino-Soper, P.B.S., Lynnes, T.C., Zhang, L., *et al.* (2018). Genetic analyses in a bonobo (*Pan paniscus*) with arrhythmogenic right ventricular cardiomyopathy. *Scientific Reports*, **8**, 4350. DOI: 10.1038/s41598-018-22334-5.

Center for Global Health (2016). *Implementing the Global Health Security Agenda: Progress and Impact from U.S. Government Investments*. Atlanta, GA: Centers for Disease Control and Prevention (CDC). Available at: https://stacks.cdc.gov/view/cdc/59125.

Centurión-Lara, A., Molini, B.J., Godornes, C., *et al.* (2006). Molecular differentiation of *Treponema pallidum* subspecies. *Journal of Clinical Microbiology*, **44**(9), 3377–80. DOI: 10.1128/JCM.00784-06.

Cerdán, P. and Kirk-Cohen, G., ed. (2020). *How to Reduce Single-Use Plastic at your Zoo or Aquarium*. Barcelona, Spain: World Association of Zoos and Aquariums (WAZA). Available at: https://www.waza.org/wp-content/uploads/2020/10/WAZA-short-guide-final-online.pdf.

Cerveny, S. and Sleeman, J. (2014). Great apes. In *Zoo Animal and Wildlife Immobilization and Anesthesia*, ed. G. West, D. Heard and N. Caulkett. Ames, IA: Wiley Blackwell, pp. 573–84. DOI: 10.1002/9781118792919.ch39.

Champion, J. (2013). *The effects of a hurricane and fire on feeding ecology, activity budget, and social patterns of spider monkeys (Ateles geoffroyi) in Central Belize*. Master's thesis. Calgary, Canada: University of Calgary. DOI: 10.11575/PRISM/28094.

Chan, B.P.L., Lo, Y.F.P., Hong, X.-J., Mak, C.F. and Ma, Z. (2020). First use of artificial canopy bridge by the world's most critically endangered primate the Hainan gibbon *Nomascus hainanus*. *Scientific Reports*, **10**, 15176. DOI: 10.1038/s41598-020-72641-z.

Chan, J.K.L., Marzuki, K.M. and Mohtar, T.M. (2021). Local community participation and responsible tourism practices in ecotourism destination: a case of Lower Kinabatangan, Sabah. *Sustainability*, **13**(23), 13302. DOI: 10.3390/su132313302.

Chancellor, R.L., Rundus, A.S. and Nyandwi, S. (2017). Chimpanzee seed dispersal in a montane forest fragment in Rwanda. *American Journal of Primatology*, **79**(3), e22624. DOI: 10.1002/ajp.22624.

Chappell, J.M. and Thorpe, S.K.S. (2021). *The Enclosure Design Tool: An Evidence-Based Framework for Improving Captive Ape Well-Being*. Unpublished data. Birmingham, UK: University of Birmingham.

Chappell, J.M. and Thorpe, S.K.S. (2022). The role of great ape behavioral ecology in One Health: implications for captive welfare and re-habilitation success. *American Journal of Primatology*, **84**(4–5), e23328. DOI: 10.1002/ajp.23328.

Charles-Dominique, P. (1977). Urine marking and territoriality in *Galago alleni* (Waterhouse, 1837 – Lorisoidea, Primates) – a field study by radio-telemetry. *Zeitschrift für Tierpsychologie*, **43**(2), 113–38. DOI: 10.1111/j.1439-0310.1977.tb00063.x.

Chelluri, G.I., Ross, S.R. and Wagner, K.E. (2013). Behavioral correlates and welfare implications of informal interactions between caretakers and zoo-housed chimpanzees and gorillas. *Applied Animal Behaviour Science*, **147**(3), 306–15. DOI: 10.1016/j.applanim.2012.06.008.

Cheng, A.C. and Currie, B.J. (2005). Melioidosis: epidemiology, pathophysiology, and management. *Clinical Microbiology Reviews*, **18**(2), 383–416. DOI: 10.1128/CMR.18.2.383-416.2005.

Cheptoris, S. (2020). *Statement on the Rising Water Levels of Lake Victoria and the Nile System*. Kampala, Uganda: Uganda Media Centre. Available at: https://www.mediacentre.go.ug/media/statement-rising-water-levels-lake-victoria-and-nile-system.

Chester Zoo (2021). *Chester Zoo Risk Assessment COVID-19 (Coronavirus)*. Chester, UK Chester Zoo. Available at: https://cdn.chesterzoo.org/2021/04/COVID-19-RISK-ASSESSMENT-CHESTER-ZOO-2021.pdf.

Cheyne, S.M. (2008a). Effects of meteorology, astronomical variables, location and human disturbance on the singing apes: *Hylobates albibarbis*. *American Journal of Primatology*, **70**(4), 386–92. DOI: 10.1002/ajp.20502.

Cheyne, S.M. (2008b). Feeding ecology, food choice and diet characteristics of gibbons in a disturbed peat-swamp forest, Indonesia. In *XXII Congress of the International Primatological Society*, ed. P. C. Lee, P. Honess, H. Buchanan-Smith, A. MaClarnon and W. I. Sellers. Edinburgh, UK, pp. 3–8.

Cheyne, S.M. (2010). Behavioural ecology of gibbons (*Hylobates albibarbis*) in a degraded peat-swamp forest. In *Indonesian Primates*, ed. S. Gursky and J. Supriatna. New York, NY: Springer, pp. 121–56. DOI: 10.1007/978-1-4419-1560-3_8.

Cheyne, S.M., Campbell, C.O. and Payne, K.L. (2012). Proposed guidelines for in situ gibbon rescue, rehabilitation and reintroduction. *International Zoo Yearbook*, **46**(1). DOI: 10.1111/j.1748-1090.2011.00149.x.

Cheyne, S.M., Gilhooly, L.J., Hamard, M.C., *et al.* (2016). Population mapping of gibbons in Kalimantan, Indonesia: correlates of gibbon density and vegetation across the species' range. *Endangered Species Research*, **30**(1), 133–43. DOI: 10.3354/esr00734.

Chi, F., Leider, M., Leendertz, F.H., *et al.* (2007). New *Streptococcus pneumoniae* clones in deceased wild chimpanzees. *Journal of Bacteriology*, **189**(16), 6085–8. DOI: 10.1128/JB.00468-07.

Chimfunshi Wildlife Orphanage (n.d.). *Chimfunshi*. Chingola, Zambia: Chimfunshi Wildlife Orphanage Trust. Available at: https://www.chimfunshi.de/en. Accessed: October, 2020.

Chimp Eden (n.d.). *Chimp Guardianship Programme*. Mpumalanga, South Africa: Chimp Eden, Jane Goodall Institute South Africa. Available at: https://www.chimpeden.com/adoptions.html. Accessed: October, 2020.

Chimpanzee Conservation Center (2020). *Annual Report 2019*. High Niger National Park, Guinea: Project Primates. Available at: https://www.projetprimates.com/wp-content/uploads/CCC-2019-Annual-Report.pdf.

ChimpCARE (n.d.-a). *Chimpanzees in the US*. Chicago, IL: Lincoln Park Zoo. Available at: http://www.chimpcare.org/map. Accessed: October, 2020.

ChimpCARE (n.d.-b). *Welfare Assessment*. Chicago, IL: ChimpCARE. Available at: https://chimpcare.org/welfare_assessment. Accessed: May, 2022.

Chivers, D.J. (1974). *The Siamang in Malaya: A Field Study of a Primate in Tropical Rainforest*. Contributions to Primatology 4. Basel, Switzerland: Karger.

Chok, S., Macbeth, J. and Warren, C. (2007). Tourism as a tool for poverty alleviation: a critical analysis of "pro-poor tourism" and implications for sustainability. *Current Issues in Tourism*, **10**(2–3), 144–65. DOI: 10.2167/cit303.

Chomel, B.B., Belotto, A. and Meslin, F.-X. (2007). Wildlife, exotic pets, and emerging zoonoses. *Emerging Infectious Diseases*, **13**(1), 6. DOI: 10.3201/eid1301.060480.

Choo, Y. (2011). *Orangutan behaviour in captivity: activity budgets, enclosure use and the visitor effect*. MSc thesis. Singapore: National University of Singapore.

Choo, Y., Todd, P.A. and Li, D. (2011). Visitor effects on zoo orangutans in two novel, naturalistic enclosures. *Applied Animal Behaviour Science*, **133**(1), 78–86. DOI: 10.1016/j.applanim.2011.05.007.

Choudhury, A. (2013). Description of a new subspecies of hoolock gibbon *Hoolock hoolock* from northeast India. *Newsletter and Journal of the Rhino Foundation for Nature in Northeast India*, **9**, 49–59.

Christmann, P., Ayuk, E.T., Pedro, A.M.A. and Kumar, S.V. (2022). Future mineral demand: the necessary transition toward sustainability. In *Routledge Handbook of the Extractive Industries and Sustainable Development*, ed. N. Yakovleva and E. Nickless. London, UK: Routledge, pp. 101–32.

Chua, L., Fair, H., Schreer, V., Stępień, A. and Thung, P.H. (2021). Only the orangutans get a life jacket. *American Ethnologist*, **48**(4), 370–85. DOI: 10.1111/amet.13045.

Chua, L., Harrison, M.E., Fair, H., *et al.* (2020). Conservation and the social sciences: beyond critique and co-optation. A case study from orangutan conservation. *People and Nature*, **2**(1), 42–60. DOI: 10.1002/pan3.10072.

Chuma, I.S., Batamuzi, E.K., Collins, D.A., *et al.* (2018). Widespread *Treponema pallidum* infection in nonhuman primates, Tanzania. *Emerging Infectious Diseases*, **24**(6), 1002–9. DOI: 10.3201/eid2406.180037.

Chuma, I.S., Roos, C., Atickem, A., *et al.* (2019). Strain diversity of *Treponema pallidum* subsp. *pertenue* suggests rare interspecies transmission in African nonhuman primates. *Scientific Reports*, **9**, 14243. DOI: 10.1038/s41598-019-50779-9.

Chumo, C. (2021). *ANAW and WFA Lead Global Initiative for Adoption of UN Animal Welfare Resolution.* Boston, MA: World Federation for Animals (WFA). Available at: https://wfa.org/wfa-anaw-resolution/.

Cibot, M., Krief, S., Philippon, J., *et al.* (2016). Feeding consequences of hand and foot disability in wild adult chimpanzees (*Pan troglodytes schweinfurthii*). *International Journal of Primatology*, **37**(4–5), 479–94. DOI: 10.1007/s10764-016-9914-0.

Cipolletta, C. (2003). Ranging patterns of a western gorilla group during habituation to humans in the Dzanga-Ndoki National Park, Central African Republic. *International Journal of Primatology*, **24**(6), 1207–26. DOI: 10.1023/B:IJOP.0000005988.52177.45.

CITES (2016). *Resolution Conf 17.8 Disposal of Illegally Traded and Confiscated Specimens of CITES-Listed Species.* Geneva, Switzerland: Convention on International Trade in Endangered Species of Wild Fauna and Flora (CITES). Available at: https://www.cites.org/sites/default/files/document/E-Res-17-08.pdf.

CITES (n.d.). *CITES Trade Database.* Cambridge, UK: United Nations Environment Programme (UNEP) World Conservation Monitoring Centre (WCMC). Available at: https://trade.cites.org/. Accessed: December, 2020.

CITES Secretariat (2017). *SC69 Doc. 34.1 Sixty-ninth Meeting of the Standing Committee Geneva (Switzerland), 27 November–1 December 2017. Interpretation and Implementation Matters. General Compliance and Enforcement. Disposal of Confiscated Specimens. Report of the Secretariat.* Geneva, Switzerland: Convention on International Trade in Endangered Species of Wild Fauna and Flora (CITES). Available at: https://stag.cites.org/sites/default/files/eng/com/sc/69/E-SC69-34-01.pdf.

CITES Secretariat (2021). *Revised Draft (2021) Model Law on International Trade in Wild Fauna and Flora.* Geneva, Switzerland: Convention on International Trade in Endangered Species of Wild Fauna and Flora (CITES). Available at: https://cites.org/sites/default/files/projects/NLP/E-Model_law-revised_Oct.2021.FINAL.DRAFT.pdf.

Clark, F.E., Fitzpatrick, M., Hartley, A., *et al.* (2012). Relationship between behavior, adrenal activity, and environment in zoo-housed western lowland gorillas (*Gorilla gorilla gorilla*). *Zoo Biology*, **31**(3), 306–21. DOI: 10.1002/zoo.20396.

Clarke, E., Reichard, U.H. and Zuberbühler, K. (2006). The syntax and meaning of wild gibbon songs. *PLoS ONE*, **1**(1), e73. DOI: 10.1371/journal.pone.0000073.

Clayton, J.B., Gomez, A., Amato, K., *et al.* (2018). The gut microbiome of nonhuman primates: lessons in ecology and evolution. *American Journal of Primatology*, **80**(6), e22867. DOI: 10.1002/ajp.22867.

Clayton, J.B., Vangay, P., Huang, H., *et al.* (2016). Captivity humanizes the primate microbiome. *Proceedings of the National Academy of Sciences*, **113**(37), 10376–81. DOI: 10.1073/pnas.1521835113.

Clegg, I. (2021). *How to Take Your Welfare Program to the Next Level: The C-Well Dolphin Assessment as an Example of a Species-Specific Tool, and AnimalCare Software's Cloud-Based Welfare Platform. EAZA Animal Welfare Webinar 1 March 2021.* Amsterdam, the Netherlands: European Association of Zoos and Aquaria (EAZA). Available at: https://www.eaza.net/about-us/areas-of-activity/animal-welfare/animal-welfare-webinars/.

Clegg, I.L.K., Borger-Turner, J.L. and Eskelinen, H.C. (2015). C-Well: the development of a welfare assessment index for captive bottlenose dolphins (*Tursiops truncatus*). *Animal Welfare*, **24**(3), 267–82. DOI: 10.7120/09627286.24.3.267.

Clifford, D.H., Yoo, S.Y., Fazekas, S. and Hardin, C.J. (1977). Surgical drainage of a submandibular air sac in an orangutan. *Journal of the American Veterinary Medical Association*, **171**(9), 862–5.

Clifford, W. and Steedman, C. (2021). Wildlife–pet markets in a One-Health context. *International Journal of One Health*, **7**(1), 42–64. DOI: 10.14202/IJOH.2021.42-64.

Clink, D.J., Crofoot, M.C. and Marshall, A.J. (2019). Application of a semi-automated vocal fingerprinting approach to monitor Bornean gibbon females in an experimentally fragmented landscape in Sabah, Malaysia. *Bioacoustics*, **28**(3), 193–209. DOI: 10.1080/09524622.2018.1426042.

Cliquet, F., Müller, T., Mutinelli, F., *et al.* (2003). Standardisation and establishment of a rabies ELISA test in European laboratories for assessing the efficacy of oral fox vaccination campaigns. *Vaccine*, **21**(21), 2986–93. DOI: 10.1016/S0264-410X(03)00102-6.

Cochrane, A. (2012). *Animal Rights Without Liberation Applied Ethics and Human Obligations*. New York, NY: Columbia University Press.

Coe, J. and Hoy, J. (2020). Choice, control and computers: empowering wildlife in human care. *Multimodal Technologies and Interaction*, **4**(4), 92. DOI: 10.3390/mti4040092.

Cohen, J. (2010). Chimpanzee research today. A matter of life and limb. *Science*, **328**(5974), 33. DOI: 10.1126/science.328.5974.33.

Cohen, S.E. (2013). Sandy marked a shift for social media use in disasters. *Government Technology*, March 7, 2013. Available at: https://www.govtech.com/em/disaster/Sandy-Social-Media-Use-in-Disasters.html.

Colditz, I.G. and Hine, B.C. (2016). Resilience in farm animals: biology, management, breeding and implications for animal welfare. *Animal Production Science*, **56**(12), 1961–83. DOI: 10.1071/AN15297.

Collins, A. (2003). Health guidelines for visiting researchers in Gombe National Park to minimize risk of disease transmission among primates (updated 8/01/03). *Pan Africa News*, **10**(1), 1–3.

Collins, A. and Goodall, J. (2008). Long-term research and conservation in Gombe National Park, Tanzania. In *Science and Conservation in African Forests: The Benefits of Longterm Research*, ed. E. Ross and R. Wrangham. Cambridge, UK: Cambridge University Press, pp. 158–72. DOI: 10.1017/CBO9780511754920.016.

Conover, M.R. and Conover, D.O. (2022). *Human–Wildlife Interactions: From Conflict to Coexistence*, 2nd edn. Boca Raton, FL: CRC Press. DOI: 10.1201/9780429401404.

Conservation Evidence (n.d.). *Conservation Evidence: Providing Evidence to Improve Practice*. Cambridge, UK: University of Cambridge. Available at: https://www.conservationevidence.com/data/index?terms=prism. Accessed: May, 2019.

Cooke, S.J., Madliger, C.L., Cramp, R.L., *et al.* (2020). Reframing conservation physiology to be more inclusive, integrative, relevant and forward-looking: reflections and a horizon scan. *Conservation Physiology*, **8**(1), coaa016. DOI: 10.1093/conphys/coaa016.

Cooper, J.E. and Hull, G., ed. (2017). *Gorilla Pathology and Health: With a Catalogue of Preserved Materials*. San Diego, CA: Academic Press. DOI: 10.1016/B978-0-12-802039-5.00021-4.

Cooper, K. (2018). What does a good response to an emergency wildlife disease look like? Presented at: *WDA-A Annual Conference, Bali, Indonesia*. Wildlife Disease Association Australasia Section (WDA-A) in association with the Asian Society of Conservation Medicine.

Corbey, R. (2005). *The Metaphysics of Apes: Negotiating the Animal–Human Boundary*. New York, NY: Cambridge University Press.

Corlett, R.T., Primack, R.B., Devictor, V., *et al.* (2020). Impacts of the coronavirus pandemic on biodiversity conservation. *Biological Conservation*, **246**, 108571. DOI: 10.1016/j.biocon.2020.108571.

Corrigan, A. (2010). *An Investigation into the Welfare Standards of Zoos in Malaysia*. Singapore: Animal Concerns Research and Education Society (ACRES). Available at: http://www.zoocheck.com/wp-content/uploads/2015/06/MalaysiaZooReport2010.pdf.

Coscollá, M., Lewin, A., Metzger, S., *et al.* (2013). Novel *Mycobacterium tuberculosis* complex isolate from a wild chimpanzee. *Emerging Infectious Diseases*, **19**(6), 969–76.

Coudrat, C.N.Z., Nanthavong, C., Ngoprasert, D., Suwanwaree, P. and Savini, T. (2015). Singing patterns of white-cheeked gibbons (*Nomascus* sp.) in the Annamite Mountains of Laos. *International Journal of Primatology*, **36**(4), 691–706. DOI: 10.1007/s10764-015-9849-x.

Cox, C., Burgess, S., Sellitto, C. and Buultjens, J. (2009). The role of user-generated content in tourists' travel planning behavior. *Journal of Hospitality Marketing & Management*, **18**(8), 743–64. DOI: 10.1080/19368620903235753.

Cox, J. and Lennkh, S. (2016). *Model Animal Welfare Act – A Comprehensive Framework Law*. Boston, MA: World Animal Net. Available at: http://worldanimal.net/images/stories/documents/Model_AWA/WAN-Model-Animal-Welfare-Act.pdf.

Cozannet, G.L. (2007). *IGOS Geohazards: Toward an Improved use of Earth Observations for Geohazards Mitigation.* Scientific and Technical Subcommittee Session. Vienna, Austria: United Nations Office for Outer Space Affairs (UNOOSA) Available at: https://www.unoosa.org/pdf/pres/stsc2007/tech-12.pdf.

Cranfield, M.R. and Minnis, R.B. (2007). An integrated health approach to the conservation of mountain gorillas *Gorilla beringei beringei. International Zoo Yearbook*, **41**, 110–21.

Crissey, S., Pribyl, L., Pruett-Jones, M. and Meehan, T. (1998). Nutritional management of Old World primates with special consideration for vitamin D. *International Zoo Yearbook*, **36**(1), 122–30. DOI: 10.1111/j.1748-1090.1998.tb02894.x.

Crockett, C.M. and Ha, R.R. (2010). Data collection in the zoo setting, emphasizing behavior. In *Wild Mammals in Captivity: Principles and Techniques*, ed. D. G. Kleiman, K. V. Thompson and C. K. Baer. Chicago, IL: University of Chicago Press, pp. 386–406.

Crunchant, A.-S., Egerer, M., Loos, A., *et al.* (2017). Automated face detection for occurrence and occupancy estimation in chimpanzees. *American Journal of Primatology*, **79**(3), e22627. DOI: 10.1002/ajp.22627.

Crutzen, P.J. (2006). The "Anthropocene". In *Earth System Science in the Anthropocene*, ed. E. Ehlers and T. Krafft. Berlin, Heidelberg, Germany: Springer, pp. 13–18. DOI: 10.1007/3-540-26590-2_3.

CSBI and TBC (2015). *A Cross-Sector Guide to Implementing the Mitigation Hierarchy.* Cambridge, UK: Cross-Sector Biodiversity Initiative (CSBI). Available at: http://www.csbi.org.uk/our-work/mitigation-hierarchy-guide/.

CTPH (n.d.-a). *Alternative Livelihoods.* Entebbe, Uganda: Conservation Through Public Health (CTPH). Available at: https://ctph.org/alternative-livelihoods-program/. Accessed: August, 2022.

CTPH (n.d.-b). *Bwindi Impenetrable National Park.* Entebbe, Uganda: Conservation Through Public Health (CTPH). Available at: https://ctph.org/conservation-locations/. Accessed: August, 2022.

CTPH (n.d.-c). *Conservation Through Public Health.* Entebbe, Uganda: Conservation through Public Health (CTPH). Available at: https://ctph.org/. Accessed: August, 2022.

CTPH (n.d.-d). *One Health.* Entebbe, Uganda: Conservation Through Public Health (CTPH). Available at: https://ctph.org/one-health-program/. Accessed: August, 2022.

Cunningham, E.P., Unwin, S. and Setchell, J.M. (2015). Darting primates in the field: a review of reporting trends and a survey of practices and their effect on the primates involved. *International Journal of Primatology*, **36**(5), 911–32. DOI: 10.1007/s10764-015-9862-0.

Curry, B.A., Drane, A.L., Atencia, R., *et al.* (2023). Body mass and growth rates in captive chimpanzees (*Pan troglodytes*) cared for in African wildlife sanctuaries, zoological institutions, and research facilities. *Zoo Biology*, **42**(1), 98–106. DOI: 10.1002/zoo.21718.

Cusick, D. (2019). Some disaster prevention spending reaps higher rewards. *Scientific American*, June 24, 2019. Available at: https://www.scientificamerican.com/article/some-disaster-prevention-spending-reaps-higher-rewards/.

CZS (n.d.). *Animal Welfare Research: WelfareTrak®.* Chicago, IL: Chicago Zoological Society (CZS). Available at: https://welfaretrak.org/. Accessed: May, 2022.

D'arc, M., Ayouba, A., Esteban, A., *et al.* (2015). Origin of the HIV-1 group O epidemic in western lowland gorillas. *Proceedings of the National Academy of Sciences*, **112**(11), E1343–52.

D'Cruze, N., Green, J., Elwin, A. and Schmidt-Burbach, J. (2020). Trading tactics: time to rethink the global trade in wildlife. *Animals*, **10**(12), 2456. DOI: 10.3390/ani10122456.

D'Cruze, N. and Macdonald, D.W. (2016). A review of global trends in CITES live wildlife confiscations. *Nature Conservation*, **15**. DOI: 10.3897/natureconservation.15.10005.

Daddoust, L., Asgary, A., McBey, K.J., Elliott, S. and Normand, A. (2021). Spontaneous volunteer coordination during disasters and emergencies: opportunities, challenges, and risks. *International Journal of Disaster Risk Reduction*, **65**, 102546. DOI: 10.1016/j.ijdrr.2021.102546.

Dalkey, N. and Helmer, O. (1963). An experimental application of the Delphi method to the use of experts. *Management Science*, **9**(3), 458–67.

Dalton, J. (2020). Coronavirus: sharp rise in poaching of Africa's mountain gorillas as people hunt more bushmeat. *Independent*, July 20, 2020. Available at: https://www.independent.co.uk/independentpremium/world/poaching-gorillas-africa-congo-uganda-bushmeat-coronavirus-apes-a9628501.html.

Dampage, U., Bandaranayake, L., Wanasinghe, R., Kottahachchi, K. and Jayasanka, B. (2022). Forest fire detection system using wireless sensor networks and machine learning. *Scientific Reports*, **12**, 46. DOI: 10.1038/s41598-021-03882-9.

Das, J., Biswas, J., Bhattacherjee, P.C. and Rao, S.S. (2009). Canopy bridges: an effective conservation tactic for supporting gibbon populations in forest fragments. In *The Gibbons: New Perspectives on Small Ape Socioecology and Population Biology*, ed. D. Whittaker and S. Lappan. New York, NY: Springer, pp. 467–75. DOI: 10.1007/978-0-387-88604-6_22.

Daszak, P., Cunningham, A.A. and Hyatt, A.D. (2000). Emerging Infectious Diseases of wildlife – threats to biodiversity and human health. *Science*, **287**(5452), 443–9. DOI: 10.1126/science.287.5452.443.

Daszak, P., Cunningham, A.A. and Hyatt, A.D. (2001). Anthropogenic environmental change and the emergence of infectious diseases in wildlife. *Acta Tropica*, **78**(2), 103–16. DOI: 10.1016/S0001-706X(00)00179-0.

Daud, Z. (2019). Sepilok centre must heed the rules. *New Straits Times*, December 7, 2019. Available at: https://www.nst.com.my/opinion/letters/2019/12/545536/sepilok-centre-must-heed-rules.

Davis, J.T., Mengersen, K., Abram, N.K., *et al.* (2013). It's not just conflict that motivates killing of orangutans. *PLoS ONE*, **8**(10), e75373. DOI: 10.1371/journal.pone.0075373.

Dawson, C.P. (2008). Ecotourism and nature-based tourism: one end of the tourism opportunity spectrum? In *Tourism, Recreation and Sustainability: Linking Culture and the Environment*, 2nd edn, ed. S. F. McCool and R. N. Moisey. Wallingford, UK: CABI International, pp. 38–50. DOI: 10.1079/9781845934705.0038.

de Haas, A. (2020). Transmission of diseases from humans to apes: why extra vigilance is now needed. *The Conversation*, March 24, 2020. Available at: https://theconversation.com/transmission-of-diseases-from-humans-to-apes-why-extra-vigilance-is-now-needed-134083.

De Santis, O., Audran, R., Pothin, E., *et al.* (2016). Safety and immunogenicity of a chimpanzee adenovirus-vectored Ebola vaccine in healthy adults: a randomised, double-blind, placebo-controlled, dose-finding, phase 1/2a study. *The Lancet Infectious Diseases*, **16**(3), 311–20. DOI: 10.1016/s1473-3099(15)00486-7.

de Silva, G.C., Regan, E.C., Pollard, E.H.B. and Addison, P.F.E. (2019). The evolution of corporate no net loss and net positive impact biodiversity commitments: understanding appetite and addressing challenges. *Business Strategy and the Environment*, **28**(7), 1481–95. DOI: 10.1002/bse.2379.

Decision Tree Writing Group (2006). Clinical response decision tree for the mountain gorilla (*Gorilla beringeii*) as a model for great apes. *American Journal of Primatology*, **68**(9), 909–27. DOI: 10.1002/ajp.20297.

Deem, S.L. (2007). Role of the zoo veterinarian in the conservation of captive and free-ranging wildlife. *International Zoo Yearbook*, **41**(1), 3–11. DOI: 10.1111/j.1748-1090.2007.00020.x.

Deem, S.L. (2016). Conservation medicine: a solution-based approach for saving nonhuman primates. In *Ethnoprimatology: Primate Conservation in the 21st Century*, ed. M. T. Waller. Cham, Switzerland: Springer, pp. 63–76. DOI: 10.1007/978-3-319-30469-4_4.

Deere, J.R., Parsons, M.B., Lonsdorf, E.V., *et al.* (2019). *Entamoeba histolytica* infection in humans, chimpanzees and baboons in the Greater Gombe Ecosystem, Tanzania. *Parasitology*, **146**(9), 1116–22. DOI: 10.1017/S0031182018001397.

DeGrazia, D. (2016). Nonhuman primates, human need, and ethical constraints. *Hastings Center Report*, **46**(4), 27–8. DOI: 10.1002/hast.601.

Deiner, K., Bik, H.M., Mächler, E., *et al.* (2017). Environmental DNA metabarcoding: transforming how we survey animal and plant communities. *Molecular Ecology*, **26**(21), 5872–95. DOI: 10.1111/mec.14350.

Delgado, R.A. (2010). Communication, culture and conservation in orangutans. In *Indonesian Primates*, ed. S. Gursky and J. Supriatna. New York, NY: Springer, pp. 23–40. DOI: 10.1007/978-1-4419-1560-3_3.

Delgado, R.A. and van Schaik, C.P. (2000). The behavioral ecology and conservation of the orangutan (*Pongo pygmaeus*): a tale of two islands. *Evolutionary Anthropology: Issues, News, and Reviews*, **9**(5), 201–18. DOI: 10.1002/1520-6505(2000)9:5<201::AID-EVAN2>3.0.CO;2-Y.

Dellatore, D.F., Waitt, C.D. and Foitovà, I. (2014). The impact of tourism on the behavior of rehabilitated orangutans (*Pongo abelii*) in Bukit Lawang, North Sumatra, Indonesia. In *Primate Tourism: A Tool for Conservation?*, ed. A. E. Russon and J. Wallis. Cambridge, UK: Cambridge University Press, pp. 98–120. DOI: 10.1017/CBO9781139087407.008.

Demetria, C., Smith, I., Tan, T., *et al.* (2018). Reemergence of Reston Ebola virus in cynomolgus monkeys, the Philippines, 2015. *Emerging Infectious Diseases*, **24**(7), 1285–91. DOI: 10.3201/eid2407.171234.

Dench, R., Sulistyo, F., Fahroni, A. and Philippa, J. (2015). Evaluation of diagnostic accuracy of the comparative tuberculin skin test in rehabilitant Bornean orangutans (*Pongo pygmaeus*). *Journal of Zoo and Wildlife Medicine*, **46**(4), 833–42. DOI: 10.1638/2014-0220.1.

Dennis, R. (1999). *A Review of Fire Projects in Indonesia (1982–1998)*. Bogor, Indonesia: Center for International Forestry Research (CIFOR). Available at: https://www.cifor.org/publications/pdf_files/firereport.pdf.

DePaoli, A. and Johnsen, D.O. (1978). Fatal strongyloidiasis in gibbons (*Hylobates lar*). *Veterinary Pathology*, **15**(1), 31–9. DOI: 10.1177/030098587801500104.

Desmond, J.S. and Desmond, J.A.Z. (2014). Evaluating the effectiveness of chimpanzee tourism. In *Primate Tourism: A Tool for Conservation?*, ed. A. E. Russon and J. Wallis. Cambridge, UK: Cambridge University Press, pp. 199–212. DOI: 10.1017/CBO9781139087407.014.

Detroit Zoological Society (n.d.). *The Great Ape Heart Project*. Royal Oak, MI: Detroit Zoological Society. Available at: https://greatapeheartproject.org/. Accessed: January, 2022.

Devaux, C.A., Mediannikov, O., Medkour, H. and Raoult, D. (2019). Infectious disease risk across the growing human–non human primate interface: a review of the evidence. *Frontiers in Public Health*, **7**, 305. DOI: 10.3389/fpubh.2019.00305.

DHS (2019). *Guidelines for Managing Spontaneous Volunteers in South Australia*. Adelaide, Australia: Department of Human Services (DHS), Government of South Australia. Available at: https://naturaldisaster.royalcommission.gov.au/system/files/2020-07/SSA.468.001.0257.pdf.

Di Giulio, D.B. and Eckburg, P.B. (2004). Human monkeypox: an emerging zoonosis. *The Lancet Infectious Diseases*, **4**(1), 15–25. DOI: 10.1016/s1473-3099(03)00856-9.

Dickman, A.J. and Hazzah, L. (2016). Money, myths and man-eaters: complexities of human–wildlife conflict. In *Problematic Wildlife: A Cross-Disciplinary Approach*, ed. F. M. Angelici. Cham, Switzerland: Springer International Publishing, pp. 339–56. DOI: 10.1007/978-3-319-22246-2_16.

Digun-Aweto, O. (2020). Taking a cue from Rwanda; replicating gorilla tourism in Nigeria. *AfricaGrowth Agenda*, **17**(2), 10–11. DOI: 10.10520/EJC-1f024f79ad.

Dinsi, S.C. and Eyebe, S.A. (2016). *Great Ape Conservation in Cameroon: Mapping Institution and Policies. Poverty and Conservation Learning Group (PCLG) Research Report*. London, UK: International Institute for Environment and Development (IIED). Available at: http://pubs.iied.org/pdfs/G04017.pdf.

Doane, C.J., Lee, D.R. and Sleeper, M.M. (2006). Electrocardiogram abnormalities in captive chimpanzees (*Pan troglodytes*). *Comparative Medicine*, **56**(6), 512–18.

Dobson, A.P., Pimm, S.L., Hannah, L., *et al.* (2020). Ecology and economics for pandemic prevention. *Science*, **369**(6502), 379–81. DOI: 10.1126/science.abc3189.

Doran-Sheehy, D., Derby, A., Greer, D. and Mongo, P. (2007). Habituation of western gorillas: the process and factors that influence it. *American Journal of Primatology*, **69**, 1–16. DOI: 10.1002/ajp.20442.

Doran-Sheehy, D., Mongo, P., Lodwick, J. and Conklin-Brittain, N.L. (2009). Male and female western gorilla diet: preferred foods, use of fallback resources, and implications for ape versus old world monkey foraging strategies. *American Journal of Physical Anthropology*, **140**(4), 727–38. DOI: 10.1002/ajpa.21118.

Dore, K.M., Hansen, M.F., Klegarth, A.R., *et al.* (2020). Review of GPS collar deployments and performance on nonhuman primates. *Primates*, **61**(3), 373–87. DOI: 10.1007/s10329-020-00793-7.

Dore, K.M., Riley, E.P. and Fuentes, A. (2017). *Ethnoprimatology: A Practical Guide to Research at the Human–Nonhuman Primate Interface*. Cambridge, UK: Cambridge University Press. DOI: 10.1017/9781316272466.

Doyle, C. (2017). Captive wildlife sanctuaries: definition, ethical considerations and public perception. *Animal Studies Journal*, **6**(2), 55–85. DOI: https://ro.uow.edu.au/asj/vol6/iss2/5.

Drane, A.L., Atencia, R., Cooper, S.-M., *et al.* (2019). Cardiac structure and function characterized across age groups and between sexes in healthy wild-born captive chimpanzees (*Pan troglodytes*) living in sanctuaries. *American Journal of Veterinary Research*, **80**(6), 547–57. DOI: 10.2460/ajvr.80.6.547.

Drane, A.L., Atencia, R., Cooper, S.-M., *et al.* (2020). Evaluation of relationships between results of electrocardiography and echocardiography in 341 chimpanzees (*Pan troglodytes*). *American Journal of Veterinary Research*, **81**(6), 488–98. DOI: 10.2460/ajvr.81.6.488.

Draper, C., Baker, L. and Ramp, D. (2015). Poster: Why compassionate conservation can improve the welfare of wild animals. Presented at: *Animal Populations – World Resources and Animal Welfare, UFAW International Animal Welfare Science Symposium, 14–15th July 2015, Zagreb, Croatia*. Wheathampstead, UK: Universities Federation for Animal Welfare (UFAW), p. 51.

Draper, C. and Harris, S. (2012). The assessment of animal welfare in British zoos by government-appointed inspectors. *Animals*, **2**(4), 507–28. DOI: 10.3390/ani2040507.

Drews, B., Harmann, L.M., Beehler, L.L., *et al.* (2011). Ultrasonographic monitoring of fetal development in unrestrained bonobos (*Pan paniscus*) at the Milwaukee County Zoo. *Zoo Biology*, **30**(3), 241–53. DOI: 10.1002/zoo.20304.

Dunay, E., Apakupakul, K., Leard, S., Palmer, J.L. and Deem, S.L. (2018). Pathogen transmission from humans to great apes is a growing threat to primate conservation. *EcoHealth*, **15**(1), 148–62. DOI: 10.1007/s10393-017-1306-1.

Dunkley, J. and Whelan, T.A. (2006). Vicarious traumatisation: current status and future directions. *British Journal of Guidance & Counselling*, **34**(1), 107–16. DOI: 10.1080/03069880500483166.

Dunlop, A.L., Logue, K.M., Vaidyanathan, L. and Isakov, A.P. (2016). Facilitators and barriers for effective academic–community collaboration for disaster preparedness and response. *Journal of Public Health Management and Practice*, **22**(3), E20–28.

Dunn, A., Bergl, R., Byler, D., *et al.* (2014). *Revised Regional Action Plan for the Conservation of the Cross River Gorilla (Gorilla gorilla diehli) 2014–2019*. New York, NY: International Union for Conservation of Nature (IUCN) Species Survival Commission (SSC) Primate Specialist Group (PSG) and Wildlife Conservation Society (WCS). Available at: https://portals.iucn.org/library/node/44661.

Durham, D. (2015). The status of captive apes. In *State of the Apes: Industrial Agriculture and Ape Conservation*, ed. Arcus Foundation. Cambridge, UK: Cambridge University Press, pp. 228–59. Available at: http://www.stateoftheapes.com/themes/the-status-of-captive-apes/.

Durham, D. (2018). The status of captive apes: a statistical update. In *State of the Apes: Infrastructure Development and Ape Conservation*, ed. Arcus Foundation. Cambridge, UK: Cambridge University Press, pp. 255–63. Available at: https://www.stateoftheapes.com/volume-3-infrastructure-development/.

Durham, D. (2020). The status of captive apes: a statistical update. In *State of the Apes: Killing, Capture, Trade and Conservation*, ed. Arcus Foundation. Cambridge, UK: Cambridge University Press, pp. 255–62. Available at: https://www.stateoftheapes.com/themes/volume-4-chapter-8-campaign-for-nonhuman-rights-and-status-of-captive-apes/.

Durham, D. and Phillipson, A. (2014). Status of captive apes across Africa and Asia: the impact of extractive industry. In *State of the Apes: Extractive Industries and Ape Conservation*, ed. Arcus Foundation. Cambridge, UK: Cambridge University Press, pp. 279–305. Available at: http://www.stateoftheapes.com/volume-1-extractive-industries/.

EAGLE (2019). *The EAGLE Network Annual Report 2019*. Eco Activists for Governance and Law Enforcement (EAGLE). Available at: https://www.eagle-enforcement.org/data/files/eagle-network-annual-report-2019.pdf.

Earth Systems (2015). *Mako Gold Project Environmental and Social Impact Assessment Report*. Dakar, Senegal: Earth Systems. Available at: https://documents.pub/document/mako-gold-project-rmlcomau-mako-gold-project-esia-non-technical-summary-final.html?page=1.

Eberle, R., Black, D.H. and Hilliard, J.K. (1989). Relatedness of glycoproteins expressed on the surface of simian herpesvirus virions and infected cells to specific HSV glycoproteins. *Archives of Virology*, **109**(3–4), 233–52. DOI: 10.1007/BF01311084.

Eberle, R. and Jones-Engel, L. (2017). Understanding primate herpesviruses. *Journal of Emerging Diseases and Virology*, **3**(1). DOI: 10.16966/2473-1846.127.

Edes, A.N. (2018). *Assessing long-term stress in great apes: allostatic load in western lowland gorillas (Gorilla gorilla gorilla)*. Doctoral thesis. Columbus, OH: The Ohio State University.

Edes, A.N., Edwards, K.L., Wolfe, B.A., Brown, J.L. and Crews, D.E. (2020). Allostatic load indices with cholesterol and triglycerides predict disease and mortality risk in zoo-housed western lowland gorillas (*Gorilla gorilla gorilla*). *Biomarker Insights*, **15**, 1177271920914585. DOI: 10.1177/1177271920914585.

Edes, A.N., Wolfe, B.A. and Crews, D.E. (2018). Evaluating allostatic load: a new approach to measuring long-term stress in wildlife. *Journal of Zoo and Wildlife Medicine*, **49**(2), 272–82, 11. DOI: 10.1638/2016-0070.1.

Edwards, D.P., Sloan, S., Weng, L., *et al.* (2014). Mining and the African environment. *Conservation Letters*, **7**(3), 302–11. DOI: 10.1111/conl.12076.

Edwards, S.J.L., Chatterjee, H.J. and Santini, J.M. (2021). Anthroponosis and risk management: a time for ethical vaccination of wildlife? *The Lancet Microbe*, **2**(6), e230–31. DOI: 10.1016/S2666-5247(21)00081-1.

Edwards, S.J.L., Norell, C.H., Illari, P., Clarke, B. and Neuhaus, C.P. (2018). A radical approach to Ebola: saving humans and other animals. *American Journal of Bioethics*, **18**(10), 35–42. DOI: 10.1080/15265161.2018.1513584.

Elder, A.A. (2009). Hylobatid diets revisited: the importance of body mass, fruit availability, and interspecific competition. In *The Gibbons: New Perspectives on Small Ape Socioecology and Population Biology*, ed. D. Whittaker and S. Lappan. New York, NY: Springer, pp. 133–59. DOI: 10.1007/978-0-387-88604-6_8.

Elichai, A. (2018). How big data can help in disaster response. *Scientific American Observations*, December 13, 2018. Available at: https://blogs.scientificamerican.com/observations/how-big-data-can-help-in-disaster-response/.

Ellwanger, J.H. and Chies, J.A.B. (2021). Zoonotic spillover: understanding basic aspects for better prevention. *Genetics and Molecular Biology*, **44**(1, S1), e20200355. DOI: 10.1590/1678-4685-GMB-2020-0355.

Ely, J.J., Bishop, M.A., Lammey, M.L., *et al.* (2010). Use of biomarkers of collagen types I and III fibrosis metabolism to detect cardiovascular and renal disease in chimpanzees (*Pan troglodytes*). *Comparative Medicine*, **60**(2), 154–8.

Ely, J.J., Zavaskis, T. and Lammey, M.L. (2013). Hypertension increases with aging and obesity in chimpanzees (*Pan troglodytes*). *Zoo Biology*, **32**(1), 79–87.

Emery Thompson, M., Jones, J.H., Pusey, A.E., *et al.* (2007). Aging and fertility patterns in wild chimpanzees provide insights into the evolution of menopause. *Current Biology*, **17**(24), 2150–6. DOI: 10.1016/j.cub.2007.11.033.

Emery Thompson, M., Muller, M.N., Machanda, Z.P., Otali, E. and Wrangham, R.W. (2020). The Kibale Chimpanzee Project: over thirty years of research, conservation, and change. *Biological Conservation*, **252**, 108857. DOI: 10.1016/j.biocon.2020.108857.

Emery Thompson, M. and Wrangham, R.W. (2008). Diet and reproductive function in wild female chimpanzees (*Pan troglodytes schweinfurthii*) at Kibale National Park, Uganda. *American Journal of Physical Anthropology*, **135**(2), 171–81. DOI: 10.1002/ajpa.20718.

Emery Thompson, M. and Wrangham, R.W. (2013). *Pan troglodytes* robust chimpanzee. In *Mammals of Africa. Volume II: Primates*, ed. T. M. Butynski, J. Kingdon and J. Kalina. London, UK: Bloomsbury Publishing, pp. 55–64.

Emery Thompson, M., Zhou, A. and Knott, C.D. (2012). Low testosterone correlates with delayed development in male orangutans. *PLoS ONE*, **7**(10), e47282. DOI: 10.1371/journal.pone.0047282.

Emmons, R.W. and Lennette, E.H. (1970). Natural herpesvirus hominis infection of a gibbon (*Hylobates lar*). *Archiv für die gesamte Virusforschung*, **31**(3), 215–18. DOI: 10.1007/BF01253755.

Endangered Asian Species Trust (2020a). Endangered Asian Species Trust. *Facebook Post*, September 22, 2020. Available at: https://www.facebook.com/permalink.php?story_fbid=pfbidorcrtXsmP3mQRoWgqC29QB1fr7KmUfbKvytBjRuGeREHQs0q92imk59GML8Sa615Vl&id=164617243557223.

Endangered Asian Species Trust (2020b). Endangered Asian Species Trust 2. *Facebook Post*, September 29, 2020. Available at: https://www.facebook.com/permalink.php?story_fbid=pfbid02ZHh1AXvejtvbbzQdpw1mgM8tonVDYdoUWm6UKYP3GHkmAh1WR8EnM1ThYkjcNTMdl&id=164617243557223.

Endangered Asian Species Trust (2020c). Golden-cheeked gibbon rehabilitation. *Facebook Post*, September 9, 2020. Available at: https://www.facebook.com/164617243557223/videos/316142682948849/.

Engelman, D., Yoshizumi, J., Hay, R.J., *et al.* (2020). The 2020 International Alliance for the Control of Scabies consensus criteria for the diagnosis of scabies. *British Journal of Dermatology*, **183**(5), 808–20. DOI: 10.1111/bjd.18943.

English, P. and Ahebwa, W.M. (2018). *How can Tourism become a Driver of Economic Growth in Uganda?* Prepared for the 2018 Economic Growth Forum and National Budget Conference. S-43437-UGA-1. London, UK: International Growth Centre (IGC).

ENISA (n.d.). *Risk Treatment*. Attiki, Greece: European Union Agency for Cybersecurity (ENISA). Available at: https://www.enisa.europa.eu/topics/threat-risk-management/risk-management/current-risk/risk-management-inventory/rm-process/risk-treatment. Accessed: July, 2022.

Equator Principles (2020). *Equator Principles EP4*. Equator Principles Association. Available at: https://equator-principles.com/app/uploads/The-Equator-Principles_EP4_July2020.pdf.

Erb, W.M., Barrow, E.J., Hofner, A.N., Utami-Atmoko, S.S. and Vogel, E.R. (2018). Wildfire smoke impacts activity and energetics of wild Bornean orangutans. *Scientific Reports*, **8**, 7606. DOI: 10.1038/s41598-018-25847-1.

Erdős, L. (2019). No one loved gorillas more – the life and legacy of Diane Fossey. In *Green Heroes: From Buddha to Leonardo DiCaprio*, ed. L. Erdős. Cham, Switzerland: Springer International Publishing, pp. 117–23. DOI: 10.1007/978-3-030-31806-2_24.

Errecaborde, K.M., Rist, C., Travis, D.A., *et al.* (2019). Evaluating One Health: the role of team science in multisectoral collaboration. *Revue Scientifique et Technique de l'Office International des Épizooties*, **38**(1), 279–89. DOI: 10.20506/rst.38.1.2960.

Estienne, V. (2022). *Nouabalé-Ndoki National Park. A.P.E.S. Wiki*. Munich, Germany: Max Planck Society for the Advancement of Science e.V. Available at: https://wiki.iucnapesportal.org/index.php/Nouabal%C3%A9-Ndoki_National_Park.

Estrada, A. (2013). Socioeconomic contexts of primate conservation: population, poverty, global economic demands, and sustainable land use. *American Journal of Primatology*, **75**(1), 30–45. DOI: 10.1002/ajp.22080.

Estrada, A. and Garber, P.A. (2022). Principal drivers and conservation solutions to the impending primate extinction crisis: introduction to the special issue. *International Journal of Primatology*, **43**(1), 1–14. DOI: 10.1007/s10764-022-00283-1.

Estrada, A., Garber, P.A., Mittermeier, R.A., *et al.* (2018). Primates in peril: the significance of Brazil, Madagascar, Indonesia and the Democratic Republic of the Congo for global primate conservation. *PeerJ*, **6**, e4869. DOI: 10.7717/peerj.4869.

Estrada, A., Garber, P.A., Rylands, A.B., *et al.* (2017). Impending extinction crisis of the world's primates: why primates matter. *Science Advances*, **3**(1), e1600946. DOI: 10.1126/sciadv.1600946.

Etieyibo, E. (2017). Ubuntu and the environment. In *The Palgrave Handbook of African Philosophy*, ed. A. Afolayan and T. Falola. New York, NY: Palgrave Macmillan US, pp. 633–57. DOI: 10.1057/978-1-137-59291-0_41.

European Commission (2021). *DG ECHO Guidance Note Disaster Preparedness*. Brussels, Belgium: European Commission. Available at: https://ec.europa.eu/echo/files/policies/sectoral/dg_echo_guidance_note_-_disaster_preparedness_en.pdf.

European Commission (n.d.). *Economics for Disaster Prevention and Preparedness*. Brussels, Belgium: European Commission. Available at: https://civil-protection-humanitarian-aid.ec.europa.eu/what/civil-protection/european-disaster-risk-management/economics-disaster-prevention-and-preparedness_en. Accessed: July, 2022.

Evans, T., Wingard, J. and Humle, T. (2021). The mitigation hierarchy in environmental impact assessment and related legislation as a tool for species conservation: a case study of western chimpanzees and mining development. *Biological Conservation*, **261**, 109237. DOI: 10.1016/j.biocon.2021.109237.

Eze, M. (2010). *Intellectual History in Contemporary South Africa*. London, UK: Palgrave MacMillan.

Fan, P.-F. (2017). The past, present, and future of gibbons in China. *Biological Conservation*, **210**, 29–39. DOI: 10.1016/j.biocon.2016.02.024.

Fan, P.-F., Fei, H., Xiang, Z., *et al.* (2010). Social structure and group dynamics of the Cao Vit gibbon (*Nomascus nasutus*) in Bangliang, Jingxi, China. *Folia Primatologica*, **81**(5), 245–53.

Fan, P.-F., He, K., Chen, X., *et al.* (2017). Description of a new species of hoolock gibbon (Primates: Hylobatidae) based on integrative taxonomy. *American Journal of Primatology*, **79**(5), e22631. DOI: 10.1002/ajp.22631.

Fan, P.-F. and Jiang, X.-L. (2008). Effects of food and topography on ranging behavior of black crested gibbon (*Nomascus concolor jingdongensis*) in Wuliang Mountain, Yunnan, China. *American Journal of Primatology*, **70**(9), 871–8. DOI: 10.1002/ajp.20577.

Fan, P.-F. and Jiang, X.-L. (2010). Maintenance of multifemale social organization in a group of *Nomascus concolor* at Wuliang Mountain, Yunnan, China. *International Journal of Primatology*, **31**(1), 1–13. DOI: 10.1007/s10764-009-9375-9.

Fan, P.-F., Turvey, S.T. and Bryant, J.V. (2020). Hoolock tianxing *(amended version of 2019 assessment). The IUCN Red List of Threatened Species 2020: e.T118355648A166597159*. Gland, Switzerland: International Union for Conservation of Nature (IUCN). DOI: 10.2305/IUCN.UK.2020-1.RLTS.T118355648A166597159.en.

FAO (2018). *The State of Food Security and Nutrition in the World 2018*. Rome, Italy: Food and Agriculture Organization of the United Nations (FAO). Available at: https://www.fao.org/3/I9553EN/i9553en.pdf.

FAO (n.d.-a). *Conservation Agriculture*. Rome, Italy: Food and Agriculture Organization of the United Nations (FAO). Available at: http://www.fao.org/conservation-agriculture/overview/what-is-conservation-agriculture/en/. Accessed: August, 2021.

FAO (n.d.-b). *Towards a New Green Revolution*. Rome, Italy: Food and Agriculture Organization of the United Nations (FAO). Available at: https://www.fao.org/3/x0262e/x0262e06.htm. Accessed: August, 2021.

FAO and NACA (2001). Contingency planning. In *Manual of Procedures for the Implementation of the Asia Regional Technical Guidelines on Health Management for the Responsible Movement of Live Aquatic Animals*. FAO Fisheries Technical Paper No. 402, Supplement 1, ed. Food and Agriculture Organization of the United Nations (FAO) and Network of Aquaculture Centres in Asia-Pacific (NACA). Rome, Italy: FAO, pp. 53–62. Available at: http://www.fao.org/3/y1238e/y1238e09.pdf.

Farmer, K.H. (2002). Pan-African Sanctuary Alliance: status and range of activities for great ape conservation. *American Journal of Primatology*, **58**(3), 117–32. DOI: 10.1002/ajp.10054.

Farmer, K.H. (2012). *Building Sustainable Sanctuaries*. Cambridge, UK: Arcus Foundation. Available at: http://www.sanctuaryfederation.org/gfas/wp-content/uploads/2013/09/Arcus_Building_Sustainable_Sanctuaries.pdf.

Farmer, K.H. (2018). *Compassionate Conservation Organizations: Challenges, Priorities and Recommended Action*. Unpublished report prepared for the Arcus Foundation.

Farrell, M., Rando, C. and Garrod, B. (2015). Lessons from the past: metabolic bone disease in historical captive primates. *International Journal of Primatology*, **36**(2), 398–411. DOI: 10.1007/s10764-015-9831-7.

Fasina, F.O., Fasanmi, O.G., Makonnen, Y.J., *et al.* (2021). The One Health landscape in Sub-Saharan African countries. *One Health*, **13**, 100325. DOI: 10.1016/j.onehlt.2021.100325.

Fauna Silvestre de Nicaragua (2020). Fauna Silvestre de Nicaragua. *Facebook Post*, April 12, 2020. Available at: https://www.facebook.com/eduszoo/videos/este-es-pipo-el-chimpance-del-zoo-nicaragua-es-un-gran-amigo-y-le-gusta-jugar-mu/215515939731881/.

Faust, C.L., McCallum, H.I., Bloomfield, L.S.P., *et al.* (2018). Pathogen spillover during land conversion. *Ecology Letters*, **21**(4), 471–83. DOI: 10.1111/ele.12904.

Faust, L.J., Cress, D., Farmer, K.H., Ross, S.R. and Beck, B.B. (2011). Predicting capacity demand on sanctuaries for African chimpanzees (*Pan troglodytes*). *International Journal of Primatology*, **32**(4), 849–64. DOI: 10.1007/s10764-011-9505-z.

Federer, K., Armua-Fernandez, M.T., Gori, F., *et al.* (2016). Detection of taeniid (*Taenia* spp., *Echinococcus* spp.) eggs contaminating vegetables and fruits sold in European markets and the risk for metacestode infections in captive primates. *International Journal for Parasitology: Parasites and Wildlife*, **5**(3), 249–53. DOI: 10.1016/j.ijppaw.2016.07.002.

Fedigan, L.M. (2010). Ethical issues faced by field primatologists: asking the relevant questions. *American Journal of Primatology*, **72**(9), 754–71. DOI: 10.1002/ajp.20814.

Feinberg, J. (1974). The rights of animals and unborn generations. In *Philosophy and Environmental Crisis*, ed. W. T. Blackstone. Athens, GA: University of Georgia Press, pp. 43–68.

Feldmann, H. and Geisbert, T.W. (2011). Ebola haemorrhagic fever. *The Lancet*, **377**(9768), 849–62. DOI: 10.1016/s0140-6736(10)60667-8.

FEMA (2017). *National Incident Management System*. Washington DC: Federal Emergency Management Agency (FEMA), US Department of Homeland Security. Available at: https://www.fema.gov/sites/default/files/2020-07/fema_nims_doctrine-2017.pdf.

Ferber, D. (2000). Human diseases threaten great apes. *Science*, **289**(5483), 1277–8. DOI: 10.1126/science.289.5483.1277.

Ferdowsian, H. (2020). The right to bodily sovereignty and its importance to mental and physical well-being. In *Neuroethics and Nonhuman Animals*, ed. L. S. M. Johnson, A. Fenton and A. Shriver. Cham, Switzerland: Springer International Publishing, pp. 255–70. DOI: 10.1007/978-3-030-31011-0_15.

Ferdowsian, H., Durham, D.L., Kimwele, C., *et al.* (2011). Signs of mood and anxiety disorders in chimpanzees. *PLoS ONE*, **6**(6), e19855. DOI: 10.1371/journal.pone.0019855.

Ferdowsian, H. and Fuentes, A. (2014). Harms and deprivation of benefits for nonhuman primates in research. *Theoretical Medicine and Bioethics*, **35**(2), 143–56. DOI: 10.1007/s11017-014-9288-2.

Ferdowsian, H., Johnson, L.S.M., Johnson, J., *et al.* (2020). A Belmont report for animals? *Cambridge Quarterly of Healthcare Ethics*, **29**(1), 19–37. DOI: 10.1017/S0963180119000732.

Fernandez-Duque, E. and Rotundo, M. (2003). Field methods for capturing and marking azarai night monkeys. *International Journal of Primatology*, **24**(5), 1113–20. DOI: 10.1023/A:1026284430453.

Fernie, A.C. (2008). *The creation and implementation of a great ape welfare index.* Doctor of Philosophy thesis. Brisbane, Australia: The University of Queensland.

Fernie, A.C., Tribe, A., Murray, P.J., Lisle, A. and Phillips, C.J.C. (2012). A survey of the attitudes of stakeholders in the zoo industry towards the husbandry requirements of captive great apes. *Animal Welfare*, **21**(2), 233–45. DOI: 10.7120/09627286.21.2.233.

Ferrie, G.M., Farmer, K.H., Kuhar, C.W., *et al.* (2014). The social, economic, and environmental contributions of Pan African Sanctuary Alliance primate sanctuaries in Africa. *Biodiversity and Conservation*, **23**(1), 187–201. DOI: 10.1007/s10531-013-0592-3.

Ferris, R.L., Ali, I.K.M. and West, G.D. (2021). Use of a human indirect immunofluorescence antibody assay for *Balamuthia mandrillaris* in a group of captive northwest Bornean orangutans (*Pongo pygmaeus pygmaeus*). *Journal of Zoo and Wildlife Medicine*, **52**, 310–14. DOI: 10.1638/2019-0018.

Figley, C.R. (1995). *Compassion Fatigue: Coping with Secondary Traumatic Stress Disorder in Those Who Treat the Traumatized.* Brunner/Mazel Psychological Stress Series No. 23. Philadelphia, PA: Brunner/Mazel.

Filippone, C., Betsem, E., Tortevoye, P., *et al.* (2015). A severe bite from a nonhuman primate is a major risk factor for HTLV-1 infection in hunters from Central Africa. *Clinical Infectious Diseases*, **60**(11), 1667–76. DOI: 10.1093/cid/civ145.

Finley, N. (2019). Out on a limb: unlikely collaboration boosts orangutans in Borneo. *Mongabay Series: Great Apes*, June 12, 2019. Available at: https://news.mongabay.com/2019/06/out-on-a-limb-unlikely-collaboration-boosts-orangutans-in-borneo/.

Fischer, C.P. and Romero, L.M. (2019). Chronic captivity stress in wild animals is highly species-specific. *Conservation Physiology*, **7**(1), coz093. DOI: 10.1093/conphys/coz093.

Fischer, J. and Lindenmayer, D.B. (2000). An assessment of the published results of animal relocations. *Biological Conservation*, **96**(1), 1–11. DOI: 10.1016/S0006-3207(00)00048-3.

Fitzgerald, K. (2022). *Mountain Gorilla Tourism Drives Economic Growth And Conservation.* Nairobi, Kenya: African Wildlife Foundation (AWF) Conservation Centre. Available at: https://www.awf.org/blog/mountain-gorilla-tourism-drives-economic-growth-and-conservation.

Fleury, E. (2017). Money for monkeys, and more: ensuring sanctuary retirement of nonhuman primates. *Animal Studies Journal*, **6**(2), 30–54. DOI: https://ro.uow.edu.au/asj/vol6/iss2/4.

Foitová, I., Civáňová, K., Baruš, V. and Nurcahyo, W. (2014). Phylogenetic relationships between pinworms (Nematoda: Enterobiinae) parasitising the critically endangered orang-utan, according to the characterisation of molecular genomic and mitochondrial markers. *Parasitology Research*, **113**(7), 2455–66. DOI: 10.1007/s00436-014-3892-y.

Foitová, I., Koubková, B., Barus, V. and Nurcahyo, W. (2008). Presence and species identification of the gapeworm *Mammomonogamus laryngeus* (Railliet, 1899) (Syngamidae: Nematoda) in a semi-wild population of Sumatran orangutan (*Pongo abelii*) in Indonesia. *Research in Veterinary Science*, **84**(2), 232–6. DOI: 10.1016/j.rvsc.2007.04.021.

Fontseré, C., Frandsen, P., Hernández-Rodríguez, J., *et al.* (2021). The genetic impact of an Ebola outbreak on a wild gorilla population. *BMC Genomics*, **22**(1), 735. DOI: 10.1186/s12864-021-08025-y.

Formenty, P.B.H., Boesch, C., Wyers, M., *et al.* (1999). Ebola virus outbreak among wild chimpanzees living in a rain forest of Côte d'Ivoire. *Journal of Infectious Diseases*, **179** (S1), S120–6. DOI: 10.1086/514296.

Fort Worth Zoo (2020). Fort Worth Zoo. *Facebook Post*, June 30, 2020. Available at: https://www.facebook.com/FortWorthZoo/posts/we-have-10-bonobos-at-zoo-i-will-give-you-a-brief-introduction-from-oldest-to-yo/10158320883452095/.

Four Paws International (2020a). *Annual Report 2020.* Vienna, Austria: Four Paws International. Available at: https://media.4-paws.org/e/3/0/e/e30e825655512c7a48099a8290231844d4a13af3/210811_FOURPAWSAnnualReport2020.pdf.

Four Paws International (2020b). Circuses without wild animals. *Four Paws International*, March 21, 2023. Available at: https://www.four-paws.org/campaigns-topics/topics/wild-animals/worldwide-circus-bans.

Fourie, C. (2017). Who is experiencing what kind of moral distress? Distinctions for moving from a narrow to a broad definition of moral distress. *AMA Journal of Ethics*, **19**(6), 578–84. DOI: 10.1001/journalofethics.2017.19.6.nlit1-1706.

Fraser, D. (2009). Assessing animal welfare: different philosophies, different scientific approaches. *Zoo Biology*, **28**(6), 507–18. DOI: 10.1002/zoo.20253.

Fraser, D. (2010). Toward a synthesis of conservation and animal welfare science. *Animal Welfare*, **19**(2), 121–4. DOI: 10.1017/S0962728600001378.

Freeland, W.J. (1976). Pathogens and the evolution of primate sociality. *Biotropica*, **8**, 12–24.

Freund, C., Rahman, E. and Knott, C. (2017). Ten years of orangutan-related wildlife crime investigation in West Kalimantan, Indonesia. *American Journal of Primatology*, **79**(11), 22620. DOI: 10.1002/ajp.22620.

Fribourg-Blanc, A. and Mollaret, H.H. (1969). Natural treponematosis of the African primate. *Primates in Medicine*, **3**(0), 113–21.

Fribourg-Blanc, A., Mollaret, H.H. and Niel, G. (1966). [Serologic and microscopic confirmation of treponemosis in Guinea baboons]. *Bulletin de la Société de Pathologie Exotique et de ses Filiales*, **59**(1), 54–9.

Friend, M., Hurley, J.W., Nol, P. and Wesenberg, K. (2006). *Disease Emergence and Resurgence – The Wildlife–Human Connection*. Circular 1285. Reston, VA: US Geological Survey. DOI: 10.3133/cir1285.

Friends of Animals (n.d.). *Chimpanzee Rehabilitation Project River Gambia National Park Visitor Information Sheet*. Darien, CT: Friends of Animals. Available at: https://www.friendsofanimals.org/program/visit-the-river-gambia-national-park/. Accessed: October, 2020.

Fröhlich, M., Kunz, J.A., Fryns, C., *et al.* (2020). Social interactions and interaction partners in infant orang-utans of two wild populations. *Animal Behaviour*, **166**, 183–91.

Fruth, B., Hickey, J.R., André, C., *et al.* (2016). Pan paniscus *(errata version published in 2016)*. The IUCN Red List of Threatened Species 2016: e.T15932A102331567. Gland, Switzerland: International Union for Conservation of Nature (IUCN). DOI: 10.2305/IUCN.UK.2016-2.RLTS.T15932A17964305.en.

Fruth, B. and Hohmann, G. (1996). Nest building behavior in the great apes: the great leap forward? In *Great Ape Societies*, ed. W. McGrew, L. Marchant and T. Nishida. Cambridge, UK: Cambridge University Press, pp. 225–40. DOI: 10.1017/CBO9780511752414.019.

Fruth, B., Tagg, N. and Stewart, F. (2018). Sleep and nesting behavior in primates: a review. *American Journal of Physical Anthropology*, **166**(3), 499–509. DOI: 10.1002/ajpa.23373.

Fruth, B., Williamson, E.A. and Richardson, M.C. (2013). Bonobo *Pan paniscus*. In *Handbook of the Mammals of the World. Volume 3: Primates*, ed. R. A. Mittermeier, A. B. Rylands and D. E. Wilson. Barcelona, Spain: Lynx Edicions, pp. 853–4.

FSC (2019). *PSU Review Report of FSC-STD-01-001 FSC Principles and Criteria for Forest Stewardship*. Bonn, Germany: Forest Stewardship Council (FSC) International Center. Available at: https://connect.fsc.org/sites/default/files/2019-10/Review%20report%20FSC-STD-01-001.pdf.

FSC (2023). *FSC Principles and Criteria for Forest Stewardship. FSC-STD-01-001 V5-3*. Bonn, Germany: Forest Stewardship Council (FSC). Available at: https://connect.fsc.org/document-centre/documents/resource/392.

FSC (n.d.). *Our History*. London, UK: Forest Stewardship Council (FSC). Available at: https://fsc.org/en/our-history. Accessed: December, 2022.

Fujita, S. (2011). Health monitoring. In *The Chimpanzees of Bossou and Nimba*, ed. T. Matsuzawa, T. Humle and Y. Sugiyama. Tokyo, Japan: Springer, pp. 353–9. DOI: 10.1007/978-4-431-53921-6_37.

Fuller, G., Margulis, S.W. and Santymire, R.M. (2011). The effectiveness of indigestible markers for identifying individual animal feces and their prevalence of use in North American zoos. *Zoo Biology*, **30**(4), 379–98.

Furuichi, T. (2009). Factors underlying party size differences between chimpanzees and bonobos: a review and hypotheses for future study. *Primates*, **50**(3), 197–209. DOI: 10.1007/s10329-009-0141-6.

Furuichi, T., Hashimoto, C., Idani, G., *et al.* (1999). Current situation of studies of bonobos (*Pan paniscus*) at Wamba, D.R. Congo. *Primate Research*, **15**(2), 115–27. DOI: 10.2354/psj.15.115.

FVE (n.d.). *Code of Good Veterinary Practice*. Brussels, Belgium: Federation of Veterinarians of Europe (FVE). Available at: https://www.eesc.europa.eu/sites/default/files/resources/docs/130-private-act.pdf. Accessed: February, 2021.

G1 (2020). Aos 57 anos, chimpanzé Bob recebe novo companheiro de recinto no Zoo de Curitiba. *G1*, April 2, 2020. Available at: https://g1.globo.com/pr/parana/noticia/2020/04/02/chimpanze-bob-recebe-novo-companheiro-de-recinto-no-zoo-de-curitiba.ghtml.

GADM (n.d.). *GADM Maps and Data*. Available at: https://gadm.org/index.html. Accessed: October, 2021.

Gagliardi, A., Totino, V., Cacciotti, F., *et al.* (2018). Rebuilding the gut microbiota ecosystem. *International Journal of Environmental Research and Public Health*, **15**(8), 1679. DOI: 10.3390/ijerph15081679.

GAHP (n.d.). *Anesthesia Info.* Royal Oak, MI: Great Ape Heart Project (GAHP), Detroit Zoological Society. Available at: https://greatapeheartproject.org/resources/anesthesia-info/. Accessed: December, 2020.

GAIN (n.d.). *Great Ape Information Network.* Kyoto, Japan: National BioResource Project (NBRP)—Great Ape Information Network (GAIN). Available at: http://www.shigen.nig.ac.jp/gain/index.jsp. Accessed: October, 2020.

GAL (2018). *UN Convention of Animal Health and Protection (UNCAHP). First Pre-Draft of the Global Animal Welfare Law Association August 23rd 2018.* Zürich, Switzerland: Global Animal Law (GAL) Association. Available at: https://www.globalanimallaw.org/downloads/Folder-UNCAHP.pdf.

Gamble, K.C., North, M.C.K., Backues, K. and Ross, S.R. (2004). Pathologic review of the chimpanzee (*Pan troglodytes*): 1990–2003. Presented at: *Proceedings of the Annual Meeting of the American Association of Zoo Veterinarians, San Diego, CA, 28 August–3 September 2004.* Jacksonville, FL: American Association of Zoo Veterinarians.

Game, E.T., Meijaard, E., Sheil, D. and McDonald-Madden, E. (2014). Conservation in a wicked complex world; challenges and solutions. *Conservation Letters*, **7**(3), 271–7. DOI: 10.1111/conl.12050.

Ganas, J., Robbins, M.M., Nkurunungi, J.B., Kaplin, B.A. and McNeilage, A. (2004). Dietary variability of mountain gorillas in Bwindi Impenetrable National Park, Uganda. *International Journal of Primatology*, **25**(5), 1043–72. DOI: 10.1023/b:ijop.0000043351.20129.44.

Gaskin, J.M. (2022). Encephalomyocarditis virus infection in animals. In *MSD Manual Veterinary Manual*, ed. MSD. Rahway, NJ: Merck & Co. Inc (MSD). Available at: https://www.msdvetmanual.com/generalized-conditions/encephalomyocarditis-virus-infection/encephalomyocarditis-virus-infection-in-animals.

GATO (2020). Historical achievements of GATO and its following objective. *GATO News*, October 26, 2020. Available at: https://gyvunuapsauga.lt/en/news/historical-achievements-of-gato-and-its-following-objectives/.

Gaveau, D.L.A., Sloan, S., Molidena, E., *et al.* (2014). Four decades of forest persistence, clearance and logging on Borneo. *PLoS ONE*, **9**(7), e101654. DOI: 10.1371/journal.pone.0101654.

GCC (n.d.). *GCC Gibbons.* Santa Clarita, CA: Gibbon Conservation Center (GCC). Available at: https://www.gibboncenter.org/list-of-gcc-gibbons.html. Accessed: October, 2020.

GDPC (n.d.). *Early Warning Systems.* Washington DC: Global Disaster Preparedness Center (GDPC). Available at: https://preparecenter.org/topic/early-warning-systems/#:~:text=Early%20warning%20system%20%E2%80%93%20The%20set,possibility%20of%20harm%20or%20loss. Accessed: July, 2022.

Geissmann, T. (1991). Reassessment of age of sexual maturity in gibbons (*Hylobates* spp.). *American Journal of Primatology*, **23**(1), 11–22. DOI: 10.1002/ajp.1350230103.

Geissmann, T., Grindley, M., Ngwe, L., *et al.* (2013). *The Conservation Status of Hoolock Gibbons in Myanmar.* Zürich, Switzerland: Gibbon Conservation Alliance. Available at: http://www.gibbonconservation.org/07_publications/book/2013_hoolock_myanmar.pdf.

Genton, C.l., Cristescu, R.H., Gatti, S., *et al.* (2017). Using demographic characteristics of populations to detect spatial fragmentation following suspected Ebola outbreaks in great apes. *American Journal of Physical Anthropology*, **164**(1), 3–10.

Genton, C.l., Pierre, A., Cristescu, R.H., *et al.* (2015). How Ebola impacts social dynamics in gorillas: a multistate modelling approach. *Journal of Animal Ecology*, **84**(1), 166–76.

Georges, A.-J., Leroy, E.M., Renaut, A., *et al.* (1999). Ebola hemorrhagic fever outbreaks in Gabon, 1994–1997: epidemiologic and health control issues. *Journal of Infectious Diseases*, **179**, S65–75.

Georges-Courbot, M.C., Sanchez, A.J., Lu, C.Y., *et al.* (1997). Isolation and phylogenetic characterization of Ebola viruses causing different outbreaks in Gabon. *Emerging Infectious Diseases*, **3**(1), 59–62.

Gevers, D., Kugathasan, S., Denson, Lee A., *et al.* (2014). The treatment-naive microbiome in new-onset Crohn's disease. *Cell Host & Microbe*, **15**(3), 382–92. DOI: 10.1016/j.chom.2014.02.005.

GFAS (2017). *Contingency Planning for Sanctuaries and Rehabilitation Centers.* Phoenix, AZ: Global Federation of Animal Sanctuaries (GFAS). Available at: https://sanctuaryfederation.org/webinars/contingency-planning-for-sanctuaries-and-rehabilitation-centers/.

GFAS (2019). *Gorilla Rehabilitation and Conservation Education (GRACE) Center.* Phoenix, AZ: Global Federation of Animal Sanctuaries (GFAS). Available at: https://sanctuaryfederation.org/sanctuaries/gorilla-rehabilitation-and-conservation-education-center-grace/.

GFAS (2020). *Recipients of the 2020 Carole Noon and Outstanding Sanctuary Awards Announced.* Phoenix, AZ: Global Federation of Animal Sanctuaries (GFAS). Available at: https://www.sanctuaryfederation.org/2020/10/23/recipients-of-the-2020-carole-noon-and-outstanding-sanctuary-awards-announced/.

GFAS (2022). *Standards Appendix for Ape Sanctuaries.* Phoenix, AZ: Global Federation of Animal Sanctuaries (GFAS). Available at: https://sanctuaryfederation.org/wp-content/uploads/2023/02/Ape-Standards-Appendix-2022.pdf.

GFAS (n.d.). *Find a Sanctuary.* Phoenix, AZ: Global Federation of Animal Sanctuaries (GFAS). Available at: https://sanctuaryfederation.org/find-a-sanctuary/. Accessed: December, 2022.

GHSA (2020). *Turning Crisis to Opportunities for Workforce Development.* Global Health Security Agenda (GHSA).

Giannetti, B.F., Agostinho, F., Almeida, C.M.V.B. and Huisingh, D. (2015). A review of limitations of GDP and alternative indices to monitor human wellbeing and to manage eco-system functionality. *Journal of Cleaner Production*, **87**, 11–25. DOI: 10.1016/j.jclepro.2014.10.051.

Gibbon Rehabilitation Project (n.d.). *Gibbon Sponsorship Program.* Phuket, Thailand: The Wild Animal Rescue Foundation of Thailand (WARF). Available at: https://www.gibbonproject.org/gibbon-sponsorship-program/. Accessed: October, 2020.

Gibbons, A. (2020). Ape researchers mobilize to save primates from coronavirus. *Science*, **368**(6491), 566. DOI: 10.1126/science.368.6491.566-a.

Gibbs, E.P.J. (2014). The evolution of One Health: a decade of progress and challenges for the future. *Veterinary Record*, **174**(4), 85–91. DOI: 10.1136/vr.g143.

GIBOP (2019). *Global Inventory of Biodiversity Offset Policies (GIBOP).* Gland, Switzerland: International Union for Conservation of Nature (IUCN). Available at: https://portals.iucn.org/offsetpolicy/.

Gibson, M. (2011). The universal declaration of animal welfare. *Deakin Law Review*, **16**(2), 539–67. DOI: 10.21153/dlr2011vol16no2art112.

Gilardi, K.V., Gillespie, T.R., Leendertz, F.H., *et al.* (2015). *Best Practice Guidelines for Health Monitoring and Disease Control in Great Ape Populations.* Gland, Switzerland: International Union for Conservation of Nature (IUCN) Species Survival Commission (SSC) Primate Specialist Group (PSG). Available at: https://www.iucngreatapes.org/health-monitoring-and-disease-prevention.

Gilardi, K.V., Nziza, J., Ssebide, B., *et al.* (2022). Endangered mountain gorillas and COVID-19: One Health lessons for prevention and preparedness during a global pandemic. *American Journal of Primatology*, **84**(4–5), e23291. DOI: 10.1002/ajp.23291.

Gilardi, K.V., Oxford, K.L., Gardner-Roberts, D., *et al.* (2014). Human herpes simplex virus type 1 in confiscated gorilla. *Emerging Infectious Diseases*, **20**(11), 1883–6. DOI: 10.3201/eid2011.140075.

Gilardi, K.V. and Uwingeli, P. (2022). Keep mountain gorillas free from pandemic virus. *Nature*, **602**(7896), 211. DOI: 10.1038/d41586-022-00331-z.

Gill, V. (2017). Endangered apes saved from pet trade. *BBC News*, November 7, 2017. Available at: https://www.bbc.com/news/science-environment-41767347.

Gillespie, T.R. (2019). Guest editorial: Protecting wild primates during the novel coronavirus pandemic and beyond. *Asian Primates Journal*, **8**(1), 1.

Gillespie, T.R. and Chapman, C.A. (2006). Prediction of parasite infection dynamics in primate metapopulations based on attributes of forest fragmentation. *Conservation Biology*, **20**(2), 441–8.

Gillespie, T.R. and Chapman, C.A. (2008). Forest fragmentation, the decline of an endangered primate, and changes in host–parasite interactions relative to an unfragmented forest. *American Journal of Primatology*, **70**(3), 222–30. DOI: 10.1002/ajp.20475.

Gillespie, T.R., Chapman, C.A. and Greiner, E.C. (2005). Effects of logging on gastrointestinal parasite infections and infection risk in African primates. *Journal of Applied Ecology*, **42**(4), 699–707. DOI: 10.1111/j.1365-2664.2005.01049.x.

Gillespie, T.R., Jones, K.E., Dobson, A.P., Clennon, J.A. and Pascual, M. (2021). COVID-clarity demands unification of health and environmental policy. *Global Change Biology*, **27**(7), 1319–21. DOI: 10.1111/gcb.15508.

Gillespie, T.R. and Leendertz, F.H. (2020). COVID-19: protect great apes during human pandemics. *Nature*, **579**(7800), 497. DOI: 10.1038/d41586-020-00859-y.

Gillespie, T.R., Lonsdorf, E.V., Canfield, E.P., *et al.* (2010). Demographic and ecological effects on patterns of parasitism in eastern chimpanzees (*Pan troglodytes schweinfurthii*) in Gombe National Park, Tanzania. *American Journal of Physical Anthropology*, **143**(4), 534–44. DOI: 10.1002/ajpa.21348.

Gillespie, T.R., Nunn, C.L. and Leendertz, F.H. (2008). Integrative approaches to the study of primate infectious disease: implications for biodiversity conservation and global health. *American Journal of Physical Anthropology*, **137**(S47), 53–69. DOI: 10.1002/ajpa.20949.

Gjeltema, J., Troan, B.V., Muehlenbachs, A., *et al.* (2016). Amoebic meningoencephalitis and disseminated infection caused by *Balamuthia mandrillaris* in a western lowland gorilla (*Gorilla gorilla gorilla*). *Journal of the American Veterinary Medical Association*, **248**(3), 315–21. DOI: 10.2460/javma.248.3.315.

Gogarten, J.F., Akoua-Koffi, C., Calvignac-Spencer, S., *et al.* (2014). The ecology of primate retroviruses: an assessment of 12 years of retroviral studies in the Taï National Park area, Côte d'Ivoire. *Virology*, **460–461**, 147–53. DOI: 10.1016/j.virol.2014.05.012.

Gogarten, J.F., Calvignac-Spencer, S., Nunn, C.L., *et al.* (2020). Metabarcoding of eukaryotic parasite communities describes diverse parasite assemblages spanning the primate phylogeny. *Molecular Ecology Resources*, **20**, 204–15. DOI: 10.1111/1755-0998.13101.

Gogarten, J.F., Davies, T.J., Benjamino, J., *et al.* (2018). Factors influencing bacterial microbiome composition in a wild non-human primate community in Taï National Park, Côte d'Ivoire. *The ISME Journal*, **12**(10), 2559–74. DOI: 10.1038/s41396-018-0166-1.

Gogarten, J.F., Düx, A., Mubemba, B., *et al.* (2019a). Tropical rainforest flies carrying pathogens form stable associations with social nonhuman primates. *Molecular Ecology*, **28**(18), 4242–58. DOI: 10.1111/mec.15145.

Gogarten, J.F., Düx, A., Schuenemann, V.J., *et al.* (2016). Tools for opening new chapters in the book of *Treponema pallidum* evolutionary history. *Clinical Microbiology and Infection*, **22**(11), 916–21.

Gogarten, J.F., Rühlemann, M.C., Archie, E.A., *et al.* (2021). Primate phageomes are structured by superhost phylogeny and environment. *Proceedings of the National Academy of Sciences*, **118**(15), e2013535118.

Gogarten, J.F., Schubert, G., Leendertz, F.H. and Calvignac-Spencer, S. (2019b). The chimpanzees of the Taï Forest as models for hominine microorganism ecology and evolution. In *The Chimpanzees of the Taï Forest: 40 Years of Research*, ed. C. Boesch, R. Wittig, C. Crockford, *et al.* Cambridge, UK: Cambridge University Press, pp. 366–84.

Goldberg, T.L., Gendron-Fitzpatrick, A., Deering, K.M., *et al.* (2014). Fatal metacestode infection in Bornean orangutan caused by unknown *Versteria* species. *Emerging Infectious Diseases*, **20**(1), 109–13. DOI: 10.3201/eid2001.131191.

Goldsmith, M. (2000). Effects of ecotourism on the behavioral ecology of Bwindi gorillas, Uganda: preliminary results. *American Journal of Physical Anthropology*, **111**(S30), 161.

Goldsmith, M.L. (2014). Mountain gorilla tourism as a conservation tool: have we tipped the balance? In *Primate Tourism: A Tool for Conservation?*, ed. A. E. Russon and J. Wallis. Cambridge, UK: Cambridge University Press, pp. 177–98. DOI: 10.1017/CBO9781139087407.013.

Gonçalves, A. and Carvalho, S. (2019). Death among primates: a critical review of non-human primate interactions towards their dead and dying. *Biological Reviews*, **94**(4), 1502–29.

Gond, V., Fayolle, A., Pennec, A., *et al.* (2013). Vegetation structure and greenness in Central Africa from Modis multi-temporal data. *Philosophical Transactions of the Royal Society B: Biological Sciences*, **368**(1625), 20120309. DOI: 10.1098/rstb.2012.0309.

Goodall, J. (1983). Population dynamics during a 15 year period in one community of free-living chimpanzees in the Gombe National Park, Tanzania. *Zeitschrift für Tierpsychologie*, **61**(1), 1–60. DOI: 10.1111/j.1439-0310.1983.tb01324.x.

Goodall, J. (1986). *The Chimpanzees of Gombe: Patterns of Behavior*. Cambridge, MA: Harvard University Press.

Goodall, J. (1998). Essays on science and society: learning from the chimpanzees: a message humans can understand. *Science*, **282**(5397), 2184–5. DOI: 10.1126/science.282.5397.2184.

Goodall, J. (2000). *In the Shadow of Man*. New York, NY: Houghton Mifflin.

Goodman, R.A., Bunnell, R. and Posner, S.F. (2014). What is "community health"? Examining the meaning of an evolving field in public health. *Preventive Medicine*, **67**, S58–61. DOI: 10.1016/j.ypmed.2014.07.028.

Goodpaster, K.E. (1978). On being morally considerable. *Journal of Philosophy*, **75**(6), 308–25.

Goodwin, H. (2007). Indigenous tourism and poverty reduction. In *Tourism and Indigenous Peoples*, ed. R. Butler and T. Hinch. Oxford, UK: Butterworth-Heinemann, pp. 84–94.

Goodwin, H. (2014). Responsible tourism and the green economy. In *Green Growth and Travelism: Concept, Policy and Practice for Sustainable Tourism*, ed. T. DeLacy, M. Jiang, G. Lipman and S. Vorster: Routledge, pp. 133–44.

Goodwin, H. (2016). *Responsible Tourism: Using Tourism for Sustainable Development*, 2nd edn. Oxford, UK: Goodfellow Publishers Ltd.

Goossens, B., Kapar, M.D., Kahar, S. and Ancrenaz, M. (2011). First sighting of Bornean orang-utan twins in the wild. *Asian Primates Journal*, **2**(1), 10–12.

Goossens, B., Setchell, J.M., James, S.S., *et al.* (2006). Philopatry and reproductive success in Bornean orang-utans (*Pongo pygmaeus*). *Molecular Ecology*, **15**(9), 2577–88. DOI: 10.1111/j.1365-294x.2006.02952.x.

Goossens, B., Setchell, J.M., Tchidongo, E., *et al.* (2005). Survival, interactions with conspecifics and reproduction in 37 chimpanzees released into the wild. *Biological Conservation*, **123**(4), 461–75. DOI: 10.1016/j.biocon.2005.01.008.

Gorilla Doctors (n.d.-a). *About Us*. Davis, CA: Gorilla Doctors. Available at: https://www.gorilladoctors.org/about-us/. Accessed: September, 2022.

Gorilla Doctors (n.d.-b). *Doctors and Staff*. Davis, CA: Gorilla Doctors. Available at: https://www.gorilladoctors.org/about-us/team/. Accessed: September, 2022.

Gorilla Doctors (n.d.-c). *Employee Health Program*. Davis, CA: Gorilla Doctors. Available at: https://www.gorilla-doctors.org/saving-lives/one-health-medicine/employee-health-program/ Accessed: September, 2022.

Gorilla Doctors (n.d.-d). *History*. Davis, CA: Gorilla Doctors. Available at: https://www.gorilladoctors.org/about-us/history-past-gorilla-doctors/. Accessed: October, 2022.

Gorilla Doctors (n.d.-e). *One Health Medicine*. Davis, CA: Gorilla Doctors. Available at: https://www.gorilladoctors.org/saving-lives/one-health-medicine/. Accessed: September, 2022.

Gorilla Doctors (n.d.-f). *UC Davis and MGVP Partnership*. Davis, CA: Gorilla Doctors. Available at: https://www.gorilladoctors.org/about-us/uc-davis/. Accessed: September, 2022.

Gormus, B.J., Xu, K., Alford, P.L., *et al.* (1991). A serologic study of naturally acquired leprosy in chimpanzees. *International Journal of Leprosy and Other Mycobacterial Diseases*, **59**(3), 450–7.

Government of Uganda (2019). *The National Environmental Act, 2019*. Uganda: National Environment Management Authority. Available at: https://nema.go.ug/sites/all/themes/nema/docs/National%20Environment%20Act,%20No.%205%20of%202019.pdf.

GRACE (2019). *GRACE Awarded Accreditation by the Global Federation of Animal Sanctuaries*. North Kivu Province, Democratic Republic of Congo: Gorilla Rehabilitation and Conservation Education (GRACE). Available at: https://gracegorillas.org/2019/09/17/grace-gfas-accredited/.

GRACE (2020). *GRACE 2019 Annual Report*. North Kivu Province, Democratic Republic of Congo: Gorilla Rehabilitation and Conservation Education (GRACE). Available at: https://gracegorillas.org/annual-reports/.

Graczyk, T.K., Mudakikwa, A.B., Cranfield, M.R. and Eilenberger, U. (2001). Hyperkeratotic mange caused by *Sarcoptes scabiei* (Acariformes: Sarcoptidae) in juvenile human-habituated mountain gorillas (*Gorilla gorilla beringei*). *Parasitology Research*, **87**, 1024–8. DOI: 10.1007/s004360100489.

Graef, A. (2021). Jane Goodall among 80+ leaders in animal advocacy & conservation calling for AP stylebook update. *In Defense of Animals Media Release*, March 25, 2021. Available at: https://www.idausa.org/campaign/guardian/latest-news/jane-goodall-joins-in-defense-of-animals-call-for-ap-stylebook-update/.

Graham, T.L., Matthews, H.D. and Turner, S.E. (2016). A global-scale evaluation of primate exposure and vulnerability to climate change. *International Journal of Primatology*, **37**(2), 158–74. DOI: 10.1007/s10764-016-9890-4.

Granjon, A.-C., Robbins, M.M., Arinaitwe, J., *et al.* (2020a). Estimating abundance and growth rates in a wild mountain gorilla population. *Animal Conservation*, **23**(4), 455–65. DOI: 10.1111/acv.12559.

Granjon, A.-C., Robbins, M., Arinaitwe, J., *et al.* (2020b). Increased survey effort and intrinsic growth contribute to the largest recorded mountain gorilla population. *Animal Conservation*, **23**(4), 455–65.

Grantham, H.S., Duncan, A., Evans, T.D., *et al.* (2020a). Anthropogenic modification of forests means only 40% of remaining forests have high ecosystem integrity. *Nature Communications*, **11**(1), 5978. DOI: 10.1038/s41467-020-19493-3.

Grantham, H.S., Shapiro, A., Bonfils, D., *et al.* (2020b). Spatial priorities for conserving the most intact biodiverse forests within Central Africa. *Environmental Research Letters*, **15**(9), 094005. DOI: 10.1088/1748-9326/ab9fae.

GRASP and IUCN (2018). *Report to the CITES Standing Committee on the Status of Great Apes.* Nairobi, Kenya, and Gland, Switzerland: United Nations Environment Programme Great Apes Survival Partnership (GRASP) and International Union for Conservation of Nature (IUCN). Available at: http://www.primate-sg.org/storage/pdf/GRASP__IUCN_2018_Report_to_CITES_on_the_Status_of_Great_Apes.pdf.

Graving, J.M., Chae, D., Naik, H., *et al.* (2019). DeepPoseKit, a software toolkit for fast and robust animal pose estimation using deep learning. *eLife*, **8**, e47994. DOI: 10.7554/eLife.47994.

Gray, C. and Favre, D. (2022). Veterinary ethics and the law. In *Ethics in Veterinary Practice: Balancing Conflicting Interests*, ed. B. Kipperman and B. E. Rollin. Hoboken, NJ: John Wiley & Sons Inc, pp. 78–99.

Gray, M., McNeilage, A., Fawcett, K., *et al.* (2010). Censusing the mountain gorillas in the Virunga Volcanoes: complete sweep method versus monitoring. *African Journal of Ecology*, **48**(3), 588–99. DOI: 10.1111/j.1365-2028.2009.01142.x.

Gray, M., Roy, J., Vigilant, L., *et al.* (2013). Genetic census reveals increased but uneven growth of a critically endangered mountain gorilla population. *Biological Conservation*, **158**(Supplement C), 230–8. DOI: 10.1016/j.biocon.2012.09.018.

Gray, S.J. (2012). *Conservation difficulties for* Hylobates lar: *effects the illegal pet trade has on white-handed gibbons' behavioral health and successful rehabilitation.* Undergraduate Honors thesis. Boulder, CO: University of Colorado Boulder. Available at: https://scholar.colorado.edu/concern/undergraduate_honors_theses/12579s59n.

Great Ape Project (n.d.). *World Declaration on Great Apes.* Great Ape Project (GAP). Available at: https://www.projetogap.org.br/en/world-declaration-on-great-primates/. Accessed: December, 2022.

Greene, M. (2005). *Jane Goodall: A Biography.* Westport, CT: Greenwood Press.

Greenpeace Africa (2020). 34 plastic bans in Africa: a reality check. *Greenpeace Africa*, May 19, 2020. Available at: https://www.greenpeace.org/africa/en/blogs/11156/34-plastic-bans-in-africa/.

Greggor, A.L., Berger-Tal, O., Blumstein, D.T., *et al.* (2016). Research priorities from animal behaviour for maximising conservation progress. *Trends in Ecology & Evolution*, **31**(12), 953–64. DOI: 10.1016/j.tree.2016.09.001.

Gresl, T.A., Baum, S.T. and Kemnitz, J.W. (2000). Glucose regulation in captive *Pongo pygmaeus abeli, P. p. pygmaeus,* and *P. p. abeli* × *P. p. pygmaeus* orangutans. *Zoo Biology*, **19**(3), 193–208.

Grimm, D. (2020). Chimpanzee sanctuaries are under fire. Can a new science-based tool improve ape welfare? *Science*, December 2, 2020. Available at: https://www.sciencemag.org/news/2020/12/chimpanzee-sanctuaries-are-under-fire-can-new-science-based-tool-improve-ape-welfare.

Gruen, L. (2015). *Entangled Empathy: An Alternative Ethic for our Relationships with Animals.* Brooklyn, NY: Lantern Books.

Gruen, L. (2018). More risky than radical. *American Journal of Bioethics*, **18**(10), 45–7. DOI: 10.1080/15265161.2018.1513606.

Gruen, L., Fultz, A. and Pruetz, J. (2013). Ethical issues in African great ape field studies. *Institute for Laboratory Animal Research (ILAR) Journal*, **54**(1), 24–32. DOI: 10.1093/ilar/ilt016.

Gruen, L., Jamieson, D. and Schlottmann, C. (2012). *Reflecting on Nature: Readings in Environmental Ethics and Philosophy*, 2nd edn. New York, NY: Oxford University Press.

Grunert, K.G., Hieke, S. and Wills, J. (2014). Sustainability labels on food products: consumer motivation, understanding and use. *Food Policy*, **44**, 177–89. DOI: 10.1016/j.foodpol.2013.12.001.

Grützmacher, K.S., Karesh, W.B., Amuasi, J.H., *et al.* (2021). The Berlin principles on one health: bridging global health and conservation. *Science of The Total Environment*, **764**, 142919. DOI: 10.1016/j.scitotenv.2020.142919.

Grützmacher, K.S., Keil, V., Leinert, V., *et al.* (2018a). Human quarantine: toward reducing infectious pressure on chimpanzees at the Taï Chimpanzee Project, Côte d'Ivoire. *American Journal of Primatology*, **80**(1), e22619. DOI: 10.1002/ajp.22619.

Grützmacher, K.S., Keil, V., Metzger, S., *et al.* (2018b). Human respiratory syncytial virus and *Streptococcus pneumoniae* infection in wild bonobos. *EcoHealth*, **15**(2), 462–6. DOI: 10.1007/s10393-018-1319-4.

Grützmacher, K.S., Köndgen, S., Keil, V., *et al.* (2016). Codetection of respiratory syncytial virus in habituated wild western lowland gorillas and humans during a respiratory disease outbreak. *EcoHealth*, **13**(3), 499–510. DOI: 10.1007/s10393-016-1144-6.

Gryseels, S., Watts, T.D., Kabongo, J.-M.M., *et al.* (2019). A near-full-length HIV-1 genome from 1966 recovered from formalin-fixed paraffin-embedded tissue. *bioRxiv*, 687863. DOI: 10.1101/687863; t.

Guagliardo, S.A.J., Monroe, B.P., Moundjoa, C., *et al.* (2020). Asymptomatic orthopoxvirus circulation in humans in the wake of a monkeypox outbreak among chimpanzees in Cameroon. *American Journal of Tropical Medicine and Hygiene*, **102**(1), 206–12. DOI: 10.4269/ajtmh.19-0467.

Guarino, B. (2016). North Korea's newest zoo attraction is a chimpanzee trained to smoke cigarettes. *The Washington Post*, October 21, 2016. Available at: https://www.4apes.com/news/item/1538-https-www-washingtonpost-com-news-morning-mix-wp-2016-10-21-north-koreas-newest-zoo-attraction-is-a-chimpanzee-trained-to-smoke-cigarettes.

Guatelli-Steinberg, D. (2000). Linear enamel hypoplasia in gibbons (*Hylobates lar carpenteri*). *American Journal of Physical Anthropology*, **112**(3), 395–410. DOI:10.1002/1096-8644(200007)112:3<395::AID-AJPA9>3.0.CO;2-H.

Guatelli-Steinberg, D., Ferrell, R.J. and Spence, J.M. (2012). Linear enamel hypoplasia as an indicator of physiological stress in great apes: reviewing the evidence in light of enamel growth variation. *American Journal of Physical Anthropology*, **148**(2), 191–204.

Guatelli-Steinberg, D. and Skinner, M.F. (2000). Prevalence and etiology of linear enamel hypoplasia in monkeys and apes from Asia and Africa. *Folia Primatologica*, **71**(3), 115–32. DOI: 10.1159/000021740.

Guerrera, W., Sleeman, J.M., Jasper, S.B., *et al.* (2003). Medical survey of the local human population to determine possible health risks to the mountain gorillas of Bwindi Impenetrable Forest National Park, Uganda. *International Journal of Primatology*, **24**(1), 197–207. DOI: 10.1023/A:1021410931928.

Guimarães, V.Y., Justo, A.A., Martins, L.L., Catão-Dias, J.L. and Sacristán, C. (2020). Emerging coronaviruses in Neotropical primates: a new threat? *Revista de Ciência Veterinária e Saúde Pública*, **7**(1). DOI: 10.4025/revcivet.v7i1.55490.

Guo, Y.-R., Cao, Q.-D., Hong, Z.-S., *et al.* (2020). The origin, transmission and clinical therapies on coronavirus disease 2019 (COVID-19) outbreak: an update on the status. *Military Medical Research*, **7**(1), 11. DOI: 10.1186/s40779-020-00240-0.

Gut Aiderbichl (n.d.). *Schimpansen*. Salzburg, Germany: Gut AIderbichl GmbH. Available at: https://www.gut-aiderbichl.com/tiere/unseretiereo/schimpansen/. Accessed: December, 2020.

Guy, A.J., Curnoe, D. and Banks, P.B. (2014). Welfare based primate rehabilitation as a potential conservation strategy: does it measure up? *Primates*, **55**(1), 139–47. DOI: 10.1007/s10329-013-0386-y.

Guyson, N. (2021). Lockdown underscores Uganda's overreliance on tourism to fund conservation. *Mongabay Series: Great Apes*, December 21, 2021. Available at: https://news.mongabay.com/2021/12/lockdown-underscores-ugandas-overreliance-on-tourism-to-fund-conservation/.

GVTC (2020). *Regional EVD and COVID-19 Contingency Plans for Mountain Gorillas*. Kigali, Rwanda: Greater Virunga Transboundary Collaboration (GVTC). Available at: https://pfbc-cbfp.org/actualites-partenaires/coll%C3%A8ge-multilat%C3%A9ral.html?file=files/docs/news/6-2020/Doc%202_GVTC%20Contingency%20Planning%20Overview.pdf.

Haberthur, K. and Messaoudi, I. (2013). Animal models of varicella zoster virus infection. *Pathogens*, **2**(2), 364–82. DOI: 10.3390/pathogens2020364.

Haggblade, M.K., Smith, W.A., Noheri, J.B., *et al.* (2019). Outcomes of snare-related injuries to endangered mountain gorillas (*Gorilla beringei beringei*) in Rwanda. *Journal of Wildlife Diseases*, **55**(2), 298–303, 6. DOI: 10.7589/2018-01-008.

Hahn, B.H., Shaw, G.M., De Cock, K.M. and Sharp, P.M. (2000). AIDS as a zoonosis: scientific and public health implications. *Science*, **287**(5453), 607–14. DOI: 10.1126/science.287.5453.607.

Halifax, J. (2011). The precious necessity of compassion. *Journal of Pain and Symptom Management*, **41**(1), 146–53. DOI: 10.1016/j.jpainsymman.2010.08.010.

Hall, C.M., Scott, D. and Gössling, S. (2020). Pandemics, transformations and tourism: be careful what you wish for. *Tourism Geographies*, **22**(3), 577–98. DOI: 10.1080/14616688.2020.1759131.

Hall, M.J., Ng, A., Ursano, R.J., *et al.* (2004). Psychological impact of the animal–human bond in disaster prepared-ness and response. *Journal of Psychiatric Practice*, **10**(6), 368–74.

Halter, C. (2018). Paul Richards and Esther Mokuwa on lessons learned during the Ebola epidemic. *KGOU*, March 16, 2018. Available at: https://www.kgou.org/post/paul-richards-and-esther-mokuwa-lessons-learned-during-ebola-epidemic.

Hamard, M., Cheyne, S.M. and Nijman, V. (2010). Vegetation correlates of gibbon density in the peat-swamp forest of the Sabangau catchment, Central Kalimantan, Indonesia. *American Journal of Primatology*, **72**(7), 607–16. DOI: 10.1002/ajp.20815.

Hamer, D.H. and Connor, B.A. (2004). Travel health knowledge, attitudes and practices among United States travelers. *Journal of Travel Medicine*, **11**(1), 23–6. DOI: 10.2310/7060.2004.13577.

Hampton, J.O., Jones, B. and McGreevy, P.D. (2020). Social license and animal welfare: developments from the past decade in Australia. *Animals*, **10**(12), 2237. DOI: 10.3390/ani10122237.

Han, X.Y., Seo, Y.H., Sizer, K.C., *et al.* (2008). A new *Mycobacterium* species causing diffuse lepromatous leprosy. *American Journal of Clinical Pathology*, **130**(6), 856–64. DOI: 10.1309/ajcpp72fjzzrrvmm.

Han, X.Y., Sizer, K.C., Thompson, E.J., *et al.* (2009). Comparative sequence analysis of *Mycobacterium leprae* and the new leprosy-causing *Mycobacterium lepromatosis*. *Journal of Bacteriology*, **191**(19), 6067–74. DOI: 10.1128/JB.00762-09.

Hanamura, S., Kiyono, M., Lukasik-Braum, M., *et al.* (2008). Chimpanzee deaths at Mahale caused by a flu-like disease. *Primates*, **49**(1), 77–80. DOI: 10.1007/s10329-007-0054-1.

Hanes, A., Kalema-Zikusoka, G., Svensson, M.S. and Hill, C.M. (2018). Assessment of health risks posed by tourists visiting mountain gorillas in Bwindi Impenetrable National Park, Uganda. *Primate Conservation*, **32**, 123–32.

Hannibal, D.L. and Guatelli-Steinberg, D. (2005). Linear enamel hypoplasia in the great apes: analysis by genus and locality. *American Journal of Physical Anthropology*, **127**(1), 13–25. DOI: 10.1002/ajpa.20141.

Hansen, B.K., Fultz, A.L., Hopper, L.M. and Ross, S.R. (2018). An evaluation of video cameras for collecting obser-vational data on sanctuary-housed chimpanzees (*Pan troglodytes*). *Zoo Biology*, **37**(3), 156–61. DOI: 10.1002/zoo.21410.

Harcourt, A.H., Fossey, D. and Sabater-Pi, J. (1981). Demography of *Gorilla gorilla*. *Journal of Zoology*, **195**(2), 215–33. DOI: 10.1111/j.1469-7998.1981.tb03460.x.

Harcourt, A.H. and Greenberg, J. (2001). Do gorilla females join males to avoid infanticide? A quantitative model. *Animal Behaviour*, **62**(5), 905–15. DOI: 10.1006/anbe.2001.1835.

Harcourt, A.H. and Stewart, K.J. (2007). *Gorilla Society: Conflict, Compromise, and Cooperation Between the Sexes.* Chicago, IL: University of Chicago Press. DOI: 10.7208/chicago/9780226316048.001.0001.

Hardgrove, E.H., Zimmerman, D.M., von Fricken, M.E. and Deem, S.L. (2021). A scoping review of rodent-borne pathogen presence, exposure, and transmission at zoological institutions. *Preventive Veterinary Medicine*, **193**, 105345. DOI: 10.1016/j.prevetmed.2021.105345.

Harper, K.N. and Knauf, S. (2013). *Treponema pallidum* infection in primates: clinical manifestations, epidemiol-ogy, and evolution of a stealthy pathogen. In *Primates, Pathogens, and Evolution*, ed. J. F. Brinkworth and K. Pechenkina. New York, NY: Springer, pp. 189–219.

Harrington, L.A., Moehrenschlager, A., Gelling, M., *et al.* (2013). Conflicting and complementary ethics of animal welfare considerations in reintroductions. *Conservation Biology*, **27**(3), 486–500. DOI: 10.1111/cobi.12021.

Harris, D.J., Ebika, S.T.N., Sanz, C.M., Madingou, M.P.N. and Morgan, D.B. (2021). Large trees in tropical rain forests require big plots. *Plants People Planet*, **3**(3), 282–94. DOI: 10.1002/ppp3.10194.

Harrison, M., Baker, J., Twinamatsiko, M. and Milner-Gulland, E.J. (2015). Profiling unauthorized natural resource users for better targeting of conservation interventions. *Conservation Biology*, **29**(6), 1636–46. DOI: 10.1111/cobi.12575.

Harrison, M.E., Cheyne, S.M., Sulistiyanto, Y. and Rieley, J.O. (2007). Biological effects of smoke from dry-season fires in non-burnt areas of the Sabangau peat swamp forest, Central Kalimantan, Indonesia. In *Carbon–Climate–Human Interaction on Tropical Peatland: Proceedings of the International Symposium and Workshop on Tropical Peatland, Yogyakarta, August 27–29, 2007. EU CARBOPEAT and RESTOPEAT Partnership*, ed. J. O. Rieley, C. J. Banks and G. Radjaguk. Sleman, Indonesia, and Leicester, UK: Gadjah Mada University, University of Leicester, pp. 1–5.

Harrison, M.E., Ottay, J.B., D'Arcy, L.J., *et al.* (2020a). Tropical forest and peatland conservation in Indonesia: challenges and directions. *People and Nature*, **2**(1), 4–28. DOI: 10.1002/pan3.10060.

Harrison, M.E., Wijedasa, L.S., Cole, L.E.S., *et al.* (2020b). Tropical peatlands and their conservation are important in the context of COVID-19 and potential future (zoonotic) disease pandemics. *PeerJ*, **8**, e10283. DOI: 10.7717/peerj.10283.

Hartel, J.A., Otali, E., Machanda, Z., *et al.* (2020). Holistic approach for conservation of chimpanzees in Kibale National Park, Uganda. In *Chimpanzees in Context: A Comparative Perspective on Chimpanzee Behavior, Cognition, Conservation, and Welfare*, ed. L. M. Hopper and S. R. Ross. Chicago, IL: University of Chicago Press, pp. 612–43. DOI: 10.7208/chicago/9780226728032.003.0026.

Hasegawa, H. and Udono, T. (2007). Chimpanzee pinworm, *Enterobius anthropopitheci* (Nematoda: Oxyuridae), maintained for more than twenty years in captive chimpanzees in Japan. *Journal of Parasitology*, **93**(4), 850–3.

Hashimoto, C. (1997). Context and development of sexual behavior of wild bonobos (*Pan paniscus*) at Wamba, Zaire. *International Journal of Primatology*, **18**(1), 1–21. DOI: 10.1023/A:1026384922066.

Hashimoto, C. (1999). Snare injuries of chimpanzees in the Kalinzu Forest, Uganda. *Pan Africa News*, **6**(2), 20–2.

Häsler, B., Cornelsen, L., Bennani, H. and Rushton, J. (2014). A review of the metrics for One Health benefits. *Revue Scientifique et Technique de l'Office International des Épizooties*, **33**(2), 453–64. DOI: 10.20506/rst.33.2.2294.

Hassan, K.H. (2016). Ensuring animal welfare in zoos? Operations: a comparative note on Malaysian and Japanese legislation. *Mediterranean Journal of Social Sciences*, **7**(1), 328. DOI: 10.5901/mjss.2016.v7n1p328.

Hassell, J.M., Zimmerman, D., Cranfield, M.R., *et al.* (2017). Morbidity and mortality in infant mountain gorillas (*Gorilla beringei beringei*): a 46-year retrospective review. *American Journal of Primatology*, **79**(10), e22686. DOI: 10.1002/ajp.22686.

Haurez, B., Daïnou, K., Tagg, N., Petre, C.-A. and Doucet, J.-L. (2015). The role of great apes in seed dispersal of the tropical forest tree species *Dacryodes normandii* (Burseraceae) in Gabon. *Journal of Tropical Ecology*, **31**(5), 395–402. DOI: 10.1017/S0266467415000322.

Haurez, B., Daïnou, K., Vermeulen, C., *et al.* (2017). A look at intact forest landscapes (IFLs) and their relevance in Central African forest policy. *Forest Policy and Economics*, **80**, 192–9. DOI: 10.1016/j.forpol.2017.03.021.

Hawkins, S.J., Struthers, J.D., Phair, K.A., *et al.* (2021). Diagnostic evaluation of fatal *Balamuthia mandrillaris* meningoencephalitis in a captive Bornean orangutan (*Pongo pygmaeus*) with identification of potential environmental source and evidence of chronic exposure. *Primates*, **62**(1), 51–61. DOI: 10.1007/s10329-020-00860-z.

He, B., Feng, Y., Zhang, H., *et al.* (2015). Filovirus RNA in fruit bats, China. *Emerging Infectious Diseases*, **21**(9), 1675–7. DOI: 10.3201/eid2109.150260.

Head, J.S., Boesch, C., Makaga, L. and Robbins, M.M. (2011). Sympatric chimpanzees (*Pan troglodytes troglodytes*) and gorillas (*Gorilla gorilla gorilla*) in Loango National Park, Gabon: dietary composition, seasonality, and intersite comparisons. *International Journal of Primatology*, **32**(3), 755–75. DOI: 10.1007/s10764-011-9499-6.

Head, J.S., Boesch, C., Robbins, M.M., *et al.* (2013). Effective sociodemographic population assessment of elusive species in ecology and conservation management. *Ecology and Evolution*, **3**(9), 2903–16. DOI: 10.1002/ece3.670.

Heinicke, S., Mundry, R., Boesch, C., *et al.* (2019). Advancing conservation planning for western chimpanzees using IUCN SSC A.P.E.S.: the case of a taxon-specific database. *Environmental Research Letters*, **14**(6), 064001. DOI: 10.1088/1748-9326/ab1379.

Heldstab, A., Rüedi, D., Sonnabend, W.F. and Deinhardt, F. (1981). Spontaneous generalized herpesvirus hominis infection of a lowland gorilla (*Gorilla gorilla gorilla*). *Journal of Medical Primatology*, **10**(2–3), 129–35. DOI: 10.1159/000460063.

HELP Congo (n.d.). *Parrainage*. Lissieu, France: HELP Congo. Available at: http://www.help-primates.org/fr/parrainage.html. Accessed: October, 2020.

Henao-Restrepo, A.M., Longini, I.M., Egger, M., *et al.* (2015). Efficacy and effectiveness of an rVSV-vectored vaccine expressing Ebola surface glycoprotein: interim results from the Guinea ring vaccination cluster-randomised trial. *The Lancet*, **386**(9996), 857–66. DOI: 10.1016/s0140-6736(15)61117-5.

Henseler, M., Maisonnave, H. and Maskaeva, A. (2022). Economic impacts of COVID-19 on the tourism sector in Tanzania. *Annals of Tourism Research Empirical Insights*, **3**(1), 100042. DOI: 10.1016/j.annale.2022.100042.

Herbinger, I., Boesch, C. and Rothe, H. (2001). Territory characteristics among three neighboring chimpanzee communities in the Taï National Park, Côte d'Ivoire. *International Journal of Primatology*, **22**(2), 143–67. DOI: 10.1023/a:1005663212997.

Hernandez, E., Fawcett, A., Brouwer, E., Rau, J. and Turner, P.V. (2018). Speaking up: veterinary ethical responsibilities and animal welfare issues in everyday practice. *Animals*, **8**(1), 15. DOI: 10.3390/ani8010015.

Hernández, P., Gangsei, D. and Engstrom, D. (2007). Vicarious resilience: a new concept in work with those who survive trauma. *Family Process*, **46**(2), 229–41. DOI: 10.1111/j.1545-5300.2007.00206.x.

Herrera, J. and Nunn, C.L. (2019). Behavioural ecology and infectious disease: implications for conservation of biodiversity. *Philosophical Transactions of the Royal Society B: Biological Sciences*, **374**(1781), 20180054. DOI: 10.1098/rstb.2018.0054.

Hewitt, G., MacLarnon, A. and Jones, K.E. (2002). The functions of laryngeal air sacs in primates: a new hypothesis. *Folia Primatologica*, **73**, 70–94.

HHS (2012). *Emergency Management and the Incident Command System*. Washington DC: United States Department of Health and Human Services (HHS).

Hickel, J. (2019). Is it possible to achieve a good life for all within planetary boundaries? *Third World Quarterly*, **40**(1), 18–35. DOI: 10.1080/01436597.2018.1535895.

Hickey, J.R., Basabose, A., Gilardi, K.V., *et al.* (2020). Gorilla beringei *ssp.* beringei *(amended version of 2018 assessment). The IUCN Red List of Threatened Species 2020: e.T39999A176396749*. Gland, Switzerland: International Union for Conservation of Nature (IUCN). DOI: 10.2305/IUCN.UK.2020-3.RLTS.T39999A176396749.en.

Hickey, J.R., Granjon, A.-C., Vigilant, L., *et al.* (2019a). *Virunga 2015–2016 Surveys: Monitoring Mountain Gorillas, Other Select Mammals, and Illegal Activities*. Kigali, Rwanda: GVTC, IGCP & Partners. Available at: http://igcp.org/wp-content/uploads/Virunga-Census-2015-2016-Final-Report-2019-with-French-summary-2019_04_24.pdf.

Hickey, J.R., Uzabaho, E., Akantorana, M., *et al.* (2019b). *Bwindi-Sarambwe 2018 Surveys: Monitoring Mountain Gorillas, Other Select Mammals, and Human Activities*. Unpublished report to Uganda Wildlife Authority. Kampala, Uganda: GVTC, IGCP & Partners. Available at: http://ir.must.ac.ug/handle/123456789/762.

Highland Farm (n.d.). *Gibbons in Need*. GoFundMe. Available at: https://www.gofundme.com/f/gibbons-in-need. Accessed: October, 2020.

HiH (n.d.). *Health in Harmony is a Climate Solution*. Portland, OR: Health in Harmony (HiH). Available at: https://healthinharmony.org/. Accessed: November, 2021.

Hill, S.P. and Broom, D.M. (2009). Measuring zoo animal welfare: theory and practice. *Zoo Biology*, **28**(6), 531–44. DOI: 10.1002/zoo.20276.

Hilser, H. (2011). *An assessment of primate health in the Sabangau peat-swamp forest, Central Kalimantan, Indonesian Borneo*. MSc thesis. Oxford, UK: Oxford Brookes University.

Himalayan News Service (2009). Central Zoo bans plastic bags. *The Himalayan*, June 6, 2009. Available at: https://thehimalayantimes.com/nepal/central-zoo-bans-plastic-bags.

Hing, S., Narayan, E.J., Thompson, R.C.A. and Godfrey, S.S. (2016). The relationship between physiological stress and wildlife disease: consequences for health and conservation. *Wildlife Research*, **43**(1), 51–60. DOI: 10.1071/WR15183.

Hingham, J.E.S. (2007). *Critical Issues in Ecotourism: Understanding a Complex Tourism Phenomenon*. Oxford, UK: Butterworth-Heinemann.

Hirata, S., Morimura, N., Watanuki, K., Ross, S.R. and Goodall, J. (2020). The establishment of sanctuaries for former laboratory chimpanzees: challenges, successes, and cross-cultural context. In *Chimpanzees in Context: A Comparative Perspective on Chimpanzee Behavior, Cognition, Conservation, and Welfare*, ed. L. M. Hopper and S. R. Ross. Chicago, IL: University of Chicago Press, pp. 208–32. DOI: 10.7208/chicago/9780226728032.003.0009.

Hobson, K. (2007). Political animals? On animals as subjects in an enlarged political geography. *Political Geography*, **26**(3), 250–67. DOI: 10.1016/j.polgeo.2006.10.010.

Hockings, K. and Humle, T. (2009). *Best Practice Guidelines for the Prevention and Mitigation of Conflict Between Humans and Great Apes*. Gland, Switzerland: International Union for Conservation of Nature (IUCN) Species Survival Commission (SSC) Primate Specialist Group (PSG). Available at: https://portals.iucn.org/library/efiles/documents/ssc-op-037.pdf.

Hockings, K.J., McLennan, M.R., Carvalho, S., *et al.* (2015). Apes in the Anthropocene: flexibility and survival. *Trends in Ecology & Evolution*, **30**(4), 215–22. DOI: 10.1016/j.tree.2015.02.002.

Hockings, K.J., Mubemba, B., Avanzi, C., *et al.* (2021). Leprosy in wild chimpanzees. *Nature*, **598**(7882), 652–6. DOI: 10.1038/s41586-021-03968-4.

Hockings, K.J., Yamakoshi, G., Kabasawa, A. and Matsuzawa, T. (2010). Attacks on local persons by chimpanzees in Bossou, Republic of Guinea: long-term perspectives. *American Journal of Primatology*, **72**(10), 887–96. DOI: 10.1002/ajp.20784.

Hockings, M., Dudley, N., Elliott, W., *et al.* (2020). Editorial essay: COVID-19 and protected and conserved areas. *Parks*, **26**(1), 7–24.

Hoffmann, C., Zimmermann, F., Biek, R., *et al.* (2017). Persistent anthrax as a major driver of wildlife mortality in a tropical rainforest. *Nature*, **548**, 82–6. DOI: 10.1038/nature23309, https://www.nature.com/articles/nature23309#supplementary-information.

Hohmann, G., Robbins, M.M. and Boesch, C., ed. (2006). *Feeding Ecology in Apes and Other Primates: Ecological, Physiological, and Behavioural Aspects*. Cambridge Studies in Biological and Evolutionary Anthropology Volume 48. Cambridge, UK: Cambridge University Press.

Homsy, J. (1999). *Ape Tourism and Human Diseases: How Close Should We Get? A Critical Review of the Rules and Regulations Governing Park Management & Tourism for the Wild Mountain Gorilla,* Gorilla gorilla beringei. Nairobi, Kenya: Report of a Consultancy for the International Gorilla Conservation Programme.

Hooper, L.V., Littman, D.R. and Macpherson, A.J. (2012). Interactions between the microbiota and the immune system. *Science*, **336**(6086), 1268–73. DOI: 10.1126/science.1223490.

Hopper, L.M. and Ross, S.R., ed. (2020). *Chimpanzees in Context: A Comparative Perspective on Chimpanzee Behavior, Cognition, Conservation, and Welfare*. Chicago, IL: University of Chicago Press. DOI: 10.7208/chicago/9780226728032.001.0001.

Horvath, L.L., Murray, C.K. and DuPont, H.L. (2003). Travel health information at commercial travel websites. *Journal of Travel Medicine*, **10**(5), 272–9. DOI: 10.2310/7060.2003.2699.

Hosey, G. (2008). A preliminary model of human–animal relationships in the zoo. *Applied Animal Behaviour Science*, **109**(2), 105–27. DOI: 10.1016/j.applanim.2007.04.013.

Hosey, G.R. and Druck, P.L. (1987). The influence of zoo visitors on the behaviour of captive primates. *Applied Animal Behaviour Science*, **18**(1), 19–29. DOI: 10.1016/0168-1591(87)90251-6.

Hosey, G., Melfi, V. and Pankhurst, S. (2013). *Zoo Animals: Behaviour, Management, and Welfare*. Oxford, UK: Oxford University Press.

Hosey, G., Melfi, V. and Ward, S.J. (2020). Problematic animals in the zoo: the issue of charismatic megafauna. In *Problematic Wildlife II: New Conservation and Management Challenges in the Human–Wildlife Interactions*, ed. F. M. Angelici and L. Rossi. Cham, Switzerland: Springer International Publishing, pp. 485–508. DOI: 10.1007/978-3-030-42335-3_15.

Hrdy, S.B. (1979). Infanticide among animals: a review, classification, and examination of the implications for the reproductive strategies of females. *Ethology and Sociobiology*, **1**(1), 13–40. DOI: 10.1016/0162-3095(79)90004-9.

Hsu, C.-C. and Sandford, B.A. (2007). The Delphi technique: making sense of consensus. *Practical Assessment, Research, and Evaluation*, **12**, 10. DOI: 10.7275/pdz9-th90.

Hu, N., Guan, Z., Huang, B., *et al.* (2018). Dispersal and female philopatry in a long-term, stable, polygynous gibbon population: evidence from 16 years field observation and genetics. *American Journal of Primatology*, **80**(9), e22922. DOI: 10.1002/ajp.22922.

Hu, T., Chitnis, N., Monos, D. and Dinh, A. (2021). Next-generation sequencing technologies: an overview. *Human Immunology*, **82**(11), 801–11. DOI: 10.1016/j.humimm.2021.02.012.

Hubálek, Z. (2003). Emerging human infectious diseases: anthroponoses, zoonoses, and sapronoses. *Emerging Infectious Diseases*, **9**(3), 403–4.

Hughes, A.C. (2019). Understanding and minimizing environmental impacts of the Belt and Road Initiative. *Conservation Biology*, **33**(4), 883–94. DOI: 10.1111/cobi.13317.

Humle, T. (2015). *The Dimensions of Ape–Human Interactions in Industrial Agricultural Landscapes. Background Paper for State of the Apes: Industrial Agriculture and Ape Conservation. Arcus Foundation.* Cambridge, UK:

Cambridge University Press. Available at: http://www.stateoftheapes.com/wp-content/uploads/2016/03/Ape-Human-Interactions-in-Industrial-Agricultural-Landscapes.pdf.

Humle, T., Boesch, C., Campbell, G., *et al.* (2016a). Pan troglodytes *ssp.* verus *(errata version published in 2016). The IUCN Red List of Threatened Species 2016: e.T15935A102327574.* Gland, Switzerland: International Union for Conservation of Nature (IUCN). DOI: 10.2305/IUCN.UK.2016-2.RLTS.T15935A17989872.en.

Humle, T., Colin, C., Laurans, M. and Raballand, E. (2011). Group release of sanctuary chimpanzees (*Pan troglodytes*) in the Haut Niger National Park, Guinea, west Africa: ranging patterns and lessons so far. *International Journal of Primatology*, **32**(2), 456–73. DOI: 10.1007/s10764-010-9482-7.

Humle, T. and Hill, C. (2016). People–primate interactions: implications for primate conservation. In *Introduction to Primate Conservation*, ed. S. A. Wich and A. J. Marshall. Oxford, UK: Oxford University Press, pp. 219–40.

Humle, T., Maisels, F., Oates, J.F., Plumptre, A. and Williamson, E.A. (2016b). Pan troglodytes *(errata version published in 2018). The IUCN Red List of Threatened Species 2016: e.T15933A129038584.* Gland, Switzerland: International Union for Conservation of Nature (IUCN). DOI: 10.2305/IUCN.UK.2016-2.RLTS.T15933A17964454.en.

Hursthouse, R. (2011). Virtue ethics and the treatment of animals. In *The Oxford Handbook of Animal Ethics*, ed. T. L. Beauchamp and R. G. Frey. Oxford, UK: Oxford University Press, pp. 119–43. DOI: 10.1093/oxford-hb/9780195371963.013.0005.

Hutchins, M., Foose, T. and Seal, U.S. (1991). The role of veterinary medicine in endangered species conservation. *Journal of Zoo and Wildlife Medicine*, **22**(3), 277–81.

Huynh, D.V., Truong, T.T.K., Duong, L.H., *et al.* (2021). The COVID-19 pandemic and its impacts on tourism business in a developing city: insight from Vietnam. *Economies*, **9**(4), 172. DOI: 10.3390/economies9040172.

Hvenegaard, G.T. (2014). Economic aspects of primate tourism associated with primate conservation. In *Primate Tourism: A Tool for Conservation?*, ed. A. E. Russon and J. Wallis. Cambridge, UK: Cambridge University Press, pp. 259–77. DOI: 10.1017/CBO9781139087407.020.

Hyeroba, D., Apell, P. and Otali, E. (2011). Managing a speared alpha male chimpanzee (*Pan troglodytes*) in Kibale National Park, Uganda. *Veterinary Record*, **169**(25), 658. DOI: 10.1136/vr.d4680.

ICCA Consortium (2021). *Territories of Life: 2021 Report.* ICCA Consortium. Available at: https://report.territoriesoflife.org/.

Idani, G. (1990). Relations between unit-groups of bonobos at Wamba, Zaire: encounters and temporary fusions. *African Study Monographs*, **11**, 153–86.

IFAW (2018). *Disrupt: Wildlife Cybercrime.* London, UK: International Fund for Animal Welfare (IFAW). Available at: https://www.ifaw.org/eu/resources/disrupt-wildlife-cybercrime.

IFC (2012). *Performance Standard 6 – Biodiversity Conservation and Sustainable Management of Living Natural Resources.* Washington DC: International Finance Corporation (IFC), World Bank Group. Available at: https://www.ifc.org/wps/wcm/connect/Topics_Ext_Content/IFC_External_Corporate_Site/Sustainability-At-IFC/Policies-Standards/Performance-Standards/PS6.

IFC (2013). *Good Practice Handbook: Cumulative Impact Assessment and Management: Guidance for the Private Sector in Emerging Markets.* Washington DC: International Finance Corporation (IFC), World Bank Group. Available at: https://www.ifc.org/wps/wcm/connect/58fb524c-3f82-462b-918f-0ca1af135334/IFC_GoodPracticeHandbook_CumulativeImpactAssessment.pdf?MOD=AJPERES&CVID=kbnYgI5.

IFC (2019). *Guidance Note 6 – Biodiversity Conservation and Sustainable Management of Living Natural Resources.* Washington DC: International Finance Corporation (IFC), World Bank Group. Available at: https://www.ifc.org/wps/wcm/connect/5e0f3c0c-0aa4-4290-a0f8-4490b61de245/GN6_English_June-27-2019.pdf?MOD=AJPERES&CVID=mRQjZva.

IFRC (2021). *Contingency Plan Guide.* Geneva, Switzerland: International Federation of Red Cross and Red Crescent Societies (IFRC). Available at: https://www.ifrc.org/document/contingency-planning-guide.

IGCP (n.d.). *Certified Gorilla Friendly TM.* Kigali, Rwanda: International Gorilla Conservation Programme (IGCP). Available at: https://igcp.org/tourism/certified-gorilla-friendly/. Accessed: April, 2022.

iHMP Research Network Consortium (2019). The Integrative Human Microbiome Project. *Nature*, **569**(7758), 641–8. DOI: 10.1038/s41586-019-1238-8.

ILRI [International Livestock Research Institute] (2019). *Meat: The Future Series. Options for the Livestock Sector in Developing and Emerging Economies to 2030 and Beyond.* Geneva, Switzerland: World Economic Forum.

Imster, E. (2018). Wildfire smoke messing with orangutans' eating and sleep. *EarthSky*, June 3, 2018. Available at: https://earthsky.org/earth/wildfire-smoke-orangutan-health-threat/.

Inclean Magazine (2019). Zoos Victoria ban all single-use plastic bottles and straws. *Inclean Magazine*, May 1, 2019. Available at: https://incleanmag.com.au/zoos-victoria-ban-all-single-use-plastic/.

Ingram, J. (2020). Nutrition security is more than food security. *Nature Food*, **1**(1), 2. DOI: 10.1038/s43016-019-0002-4.

Inogwabini, B.I. and Leader-Williams, N. (2012). Effects of epidemic diseases on the distribution of bonobos. *PLoS ONE*, **7**(12), e51112.

Inoue, E., Tashiro, Y., Ogawa, H., *et al.* (2013). Gene flow and genetic diversity of chimpanzees in Tanzanian habitats. *Primate Conservation*, **26**(1), 67–74. DOI: 10.1896/052.026.0105.

International Animal Rescue (2020). *Report of the Trustees and Financial Statements for the Year Ended 31 December 2019*. Uckfield, UK: International Animal Rescue. Available at: https://register-of-charities.charitycommission.gov.uk/charity-search/-/charity-details/4029510/accounts-and-annual-returns.

IOC-UNESCO (n.d.). *Global Tsunami Early Warning and Mitigation Programme*. Paris, France: United Nations Educational, Scientific and Cultural Organization (UNESCO) Intergovernmental Oceanographic Commission (IOC). Available at: https://www.ioc.unesco.org/en/global-tsunami-early-warning-and-mitigation-programme. Accessed: October, 2022.

IPBES (2019). Nature's dangerous decline "unprecedented"; species extinction rates "accelerating". *IPBES* [Intergovernmental Science-Policy Platform on Biodiversity and Ecosystem Services] *Media Release*, May 5, 2019. Available at: https://www.ipbes.net/news/Media-Release-Global-Assessment.

IPBES (2020). *Workshop Report on Biodiversity and Pandemics*. Bonn, Germany: Intergovernmental Platform on Biodiversity and Ecosystem Services (IPBES). DOI: 10.5281/zenodo.4147317.

IPCC (2023). *AR6 Synthesis Report Climate Change 2023*. Geneva, Switzerland: Intergovernmental Panel on Climate Change (IPCC). Available at: https://www.ipcc.ch/report/ar6/syr/.

IPPL (n.d.). *Our Gibbon Sanctuary*. Summerville, SC: International Primate Protection League (IPPL). Available at: https://www.ippl.org/gibbon/ippls-gibbon-sanctuary/. Accessed: October, 2020.

IRMA (2018). *IRMA Standard for Responsible Mining IRMA-STD-001*. Initiative for Responsible Mining Assurance (IRMA). Available at: https://responsiblemining.net/wp-content/uploads/2018/07/IRMA_STANDARD_v.1.0_FINAL_2018.pdf.

Isakov, A., O'Neal, P., Prescott, J., *et al.* (2014). Academic–community partnerships for sustainable preparedness and response systems. *American Journal of Disaster Medicine*, **9**(2), 97–106. DOI: 10.5055/ajdm.2014.0146.

Ishizuka, S., Toda, K. and Furuichi, T. (2020). Genetic analysis of migration pattern of female bonobos (*Pan paniscus*) among three neighboring groups. *International Journal of Primatology*, **41**, 401–14. DOI: 10.1007/s10764-019-00106-w.

IUCN (2012). *IUCN Red List Categories and Criteria, Version 3.1*, 2nd edn. Gland, Switzerland, and Cambridge, UK: International Union for Conservation of Nature (IUCN) Species Survival Commission (SSC). Available at: https://www.iucnredlist.org/resources/categories-and-criteria.

IUCN (2014). *Regional Action Plan for the Conservation of Western Lowland Gorillas and Central Chimpanzees 2015–2025*. Gland, Switzerland: International Union for Conservation of Nature (IUCN) Species Survival Commission (SSC) Primate Specialist Group (PSG). Available at: https://portals.iucn.org/library/node/45060.

IUCN (2019a). *Guidelines for the Management of Confiscated, Live Organisms*. Gland, Switzerland: International Union for Conservation of Nature (IUCN). Available at: https://portals.iucn.org/library/sites/library/files/documents/2019-005-En.pdf.

IUCN (2019b). *The IUCN Red List of Threatened Species. Version 2019-1*. Gland, Switzerland: International Union for Conservation of Nature (IUCN). Available at: https://www.iucnredlist.org.

IUCN (2020a). *Bwindi Impenetrable National Park: 2020 Conservation Outlook Assessment*. International Union for Conservation of Nature (IUCN) World Heritage Outlook. Available at: https://rris.biopama.org/node/20652?language=en.

IUCN (2020b). *Great Apes, COVID-19 and the SARS CoV-2: Joint Statement of the IUCN SSC Wildlife Health Specialist Group and the Primate Specialist Group, Section on Great Apes. March 15 2020*. Geneva, Switzerland: International Union for Conservation of Nature (IUCN). Available at: http://www.primate-sg.org/storage/SARSCoV-2.pdf.

IUCN (2022). *The IUCN Red List of Threatened Species. Version 2022-1.* Gland, Switzerland: International Union for Conservation of Nature (IUCN). Available at: https://www.iucnredlist.org.

IUCN (2023). *The IUCN Red List of Threatened Species. Version 2022-2.* Gland, Switzerland: International Union for Conservation of Nature (IUCN). Available at: https://www.iucnredlist.org.

IUCN SSC Human–Wildlife Conflict & Coexistence Specialist Group (n.d.). *IUCN SSC Human–Wildlife Conflict & Coexistence Specialist Group.* International Union for Conservation of Nature (IUCN) Species Survival Commission (SSC). Available at: https://www.hwctf.org. Accessed: September, 2022.

IUCN SSC PSG (2020a). *COVID-19 and Great Apes: Advisory for Extractive Industry Personnel, Applicable to Energy, Extractives, Transport Infrastructure, Agro-Industry and Other Projects Operating in Great Ape Habitats.* International Union for Conservation of Nature (IUCN) Species Survival Commission (SSC) Primate Specialist Group (PSG). Available at: http://static1.1.sqspcdn.com/static/f/1200343/28297087/1589210933010/COVID-19_advisory_for_extractive_projects.pdf?token=vXaQFVpBmg%2FmT%2B8G%2F8vBcZ7O0Q8%3D.

IUCN SSC PSG (2020b). *Regional Action Plan for the Conservation of Western Chimpanzees (*Pan troglodytes verus*) 2020–2030.* Gland, Switzerland: International Union for Conservation of Nature (IUCN) Species Survival Commission (SSC) Primate Specialist Group (PSG). DOI: 10.2305/IUCN.CH.2020.SSC-RAP.2.en.

IUCN SSC PSG (n.d.). *SARS-CoV-2 & COVID-19.* Geneva, Switzerland: International Union for Conservation of Nature (IUCN) Species Survival Commission (SSC) Primate Specialist Group (PSG). Available at: http://www.primate-sg.org/covid-19. Accessed: March, 2022.

IUCN SSC PSG SGA (n.d.-a). *COVID-19 Resources.* International Union for Conservation of Nature (IUCN) Species Survival Commission (SSC) Primate Specialist Group (PSG) Section on Great Apes (SGA). Available at: https://www.iucngreatapes.org/covid-19. Accessed: September, 2022.

IUCN SSC PSG SGA (n.d.-b). *IUCN SSC A.P.E.S. Database.* International Union for Conservation of Nature (IUCN) Species Survival Commission (SSC) Primate Specialist Group (PSG) Section on Great Apes (SGA). Available at: https://www.iucngreatapes.org/apes-database. Accessed: January, 2023.

Ives, C.D. and Bekessy, S.A. (2015). The ethics of offsetting nature. *Frontiers in Ecology and the Environment*, **13**(10), 568–73. DOI: 10.1890/150021.

J.A.C.K. Sanctuary (n.d.). *J.A.C.K. Primate Rehabilitation Centre.* Lubumbashi, DRC: Jeunes Animaux Confisqués au Katanga (J.A.C.K.). Available at: http://www.jacksanctuary.org/. Accessed: October, 2020.

Jacob, S.T., Crozier, I., Fischer, W.A. II, *et al.* (2020). Ebola virus disease. *Nature Reviews Disease Primers*, **6**(1), 13. DOI: 10.1038/s41572-020-0147-3.

Jacobson, S.L., Ross, S.R. and Bloomsmith, M.A. (2016). Characterizing abnormal behavior in a large population of zoo-housed chimpanzees: prevalence and potential influencing factors. *PeerJ*, **4**, e2225. DOI: 10.7717/peerj.2225.

Jakob-Hoff, R.M., MacDiarmid, S.C., Lees, C., *et al.* (2014). *Manual of Procedures for Wildlife Disease Risk Analysis.* Paris, France: World Organisation for Animal Health (OIE), in association with the International Union for Conservation of Nature (IUCN) and the Species Survival Commission (SSC). Available at: https://portals.iucn.org/library/sites/library/files/documents/2014-007.pdf.

Jameton, A. (1984). *Nursing Practice: The Ethical Issues.* Englewood Cliffs, NJ: Prentice-Hall.

Jane Goodall Institute (n.d.). *Tchimpounga Chimpanzee Rehabilitation Centre.* Lymington, UK: The Jane Goodall Institute UK. Available at: https://www.janegoodall.co.uk/our-programmes/tchimpounga-chimpanzee-rehabilitation-centre. Accessed: October, 2020.

Janson, C.H. and Goldsmith, M.L. (1995). Predicting group size in primates: foraging costs and predation risks. *Behavioral Ecology*, **6**(3), 326–36. DOI: 10.1093/beheco/6.3.326.

Jayme, S.I., Field, H.E., de Jong, C.E., *et al.* (2015). Molecular evidence of Ebola Reston virus infection in Philippine bats. *Virology Journal*, **12**(1), 107. DOI: 10.1186/s12985-015-0331-3.

Jean, S.M., Preuss, T.M., Sharma, P., *et al.* (2012). Cerebrovascular accident (stroke) in captive, group-housed, female chimpanzees. *Comparative Medicine*, **62**(4), 322–9.

JGI (n.d.). *Project Snare Removal.* Washington DC: Jane Goodall Institute (JGI). Available at: https://www.janegoodall.org/snare-removal-2/. Accessed: September, 2022.

Jim, C.Y. (1999). The forest fires in Indonesia 1997–98: possible causes and pervasive consequences. *Geography*, **84**(3), 251–60. DOI: http://www.jstor.org/stable/40573309.

John, D.A. and Babu, G.R. (2021). Lessons from the aftermaths of green revolution on food system and health. *Frontiers in Sustainable Food Systems*, **5**. DOI: 10.3389/fsufs.2021.644559.

Johns, B.G. (1996). Responses of chimpanzees to habituation and tourism in the Kibale Forest, Uganda. *Biological Conservation*, **78**(3), 257–62. DOI: 10.1016/S0006-3207(96)00044-4.

Johnson, C.K., Hitchens, P.L., Pandit, P.S., *et al.* (2020). Global shifts in mammalian population trends reveal key predictors of virus spillover risk. *Proceedings of the Royal Society B: Biological Sciences*, **287**(1924), 20192736. DOI: 10.1098/rspb.2019.2736.

Johnson, D.F., Druce, J.D., Birch, C. and Grayson, M.L. (2009). A quantitative assessment of the efficacy of surgical and N95 masks to filter influenza virus in patients with acute influenza infection. *Clinical Infectious Diseases*, **49**(2), 275–7. DOI: 10.1086/600041.

Jones, I.J., MacDonald, A.J., Hopkins, S.R., *et al.* (2020). Improving rural health care reduces illegal logging and conserves carbon in a tropical forest. *Proceedings of the National Academy of Sciences*, **117**(45), 28515–24. DOI: 10.1073/pnas.2009240117.

Jones, K.E., Patel, N.G., Levy, M.A., *et al.* (2008). Global trends in emerging infectious diseases. *Nature*, **451**(7181), 990–3. DOI: 10.1038/nature06536.

Jones, R.C. and Greek, R. (2014). A review of the Institute of Medicine's analysis of using chimpanzees in bio-medical research. *Science and Engineering Ethics*, **20**(2), 481–504. DOI: 10.1007/s11948-013-9442-7.

Jones, R.M. and Brosseau, L.M. (2015). Aerosol transmission of infectious disease. *Journal of Occupational and Environmental Medicine*, **57**(5), 501–8.

Jong, H.N. (2020). Indonesia forest fires push orangutans into starvation mode, study finds. *Mongabay*, January 24, 2020. Available at: https://news.mongabay.com/2020/01/indonesia-forest-fires-burning-orangutans-starvation-mode-borneo/.

Jong, H.N. (2022). Plantations threaten Indonesia's orangutans, but they're not oil palm. *Mongabay*, July 15, 2022. Available at: https://news.mongabay.com/2022/07/plantations-threaten-indonesias-orangutans-but-theyre-not-oil-palm/.

Joppa, L.N. (2015). Technology for nature conservation: an industry perspective. *Ambio*, **44**(4), 522–6. DOI: 10.1007/s13280-015-0702-4.

Juarez, C.P., Rotundo, M.A., Berg, W. and Fernández-Duque, E. (2011). Costs and benefits of radio-collaring on the behavior, demography, and conservation of owl monkeys (*Aotus azarai*) in Formosa, Argentina. *International Journal of Primatology*, **32**(1), 69–82. DOI: 10.1007/s10764-010-9437-z.

Junge, R.E., Gannon, F.H., Porton, I., McAlister, W.H. and Whyte, M.P. (2000). Management and prevention of vitamin D deficiency rickets in captive-born juvenile chimpanzees (*Pan troglodytes*). *Journal of Zoo and Wildlife Medicine*, **31**(3), 361–9.

Junker, J., Blake, S., Boesch, C., *et al.* (2012). Recent decline in suitable environmental conditions for African great apes. *Diversity and Distributions*, **18**(11), 1077–91. DOI: 10.1111/ddi.12005.

Junker, J., Kühl, H.S., Orth, L., *et al.* (2017). *Primate Conservation: Global Evidence for the Effects of Interventions.* Cambridge, UK: University of Cambridge. Available at: https://www.conservationevidence.com/synopsis/pdf/14.

Junker, J., Petrovan, S.O., Arroyo-Rodríguez, V., *et al.* (2020). A severe lack of evidence limits effective conservation of the world's primates. *BioScience*, **70**(9), 794–803. DOI: 10.1093/biosci/biaa082.

Justice, W.S.M., O'Brien, M.F., Szyszka, O., *et al.* (2017). Adaptation of the animal welfare assessment grid (AWAG) for monitoring animal welfare in zoological collections. *Veterinary Record*, **181**(6), 143. DOI: 10.1136/vr.104309.

Kabano, P., Arinaitwe, J. and Robbins, M.M. (2014). A brief history of habituated gorillas in Bwindi Impenetrable National Park. *Gorilla Journal*, **48**, 7–10.

Kagan, R., Carter, S. and Allard, S. (2015). A universal animal welfare framework for zoos. *Journal of Applied Animal Welfare Science*, **18**(S1), S1–10. DOI: 10.1080/10888705.2015.1075830.

Kahn, M. (1992). The passive voice of science: language abuse in the wildlife profession. *The Trumpeter Journal of Ecosophy*, **9**(4), 152–4.

Kalan, A.K., Piel, A.K., Mundry, R., *et al.* (2016). Passive acoustic monitoring reveals group ranging and territory use: a case study of wild chimpanzees (*Pan troglodytes*). *Frontiers in Zoology*, **13**(1), 34. DOI: 10.1186/s12983-016-0167-8.

Kalema-Zikusoka, G. and Byonanebye, J. (2019). Scaling up a one-health model of conservation through public health: experiences in Uganda and the Democratic Republic of the Congo. *The Lancet Global Health*, **7**, S34. DOI: 10.1016/S2214-109X(19)30119-6.

Kalema-Zikusoka, G., Kock, R.A. and Macfie, E.J. (2002). Scabies in free-ranging mountain gorillas (*Gorilla beringei beringei*) in Bwindi Impenetrable National Park, Uganda. *Veterinary Record*, **150**, 12–15.

Kalema-Zikusoka, G., Rubanga, S., Mutahunga, B. and Sadler, R. (2018). Prevention of *Cryptosporidium* and *Giardia* at the human/gorilla/livestock interface. *Frontiers in Public Health*, **6**. DOI: 10.3389/fpubh.2018.00364.

Kalema-Zikusoka, G., Rubanga, S., Ngabirano, A. and Zikusoka, L. (2021). Mitigating impacts of the COVID-19 pandemic on gorilla conservation: lessons from Bwindi Impenetrable Forest, Uganda. *Frontiers in Public Health*, **9**, December 14, 2018. DOI: 10.3389/fpubh.2021.655175.

Kalema-Zikusoka, G. and Rwego, I.B. (2016). Mountain gorillas, tourism, and conflicts with people living adjacent to Bwindi Impenetrable National Park. In *Tropical Conservation, Perspectives on Local and Global Priorities*, ed. A. Aguirre and R. Sukumar. Oxford, UK: Oxford University Press, pp. 136–9.

Kalter, S.S. (1989). Infectious diseases of nonhuman primates in a zoo setting. *Zoo Biology*, **8**(S1), 61–76. DOI: 10.1002/zoo.1430080508.

Kanamori, T., Kuze, N., Bernard, H., Malim, T.P. and Kohshima, S. (2012). Fatality of a wild Bornean orangutan (*Pongo pygmaeus morio*): behavior and death of a wounded juvenile in Danum Valley, North Borneo. *Primates*, **53**(3), 221–6. DOI: 10.1007/s10329-012-0297-3.

Kaplan, G. and Rogers, L.J. (2000). *The Orangutans: Their Evolution, Behaviour and Future*. Philadelphia, PA: Perseus Running Press.

Kappeler, P.M. and Watts, D.P. (2012). *Long-Term Field Studies of Primates*. Berlin, Germany: Springer-Verlag.

Karesh, W.B. and Cook, R.A. (2009). One world – one health. *Clinical Medicine*, **9**(3), 259–60. DOI: 10.7861/clinmedicine.9-3-259.

Karlsson, M. and Edvardsson Björnberg, K. (2021). Ethics and biodiversity offsetting. *Conservation Biology*, **35**(2), 578–86. DOI: 10.1111/cobi.13603.

Karokaro, A.S., Gokkon, B. and Suriyani, L.D. (2017). Indonesia is running out of places to put rescued animals. *Mongabay*, July 3, 2017. Available at: https://news.mongabay.com/2017/07/indonesia-is-running-out-of-places-to-put-rescued-animals/.

Kaur, T., Singh, J., Tong, S., *et al.* (2008). Descriptive epidemiology of fatal respiratory outbreaks and detection of a human-related metapneumovirus in wild chimpanzees (*Pan troglodytes*) at Mahale Mountains National Park, Western Tanzania. *American Journal of Primatology*, **70**(8), 755–65. DOI: 10.1002/ajp.20565.

Kavanagh, M. and Caldecott, J.O. (2013). Strategic guidelines for the translocation of primates and other animals. *The Raffles Bulletin of Zoology*, **29**, 203–9.

KCP (n.d.). *Kibale Chimpanzee Project: Research, Conservation, and Education. Veterinary Intervention*. Kibale Chimpanzee Project (KCP). Available at: https://kibalechimpanzees.wordpress.com/veterinary-intervention/. Accessed: August, 2022.

Keele, B.F., Jones, J.H., Terio, K.A., *et al.* (2009). Increased mortality and AIDS-like immunopathology in wild chimpanzees infected with SIVcpz. *Nature*, **460**(7254), 515–19. DOI: 10.1038/nature08200.

Keele, B.F., Van Heuverswyn, F., Li, Y., *et al.* (2006). Chimpanzee reservoirs of pandemic and nonpandemic HIV-1. *Science*, **313**(5786), 523–6.

Keesing, F., Belden, L.K., Daszak, P., *et al.* (2010). Impacts of biodiversity on the emergence and transmission of infectious diseases. *Nature*, **468**(7324), 647–52. DOI: 10.1038/nature09575.

Keita, M.B., Hamad, I. and Bittar, F. (2014). Looking in apes as a source of human pathogens. *Microbial Pathogenesis*, **77**, 149–54. DOI: 10.1016/j.micpath.2014.09.003.

Kelly, A., Osburn, B. and Salman, M. (2014). Veterinary medicine's increasing role in global health. *The Lancet Global Health*, **2**(7), e379–80. DOI: 10.1016/S2214-109X(14)70255-4.

Kelly, T.R., Machalaba, C., Karesh, W.B., *et al.* (2020). Implementing One Health approaches to confront emerging and re-emerging zoonotic disease threats: lessons from PREDICT. *One Health Outlook*, **2**(1), 1. DOI: 10.1186/s42522-019-0007-9.

Kernbach, M., Ramsay, C., Rohr, J.R. and Martin, L.B. (2019). Eco-immunology: past, present, and future. In *Encyclopedia of Ecology*, 2nd edn, ed. B. Fath. Oxford, UK: Elsevier, pp. 64–71. DOI: 10.1016/B978-0-12-409548-9.10890-5.

Kik, M.J.L., Bos, J.H., Groen, J. and Dorrestein, G.M. (2005). Herpes simplex infection in a juvenile orangutan (*Pongo pygmaeus pygmaeus*). *Journal of Zoo and Wildlife Medicine*, **36**(1), 131–4.

Kilbourn, A.M., Bosi, E.J., Karesh, W.B., Landau, M. and Tambing, E. (1997). Disease evaluation of free-ranging orangutans (*Pongo pygmaeus pygmaeus*) in Sabah, Malaysia. Presented at: *Proceedings of the Annual American Association of Zoo Veterinarians Conference 1998, Houston, TX*. Jacksonville, FL: American Association of Zoo Veterinarians.

Kilbourn, A.M., Karesh, W.B., Wolfe, N.D., *et al.* (2003). Health evaluation of free-ranging and semi-captive orangutans (*Pongo pygmaeus pygmaeus*) in Sabah, Malaysia. *Journal of Wildlife Diseases*, **39**(1), 73–83. DOI: 10.7589/0090-3558-39.1.73.

Kimbrough, L. (2020). Around the world, a fire crisis flares up, fueled by human actions. *Mongabay*, September 4, 2020. Available at: https://news.mongabay.com/2020/09/around-the-world-a-fire-crisis-flares-up-fueled-by-human-actions/.

King, T., Chamberlan, C. and Courage, A. (2006). *Gorilla Reintroduction, Republic of Congo. A Report for the PASA/IUCN African Primate Reintroduction Workshop, 20–22 April 2006, Apeldoorn, the Netherlands*. Brazzaville, Republic of Congo: The John Aspinall Foundation. Available at: https://www.ppgcongo.org/reintrodution/ppg-congo-2006-gorilla-reintro-pasa-en.pdf.

King, T., Chamberlan, C. and Courage, A. (2012). Assessing initial reintroduction success in long-lived primates by quantifying survival, reproduction and dispersal parameters: western lowland gorillas (*Gorilla gorilla gorilla*) in Congo and Gabon. *International Journal of Primatology*, **33**(1), 134–49. DOI: 10.1007/s10764-011-9563-2.

Kiran, D., Sander, W.E. and Duncan, C. (2022). Empowering veterinarians to be planetary health stewards through policy and practice. *Frontiers in Veterinary Science*, **9**, 775411. DOI: 10.3389/fvets.2022.775411.

Kirby, J.N., Steindl, S.R. and Doty, J.R. (2017). Compassion as the highest ethic. In *Practitioner's Guide to Ethics and Mindfulness-Based Interventions*, ed. L. M. Monteiro, J. F. Compson and F. Musten. Cham, Switzerland: Springer International Publishing, pp. 253–77. DOI: 10.1007/978-3-319-64924-5_10.

Klailova, M., Casanova, C., Henschel, P., *et al.* (2013). Non-human predator interactions with wild great apes in Africa and the use of camera traps to study their dynamics. *Folia Primatologica*, **83**(3–6), 312–28. DOI: 10.1159/000342143.

Klee, S.R., Brzuszkiewicz, E.B., Nattermann, H., *et al.* (2010). The genome of a *Bacillus* isolate causing anthrax in chimpanzees combines chromosomal properties of *B. cereus* with *B. anthracis* virulence plasmids. *PLoS ONE*, **5**(7), e10986. DOI: 10.1371/journal.pone.0010986.

Kleinschmidt, L.M., Kinney, M.E. and Hanley, C.S. (2018). Treatment of disseminated *Strongyloides* spp. infection in an infant Sumatran orangutan (*Pongo abelii*). *Journal of Medical Primatology*, **47**(3), 201–4. DOI: 10.1111/jmp.12338.

Knauf, S., Gogarten, J.F., Schuenemann, V.J., *et al.* (2018). Nonhuman primates across sub-Saharan Africa are infected with the yaws bacterium *Treponema pallidum* subsp. *pertenue*. *Emerging Microbes & Infections*, **7**(1), 1–4. DOI: 10.1038/s41426-018-0156-4.

Knauf, S., Liu, H. and Harper, K.N. (2013). Treponemal infection in nonhuman primates as possible reservoir for human yaws. *Emerging Infectious Diseases*, **19**(12), 2058–60. DOI: 10.3201/eid1912.130863.

Knight, A. (2008). The beginning of the end for chimpanzee experiments? *Philosophy, Ethics, and Humanities in Medicine*, **3**(1), 16. DOI: 10.1186/1747-5341-3-16.

Knight, J. (2009). Making wildlife viewable: habituation and attraction. *Society & Animals*, **17**(2), 167–84. DOI: 10.1163/156853009X418091.

Knott, C.D. (1998). Orangutan in the wild. *National Geographic Magazine*, **2**(2), 30–57.

Knott, C.D. (2001). Female reproductive ecology of the apes: implications for human evolution. In *Reproductive Ecology and Human Evolution*, ed. P. T. Ellison. Hawthorne, NY: Walter de Gruyter, pp. 429–63. Available at: https://cherylknott.files.wordpress.com/2011/06/knott-2001-female-reproductive-ecology-of-the-apes.pdf.

Knott, C.D. (2005). Energetic responses to food availability in the great apes: implications for hominin evolution. In *Seasonality in Primates: Studies of Living and Extinct Human and Non-Human Primates*, ed. D. K. Brockman and C. P. van Schaik. New York, NY: Cambridge University Press, pp. 351–78.

Knott, C.D., Beaudrot, L., Snaith, T.V., *et al.* (2008). Female–female competition in Bornean orangutans. *International Journal of Primatology*, **29**, 975–97.

Knott, C.D., Kane, E.E., Achmad, M., *et al.* (2021). The Gunung Palung Orangutan Project: twenty-five years at the intersection of research and conservation in a critical landscape in Indonesia. *Biological Conservation*, **255**, 108856. DOI: 10.1016/j.biocon.2020.108856.

Knott, C.D., Scott, A.M., O'Connell, C.A., *et al.* (2019). Possible male infanticide in wild orangutans and a re-evaluation of infanticide risk. *Scientific Reports*, **9**, 7806. DOI: 10.1038/s41598-019-42856-w.

Knott, K. (2021). Hong Kong's leading role in the global extinction crisis, as hub of illegal wildlife trade, and the legal amendment that could change that. *South China Morning Post, Lifestyle*, April 23, 2021. Available at: https://www.scmp.com/lifestyle/article/3130438/hong-kongs-leading-role-global-extinction-crisis-hub-illegal-wildlife.

Kock, R., Michel, A.L., Yeboah-Manu, D., *et al.* (2021). Zoonotic tuberculosis: the changing landscape. *International Journal of Infectious Diseases*, **113**(S1), S68–72. DOI: 10.1016/j.ijid.2021.02.091.

Kock, R.A., Woodford, M.H. and Rossiter, P.B. (2010). Disease risks associated with the translocation of wildlife. *Revue Scientifique et Technique de l'Office International des Épizooties*, **29**(2), 329–50. DOI: 10.20506/rst.29.2.1980.

Koeppel, L., Siems, T., Fischer, M. and Lentz, H.H.K. (2018). Automatic classification of farms and traders in the pig production chain. *Preventive Veterinary Medicine*, **150**, 86–92. DOI: 10.1016/j.prevetmed.2017.12.003.

Köndgen, S., Calvignac-Spencer, S., Grützmacher, K., *et al.* (2017). Evidence for human *Streptococcus pneumoniae* in wild and captive chimpanzees: a potential threat to wild populations. *Scientific Reports*, **7**, 14581. DOI: 10.1038/s41598-017-14769-z.

Köndgen, S., Kühl, H., N'Goran, P.K., *et al.* (2008). Pandemic human viruses cause decline of endangered great apes. *Current Biology*, **18**(4), 260–4. DOI: 10.1016/j.cub.2008.01.012.

Köndgen, S., Leider, M., Lankester, F., *et al.* (2011). *Pasteurella multocida* involved in respiratory disease of wild chimpanzees. *PLoS ONE*, **6**(9), e24236. DOI: 10.1371/journal.pone.0024236.

Köndgen, S., Schenk, S., Pauli, G., Boesch, C. and Leendertz, F.H. (2010). Noninvasive monitoring of respiratory viruses in wild chimpanzees. *EcoHealth*, **7**(3), 332–41. DOI: 10.1007/s10393-010-0340-z.

Kooriyama, T., Okamoto, M., Yoshida, T., *et al.* (2013). Epidemiological study of zoonoses derived from humans in captive chimpanzees. *Primates*, **54**(1), 89–98. DOI: 10.1007/s10329-012-0320-8.

Kormos, R., Boesch, C., Bakarr, M.I. and Butynski, T.M. (2003). *West African Chimpanzees: Status, Survey and Conservation Action Plan*. Gland, Switzerland: International Union for Conservation of Nature (IUCN) World Conservation Union. Available at: https://portals.iucn.org/library/sites/library/files/documents/2003-059.pdf.

Kormos, R., Kormos, C.F., Humle, T., *et al.* (2014). Great apes and biodiversity offset projects in Africa: the case for national offset strategies. *PLoS ONE*, **9**(11), e111671. DOI: 10.1371/journal.pone.0111671.

Köster, P.C., Lapuente, J., Cruz, I., Carmena, D. and Ponce-Gordo, F. (2022). Human-borne pathogens: are they threatening wild great ape populations? *Veterinary Sciences*, **9**(7), 356. DOI: 10.3390/vetsci9070356.

Kralik, P. and Ricchi, M. (2017). A basic guide to real time PCR in microbial diagnostics: definitions, parameters, and everything. *Frontiers in Microbiology*, **8**, 108. DOI: 10.3389/fmicb.2017.00108.

Krebs, B.L., Marrin, D., Phelps, A., Krol, L. and Watters, J.V. (2018). Managing aged animals in zoos to promote positive welfare: a review and future directions. *Animals*, **8**(7), 116. DOI: 10.3390/ani8070116.

Krief, S., Berny, P., Gumisiriza, F., *et al.* (2017). Agricultural expansion as risk to endangered wildlife: pesticide exposure in wild chimpanzees and baboons displaying facial dysplasia. *Science of The Total Environment*, **598**(4), 647–56. DOI: 10.1016/j.scitotenv.2017.04.113.

Krief, S., Escalante, A.A., Pacheco, M.A., *et al.* (2010). On the diversity of malaria parasites in African apes and the origin of *Plasmodium falciparum* from bonobos. *PLoS Pathogens*, **6**(2), e1000765. DOI: 10.1371/journal.ppat.1000765.

Krüger, O. (2005). The role of ecotourism in conservation: panacea or Pandora's box? *Biodiversity and Conservation*, **14**(3), 579–600. DOI: 10.1007/s10531-004-3917-4.

Kühl, H.S., Boesch, C., Kulik, L., *et al.* (2019). Human impact erodes chimpanzee behavioral diversity. *Science*, **363**(6434), 1453. DOI: 10.1126/science.aau4532.

Kühl, H.S., Sop, T., Williamson, E.A., *et al.* (2017). The critically endangered western chimpanzee declines by 80%. *American Journal of Primatology*, **79**(9), e22681. DOI: 10.1002/ajp.22681.

Kühl, H., Williamson, L., Sanz, C., Morgan, D. and Boesch, C. (2007). Launch of the A.P.E.S. database. *Gorilla Journal*, **34**, 20–1.

Kuisma, E., Olson, S.H., Cameron, K.N., *et al.* (2019). Long-term wildlife mortality surveillance in northern Congo: a model for the detection of Ebola virus disease epizootics. *Philosophical Transactions of the Royal Society B: Biological Sciences*, **374**(1782), 20180339. DOI: 10.1098/rstb.2018.0339.

Kumar, S., Fox, B., Owston, M., Hubbard, G.B. and Dick, E.J. Jr. (2012). Pathology of spontaneous air sacculitis in 37 baboons and seven chimpanzees and a brief review of the literature. *Journal of Medical Primatology*, **41**(4), 266–77. DOI: 10.1111/j.1600-0684.2012.00547.x.

Kumar, S., Laurence, H., Owston, M.A., *et al.* (2017). Natural pathology of the captive chimpanzee (*Pan troglodytes*): a 35-year review. *Journal of Medical Primatology*, **46**(5), 271–90. DOI: 10.1111/jmp.12277.

Kumareswaran, K. and Jayasinghe, G.Y. (2022). Systematic review on ensuring the global food security and covid-19 pandemic resilient food systems: towards accomplishing sustainable development goals targets. *Discover Sustainability*, **3**(1), 29. DOI: 10.1007/s43621-022-00096-5.

Kumm, H.W. and Turner, T.B. (1936). The transmission of yaws from man to rabbits by an insect vector, *Hippelates pallipes* Loew. *American Journal of Tropical Medicine and Hygiene*, **S1–16**(3), 245–71. DOI: 10.4269/ajtmh.1936. s1-16.245.

Kuze, N., Dellatore, D., Banes, G.L., *et al.* (2012). Factors affecting reproduction in rehabilitant female orangutans: young age at first birth and short inter-birth interval. *Primates*, **53**(2), 181–92. DOI: 10.1007/s10329-011-0285-z.

Labes, E.M., Hegglin, D., Grimm, F., *et al.* (2010). Intestinal parasites of endangered orangutans (*Pongo pygmaeus*) in Central and East Kalimantan, Borneo, Indonesia. *Parasitology*, **137**(1), 123–35. DOI: 10.1017/s0031182009991120.

Labes, E.M., Nurcahyo, W., Deplazes, P. and Mathis, A. (2011). Genetic characterization of *Strongyloides* spp. from captive, semi-captive and wild Bornean orangutans (*Pongo pygmaeus*) in Central and East Kalimantan, Borneo, Indonesia. *Parasitology*, **138**, 1417–22.

Lahm, S.A., Kombila, M., Swanepoel, R. and Barnes, R.F.W. (2007). Morbidity and mortality of wild animals in relation to outbreaks of Ebola haemorrhagic fever in Gabon, 1994–2003. *Transactions of the Royal Society of Tropical Medicine and Hygiene*, **101**(1), 64–78. DOI: 10.1016/j.trstmh.2006.07.002.

Lainé, N. and Morand, S. (2020). Linking humans, their animals, and the environment again: a decolonized and more-than-human approach to "One Health". *Parasite*, **27**, 55. DOI: 10.1051/parasite/2020055.

Lambeth, S.P., Bloomsmith, M.A. and Alford, P.L. (1997). Effects of human activity on chimpanzee wounding. *Zoo Biology*, **16**(4), 327–33. DOI: 10.1002/(SICI)1098-2361(1997)16:4<327::AID-ZOO4>3.0.CO;2-C.

Lammey, M.L., Baskin, G.B., Gigliotti, A.P., *et al.* (2008). Interstitial myocardial fibrosis in a captive chimpanzee (*Pan troglodytes*) population. *Comparative Medicine*, **58**(4), 389–94.

Landolfi, J.A., Wellehan, J.F., Johnson, A.J. and Kinsel, M.J. (2005). Fatal human herpesvirus type 1 infection in a white-handed gibbon (*Hylobates lar*). *Journal of Veterinary Diagnostic Investigation*, **17**(4), 369–71. DOI: 10.1177/104063870501700412.

Langford, D.J., Bailey, A.L., Chanda, M.L., *et al.* (2010). Coding of facial expressions of pain in the laboratory mouse. *Nature Methods*, **7**(6), 447–9. DOI: 10.1038/nmeth.1455.

Lankester, F., Mätz-Rensing, K., Kiyang, J., *et al.* (2008). Fatal ulcerative colitis in a western lowland gorilla (*Gorilla gorilla gorilla*). *Journal of Medical Primatology*, **37**(6), 297–302. DOI: 10.1111/j.1600-0684.2008.00287.x.

Lappan, S. (2008). Male care of infants in a siamang (*Symphalangus syndactylus*) population including socially monogamous and polyandrous groups. *Behavioral Ecology and Sociobiology*, **62**(8), 1307–17. DOI: 10.1007/s00265-008-0559-7.

Lappan, S. (2009). Flowers are an important food for small apes in southern Sumatra. *American Journal of Primatology*, **71**(8), 624–35. DOI: 10.1002/ajp.20691.

Lappan, S., Malaivijitnond, S., Radhakrishna, S., Riley, E.P. and Ruppert, N. (2020). The human–primate interface in the new normal: challenges and opportunities for primatologists in the COVID-19 era and beyond. *American Journal of Primatology*, **82**(8), e23176. DOI: 10.1002/ajp.23176.

Latip, N.A., Marzuki, A., Marcela, P. and Umar, M.U. (2015). The involvement of indigenous peoples in promoting conservation and sustainable tourism at Lower Kinabatangan Sabah: common issues and challenges. *Australian Journal of Basic and Applied Science*, **9**(7), 323–5.

Laule, G.E., Bloomsmith, M.A. and Schapiro, S.J. (2003). The use of positive reinforcement training techniques to enhance the care, management, and welfare of primates in the laboratory. *Journal of Applied Animal Welfare Science*, **6**(3), 163–73. DOI: 10.1207/S15327604JAWS0603_02.

Laurance, W.F. (2013). Does research help to safeguard protected areas? *Trends in Ecology & Evolution*, **28**(5), 261–6. DOI: 10.1016/j.tree.2013.01.017.

Laurance, W.F., Croes, B.M., Tchignoumba, L., *et al.* (2006). Impacts of roads and hunting on central African rainforest mammals. *Conservation Biology*, **20**(4), 1251–61. DOI: 10.1111/j.1523-1739.2006.00420.x.

Laurance, W.F., Sloan, S., Weng, L. and Sayer, J.A. (2015). Estimating the environmental costs of Africa's massive "development corridors". *Current Biology*, **25**(24), 3202–8. DOI: 10.1016/j.cub.2015.10.046.

Laurance, W.F., Wich, S.A., Onrizal, O., *et al.* (2020). Tapanuli orangutan endangered by Sumatran hydropower scheme. *Nature Ecology & Evolution*, **4**(11), 1438–9. DOI: 10.1038/s41559-020-1263-x.

Laurence, H., Kumar, S., Owston, M.A., *et al.* (2017). Natural mortality and cause of death analysis of the captive chimpanzee (*Pan troglodytes*): a 35-year review. *Journal of Medical Primatology*, **46**(3), 106–15. DOI: 10.1111/jmp.12267.

Lavergne, A., Donato, D., Gessain, A., *et al.* (2014). African great apes are naturally infected with roseoloviruses closely related to human herpesvirus 7. *Journal of Virology*, **88**(22), 13212–20. DOI: 10.1128/jvi.01490-14.

Lawson, B., Garriga, R. and Galdikas, B.M. (2006). Airsacculitis in fourteen juvenile southern Bornean orangutans (*Pongo pygmaeus wurmbii*). *Journal of Medical Primatology*, **35**(3), 149–54. DOI: 10.1111/j.1600-0684.2006.00153.x.

Lécu, A. and Ball, R. (2011). Mycobacterial infections in zoo animals: relevance, diagnosis and management. *International Zoo Yearbook*, **45**(1), 183–202. DOI: 10.1111/j.1748-1090.2011.00141.x.

Ledger, E. (2020). Mountain gorillas face extinction due to threats of both coronavirus and poaching. *Independent*, August 20, 2020. Available at: https://www.independent.co.uk/news/world/coronavirus-mountain-gorillas-poaching-uganda-wildlife-crime-trade-a9633586.html.

Lee, A., Leong, M. and Dzar, A. (2020). Mariani Ramli's life mission to help gibbons sing again. *Malaysiakini*, June 8, 2020. Available at: https://www.malaysiakini.com/news/529224.

Lee, E.A.A. (2012). *Social interaction and occupational enrichment in captive Bonnet macaques (*Macaca radiata*)*. Undergraduate thesis. Serdang, Malaysia: Universiti Putra Malaysia.

Lee, K. and Brumme, Z.L. (2013). Operationalizing the One Health approach: the global governance challenges. *Health Policy and Planning*, **28**(7), 778–85. DOI: 10.1093/heapol/czs127.

Lee, R.V., Allan, W.P., Sidney, A., *et al.* (1990). Typhlitis due to *Balantidium coli* in captive lowland gorillas. *Reviews of Infectious Diseases*, **12**(6), 1052–9. DOI: 10.1093/clinids/12.6.1052.

Leeds, A., Elsner, R. and Lukas, K.E. (2016). The effect of positive reinforcement training on an adult female western lowland gorilla's (*Gorilla gorilla gorilla*) rate of abnormal and aggressive behavior. *Animal Behavior and Cognition*, **3**(2), 78–87.

Leempoel, K., Hebert, T. and Hadly, E.A. (2020). A comparison of eDNA to camera trapping for assessment of terrestrial mammal diversity. *Proceedings of the Royal Society B: Biological Sciences*, **287**(1918), 20192353. DOI: 10.1098/rspb.2019.2353.

Leendertz, F.H. and Kalema-Zikusoka, G. (2021). Vaccinate in biodiversity hotspots to protect people and wildlife from each other. *Nature*, **591**(7850), 369. DOI: 10.1038/d41586-021-00690-z.

Leendertz, F.H., Ellerbrok, H., Boesch, C., *et al.* (2004). Anthrax kills wild chimpanzees in a tropical rainforest. *Nature*, **430**(6998), 451–2. DOI: 10.1038/nature02722.

Leendertz, F.H., Lankester, F., Guislain, P., *et al.* (2006a). Anthrax in western and central African great apes. *American Journal of Primatology*, **68**(9), 928–33. DOI: 10.1002/ajp.20298.

Leendertz, F.H., Pauli, G., Maetz-Rensing, K., *et al.* (2006b). Pathogens as drivers of population declines: the importance of systematic monitoring in great apes and other threatened mammals. *Biological Conservation*, **131**(2), 325–37. DOI: 10.1016/j.biocon.2006.05.002.

Leendertz, S.A.J., Gogarten, J.F., Düx, A., Calvignac-Spencer, S. and Leendertz, F.H. (2016). Assessing the evidence supporting fruit bats as the primary reservoirs for Ebola viruses. *EcoHealth*, **13**(1), 18–25. DOI: 10.1007/s10393-015-1053-0.

Leendertz, S.A.J., Locatelli, S., Boesch, C., *et al.* (2011). No evidence for transmission of SIVwrc from western red colobus monkeys (*Piliocolobus badius badius*) to wild West African chimpanzees (*Pan troglodytes verus*) despite high exposure through hunting. *BMC Microbiology*, **11**(1), 24. DOI: 10.1186/1471-2180-11-24.

Leendertz, S.A.J., Wich, S.A., Ancrenaz, M., *et al.* (2017). Ebola in great apes: current knowledge, possibilities for vaccination, and implications for conservation and human health. *Mammal Review*, **47**(2), 98–111. DOI: 10.1111/mam.12082.

Lehmann, J., Korstjens, A.H. and Dunbar, R.I.M. (2010). Apes in a changing world – the effects of global warming on the behaviour and distribution of African apes. *Journal of Biogeography*, **37**(12), 2217–31. DOI: 10.1111/j.1365-2699.2010.02373.x.

Leighton, D.S.R. (1987). Gibbons: territoriality and monogamy. In *Primate Societies*, ed. B. B. Smuts, D. L. Cheyney, R. M. Seyfarth, R. W. Wrangham and T. T. Struhsaker. Chicago, IL: University of Chicago Press, pp. 135–145.

Leighty, K.A., Valuska, A.J., Grand, A.P., *et al.* (2015). Impact of visual context on public perceptions of non-human primate performers. *PLoS ONE*, **10**(2), e0118487. DOI: 10.1371/journal.pone.0118487.

Lempena, J. and Sal, A. (2018). *Captive Care Standards. An Overview of Standards for the Keeping of Wild Animals in Captive Care Settings in Malawi*. Lilongwe, Malawi: Lilongwe Wildlife Trust.

Lerner, H. and Berg, C. (2017). A comparison of three holistic approaches to health: One health, ecohealth, and planetary health. *Frontiers in Veterinary Science*, **4**, September 29, 2017. DOI: 10.3389/fvets.2017.00163.

Leroy, E.M., Kumulungui, B., Pourrut, X., *et al.* (2005). Fruit bats as reservoirs of Ebola virus. *Nature*, **438**(7068), 575–6. DOI: 10.1038/438575a.

Leroy, E.M., Rouquet, P., Formenty, P., *et al.* (2004). Multiple *Ebola* virus transmission events and rapid decline of central African wildlife. *Science*, **303**(5656), 387–90. DOI: 10.1126/science.1092528.

Lestari, A. and Puspita Ayu, K. (2020). Engaging palm oil and hot spot area to mitigate forest fires. *BIO Web of Conferences*, **20**, 01003. DOI: 10.1051/bioconf/20202001003.

Lewis, B. and Nogueira, M. (2021). Timeline: The battle for Simandou. *MINING.COM*, January 22, 2021. Available at: https://www.mining.com/web/timeline-the-battle-for-simandou/.

Liberia Chimpanzee Rescue & Protection (n.d.). *LCRP's Sanctuary and Conservation Center*. Monrovia, Liberia: Liberia Chimpanzee Rescue and Protection. Available at: https://www.liberiachimpanzeerescue.org/our-home.html. Accessed: March, 2021.

Limbe Wildlife Center (2020). *Limbe Wildlife Centre Annual Report 2019*. Limbe, Republic of Cameroon: Limbe Wildlife Center. Available at: https://issuu.com/limbewildlife/docs/lwc_annualreport_2019.

Linden, B., Foord, S., Horta-Lacueva, Q.J.B. and Taylor, P.J. (2020). Bridging the gap: how to design canopy bridges for arboreal guenons to mitigate road collisions. *Biological Conservation*, **246**, 108560. DOI: 10.1016/j.biocon.2020.108560.

Lindshield, S., Bogart, S.L., Gueye, M., Ndiaye, P.I. and Pruetz, J.D. (2019). Informing protection efforts for critically endangered chimpanzees (*Pan troglodytes verus*) and sympatric mammals amidst rapid growth of extractive industries in Senegal. *Folia Primatologica*, **90**(2), 124–36. DOI: 10.1159/000496145.

Lindshield, S., Hernandez-Aguilar, R.A., Korstjens, A.H., *et al.* (2021). Chimpanzees (*Pan troglodytes*) in savanna landscapes. *Evolutionary Anthropology: Issues, News, and Reviews*, **30**(6), 399–420. DOI: 10.1002/evan.21924.

Liptovszky, M., Poitier, R., Redrobe, S., Schüle, A. and Steinmetz, H.W. (2019). *EAZA Great Ape TAG Veterinary Guidelines (July 2019)*. Amsterdam, the Netherlands: European Association of Zoos and Aquaria (EAZA) Great Ape Taxon Advisory Group (TAG).

Listín Diario (2019). María y Linda, las chimpancés diferentes que llenan de emociones el zoológico Nacional. *Listín Diario YouTube*, December 26, 2019. Available at: https://www.youtube.com/watch?v=_UVi4ugqt30.

Liswanti, N., Indawan, A., Sumardjo, D. and Sheil, D. (2004). Persepsi Masyarakat Dayak Merap Dan Punan Tentang Pentingnya Hutan Di Lansekap Hutan Tropis, Kabupaten Malinau, Kalimantan Timur [Dayak Merap and Punan People's perception of the importance of forest in a tropical landscape, Malinau, East Kalimantan]. *Jurnal Manajemen Hutan Tropika*, **10**(2), 1–3.

Litchfield, C.A. (2008). Responsible tourism: a conservation tool or conservation threat? In *Conservation in the 21st Century: Gorillas as a Case Study*, ed. T. S. Stoinski, H. D. Steklis and P. T. Mehlman. Boston, MA: Springer, pp. 107–27. DOI: 10.1007/978-0-387-70721-1_4.

Liu, W., Li, Y., Learn, G.H., *et al.* (2010). Origin of the human malaria parasite *Plasmodium falciparum* in gorillas. *Nature*, **467**(7314), 420–5. DOI: 10.1038/nature09442.

Lochmiller, R.L. (1996). Immunocompetence and animal population regulation. *Oikos*, **76**(3), 594–602. DOI: 10.2307/3546356.

Löhrich, T., Behringer, V., Wittig, R.M., Deschner, T. and Leendertz, F.H. (2018). The use of neopterin as a non-invasive marker in monitoring diseases in wild chimpanzees. *EcoHealth*, **15**(4), 792–803. DOI: 10.1007/s10393-018-1357-y.

Loken, B., Boer, C. and Kasyanto, N. (2015). Opportunistic behaviour or desperate measure? Logging impacts may only partially explain terrestriality in the Bornean orang-utan *Pongo pygmaeus morio*. *Oryx*, **49**(3), 461–4. DOI: 10.1017/S0030605314000969.

Loken, B., Spehar, S. and Rayadin, Y. (2013). Terrestriality in the Bornean orangutan (*Pongo pygmaeus morio*) and implications for their ecology and conservation. *American Journal of Primatology*, **75**(11), 1129–38. DOI: 10.1002/ajp.22174.

Long, K. and Robley, A. (2004). *Cost Effective Feral Animal Exclusion Fencing for Areas of High Conservation Value in Australia*. Canberra, Australia: Commonwealth of Australia. Available at: https://www.environment.gov.au/biodiversity/invasive-species/publications/cost-effective-feral-animal-exclusion-fencing.

Lonsdorf, E.V., Ross, S.R. and Matsuzawa, T. (2010). *The Mind of the Chimpanzee: Experimental and Ecological Perspectives*. Chicago, IL: University of Chicago Press.

Lonsdorf, E.V., Travis, D., Pusey, A.E. and Goodall, J. (2006). Using retrospective health data from the Gombe chimpanzee study to inform future monitoring efforts. *American Journal of Primatology*, **68**(9), 897–908. DOI: 10.1002/ajp.20296.

Lonsdorf, E.V., Travis, D.A., Raphael, J., *et al.* (2022). The Gombe Ecosystem Health Project: 16 years of program evolution and lessons learned. *American Journal of Primatology*, **84**(4–5), e23300. DOI: 10.1002/ajp.23300.

Lonsdorf, E.V., Travis, D., Ssuna, R., *et al.* (2014). Field immobilization for treatment of an unknown illness in a wild chimpanzee (*Pan troglodytes schweinfurthii*) at Gombe National Park, Tanzania: findings, challenges, and lessons learned. *Primates*, **55**(1), 89–99. DOI: 10.1007/s10329-013-0372-4.

Loos, A. and Ernst, A. (2013). An automated chimpanzee identification system using face detection and recognition. *EURASIP Journal on Image and Video Processing*, **2013**(1), 49. DOI: 10.1186/1687-5281-2013-49.

Loos, A. and Kalyanasundaram, T.A.M. (2015). Face recognition for great apes: identification of primates in videos. Presented at: *2015 IEEE International Conference on Acoustics, Speech and Signal Processing (ICASSP), 19–24 April 2015, Brisbane, Queensland Australia*. Piscataway, NJ: Institute of Electrical and Electronics Engineers Signal Processing Society (IEEE), pp. 1548–1552. DOI: 10.1109/ICASSP.2015.7178230.

Lovejoy, T.E. and Nobre, C. (2019). Amazon tipping point: last chance for action. *Science Advances*, **5**(12), eaba2949. DOI: 10.1126/sciadv.aba2949.

Lowenstine, L.J., McManamon, R., Bonar, C. and Perkins, L. (2008). Preliminary results of a survey of United States and Canadian orangutan mortalities in the North American SSP population from 1980 to March 2008. Presented at: *Proceedings of the Annual Meeting of the American Association of Zoo Veterinarians, Los Angeles, CA, 11–17 October 2008*, Jacksonville, FL: American Association of Zoo Veterinarians.

Lowenstine, L.J., McManamon, R. and Terio, K.A. (2016). Comparative pathology of aging great apes: bonobos, chimpanzees, gorillas, and orangutans. *Veterinary Pathology*, **53**(2), 250–76. DOI: 10.1177/0300985815612154.

Lowenstine, L.J., McManamon, R. and Terio, K.A. (2018). Apes. In *Pathology of Wildlife and Zoo Animals*, ed. K. A. Terio, D. McAloose and J. St Leger. Cambridge, UK: Elsevier Inc., pp. 375–412.

Lu, H., Giordano, F. and Ning, Z. (2016). Oxford nanopore MinION sequencing and genome assembly. *Genomics, Proteomics & Bioinformatics*, **14**(5), 265–79. DOI: 10.1016/j.gpb.2016.05.004.

Lu, J., Bayne, K. and Wang, J. (2013). Current status of animal welfare and animal rights in China. *Alternatives to Laboratory Animals*, **41**(5), 351–7. DOI: 10.1177/026119291304100505.

Lucchesi, S., Cheng, L., Janmaat, K.R.L., *et al.* (2020). Beyond the group: how food, mates, and group size influence intergroup encounters in wild bonobos. *Behavioral Ecology*, **31**(2), 519–32.

Lundmark, F., Berg, C. and Röcklinsberg, H. (2018). Private animal welfare standards – opportunities and risks. *Animals*, **8**(1), 4. DOI: 10.3390/ani8010004.

Lyons, M., Smuts, C. and Stephens, A. (2001). Participation, empowerment and sustainability: (how) do the links work? *Urban Studies*, **38**(8), 1233–-51. DOI: 10.1080/00420980125039.

Lyra, T.M. (2006). La erradicación de la peste porcina africana en el Brasil, 1978–1984 [The eradication of African swine fever in Brazil, 1978–1984]. *Revue Scientifique et Technique de l'Office International des Épizooties*, **25**(1), 93–103.

Mabano, A. (2013). *Impact of tourists on mountain gorilla behavior*. BSc thesis. Huye, Rwanda: National University of Rwanda.

Macfie, E.J. and Williamson, E.A. (2010). *Best Practice Guidelines for Great Ape Tourism*. Gland, Switzerland: International Union for Conservation of Nature (IUCN) Species Survival Commission (SSC) Primate Specialist Group (PSG). Available at: https://portals.iucn.org/library/efiles/documents/SSC-OP-038.pdf.

Mackenzie, J.S. and Jeggo, M. (2019). The One Health approach: why is it so important? *Tropical Medicine and Infectious Disease*, **4**(2), 88. DOI: 10.3390/tropicalmed4020088.

MacQueen, K.M., McLellan, E., Metzger, D.S., *et al.* (2001). What is community? An evidence-based definition for participatory public health. *American Journal of Public Health*, **91**(12), 1929–38. DOI: 10.2105/ajph.91.12.1929.

Maddox, T., Howard, P., Knox, J. and Jenner, N. (2019). *Forest-Smart Mining: Identifying Factors Associated with the Impacts of Large-Scale Mining on Forests*. Washington DC: World Bank. DOI: 10.1596/32025.

Madliger, C.L., Love, O.P., Hultine, K.R. and Cooke, S.J. (2018). The conservation physiology toolbox: status and opportunities. *Conservation Physiology*, **6**(1). DOI: 10.1093/conphys/coy029.

Maekawa, M., Lanjouw, A., Rutagarama, E. and Sharp, D. (2013). Mountain gorilla tourism generating wealth and peace in post-conflict Rwanda. *Natural Resources Forum*, **37**(2), 127–37. DOI: 10.1111/1477-8947.12020.

Maekawa, M., Lanjouw, A., Rutagarama, E. and Sharp, D. (2015). Mountain gorilla ecotourism: supporting macro-economic growth and providing local livelihoods. In *Livelihoods, Natural Resources, and Post-Conflict Peacebuilding*, ed. H. Young and L. Goldman. Abingdon, UK: Taylor & Francis, pp. 167–86.

Maertens, B., Gagnaire, A., Paerewijck, O., De Bosscher, K. and Geldhof, P. (2021). Regulatory role of the intestinal microbiota in the immune response against *Giardia*. *Scientific Reports*, **11**, 10601. DOI: 10.1038/s41598-021-90261-z.

Maisels, F., Bergl, R.A. and Williamson, E.A. (2018). Gorilla gorilla *(amended version of 2016 assessment)*. *The IUCN Red List of Threatened Species 2018: e.T9404A136250858*. Gland, Switzerland: International Union for Conservation of Nature (IUCN). DOI: 10.2305/IUCN.UK.2018-2.RLTS.T9404A136250858.en.

Maisels, F., Plumptre, A.J. and Strindberg, S. (2021). New Grauer's gorilla estimate. *Gorilla Journal*, **63**, 6–7. Available at https://www.berggorilla.org/en/gorillas/gorilla-numbers/gorilla-numbers/new-grauers-gorilla-population-estimate/.

Maisels, F., Strindberg, S., Greer, D., *et al.* (2016). Pan troglodytes *ssp.* troglodytes *(errata version published in 2016)*. *The IUCN Red List of Threatened Species 2016: e.T15936A102332276*. Gland, Switzerland: International Union for Conservation of Nature (IUCN). DOI: 10.2305/IUCN.UK.2016-2.RLTS.T15936A17990042.en.

Maki, S., Alford, P. and Bramblett, C. (1987). The effects of unfamiliar humans on aggression in captive chimpanzee groups. *American Journal of Primatology*, **12**(3), 358.

Maldonado, O., Aveling, C., Cox, D., *et al.* (2012). *Grauer's Gorillas and Chimpanzees in Eastern Democratic Republic of Congo (Kahuzi-Biega, Maiko, Tayna and Itombwe Landscape): Conservation Action Plan 2012–2022*. Gland, Switzerland: International Union for Conservation of Nature (IUCN) Species Survival Commission (SSC) Primate Specialist Group (PSG), Ministry of Environment, Nature Conservation & Tourism, Institut Congolais pour la Conservation de la Nature & the Jane Goodall Institute. Available at: https://www.iucn.org/content/grauers-gorillas-and-chimpanzees-eastern-democratic-republic-congo-kahuzi-biega-maiko-tayna-and-itombwe-landscape-conservation-action-plan-2012-2022.

Malhi, Y., Roberts, J.T., Betts, R.A., *et al.* (2008). Climate change, deforestation, and the fate of the Amazon. *Science*, **319**(5860), 169–72. DOI: 10.1126/science.1146961.

Manansang, W. (2020). Roadmap to the 2023 goal. Progressing the SEAZA animal welfare certification program. *WAZA News*, **3** 16–17. Available at https://www.waza.org/wp-content/uploads/2020/10/WAZA-magazine3-2020-FINAL-web.pdf.

Marí Saéz, A., Weiss, S., Nowak, K., *et al.* (2015). Investigating the zoonotic origin of the West African Ebola epidemic. *EMBO Molecular Medicine*, **7**(1), 17–23. DOI: 10.15252/emmm.201404792.

Marks, M., Solomon, A.W. and Mabey, D.C. (2014). Endemic treponemal diseases. *Transactions of the Royal Society of Tropical Medicine and Hygiene*, **108**(10), 601–7. DOI: 10.1093/trstmh/tru128.

Maron, M., Hobbs, R.J., Moilanen, A., *et al.* (2012). Faustian bargains? Restoration realities in the context of biodiversity offset policies. *Biological Conservation*, **155**(Supplement C), 141–8. DOI: 10.1016/j.biocon.2012.06.003.

Marques, A., Martins, I.S., Kastner, T., *et al.* (2019). Increasing impacts of land use on biodiversity and carbon sequestration driven by population and economic growth. *Nature Ecology & Evolution*, **3**(4), 628–37. DOI: 10.1038/s41559-019-0824-3.

Marrana, M. (2022). Chapter 3. Epidemiology of disease through the interactions between humans, domestic animals, and wildlife. In *One Health*, ed. J. C. Prata, A. I. Ribeiro and T. Rocha-Santos. Cambridge, MA: Academic Press, pp. 73–111. DOI: 10.1016/B978-0-12-822794-7.00001-0.

Marshall, A.J., Ancrenaz, M., Brearley, F.Q., *et al.* (2009). The effects of forest phenology and floristics on populations of Bornean and Sumatran orangutans: are Sumatran forests more productive than Bornean forests? In *Orangutans: Geographic Variation in Behavioral Ecology and Conservation*, ed. S. A. Wich, S. Utami-Atmoko, T. Mitra Setia and C. P. van Schaik. Oxford, UK: Oxford University Press, pp. 97–117.

Marshall, A.J., Cannon, C.H. and Leighton, M. (2009). Competition and niche overlap between gibbons (*Hylobates albibarbis*) and other frugivorous vertebrates in Gunung Palung National Park, West Kalimantan, Indonesia. In *The Gibbons: New Perspectives on Small Ape Socioecology and Population Biology*, ed. D. Whittaker and S. Lappan. New York, NY: Springer, pp. 161–88. DOI: 10.1007/978-0-387-88604-6_9.

Marshall, A.J. and Leighton, M. (2006). How does food availability limit the population density of white-bearded gibbons? In *Feeding Ecology in Apes and Other Primates: Ecological, Physiological and Behavioural Aspects*. Cambridge Studies in Biological and Evolutionary Anthropology Volume 48, ed. G. Hohmann, M. Robbins and C. Boesch. Cambridge, UK: Cambridge University Press, pp. 313–35.

Martinsen, S. and Jukes, N. (2005). Towards a humane veterinary education. *Journal of Veterinary Medical Education*, **32**(4), 454–60. DOI: 10.3138/jvme.32.4.454.

Marx, V. (2015). PCR heads into the field. *Nature Methods*, **12**(5), 393–7. DOI: 10.1038/nmeth.3369.

Marzec, A.M., Kunz, J.A., Falkner, S., *et al.* (2016). The dark side of the red ape: male-mediated lethal female competition in Bornean orangutans. *Behavioral Ecology and Sociobiology*, **70**(4), 459–66. DOI: 10.1007/s00265-015-2053-3.

Marzi, A., Murphy, A.A., Feldmann, F., *et al.* (2016). Cytomegalovirus-based vaccine expressing Ebola virus glycoprotein protects nonhuman primates from Ebola virus infection. *Scientific Reports*, **6**, 21674. DOI: 10.1038/srep21674.

Maseyk, F.J.F., Maron, M., Gordon, A., Bull, J.W. and Evans, M.C. (2021). Improving averted loss estimates for better biodiversity outcomes from offset exchanges. *Oryx*, **55**(3), 393–403. DOI: 10.1017/S0030605319000528.

Masi, S., Chauffour, S., Bain, O., *et al.* (2012). Seasonal effects on great ape health: a case study of wild chimpanzees and western gorillas. *PLoS ONE*, **7**(12), e49805. DOI: 10.1371/journal.pone.0049805.

Masi, S., Cipolletta, C. and Robbins, M.M. (2009). Western lowland gorillas (*Gorilla gorilla gorilla*) change their activity patterns in response to frugivory. *American Journal of Primatology*, **71**(2), 91–100. DOI: 10.1002/ajp.20629.

Matos Mendes, D. (2020). Goiânia Zoo – siamangs in their islands. *Zoochat*, September 14, 2020. Available at: https://www.zoochat.com/community/media/goiania-zoo-siamangs-in-their-islands.501054/.

Mätz-Rensing, K., Kunze, M., Zöller, M., *et al.* (2011). Fatal *Balamuthia mandrillaris* infection in a gorilla: first case of balamuthiasis in Germany. *Journal of Medical Primatology*, **40**(6), 437–40. DOI: 10.1111/j.1600-0684.2011.00479.x.

Mazet, J.A.K., Genovese, B.N., Harris, L.A., *et al.* (2020). Human respiratory syncytial virus detected in mountain gorilla respiratory outbreaks. *EcoHealth*, **17**, 449–60.

Mazimhaka, J. (2006). *The potential impact of domestic tourism on Rwanda's tourism economy*. MA thesis. Johannesburg, South Africa: University of the Witwatersrand.

Mbaya, A.W. and Udendeye, U.J. (2011). Gastrointestinal parasites of captive and free-roaming primates at the Afi Mountain Primate Conservation Area in Calabar, Nigeria and their zoonotic implications. *Pakistan Journal of Biological Sciences*, **14**(13), 709–14. DOI: 10.3923/pjbs.2011.709.714.

Mbayahi, A. and Kalema-Zikusoka, G. (2020). *COVID-19 and Africa's Great Apes. Challenges and Threats Amidst the COVID-19 Pandemic for Sustaining Conservation Through Responsible Great Ape Tourism. Policy Brief*. Nairobi,

Kenya: African Civil Society Biodiversity Alliance (ACBA). Available at: https://africancba.org/download-resource-file/POLICY%20BRIEF%20ON%20AFRICA%20GREAT%20APES%20AND%20COVID-19%20ENGLISH%20%282%29_compressed.pdf.

McBean, G. (2004). Climate change and extreme weather: a basis for action. *Natural Hazards*, **31**(1), 177–90. DOI: 10.1023/B:NHAZ.0000020259.58716.0d.

McCarthy, M., Bigelow, J. and Taylor, M. (2018). Emergency preparedness and planning for animals: a case study in the Blue Mountains, New South Wales (NSW). *Australian Journal of Emergency Management*, **33**(4), 50–6.

McCarthy, M.S., Lester, J.D., Howe, E.J., *et al.* (2015). Genetic censusing identifies an unexpectedly sizeable population of an endangered large mammal in a fragmented forest landscape. *BMC Ecology*, **15**(1), 21. DOI: 10.1186/s12898-015-0052-x.

McConkey, K.R. (2018). Seed dispersal by primates in Asian habitats: from species, to communities, to conservation. *International Journal of Primatology*, **39**(3), 466–92. DOI: 10.1007/s10764-017-0013-7.

McConkey, K.R., Nathalang, A., Brockelman, W.Y., *et al.* (2018). Different megafauna vary in their seed dispersal effectiveness of the megafaunal fruit *Platymitra macrocarpa* (Annonaceae). *PLoS ONE*, **13**(7), e0198960. DOI: 10.1371/journal.pone.0198960.

McCormick, G.L., Shea, K. and Langkilde, T. (2015). How do duration, frequency, and intensity of exogenous CORT elevation affect immune outcomes of stress? *General and Comparative Endocrinology*, **222**, 81–7. DOI: 10.1016/j.ygcen.2015.07.008.

McCullers, J.A. (2014). The co-pathogenesis of influenza viruses with bacteria in the lung. *Nature Reviews Microbiology*, **12**(4), 252–62. DOI: 10.1038/nrmicro3231.

McDonald, M. and Johnson, S. (2014). "There's an app for that": a new program for the collection of behavioural field data. *Animal Behaviour*, **95**, 81–7. DOI: 10.1016/j.anbehav.2014.06.009.

McInturff, A., Xu, W., Wilkinson, C.E., Dejid, N. and Brashares, J.S. (2020). Fence ecology: frameworks for understanding the ecological effects of fences. *BioScience*, **70**(11), 971–85. DOI: 10.1093/biosci/biaa103.

McLennan, M.R. and Hockings, K.J. (2016). The aggressive apes? Causes and contexts of great ape attacks on local persons. In *Problematic Wildlife: A Cross-Disciplinary Approach*, ed. F. M. Angelici. Cham, Switzerland: Springer, pp. 373–94. DOI: 10.1007/978-3-319-22246-2_18.

McManamon, R. and Lowenstine, L. (2012). Cardiovascular disease in great apes. In *Fowler's Zoo and Wild Animal Medicine, Current Therapy, Volume 7*, ed. R. E. Miller and M. Fowler. St Louis, MO: Elsevier Saunders, pp. 408–15. DOI: 10.1016/B978-1-4377-1986-4.00053-6.

McManamon, R., Swenson, R.B. and Lowenstine, L.J. (1994). Update on diagnostic and therapeutic approaches to airsacculitis in orangutans. Presented at: *Proceedings of the Annual Meeting of the American Association of Zoo Veterinarians, Pittsburgh, PA, 23–28 October 1994*. Jacksonville, FL: American Association of Zoo Veterinarians, pp. 219–20.

McTighe, M.S., Hansen, B.C., Ely, J.J. and Lee, D.R. (2011). Determination of hemoglobin A1c and fasting blood glucose reference intervals in captive chimpanzees (*Pan troglodytes*). *Journal of the American Association for Laboratory Animal Science*, **50**(2), 165–70.

Medkour, H., Amona, I., Laidoudi, Y., *et al.* (2020). Parasitic infections in African humans and non-human primates. *Pathogens*, **9**(7), 561. DOI: 10.3390/pathogens9070561.

Meehan, T.P. and Lowenstine, L.J. (1994). Causes of mortality in captive lowland gorillas: a survey of the SSP population. Presented at: *Proceedings of the Annual Meeting of the American Association of Zoo Veterinarians, Pittsburgh, PA, 23–28 October 1994*. Jacksonville, FL: American Association of Zoo Veterinarians.

Meehl, G.A., Zwiers, F., Evans, J., *et al.* (2000). Trends in extreme weather and climate events: issues related to modeling extremes in projections of future climate change. *Bulletin of the American Meteorological Society*, **81**(3), 427–36. DOI: 10.1175/1520-0477(2000)081<0427:TIEWAC>2.3.CO;2.

Meijaard, E. (2015). Indonesia's silent tragedy in a connected world. *Jakarta Globe, reproduced by the Ape Alliance*, October 30, 2015. Available at: https://www.4apes.com/news/general/item/1318-erik-meijaard-indonesia-s-silent-tragedy-in-a-connected-world.

Meijaard, E., Albar, G., Nardiyono, *et al.* (2010). Unexpected ecological resilience in Bornean orangutans and implications for pulp and paper plantation management. *PLoS ONE*, **5**(9), e12813. DOI: 10.1371/journal.pone.0012813.

Meijaard, E., Ni'matullah, S., Dennis, R., *et al.* (2021). The historical range and drivers of decline of the Tapanuli orangutan. *PLoS ONE*, **16**(1), e0238087. DOI: 10.1371/journal.pone.0238087.

Meijaard, E. and Wich, S.A. (2014). *Extractive Industries and Orangutans. Occasional Paper for State of the Apes, Volume 1.* Cambridge, UK: Arcus Foundation. Available at: https://www.stateoftheapes.com/wp-content/uploads/2014/07/Extractive-Industries-and-Orangutans1.pdf.

Meijaard, E., Wich, S., Ancrenaz, M. and Marshall, A.J. (2012). Not by science alone: why orangutan conservationists must think outside the box. *Year in Ecology and Conservation Biology*, **1249**, 29–44. DOI: 10.1111/j.1749-6632.2011.06288.x.

Melin, A.D., Janiak, M.C., Marrone, F., Arora, P.S. and Higham, J.P. (2020). Comparative ACE2 variation and primate COVID-19 risk. *Communications Biology*, **3**(1), 641. DOI: 10.1038/s42003-020-01370-w.

Mellor, D.J. (2017). Operational details of the five domains model and its key applications to the assessment and management of animal welfare. *Animals*, **7**(8), 60. DOI: 10.3390/ani7080060.

Mellor, D.J., Beausoleil, N.J., Littlewood, K.E., *et al.* (2020). The 2020 Five Domains Model: including human–animal interactions in assessments of animal welfare. *Animals*, **10**(10), 1870. DOI: 10.3390/ani10101870.

Mendez, A. and Houghton, D.P. (2020). Sustainable banking: the role of multilateral development banks as norm entrepreneurs. *Sustainability*, **12**(3), 972. DOI: 10.3390/su12030972.

Meyers, W.M., Walsh, G.P., Brown, H.L., *et al.* (1985). Leprosy in a mangabey monkey: naturally acquired infection. *International Journal of Leprosy and Other Mycobacterial Diseases*, **53**(1), 1–14.

Michel, A.L., Venter, L., Espie, I.W. and Coetzee, M.L. (2003). *Mycobacterium tuberculosis* infections in eight species at the National Zoological Gardens of South Africa, 1991–2001. *Journal of Zoo and Wildlife Medicine*, **34**(4), 364–70. DOI: 10.1638/02-063.

Millar, K., Thorstensen, E., Tomkins, S., Mepham, B. and Kaiser, M. (2007). Developing the ethical Delphi. *Journal of Agricultural and Environmental Ethics*, **20**(1), 53–63. DOI: 10.1007/s10806-006-9022-9.

Miller, R.E., Calle, P.P. and Lamberski, N., ed. (2023). *Fowler's Zoo and Wild Animal Medicine Current Therapy, Volume 10.* St Louis, MO: Elsevier. DOI: 10.1016/B978-0-323-82852-9.00114-3.

Miller, R.E. and Fowler, M.E., ed. (2015). *Fowler's Zoo and Wild Animal Medicine, Volume 8.* Philadelphia, PA: Elsevier Saunders.

Miller, R.E., Lamberski, N. and Calle, P.P., ed. (2019). *Fowler's Zoo and Wild Animal Medicine Current Therapy, Volume 9.* St Louis, MO: Elsevier. DOI: 10.1016/B978-0-323-55228-8.00148-X.

Milstein, M.S., Shaffer, C.A., Suse, P., *et al.* (2020). An ethnographic approach to characterizing potential pathways of zoonotic disease transmission from wild meat in Guyana. *EcoHealth*, **17**(4), 424–36. DOI: 10.1007/s10393-021-01513-3.

MINAE (2017a). *Regulations for the Implementation of the Wildlife Conservation Law No. 7317.* San José, Costa Rica: Ministerio del Ambiente y Energía (MINAE). Available at: https://stopanimalselfies.org/wp-content/uploads/2019/10/ENGLISH-Regulation-Wildlife-Conservation-Law-Costa-Rica.pdf.

MINAE (2017b). *Wildlife Conservation Law No. 7317.* San José, Costa Rica: Ministerio del Ambiente y Energía (MINAE). Available at: https://stopanimalselfies.org/wp-content/uploads/2019/10/Ley-CVS-ingles.pdf.

Ministry of Natural Resources Energy and Mining (2017). *National Parks and Wildlife Act.* Lilongwe, Malawi: Ministry of Natural Resources, Energy and Mining. Available at: https://www.lilongwewildlife.org/wp-content/uploads/Malawi.Wildlife.Legislation.Digital.pdf.

Mirza, M.M.Q. (2003). Climate change and extreme weather events: can developing countries adapt? *Climate Policy*, **3**(3), 233–48. DOI: 10.1016/S1469-3062(03)00052-4.

Mitani, J.C. (2009). Male chimpanzees form enduring and equitable social bonds. *Animal Behaviour*, **77**(3), 633–40. DOI: 10.1016/j.anbehav.2008.11.021.

Mitani, J.C., Watts, D.P. and Amsler, S.J. (2010). Lethal intergroup aggression leads to territorial expansion in wild chimpanzees. *Current Biology*, **20**(12), R507–8. DOI: 10.1016/j.cub.2010.04.021.

Mitchell, G., Tromborg, C.T., Kaufman, J., *et al.* (1992). More on the "influence" of zoo visitors on the behaviour of captive primates. *Applied Animal Behaviour Science*, **35**(2), 189–98. DOI: 10.1016/0168-1591(92)90009-Z.

Mitman, S., Rosenbaum, M., Bello, R., *et al.* (2021). Challenges to IUCN guideline implementation in the rehabilitation and release of trafficked primates in Peru. *Primate Conservation*, **35**, 1–16.

Mitra Setia, T., Delgado, R.A., Utami-Atmoko, S.S., Singleton, I. and van Schaik, C.P. (2009). Social organization and male–female relationships. In *Orangutans: Geographic Variation in Behavioral Ecology and Conservation*, ed. S. A. Wich, S. S. Utami-Atmoko, T. M. Setia and C. P. van Schaik. Oxford, UK: Oxford University Press, pp. 245–54. DOI: 10.1093/acprof:oso/9780199213276.003.0017.

Mittermeier, R.A., Rylands, A.B. and Wilson, D.E., ed. (2013). *Handbook of the Mammals of the World. Volume 3: Primates*. Barcelona, Spain: Lynx Edicions.

Mlengeya, T. (2000). *TANAPA Veterinary Department Annual Report 2000/2001. Respiratory Disease Outbreak in the Chimpanzee Population of Gombe National Park*. Arusha, Tanzania: Tanzania National Parks (TANAPA).

Moberg, G.P. (1985). Influence of stress on reproduction: measure of well-being. In *Animal Stress*, ed. G. P. Moberg. New York, NY: Springer, pp. 245–67. DOI: 10.1007/978-1-4614-7544-6_14.

Moberg, G.P. (2000). Biological response to stress: implications for animal welfare. In *The Biology of Animal Stress: Basic Principles and Implications for Animal Welfare*, ed. G. P. Moberg and J. A. Mench. Wallingford, UK: CABI International, pp. 1–21. DOI: 10.1079/9780851993591.0001.

Modry, D., Pafćo, B., Petrželková, K.J. and Hasegawa, H., ed. (2018). *Parasites of Apes: An Atlas of Coproscopic Diagnostics*. Frankfurt Contributions to Natural History, Volume 76. Frankfurt, Germany: Editions Chimaira. Available at: https://www.chimaira.de.

Moeller, A.H. (2017). The shrinking human gut microbiome. *Current Opinion in Microbiology*, **38**, 30–5. DOI: 10.1016/j.mib.2017.04.002.

Mokuwa, E. and Richards, P. (2020). How should public health officials respond when important local rituals increase risk of contagion? Case and commentary. *AMA Journal of Ethics*, **22**(1), E5–9.

Moloney, G.K., Tuke, J., Dal Grande, E., Nielsen, T. and Chaber, A.-L. (2021). Is YouTube promoting the exotic pet trade? Analysis of the global public perception of popular YouTube videos featuring threatened exotic animals. *PLoS ONE*, **16**(4), e0235451. DOI: 10.1371/journal.pone.0235451.

Molyneaux, A., Hankinson, E., Kaban, M., *et al.* (2021). Primate selfies and anthropozoonotic diseases: lack of rule compliance and poor risk perception threatens orangutans. *Folia Primatologica*, **92**(5–6), 296–305. DOI: 10.1159/000520371.

Monkey World (n.d.). *Meet the Primates*. Wareham, UK: Monkey World. Available at: https://monkeyworld.org/our-primates/primate-groups/. Accessed: October, 2020.

Montali, R.J., Mikota, S.K. and Cheng, L.I. (2001). *Mycobacterium tuberculosis* in zoo and wildlife species. *Revue Scientifique et Technique de l'Office International des Épizooties*, **20**(1), 291–303. DOI: 10.20506/rst.20.1.1268.

Monte Adone (n.d.). *Scimpanzè*. Sasso Marconi, Italy: Centro Tutela e Ricerca Fauna Esotica e Selvatica Monte Adone ODV. Available at: https://centrotutelafauna.org/scimpanze/. Accessed: December, 2020.

Monto, A.S. (2002). Epidemiology of viral respiratory infections. *The American Journal of Medicine*, **112**(6), 4–12. DOI: 10.1016/S0002-9343(01)01058-0.

Moon, S., Sridhar, D., Pate, M.A., *et al.* (2015). Will Ebola change the game? Ten essential reforms before the next pandemic. The report of the Harvard-LSHTM Independent Panel on the Global Response to Ebola. *The Lancet*, **386**(10009), 2204–21. DOI: 10.1016/S0140-6736(15)00946-0.

Moore, J., Black, J., Hernandez-Aguilar, R.A., *et al.* (2017). Chimpanzee vertebrate consumption: savanna and forest chimpanzees compared. *Journal of Human Evolution*, **112**, 30–40. DOI: 10.1016/j.jhevol.2017.09.004.

Moore, P., Prompinchompoo, C. and Beastall, C.A. (2016). *CITES Implementation in Thailand: A Review of the Legal Regime Governing the Trade in Great Apes and Gibbons and Other CITES-Listed Species*. Selangor, Malaysia: TRAFFIC. Available at: http://www.trafficj.org/publication/16_CITES_Implementation_in_Thailand.pdf.

Moorhouse, T.P., Dahlsjö, C.A.L., Baker, S.E., D'Cruze, N.C. and Macdonald, D.W. (2015). The customer isn't always right – conservation and animal welfare implications of the increasing demand for wildlife tourism. *PLoS ONE*, **10**(10), e0138939. DOI: 10.1371/journal.pone.0138939.

Mootnick, A.R., Reingold, M., Holshuh, H.J. and Mirkovic, R.R. (1998). Isolation of a herpes simplex virus type 1-like agent from the brain of a mountain agile gibbon (*Hylobates agilis agilis*) with encephalitis. *Journal of Zoo and Wildlife Medicine*, **29**(1), 61–4.

Morbeck, M.E., Zihlman, A.L., Sumner, D.R. and Galloway, A. (1991). Poliomyelitis and skeletal asymmetry in Gombe chimpanzees. *Primates*, **32**(1), 77–91. DOI: 10.1007/BF02381602.

Morcillo, D.O., Steiner, U.K., Grayson, K.L., Ruiz-Lambides, A.V. and Hernández-Pacheco, R. (2020). Hurricane-induced demographic changes in a non-human primate population. *Royal Society Open Science*, **7**(8), 200173. DOI: 10.1098/rsos.200173.

Morgan, D., Mundry, R., Sanz, C., *et al.* (2018). African apes coexisting with logging: comparing chimpanzee (*Pan troglodytes troglodytes*) and gorilla (*Gorilla gorilla gorilla*) resource needs and responses to forestry activities. *Biological Conservation*, **218**, 277–86. DOI: 10.1016/j.biocon.2017.10.026.

Morgan, D. and Sanz, C. (2003). Naïve encounters with chimpanzees in the Goualougo Triangle, Republic of Congo. *International Journal of Primatology*, **24**(2), 369–81. DOI: 10.1023/A:1023005417897.

Morgan, D. and Sanz, C. (2006). Chimpanzee feeding ecology and comparisons with sympatric gorillas in the Goualougo Triangle, Republic of Congo. In *Feeding Ecology in Apes and Other Primates: Ecological, Physiological and Behavioural Aspects*. Cambridge Studies in Biological and Evolutionary Anthropology Volume 48, ed. G. Hohmann, M. Robbins and C. Boesch. Cambridge, UK: Cambridge University Press, pp. 97–122.

Morgan, D. and Sanz, C. (2007). *Best Practice Guidelines for Reducing the Impact of Commercial Logging on Great Apes in Western Equatorial Africa*. Gland, Switzerland: International Union for Conservation of Nature (IUCN) Species Survival Commission (SSC) Primate Specialist Group (PSG). Available at: https://portals.iucn.org/library/node/9059.

Morgan, D. and Sanz, C. (2020). *Rapport sur la biodiversity et l'importance ecologique du Triangle de Djeke, Republic du Congo*. Brazzaville, Republic of Congo: Foundation Nouabalé-Ndoki, Wildlife Conservation Society and Goualougo Triangle Ape Project.

Morgan, D., Strindberg, S., Winston, W., *et al.* (2019). Impacts of selective logging and associated anthropogenic disturbance on intact forest landscapes and apes of northern Congo. *Frontiers in Forests and Global Change*, **2**, July 3, 2019. DOI: 10.3389/ffgc.2019.00028.

Morgan, D.B., Winston, W., Ayina, C.E., *et al.* (2020). Forest certification and the high conservation value concept: protecting great apes in the Sangha Trinational Landscape in an era of industrial logging. In *Chimpanzees in Context: A Comparative Perspective on Chimpanzee Behavior, Cognition, Conservation, and Welfare*, ed. L. M. Hopper and S. R. Ross. Chicago, IL: University of Chicago Press, pp. 644–70. DOI: 10.7208/chicago/9780226728032.003.0027.

Morgan, K.N. and Tromborg, C.T. (2007). Sources of stress in captivity. *Applied Animal Behaviour Science*, **102**(3), 262–302. DOI: 10.1016/j.applanim.2006.05.032.

Morgans, C.L., Meijaard, E., Santika, T., *et al.* (2018). Evaluating the effectiveness of palm oil certification in delivering multiple sustainability objectives. *Environmental Research Letters*, **13**(6), 064032. DOI: 10.1088/1748-9326/aac6f4.

Mori Junior, R., Franks, D.M. and Ali, S.H. (2015). *Designing Sustainability Certification for Impact: Analysis of the Design Characteristics of 15 Sustainability Standards in the Mining Industry*. Brisbane, Australia: Centre for Social Responsibility in Mining, University of Queensland.

Morimura, N., Idani, G. and Matsuzawa, T. (2011). The first chimpanzee sanctuary in Japan: an attempt to care for the "surplus" of biomedical research. *American Journal of Primatology*, **73**(3), 226–32. DOI: 10.1002/ajp.20887.

Mörner, T., Obendorf, D.L., Artois, M. and Woodford, M.H. (2002). Surveillance and monitoring of wildlife diseases. *Revue Scientifique et Technique de l'Office International des Épizooties*, **21**(1), 67–76. DOI: 10.20506/rst.21.1.1321.

Morocco World News (2018). Animal trafficking in UAE still a major problem. *Morocco World News*, 2 August 2018. Available at: https://www.moroccoworldnews.com/2018/08/251652/animal-trafficking-uae-major-problem.

Morris, D.E., Cleary, D.W. and Clarke, S.C. (2017). Secondary bacterial infections associated with influenza pandemics. *Frontiers in Microbiology*, **8**, June 23, 2017. DOI: 10.3389/fmicb.2017.01041.

Morton, F.B., Todd, A.F., Lee, P. and Masi, S. (2013). Observational monitoring of clinical signs during the last stage of habituation in a wild western gorilla group at Bai Hokou, Central African Republic. *Folia Primatologica*, **84**(2), 118–33. DOI: 10.1159/000350916.

Mubemba, B., Chanove, E., Mätz-Rensing, K., *et al.* (2020). Yaws disease caused by *Treponema pallidum* subspecies *pertenue* in wild chimpanzee, Guinea, 2019. *Emerging Infectious Diseases*, **26**, 1283–6.

Muegge, B.D., Kuczynski, J., Knights, D., *et al.* (2011). Diet drives convergence in gut microbiome functions across mammalian phylogeny and within humans. *Science*, **332**(6032), 970–4. DOI: 10.1126/science.1198719.

Muehlenbein, M.P. (2013). Human–wildlife contact and emerging infectious diseases. In *Human–Environment Interactions: Current and Future Directions*, ed. E. S. Brondízio and E. F. Moran. Dordrecht, the Netherlands: Springer, pp. 79–94. DOI: 10.1007/978-94-007-4780-7_4.

Muehlenbein, M.P. and Ancrenaz, M. (2009). Minimizing pathogen transmission at primate ecotourism destinations: the need for input from travel medicine. *Journal of Travel Medicine*, **16**(4), 229–32. DOI: 10.1111/j.1708-8305.2009.00346.x.

Muehlenbein, M.P., Ancrenaz, M., Sakong, R., *et al.* (2012). Ape conservation physiology: fecal glucocorticoid responses in wild *Pongo pygmaeus morio* following human visitation. *PLoS ONE*, **7**(3), e33357. DOI: 10.1371/journal.pone.0033357.

Muehlenbein, M.P., Martinez, L.A., Lemke, A.A., *et al.* (2008). Perceived vaccination status in ecotourists and risks of anthropozoonoses. *EcoHealth*, **5**(3), 371–8. DOI: 10.1007/s10393-008-0192-y.

Muehlenbein, M.P., Martinez, L.A., Lemke, A.A., *et al.* (2010). Unhealthy travelers present challenges to sustainable primate ecotourism. *Travel Medicine and Infectious Disease*, **8**(3), 169–75. DOI: 10.1016/j.tmaid.2010.03.004.

Muehlenbein, M.P. and Wallis, J. (2014). Considering risks of pathogen transmission associated with primate-based tourism. In *Primate Tourism: A Tool for Conservation?*, ed. A. E. Russon and J. Wallis. Cambridge, UK: Cambridge University Press, pp. 278–91. DOI: 10.1017/CBO9781139087407.021.

Mugisha, L., Pauli, G., Opuda-Asibo, J., *et al.* (2010). Evaluation of poliovirus antibody titers in orally vaccinated semi-captive chimpanzees in Uganda. *Journal of Medical Primatology*, **39**(2), 123–8. DOI: 10.1111/j.1600-0684.2010.00400.x.

Muhangi, D., Gardiner, C.H., Ojok, L., *et al.* (2021). Pathological lesions of the digestive tract in free-ranging mountain gorillas (*Gorilla beringei beringei*). *American Journal of Primatology*, **83**(8), e23290. DOI: 10.1002/ajp.23290.

Mukanjari, S., Bednar-Friedl, B., Muchapondwa, E. and Zikhali, P. (2013). Evaluating the prospects of benefit sharing schemes in protecting mountain gorillas in Central Africa. *Natural Resource Modeling*, **26**(4), 455–79. DOI: 10.1111/nrm.12010.

Mukherjee, N., Hugé, J., Sutherland, W.J., *et al.* (2015). The Delphi technique in ecology and biological conservation: applications and guidelines. *Methods in Ecology and Evolution*, **6**(9), 1097–1109. DOI: 10.1111/2041-210X.12387.

Mul, I.F., Paembonan, W., Singleton, I., Wich, S.A. and van Bolhuis, H. (2007). Intestinal parasites of free-ranging, semicaptive, and captive *Pongo abelii* in Sumatra, Indonesia. *International Journal of Primatology*, **28**, 407–20. DOI: 10.1007/s10764-007-9119-7.

Mulero-Pázmány, M. (2021). The future of technology in conservation. In *Conservation Technology*, ed. S. A. Wich and A. K. Piel. Oxford, UK: Oxford University Press, pp. 255–73. DOI: 10.1093/oso/9780198850243.003.0013.

Munanura, I.E., Backman, K.F., Hallo, J.C. and Powell, R.B. (2016). Perceptions of tourism revenue sharing impacts on Volcanoes National Park, Rwanda: a Sustainable Livelihoods framework. *Journal of Sustainable Tourism*, **24**(12), 1709–26. DOI: 10.1080/09669582.2016.1145228.

Munanura, I.E., Backman, K.F. and Sabuhoro, E. (2013). Managing tourism growth in endangered species' habitats of Africa: Volcanoes National Park in Rwanda. *Current Issues in Tourism*, **16**(7–8), 700–18. DOI: 10.1080/13683500.2013.785483.

Munanura, I.E., Backman, K.F., Sabuhoro, E. and Bernhard, K.P. (2020). The potential of tourism benefits to reduce forest dependence behavior of impoverished residents adjacent to Volcanoes National Park in Rwanda. *Tourism Planning & Development*, **17**(5), 475–96. DOI: 10.1080/21568316.2019.1640282.

Munn, J. (2006). Effects of injury on the locomotion of free-living chimpanzees in the Budongo Forest Reserve, Uganda. In *Primates of Western Uganda*, ed. N. E. Newton-Fisher, H. Notman, J. D. Paterson and V. Reynolds. New York, NY: Springer, pp. 259–317.

Munson, L. and Montali, R.J. (1990). Pathology and diseases of great apes at the National Zoological Park. *Zoo Biology*, **9**, 99–105.

Murata, K., Hasegawa, H., Nakano, T., Noda, A. and Yanai, T. (2002). Fatal infection with human pinworm, *Enterobius vermicularis*, in a captive chimpanzee. *Journal of Medical Primatology*, **31**(2), 104–8. DOI: 10.1034/j.1600-0684.2002.01017.x.

Murguía, D.I., Bringezu, S. and Schaldach, R. (2016). Global direct pressures on biodiversity by large-scale metal mining: spatial distribution and implications for conservation. *Journal of Environmental Management*, **180**, 409–20. DOI: 10.1016/j.jenvman.2016.05.040.

Murphy, H.W., Dennis, P., Devlin, W., Meehan, T. and Kutinsky, I. (2011). Echocardiographic parameters of captive western lowland gorillas (*Gorilla gorilla gorilla*). *Journal of Zoo and Wildlife Medicine*, **42**(4), 572–9. DOI: 10.1638/2010-0139.1.

Murray, J.S. (2010). Moral courage in healthcare: acting ethically even in the presence of risk. *OJIN: The Online Journal of Issues in Nursing*, **15**(3), 2. DOI: 10.3912/OJIN.Vol15No03Man02.

Mutombo, M., Arita, I. and Jezek, Z. (1983). Human monkeypox transmitted by a chimpanzee in a tropical rainforest area of Zaire. *The Lancet*, **321**(8327), 735–7. DOI: 10.1016/S0140-6736(83)92027-5.

Muyambi, F. (2005). The impact of tourism on the behaviour of mountain gorillas. *Gorilla Journal*, **30**, 14–15.

Myers, B. and Zrinski, U. (2022). Resilient and inclusive public financial management systems enable governments to better respond to disasters. *World Bank Blogs*, March 17, 2022. Available at: https://blogs.worldbank.org/governance/resilient-and-inclusive-public-financial-management-systems-enable-governments-better.

Nadler, Y. (2019). Contingency planning for all hazards and foreign animal disease. In *Fowler's Zoo and Wild Animal Medicine Current Therapy, Volume 9*, ed. R. E. Miller, N. Lamberski and P. P. Calle. St Louis, MO: W.B. Saunders, pp. 45–52. DOI: 10.1016/B978-0-323-55228-8.00009-6.

Nagpal, R., Shively, C.A., Appt, S.A., *et al.* (2018). Gut microbiome composition in non-human primates consuming a western or Mediterranean diet. *Frontiers in Nutrition*, **5**. DOI: 10.3389/fnut.2018.00028.

Nakamara, M., Hosaka, K., Itoh, N. and Zamma, K. (2015). *Mahale Chimpanzees: 50 Years of Research*. Cambridge, UK: Cambridge University Press.

Nash, L.T., Fritz, J., Alford, P.A. and Brent, L. (1999). Variables influencing the origins of diverse abnormal behaviors in a large sample of captive chimpanzees (*Pan troglodytes*). *American Journal of Primatology*, **48**(1), 15–29. DOI: 10.1002/(sici)1098-2345(1999)48:1<15::Aid-ajp2>3.0.Co;2-r.

Nash, R., Johnston, H., Robbins, A. and Descovich, K. (2021). The effect of enrichment filling and engagement time on regurgitation and reingestion behaviour in three zoo-housed orangutans. *Journal of Zoological and Botanical Gardens*, **2**(1), 10–20. DOI: 10.3390/jzbg2010002.

Nasution, A., Perwitasari-Farajallah, D. and Utami-Atmoko, S.S. (2018). Declining orangutans population in the unprotected forest of Batang Toru. *Tropical Life Science Research*, **29**(2), 77–87. DOI: 10.21315/tlsr2018.29.2.6.

Nasution, A., Perwitasari-Farajallah, D. and Utami-Atmoko, S.S. (2020). The distribution and density of Tapanuli orangutans (*Pongo tapanuliensis*) at potential corridor locations between forest fragments in Batang Toru, North Sumatra, Indonesia. *Biodiversitas*, **21**, 5382–8.

Nater, A., Mattle-Greminger, M.P., Nurcahyo, A., *et al.* (2017). Morphometric, behavioral, and genomic evidence for a new orangutan species. *Current Biology*, **27**(22), 3487–98.e10. DOI: 10.1016/j.cub.2017.09.047.

Natesan, M., Jensen, S.M.R., Keasey, S.L., *et al.* (2016). Human survivors of disease outbreaks caused by Ebola or Marburg virus exhibit cross-reactive and long-lived antibody responses. *Clinical and Vaccine Immunology*, **23**, 717–24.

Nathan, S., Chieng, S., Kingsley, P.V., *et al.* (2018). Melioidosis in Malaysia: incidence, clinical challenges, and advances in understanding pathogenesis. *Tropical Medicine and Infectious Disease*, **3**(1), 25. DOI: 10.3390/tropicalmed3010025.

National Research Council (US) (2001). Towards the development of disease early warning systems. In *Under the Weather: Climate, Ecosystems, and Infectious Disease*, ed. Committee on Climate, Infectious Diseases, and Human Health. Washington DC: National Academy of Sciences, pp. 86–102. Available at: https://www.ncbi.nlm.nih.gov/books/NBK222241/.

NBI (2020). Unprecedented rise in water levels of Lake Victoria. *Nile Basin Initiative News and Events*, July 20, 2020. Available at: https://nilebasin.org/new-and-events/307-unprecedented-rise-in-water-levels-of-lake-victoria.

Negrey, J.D., Reddy, R.B., Scully, E.J., *et al.* (2019). Simultaneous outbreaks of respiratory disease in wild chimpanzees caused by distinct viruses of human origin. *Emerging Microbes & Infections*, **8**(1), 139–49. DOI: 10.1080/22221751.2018.1563456.

Nellemann, C. and Newton, A. (2002). *The Great Apes, The Road Ahead – A GLOBIO Perspective on the Impacts of Infrastructure Development on the Great Apes*. United Nations Environment Programme (UNEP), GRID-Arendal, World Conservation Monitoring Centre. Available at: https://wedocs.unep.org/handle/20.500.11822/7485.

Nelson, C., Lurie, N., Wasserman, J. and Zakowski, S. (2007). Conceptualizing and defining public health emergency preparedness. *American Journal of Public Health*, **97**(S1), S9–11. DOI: 10.2105/ajph.2007.114496.

Nepal, S.K. and Weber, K.E. (1994). A buffer zone for biodiversity conservation: viability of the concept in Nepal's Royal Chitwan National Park. *Environmental Conservation*, **21**(4), 333–41. DOI: 10.1017/S0376892900033646.

Newton-Fisher, N.E. (2003). The home range of the Sonso community of chimpanzees from the Budongo Forest, Uganda. *African Journal of Ecology*, **41**(2), 150–6. DOI: 10.1046/j.1365-2028.2003.00408.x.

Ng, L.S., Campos-Arceiz, A., Sloan, S., *et al.* (2020). The scale of biodiversity impacts of the Belt and Road Initiative in Southeast Asia. *Biological Conservation*, **248**, 108691. DOI: 10.1016/j.biocon.2020.108691.

Ngamassi, L., Shahriari, H., Ramakrishnan, T. and Rahman, S. (2022). Text mining hurricane Harvey tweet data: lessons learned and policy recommendations. *International Journal of Disaster Risk Reduction*, **70**, 102753. DOI: 10.1016/j.ijdrr.2021.102753.

Ngamba Island Chimpanzee Sanctuary (2020). Ngamba Island Chimpanzee Sanctuary. *Facebook Post*, October 13, 2020. Available at: https://www.facebook.com/friendsofchimps/posts/3742243585809513.

Nicholls, H. (2015). Conservation biology: wild at heart. *Nature*, **528**(7583), 474–5. DOI: 10.1038/528474a.

Nicholson, L.B. (2016). The immune system. *Essays in Biochemistry*, **60**(3), 275–301. DOI: 10.1042/ebc20160017.

Nidom, C.A., Nakayama, E., Nidom, R.V., *et al.* (2012). Serological evidence of Ebola virus infection in Indonesian orangutans. *PLoS ONE*, **7**(7), e40740. DOI: 10.1371/journal.pone.0040740.

Nielsen, H. and Spenceley, A. (2010). *The Success of Tourism in Rwanda: Gorillas and More.* World Development Report 2011 Background paper. Washington DC, and the Hague, the Netherlands: World Bank and the Netherlands Development Organization. Available at: https://openknowledge.worldbank.org/server/api/core/bitstreams/27f8459f-f3b4-5e13-8cbc-bb17f7254cea/content.

Nielsen, H. and Spenceley, A. (2011). The success of tourism in Rwanda: gorillas and more. In *Yes Africa Can: Success Stories from a Dynamic Continent*, ed. P. Chuhan-Pole and M. Angwafo. Washington DC: The World Bank, pp. 231–49.

Nieuwland, J. (2020). *Towards an interspecies health policy: great apes and the right to health.* PhD thesis. Leiden, the Netherlands: Leiden University.

Nijboer, J. (2020). *Nutrition in Primates.* Rahway, NJ: MSD Veterinary Manual. Available at: https://www.msdvetmanual.com/management-and-nutrition/nutrition-exotic-and-zoo-animals/nutrition-in-primates.

Nijman, V. (2017). Orangutan trade, confiscations, and lack of prosecutions in Indonesia. *American Journal of Primatology*, **79**(11), 22652. DOI: 10.1002/ajp.22652.

Nijman, V. (2021). Illegal and legal wildlife trade spreads zoonotic diseases. *Trends in Parasitology*, **37**(5), 359–60. DOI: 10.1016/j.pt.2021.02.001.

Nijman, V., Geissmann, T., Traeholt, C., Roos, C. and Nowak, M.G. (2020). Symphalangus syndactylus. *The IUCN Red List of Threatened Species 2020: e.T39779A17967873*. Gland, Switzerland: International Union for Conservation of Nature (IUCN). DOI: 10.2305/IUCN.UK.2020-2.RLTS.T39779A17967873.en.

Nishida, A.H. and Ochman, H. (2019). A great-ape view of the gut microbiome. *Nature Reviews Genetics*, **20**(4), 195–206. DOI: 10.1038/s41576-018-0085-z.

Nishida, T. (1968). The social group of wild chimpanzees in the Mahali Mountains. *Primates*, **9**(3), 167–224. DOI: 10.1007/BF01730971.

Nishida, T., Matsusaka, T. and McGrew, W.C. (2009). Emergence, propagation or disappearance of novel behavioral patterns in the habituated chimpanzees of Mahale: a review. *Primates*, **50**(1), 23–36. DOI: 10.1007/s10329-008-0109-y.

Nizamuddin, Q. and Rahman, S.A. (2019). Animal welfare in Asia: specific flaws and strengths, future trends and objectives. In *Animal Welfare: from Science to Law*, ed. S. Hild and L. Schweitzer. Paris, France: La foundation Droit Animal, Ethique et Sciences, pp. 109–18. Available at: https://www.fondation-droit-animal.org/documents/AnimalWelfare2019.v1.pdf.

Nizeyi, J.B., Innocent, R.B., Erume, J., *et al.* (2001). Campylobacteriosis, salmonellosis, and shigellosis in free-ranging human-habituated mountain gorillas of Uganda. *Journal of Wildlife Diseases*, **37**(2), 239–44. DOI: 10.7589/0090-3558-37.2.239.

Nkuringo Safaris (2021). *The History of Gorilla Tourism.* Entebbe, Uganda: Nkuringo Safaris Uganda Ltd. Available at: https://www.nkuringosafaris.com/the-history-of-gorilla-tourism/.

Nobel, E., Rybicki, D. and Martin, S. (2020). Wallaby Airlines hops in to help evacuate endangered wildlife from ACT fires and hot weather. *ABC News*, February 10, 2020. Available at: https://www.abc.net.au/news/2020-02-11/wallaby-airlines-hops-in-to-help-evacuate-endangered-wildlife/11949762.

Nolen, R.S. (2006). Gorilla conservation project takes "one-health" approach. Benefits extend beyond endangered apes. *Journal of the American Veterinary Medical Association*, **229**(10), 1546–8.

Norder, H., Ebert, J.W., Fields, H.A., Mushahwar, I.K. and Magnius, L.O. (1996). Complete sequencing of a gibbon hepatitis B virus genome reveals a unique genotype distantly related to the chimpanzee hepatitis B virus. *Virology*, **218**(1), 214–23. DOI: 10.1006/viro.1996.0181.

Normand, E. and Boesch, C. (2009). Sophisticated Euclidean maps in forest chimpanzees. *Animal Behaviour*, **77**(5), 1195–201. DOI: 10.1016/j.anbehav.2009.01.025.

Nowak, M.G., Rianti, P., Wich, S.A., Meijaard, E. and Fredriksson, G.M. (2017). Pongo tapanuliensis. *The IUCN Red List of Threatened Species 2017: e.T120588639A120588662*. Gland, Switzerland: International Union for Conservation of Nature (IUCN). DOI: 10.2305/IUCN.UK.2017-3.RLTS.T120588639A120588662.en.

NPA (2020). *Third National Development Plan (NDPIII) 2020/21–2024/25*. Kampala, Uganda: National Planning Authority (NPA).

NSW Rural Fire Service (n.d.-a). *NSW Bushfire Prone Land*. Sydney, Australia: NSW Government. Available at: https://datasets.seed.nsw.gov.au/dataset/bush-fire-prone-land. Accessed: July, 2022.

NSW Rural Fire Service (n.d.-b). *Standards for Asset Protection Zones*. Granville, Australia: NSW Rural Fire Service. Available at: https://www.rfs.nsw.gov.au/__data/assets/pdf_file/0010/13321/Standards-for-Asset-Protection-Zones.pdf. Accessed: July, 2022.

Nunamaker, E.A., Lee, D.R. and Lammey, M.L. (2012). Chronic diseases in captive geriatric female chimpanzees (*Pan troglodytes*). *Comparative Medicine*, **62**(2), 131–6.

Nunn, C. and Altizer, S. (2006). *Infectious Diseases in Primates: Behavior, Ecology and Evolution*. Oxford, UK: Oxford University Press. DOI: 10.1093/acprof:oso/9780198565857.001.0001.

Nunn, C.L., Altizer, S., Jones, K.E. and Sechrest, W. (2003). Comparative tests of parasite species richness in primates. *The American Naturalist*, **162**(5), 597–614. DOI: 10.1086/378721.

Nuno, A., Chesney, C., Wellbelove, M., *et al.* (2022). Protecting great apes from disease: compliance with measures to reduce anthroponotic disease transmission. *People and Nature*, **4**(5), 1387–400. DOI: 10.1002/pan3.10396.

Nurcahyo, W., Konstanzová, V. and Foitová, I. (2017). Parasites of orangutans (primates: Ponginae): an overview. *American Journal of Primatology*, **79**(6), e22650. DOI: 10.1002/ajp.22650.

Nutter, F.B. (1996). Respiratory disease claims the lives of at least seven Gombe chimps. *Pan Africa News*, **31**(3), 3. DOI: 10.5134/143337.

Nutter, F.B., Whittier, C.A., Cranfield, M.R. and Lowenstine, L.J. (2005). Causes of death for mountain gorillas (*Gorilla beringei beringei* and *G. b. undecided*) from 1968–2004: an aid to conservation programs. Presented at: *Wildlife Health in a Shrinking World: Ecology, Management and Conservation. Proceedings of the Wildlife Disease Association International Conference, Cairns, Queensland, Australia, 26 June–1 July 2005*. Lawrence, KS: Wildlife Disease Association, pp. 200–1.

Nyhus, P.J. (2016). Human–wildlife conflict and coexistence. *Annual Review of Environment and Resources*, **41**(1), 143–71. DOI: 10.1146/annurev-environ-110615-085634.

O'Riordan, T. and Lenton, T., ed. (2013). *Addressing Tipping Points for a Precarious Future*. London, UK: British Academy. DOI: 10.5871/bacad/9780197265536.001.0001.

Oates, J.F., Doumbe, O., Dunn, A., *et al.* (2016). Pan troglodytes ssp. ellioti. *The IUCN Red List of Threatened Species 2016: e.T40014A17990330*. Gland, Switzerland: International Union for Conservation of Nature (IUCN). DOI: 10.2305/IUCN.UK.2016-2.RLTS.T40014A17990330.en.

Odhiambo, N.M. (2021). Health expenditure and economic growth in sub-Saharan Africa: an empirical investigation. *Development Studies Research*, **8**(1), 73–81. DOI: 10.1080/21665095.2021.1892500.

Ogie, R.I., Forehead, H., Clarke, R.J. and Perez, P. (2018). Participation patterns and reliability of human sensing in crowd-sourced disaster management. *Information Systems Frontiers*, **20**(4), 713–28. DOI: 10.1007/s10796-017-9790-y.

Ohashi, G. and Matsuzawa, T. (2011). Deactivation of snares by wild chimpanzees. *Primates*, **52**(1), 1–5. DOI: 10.1007/s10329-010-0212-8.

OHHLEP, Adisasmito, W.B., Almuhairi, S., *et al.* (2022). One Health: a new definition for a sustainable and healthy future. *PLoS Pathogens*, **18**(6), e1010537. DOI: 10.1371/journal.ppat.1010537.

Olhar Animal (2020). MP investiga fuga de chimpanzé que mobilizou 40 pessoas em zoológico de Sorocaba, SP. *Olhar Animal*, November 14, 2020. Available at: https://olharanimal.org/mp-investiga-fuga-de-chimpanze-que-mobilizou-40-pessoas-em-zoologico-de-sorocaba-sp/.

Olival, K.J. and Hayman, D.T.S. (2014). Filoviruses in bats: current knowledge and future directions. *Viruses*, **6**(4), 1759–88.

Ondoua, O.G., Beodo Moundjim, E., Mambo Marindo, J.C., *et al.* (2017). *An Assessment of Poaching and Wildlife Trafficking in the Garamba-Bili-Chinko Transboundary Landscape.* Cambridge, UK: TRAFFIC. Available at: https://www.traffic.org/site/assets/files/1591/garamba-bili-chinko-xxs.pdf.

Ontl, K.M.B. (2017). *Chimpanzees in the Island of Gold: impacts of artisanal small-scale gold mining on chimpanzees (Pan troglodytes verus) in Fongoli, Senegal.* PhD thesis. Armes, IA: Iowa State University. DOI: 10.31274/etd-180810-5211.

Oosterhoff, P., Mokuwa, E.Y. and Wilkinson, A. (2015). *Community-Based Ebola Care Centers: A Formative Evaluation.* Ebola Response Anthropology Platform. Available at: http://www.ebola-anthropology.net/wp-content/uploads/2015/07/Community-Based-Ebola-Care-Centres_A-Formative-Evaluation1.pdf.

OpenStreetMap (n.d.). *OpenStreetMap.* Available at: https://www.openstreetmap.org/#map=5/54.910/-3.432. Accessed: October, 2021.

Oppenheimer, P., Clarke, E., Cupit, O., *et al.* (2021). The SPOTT index: a proof-of-concept measure for tracking public disclosure in the palm oil industry. *Current Research in Environmental Sustainability*, **3**, 100042. DOI: 10.1016/j.crsust.2021.100042.

Oram, F. (2018). *Abundance, feeding and behavioural ecology of orangutans (Pongo pygmaeus morio) in the fragmented forests of the Kinabatangan floodplain.* PhD thesis. Kota Kinabalu, Malaysia: Institute for Tropical Biology and Conservation, University Malaysia Sabah.

Orams, M.B. (2002). Feeding wildlife as a tourism attraction: a review of issues and impacts. *Tourism Management*, **23**(3), 281–93. DOI: 10.1016/S0261-5177(01)00080-2.

Orangutan Appeal UK (n.d.). *Sepilok Orangutan Rehabilitation Centre.* Orangutan Appeal UK. Available at: https://www.orangutan-appeal.org.uk/about-us/sepilok-orangutan-rehabilitation-centre. Accessed: October, 2020.

Orangutan Foundation (2020). Great apes also under threat from COVID-19. *Orangutan Foundation*, March 27, 2020. Available at: https://www.orangutan.org.uk/blog/great-apes-also-under-threat-from-covid-19.

Orangutan Foundation International (n.d.). *Orangutan Care Center and Quarantine.* Los Angeles, CA: Orangutan Foundation International. Available at: https://orangutan.org/occq. Accessed: October, 2020.

Orenstein, W.A. and Ahmed, R. (2017). Simply put: vaccination saves lives. *Proceedings of the National Academy of Sciences*, **114**(16), 4031–3. DOI: 10.1073/pnas.1704507114.

Osofsky, S. (2016). Plan it for the apes: sound science must inform any plans to vaccinate gorillas or chimps against Ebola. *LinkedIn: Pulse*, September 2020. Available at: https://www.linkedin.com/pulse/plan-apes-sound-science-must-inform-any-plans-gorillas-steve-osofsky.

Ostrom, E. and Cox, M. (2010). Moving beyond panaceas: a multi-tiered diagnostic approach for social-ecological analysis. *Environmental Conservation*, **37**(4), 451–63. DOI: 10.1017/S0376892910000834.

Otsuka, R. and Yamakoshi, G. (2020). Analyzing the popularity of YouTube videos that violate mountain gorilla tourism regulations. *PLoS ONE*, **15**(5), e0232085. DOI: 10.1371/journal.pone.0232085.

OVAG (2020a). *COVID-19 Pandemic Guidelines.* Orangutan Veterinary Advisory Group (OVAG) Non Human Primate COVID-19 Information Hub. Available at: https://www.ovag.org/.

OVAG (2020b). *COVID-19 Preparedness and Response Plan.* Orangutan Veterinary Advisory Group (OVAG) Non Human Primate COVID-19 Information Hub. Available at: https://www.ovag.org/.

OVAG (n.d.). *Orangutan Veterinary Advisory Group.* Orangutan Veterinary Advisory Group (OVAG). Available at: https://www.ovag.org. Accessed: November, 2022.

Owens, L.A., Colitti, B., Hirji, I., *et al.* (2021). A *Sarcina* bacterium linked to lethal disease in sanctuary chimpanzees in Sierra Leone. *Nature Communications*, **12**(1), 763. DOI: 10.1038/s41467-021-21012-x.

P-WAC (2020). Live Insta avec Maurice Barthélémy. *Facebook Post*, April 2, 2020. Available at: https://web.facebook.com/239308109568845/videos/678742962887867/.

Palacios, G.F., Lowenstine, L.J., Cranfield, M.R., *et al.* (2011). Human metapneumovirus infection in wild mountain gorillas, Rwanda. *Emerging Infectious Diseases*, **17**(4), 711–13.

Pallisco (2019). *Pallisco Wildlife Team Annual Report 2019*. Internal company document seen by authors. Douala, Cameroon: Pallisco and CIFM.

Palmer, A. (2018). Kill, incarcerate, or liberate? Ethics and alternatives to orangutan rehabilitation. *Biological Conservation*, **227**, 181–8. DOI: 10.1016/j.biocon.2018.09.012.

Palmer, A. (2020). *Ethical Debates in Orangutan Conservation*. London, UK: Routledge. DOI: 10.4324/9780429060533.

Palmer, C. (2010). *Animal Ethics in Context*. New York, NY: Columbia University Press.

Palombit, R.A. (1992). *Pair bonds and monogamy in wild siamang (*Hylobates syndactylus*) and white-handed gibbon (*Hylobates lar*) in northern Sumatra*. PhD thesis. Davis, CA: University of California Davis.

Palombit, R.A. (1994). Dynamic pair bonds in Hylobatids: implications regarding monogamous social systems. *Behaviour*, **128**(1), 65–101. DOI: 10.1163/156853994X00055.

Palombit, R.A. (1997). Inter- and intraspecific variation in the diets of sympatric siamang (*Hylobates syndactylus*) and Lar gibbons (*Hylobates lar*). *Folia Primatologica*, **68**(6), 321–37. DOI: 10.1159/000157260.

Panayotova-Pencheva, M.S. (2013). Parasites in captive animals: a review of studies in some European zoos. *Der Zoologische Garten*, **82**(1), 60–71. DOI: 10.1016/j.zoolgart.2013.04.005.

PanEco (2020). *PanEco Foundation Annual Report 2019*. Berg am Irchel, Switzerland: PanEco Foundation. Available at: https://issuu.com/stiftungpaneco1/docs/engl_paneco_annual_report_2019_web_single_pages_2.

Parc National des Virungas (n.d.). *Mountain Gorilla Sanctuary*. Virungas, DRC: Parc National des Virungas. Available at: https://virunga.org/wildlife/primates/mountain-gorillas/gorilla-orphans/. Accessed: October, 2022.

Park, C. (2022). Lessons learned from the World Health Organization's late initial response to the 2014–2016 Ebola outbreak in West Africa. *Journal of Public Health in Africa*, **13**(1), 1254. DOI: 10.4081/jphia.2022.1254.

Parsons, M.B., Gillespie, T.R., Lonsdorf, E.V., *et al.* (2014). Global positioning system data-loggers: a tool to quantify fine-scale movement of domestic animals to evaluate potential for zoonotic transmission to an endangered wildlife population. *PLoS ONE*, **9**(11), e110984. DOI: 10.1371/journal.pone.0110984.

Parsons, M.B., Travis, D., Lonsdorf, E.V., *et al.* (2015). Epidemiology and molecular characterization of *Cryptosporidium* spp. in humans, wild primates, and domesticated animals in the Greater Gombe Ecosystem, Tanzania. *PLoS Neglected Tropical Diseases*, **9**(2), e0003529. DOI: 10.1371/journal.pntd.0003529.

Parsons, M.B., Travis, D.A., Lonsdorf, E.V., *et al.* (2021). Antimicrobial resistance creates threat to chimpanzee health and conservation in the wild. *Pathogens*, **10**(4), 477. DOI: 10.3390/pathogens10040477.

PASA (2009). *Primate Veterinary Manual*, 2nd edn. Portland, OR: Pan African Sanctuary Alliance (PASA). Available at: https://pasa.org/wp-content/uploads/2016/05/PASA_Vet_Manual_2009_2nd_ed_677pp.pdf.

PASA (2016). *Operations Manual*, 2nd edn, December 2016. Portland, OR: Pan African Sanctuary Alliance (PASA). Available at: https://pasa.org/wp-content/uploads/2016/04/PASA_Operations_Manual_2016.pdf.

PASA (n.d.-a). *Donate to Drill Ranch*. Beaverton, OR: Pan African Sanctuary Alliance (PASA). Available at: https://pasa.org/donate-to-drill-ranch/. Accessed: October, 2020.

PASA (n.d.-b). *Pan African Sanctuary Alliance*. Beaverton, OR: Pan African Sanctuary Alliance (PASA). Available at: https://pasa.org. Accessed: November, 2022.

Patrono, L.V., Pléh, K., Samuni, L., *et al.* (2020). Monkeypox virus emergence in wild chimpanzees reveals distinct clinical outcomes and viral diversity. *Nature Microbiology*, **5**(7), 955–65. DOI: 10.1038/s41564-020-0706-0.

Patrono, L.V., Röthemeier, C., Kouadio, L., *et al.* (2022). Non-invasive genomics of respiratory pathogens infecting wild great apes using hybridisation capture. *Influenza and Other Respiratory Viruses*, **16**(5), 858–61.

Patrono, L.V., Samuni, L., Corman, V.M., *et al.* (2018). Human coronavirus OC43 outbreak in wild chimpanzees, Côte d'Ivoire, 2016. *Emerging Microbes & Infections*, **7**(1), 2–5. DOI: 10.1038/s41426-018-0121-2.

Patz, J.A., Daszak, P., Tabor, G.M., *et al.* (2004). Unhealthy landscapes: policy recommendations on land use change and infectious disease emergence. *Environmental Health Perspectives*, **112**(10), 1092–8. DOI: 10.1289/ehp.6877.

Pauly, B.M., Varcoe, C. and Storch, J. (2012). Framing the issues: moral distress in health care. *HEC Forum*, **24**(1), 1–11. DOI: 10.1007/s10730-012-9176-y.

Payne, J. (1988). *Orang-utan Conservation in Sabah. Report 3759.* Kuala Lumpur, Malaysia: World Wide Fund for Nature (WWF), Malaysia International.

PCI (2022). *Reputational Risk Assessment for Animal Sanctuaries and Crisis Communications Planning Workbook.* Chicago, IL: Public Communications Inc (PCI). Available at: https://www.pcipr.com/resources-download/.

Peacock, L.J. and Rogers, C.M. (1959). Gestation period and twinning in chimpanzees. *Science*, **129**(3354), 959. DOI: 10.1126/science.129.3354.959.

Pearlman, L.A. and Saakvitne, K.W. (1995). Treating therapists with vicarious traumatization and secondary traumatic stress disorders. In *Compassion Fatigue: Coping with Secondary Traumatic Stress Disorder in those who Treat the Traumatized.* Philadelphia, PA: Brunner/Mazel, pp. 150–77.

Pedersen, J., Sorensen, K., Lupo, B. and Marx, L. (2019). Human–ape interactions in a zoo setting: gorillas and orangutans modify their behavior depending upon human familiarity. *Anthrozoös*, **32**(3), 319–32. DOI: 10.1080/08927936.2019.1598651.

Pederson, A.K., King, J.E. and Landau, V.I. (2005). Chimpanzee (*Pan troglodytes*) personality predicts behavior. *Journal of Research in Personality*, **39**(5), 534–49. DOI: 10.1016/j.jrp.2004.07.002.

Pence, D.B. and Ueckermann, E.A. (2002). Sarcoptic mange in wildlife. *Revue Scientifique et Technique de l'Office International des Épizooties*, **21**(2), 385–98.

Penner, L.R. (1981). Concerning threadworm (*Strongyloides stercoralis*) in great apes: lowland gorillas (*Gorilla gorilla*) and chimpanzees (*Pan troglodytes*). *Journal of Zoo Animal Medicine*, **12**(4), 128–31. DOI: 10.2307/20094543.

Pepin, J. (2021). *The Origins of AIDS*, 2nd edn. Cambridge, UK: Cambridge University Press. DOI: 10.1017/9781108767019.

PETA [People for the Ethical Treatment of Animals] (2020). Victory for animal rights groups in "USDA blackout" lawsuits. *PeTA News Releases* July 20, 2020. Available at: https://www.peta.org/media/news-releases/victory-for-animal-rights-groups-in-usda-blackout-lawsuits/.

Peters, J.C. (1966). An epizootic of monkey pox at Rotterdam Zoo. *International Zoo Yearbook*, **6**(1), 274–5. DOI: 10.1111/j.1748-1090.1966.tb01794.x.

Petrovan, S.O., Junker, J., Wordley, C.F.R., *et al.* (2018). Evidence-based synopsis of interventions, a new tool in primate conservation and research. *International Journal of Primatology*, **39**(1), 1–4. DOI: 10.1007/s10764-018-0017-y.

Phalan, B., Hayes, G., Brooks, S., *et al.* (2018). Avoiding impacts on biodiversity through strengthening the first stage of the mitigation hierarchy. *Oryx*, **52**(2), 316–24. DOI: 10.1017/S0030605316001034.

Phelps, J., Aravind, S., Cheyne, S., *et al.* (2021a). Environmental liability litigation could remedy biodiversity loss. *Conservation Letters*, **14**(6), e12821. DOI: 10.1111/conl.12821.

Phelps, J., Fajrini, R., Nagara, G. and Saputra, R. (2021b). *Pioneering Civil Lawsuits for Harm to Threatened Species: A Guide to Claims with Examples from Indonesia.* UK AID, Lancaster University, Indonesian Environmental Law Institute, Auriga Nusantara, LIPI. Available at: https://www.conservation-litigation.org/resources.

Phelps, J., Fajrini, R., Nagara, G. and Saputra, R. (2021c). *Policy Brief. Civil Lawsuits: A Novel Response to Illegal Wildlife Trade.* UK AID, Lancaster University, Indonesian Environmental Law Institute, Auriga Nusantara, LIPI. Available at: https://www.conservation-litigation.org/resources.

Phelps, K.L. and Kingston, T. (2018). Environmental and biological context modulates the physiological stress response of bats to human disturbance. *Oecologia*, **188**(1), 41–52. DOI: 10.1007/s00442-018-4179-2.

Philippa, J. and Dench, R.J. (2019). Infectious diseases of orangutans in their home ranges and in zoos. In *Fowler's Zoo and Wild Animal Medicine Current Therapy, Volume 9*, ed. R. E. Miller, N. Lamberski and P. Calle. St Louis, MO: Elsevier, pp. 565–73.

Pierce, J. and Bekoff, M. (2018). A postzoo future: why welfare fails animals in zoos. *Journal of Applied Animal Welfare Science*, **21**(S1), 43–8. DOI: 10.1080/10888705.2018.1513838.

Pigott, D.M., Golding, N., Mylne, A., *et al.* (2014). Mapping the zoonotic niche of Ebola virus disease in Africa. *eLife*, **3**, e04395. DOI: 10.7554/eLife.04395.

Pigott, D.M., Millear, A.I., Earl, L., *et al.* (2016). Updates to the zoonotic niche map of Ebola virus disease in Africa. *eLife*, **5**, e16412. DOI: 10.7554/eLife.16412.

Pinillos, R.G., Appleby, M.C., Manteca, X., *et al.* (2016). One Welfare: a platform for improving human and animal welfare. *Veterinary Record*, **179**(16), 412–13. DOI: 10.1136/vr.i5470.

Plantier, J.C., Leoz, M., Dickerson, J.E., *et al.* (2009). A new human immunodeficiency virus derived from gorillas. *Nature Medicine*, **15**(8), 871–2. DOI: 10.1038/nm.2016.

Plowright, R.K., Peel, A.J., Streicker, D.G., *et al.* (2016). Transmission or within-host dynamics driving pulses of zoonotic viruses in reservoir–host populations. *PLoS Neglected Tropical Diseases*, **10**(8), e0004796. DOI: 10.1371/journal.pntd.0004796.

Plowright, R.K., Sokolow, S.H., Gorman, M.E., Daszak, P. and Foley, J.E. (2008). Causal inference in disease ecology: investigating ecological drivers of disease emergence. *Frontiers in Ecology and the Environment*, **6**(8), 420–9. DOI: 10.1890/070086.

Plumptre, A., Hart, J.A., Hicks, T.C., *et al.* (2016a). Pan troglodytes *ssp.* schweinfurthii *(errata version published in 2016)*. *The IUCN Red List of Threatened Species 2016: e.T15937A102329417*. Gland, Switzerland: International Union for Conservation of Nature (IUCN). DOI: 10.2305/IUCN.UK.2016-2.RLTS.T15937A17990187.en.

Plumptre, A., Kayitare, A., Rainer, H., *et al.* (2004). *The Socio-Economic Status of People Living Near Protected Areas in the Central Albertine Rift*. Albertine Technical Reports 4. Wildlife Conservation Society (WCS), International Gorilla Conservation Programme (IGCP) and CARE International.

Plumptre, A.J., Kirkby, A., Spira, C., *et al.* (2021). Changes in Grauer's gorilla (*Gorilla beringei graueri*) and other primate populations in the Kahuzi-Biega National Park and Oku Community Reserve, the heart of Grauer's gorilla global range. *American Journal of Primatology*, **83**(7), e23288. DOI: 10.1002/ajp.23288.

Plumptre, A., Nixon, S., Caillaud, D., *et al.* (2016b). Gorilla beringei *ssp.* graueri *(errata version published in 2016)*. *The IUCN Red List of Threatened Species 2016: e.T39995A102328430*. Gland, Switzerland: International Union for Conservation of Nature (IUCN). DOI: 10.2305/IUCN.UK.2016-2.RLTS.T39995A17989838.en.

Plumptre, A., Robbins, M.M. and Williamson, E.A. (2019). Gorilla beringei. *The IUCN Red List of Threatened Species 2019: e.T39994A115576640*. Gland, Switzerland: International Union for Conservation of Nature (IUCN). DOI: 10.2305/IUCN.UK.2019-1.RLTS.T39994A115576640.en.

Plumptre, A.J., Rose, R., Nangendo, G., *et al.* (2010). *Eastern Chimpanzee (*Pan troglodytes schweinfurthii*): Status Survey and Conservation Action Plan 2010–2020*. Gland, Switzerland: International Union for Conservation of Nature (IUCN). Available at: https://portals.iucn.org/library/sites/library/files/documents/2010-023.pdf.

Plumptre, A.J. and Williamson, E.A. (2001). Conservation-oriented research in the Virunga region. In *Mountain Gorillas: Three Decades of Research at Karisoke*, ed. K. J. Stewart, M. M. Robbins and P. Sicotte. Cambridge, UK: Cambridge University Press, pp. 361–90. DOI: 10.1017/CBO9780511661631.015.

PMP (n.d.). *Primate Microbiome Project*. Primate Microbiome Project (PMP). Available at: https://www.primatemicrobiome.org/. Accessed: September, 2022.

Polygeia (2016). *Lessons from Ebola Affected Communities: Being Prepared for Future Health Crises*. London, UK: Africa All Party Parliamentary Group (Africa APPG). Available at: https://research.monash.edu/en/publications/lessons-from-ebola-affected-communities-being-prepared-for-future.

Pomerantz, O. and Terkel, J. (2009). Effects of positive reinforcement training techniques on the psychological welfare of zoo-housed chimpanzees (*Pan troglodytes*). *American Journal of Primatology*, **71**(8), 687–95. DOI: 10.1002/ajp.20703.

Pontzer, H., Brown, M.H., Raichlen, D.A., *et al.* (2016). Metabolic acceleration and the evolution of human brain size and life history. *Nature*, **533**(7603), 390–2. DOI: 10.1038/nature17654.

Potapov, P., Hansen, M.C., Laestadius, L., *et al.* (2017). The last frontiers of wilderness: tracking loss of intact forest landscapes from 2000 to 2013. *Science Advances*, **3**(1), e1600821. DOI: 10.1126/sciadv.1600821.

Power, M. (1986). The foraging adaptation of chimpanzees, and the recent behaviors of the provisioned apes in Gombe and Mahale National Parks, Tanzania. *Human Evolution*, **1**(3), 251–65. DOI: 10.1007/BF02436583.

Pozo, A.A. (2020). Las multas por maltrato animal en Castilla-La Mancha se multiplican por diez tras renovar su ley de hace 30 años. *El Diario.es*, August 3, 2020. Available at: https://www.eldiario.es/castilla-la-mancha/multas-maltrato-animal-castilla-multiplican-diez-ley-30-anos_1_6144274.html.

Prado-Martinez, J., Sudmant, P.H., Kidd, J.M., *et al.* (2013). Great ape genetic diversity and population history. *Nature*, **499**(7459), 471–5. DOI: 10.1038/nature12228.

Prak, D. (2020). Association Papaye International. *LinkedIn*, September 17, 2020. Available at: https://www.linked-in.com/pulse/papaye-nternational-doroth%25C3%25A9e-prak?fbclid=IwAR0dfFpIpQXETU8-pLVVQS6H-6FCUU7_HhdJUhGusFeMmcPwvfydFc8i6BME.

Prasetyo, D., Ancrenaz, M., Morrogh-Bernard, H.C., *et al.* (2009). Nest building in orangutans. In *Geographic Variation in Behavioral Ecology and Conservation*, ed. S. Wich, S. Utami, T. Setia and C. van Schaik. Oxford, UK: Oxford University Press, pp. 269–78.

Prinz, J. (2007). *The Emotional Construction of Morals.* Oxford, UK: Oxford University Press.

Prisner-Levyne, Y. (2020). Trophy hunting, canned hunting, tiger farming, and the questionable relevance of the conservation narrative grounding international wildlife law. *Journal of International Wildlife Law & Policy*, **23**(4), 239–85. DOI: 10.1080/13880292.2020.1866236.

Project Chimps (2020). *Dr Steve Ross' Project ChimpCARE Chimpanzee Welfare Assessment and Project Chimps' Response.* Morganton, GA: Project Chimps. Available at: https://projectchimps.org/wp-content/uploads/2020/11/Ross-Assessment-Response-Final.pdf.

Projet Gorille Fernan-Vaz (n.d.). *Our Gorillas.* Omboué, Gabon: Projet Gorille Fernan-Vaz. Available at: https://gorillasgabon.org/gorillas/our-gorillas/. Accessed: October, 2020.

Pruetz, J.D. and Bertolani, P. (2009). Chimpanzee (*Pan troglodytes verus*) behavioral responses to stresses associated with living in a savanna-mosaic environment: Implications for hominin adaptations to open habitats. *PaleoAnthropology*, 252–62.

Pruetz, J.D. and Herzog, N.M. (2017). Savanna chimpanzees at Fongoli, Senegal, navigate a fire landscape. *Current Anthropology*, **58**(S16), S337–50. DOI: 10.1086/692112.

Pusey, A.E., Wilson, M.L. and Collins, D.A. (2008). Human impacts, disease risk, and population dynamics in the chimpanzees of Gombe National Park, Tanzania. *American Journal of Primatology*, **70**(8), 738–44. DOI: 10.1002/ajp.20567.

Qin, S., Golden Kroner, R.E., Cook, C., *et al.* (2019). Protected area downgrading, downsizing, and degazettement as a threat to iconic protected areas. *Conservation Biology*, **33**(6), 1275–85. DOI: 10.1111/cobi.13365.

QRA (2011). *Rebuilding Grantham Together.* Brisbane, Australia: Queensland Reconstruction Authority (QRA). Available at: https://www.qra.qld.gov.au/news-case-studies/case-studies/case-study-rebuilding-grantham-together-2011.

Quick, J., Loman, N.J., Duraffour, S., *et al.* (2016). Real-time, portable genome sequencing for Ebola surveillance. *Nature*, **530**(7589), 228–32. DOI: 10.1038/nature16996.

Quijano, L., Keeney, A., Schnackenberg, D., *et al.* (2016). *Creating a Community Animal Disaster Plan: A Step-By-Step Guide to Building an Animal Disaster Plan and Developing the Necessary Response Capacity for Your Community.* Fort Collins, CO: Colorado State University. Available at: https://hdl.handle.net/10217/198690.

Rabinowitz, P.M., Pappaioanou, M., Bardosh, K.L. and Conti, L. (2018). A planetary vision for one health. *BMJ Global Health*, **3**(5), e001137. DOI: 10.1136/bmjgh-2018-001137.

Rack, J., Wichmann, O., Kamara, B., *et al.* (2005). Risk and spectrum of diseases in travelers to popular tourist destinations. *Journal of Travel Medicine*, **12**(5), 248–53. DOI: 10.2310/7060.2005.12502.

Radonić, A., Metzger, S., Dabrowski, P.W., *et al.* (2014). Fatal monkeypox in wild-living sooty mangabey, Côte d'Ivoire, 2012. *Emerging Infectious Diseases*, **20**(6), 1009–11. DOI: 10.3201/eid2006.13-1329.

Rainer, H., Lanjouw, A., Llano Sánchez, K. and Banes, G.L. (2020). Drivers of the illegal trade in great apes. In *State of the Apes: Killing, Capture, Trade and Conservation*, ed. Arcus Foundation. Cambridge UK: Cambridge University Press, pp. 96–129. Available at: https://www.stateoftheapes.com/volume-4-killing-capture-trade/.

Rainfer (n.d.). *Los Primates.* Madrid, Spain: Centro de Rescate de Primates Rainfer. Available at: http://rainfer.org/los-primates-2/. Accessed: October, 2020.

Rakotonanahary, R.J.L., Andriambolamanana, H., Razafinjato, B., *et al.* (2021). Integrating health systems and science to respond to COVID-19 in a model district of rural Madagascar. *Frontiers in Public Health*, **9**, July 21, 2021. DOI: 10.3389/fpubh.2021.654299.

Ramsay, E.C., Stair, E.L., Castro, A.E. and Marks, M.I. (1982). Fatal herpesvirus hominis encephalitis in a white-handed gibbon. *Journal of the American Veterinary Medical Association*, **181**(11), 1429–30.

Rasmussen, E.B., Newland, M.C. and Hemmelman, E. (2020). The relevance of operant behavior in conceptualizing the psychological well-being of captive animals. *Perspectives on Behavior Science*, **43**(3), 617–54.

Razanatsoa, E., Andriantsaralaza, S., Holmes, S.M., *et al.* (2021). Fostering local involvement for biodiversity conservation in tropical regions: lessons from Madagascar during the COVID-19 pandemic. *Biotropica*, **53**(4), 994–1003. DOI: 10.1111/btp.12967.

Read, J. (2020). Uganda reopens with extra Covid precautions to protect its mountain gorillas. *Forbes*, October 5, 2020. Available at: https://www.forbes.com/sites/johannaread/2020/10/05/uganda-reopens-with-extra-covid-precautions-to-protect-its-mountain-gorillas/?sh=32518443efea.

Reddacliff, L.A., Kirkland, P.D., Hartley, W.J. and Reece, R.L. (1997). Encephalomyocarditis virus infections in an Australian Zoo. *Journal of Zoo and Wildlife Medicine*, **28**(2), 153–7.

Redshaw, S., Ingham, V., Hicks, J. and Millynn, J. (2017). Emergency preparedness through community sector engagement in the Blue Mountains. *Australian Journal of Emergency Management*, **32** (2), 35–40. DOI: https://knowledge.aidr.org.au/media/3657/ajem-32-02-17.pdf.

Refisch, J. (2021). COVID-19, climate change threaten last refuge of the mountain gorilla. *UN Environment Programme News and Stories*, September 23, 2021. Available at: https://www.unep.org/news-and-stories/story/covid-19-climate-change-threaten-last-refuge-mountain-gorilla.

Refisch, J. and Jenson, J. (2016). Transboundary collaboration in the Greater Virunga Landscape: from gorilla conservation to conflict-sensitive transboundary landscape management. In *Governance, Natural Resources and Post-Conflict Peacebuilding*, ed. C. Bruch, C. Muffett and S. Nichols. London, UK: Routledge, pp. 825–41. Available at: https://www.taylorfrancis.com/chapters/edit/10.4324/9780203109793-39/transboundary-collaboration-greater-virunga-landscape-gorilla-conservation-conflict-sensitive-transboundary-landscape-management-johannes-refisch-johann-jenson.

Reichard, U. (1995). Extra-pair copulations in a monogamous gibbon (*Hylobates lar*). *Ethology*, **100**(2), 99–112. DOI: 10.1111/j.1439-0310.1995.tb00319.x.

Reid, M.J.C. (2020). Is 2020 the year when primatologists should cancel fieldwork? *American Journal of Primatology*, **82**(8), e23161. DOI: 10.1002/ajp.23161.

Reinartz, G., Ingmanson, E.J. and Vervaecke, H. (2013). *Pan paniscus gracile* chimpanzee (bonobo, pygmy chimpanzee). In *Mammals of Africa. Volume II: Primates*, ed. T. M. Butynski, J. Kingdon and J. Kalina. London, UK: Bloomsbury Publishing, pp. 64–9.

ReliefWeb (2015). *Joint Statement on Ebola Response and WHO Reforms*. Geneva, Switzerland: United Nations Office for the Coordination of Humanitarian Affairs (OCHA). Available at: https://reliefweb.int/report/sierra-leone/joint-statement-ebola-response-and-who-reforms.

Research Animal Resources (n.d.). *Anesthesia Guidelines: Non-Human Primates*. St Paul, MN: University of Minnesota. Available at: https://research.umn.edu/units/rar/guidelines/anesthesia-non-human-primates. Accessed: May, 2023.

Resolute (2019). *Mine Gold. Create Value. 2019 Annual Report*. Perth, Australia: Resolute Mining Ltd. Available at: https://www.rml.com.au/investors/reports/annual-reports/.

ResponsibleSteel (2022). *ResponsibleSteel International Standard: Version 2.0*. Newcastle West, Australia: ResponsibleSteel. Available at: https://www.responsiblesteel.org/wp-content/uploads/2022/09/ResponsibleSteel-Standard-2.0.pdf.

Reuter, K.E., Andriantsaralaza, S., Hansen, M.F., *et al.* (2022). Impact of the COVID-19 pandemic on primate research and conservation. *Animals*, **12**(9), 1214. DOI: 10.3390/ani12091214.

Reuters and Gorman, S. (2021). Gorillas at San Diego Zoo Safari Park diagnosed with COVID-19. *Reuters*, January 11, 2021. Available at: https://www.reuters.com/business/healthcare-pharmaceuticals/two-gorillas-san-diego-zoo-test-positive-covid-19-2021-01-11/.

Reuters Staff (2021). Gorilla loses appetite, lions develop cough after catching COVID-19 at Prague Zoo. *Reuters*, February 25, 2021. Available at: https://www.reuters.com/article/us-health-coronavirus-czech-zoo-idUSK-BN2AP2GI.

Richards, P. (2016). *Ebola: How a People's Science Helped End an Epidemic*. London, UK: Zed Books. DOI: 10.5040/9781350219779.

Richardson, H. (2021). For Africa's great apes, a post pandemic future looks beyond tourism. *Mongabay*, June 9, 2021. Available at: https://news.mongabay.com/2021/06/for-africas-great-apes-a-post-pandemic-future-looks-beyond-tourism.

Richeson, J.T., Hughes, H.D., Broadway, P.R. and Carroll, J.A. (2019). Vaccination management of beef cattle: delayed vaccination and endotoxin stacking. *Veterinary Clinics of North America: Food Animal Practice*, **35**(3), 575–92. DOI: 10.1016/j.cvfa.2019.07.003.

Rideout, B.A., Gardiner, C., Stalis, I.H., *et al.* (1997). Fatal Infections with *Balamuthia mandrillaris* (a free-living amoeba) in gorillas and other Old World primates. *Veterinary Pathology*, **34**(1), 15–22. DOI: 10.1177/030098589703400103.

Riede, T., Tokuda, I.T., Munger, J.B. and Thomson, S.L. (2008). Mammalian laryngeal air sacs add variability to the vocal tract impedance: physical and computational modeling. *Journal of the Acoustical Society of America*, **124**(1), 634–47. DOI: 10.1121/1.2924125.

Rietkerk, F. and Pereboom, J.J.M. (2018). Editorial: Conservation of great apes. Zoo contributions towards improving management and well-being of great apes: augmenting knowledge to safeguard our closest relative. *International Zoo Yearbook*, **52**(1), 9–15. DOI: 10.1111/izy.12202.

Rijksen, H.D. (1978). *A field study on Sumatran orangutans (*Pongo pygmaeus abelii *Lesson 1827). Ecology, behaviour and conservation*. PhD thesis. Wageningen, the Netherlands: Nature Conservation Department, Agricultural University Wageningen. Available at: https://library.wur.nl/WebQuery/wurpubs/fulltext/209957.

Rijksen, H.D. and Meijaard, E. (1999). *Our Vanishing Relative? The Status of Wild Orangutans at the Close of the Twentieth Century*. Dordrecht, the Netherlands: Kluwer Academic.

Rima, B., Collins, P., Easton, A., *et al.* (2017). ICTV virus taxonomy profile: Pneumoviridae. *Journal of General Virology*, **98**(12), 2912–13. DOI: 10.1099/jgv.0.000959.

Ringer, G.D. (2002). Gorilla tourism: Uganda uses tourism to recover from decades of violent conflict. *Alternatives Journal: Canadian Environmental Ideas and Action*, **28**(4), 16–19.

Rio Tinto Simfer S.A. (2012a). *Simandou Social and Environmental Impact Assessment (SEIA). Volume I. Mine. Chapter 1: Introduction*. Conakry, Republic of Guinea, and London, UK: Rio Tinto Simfer S.A. Available at: https://icsid.worldbank.org/sites/default/files/parties_publications/C3765/Respondent%27s%20Counter-Memorial/Pi%C3%A8ces%20factuelles/R-0140.pdf.

Rio Tinto Simfer S.A. (2012b). *Simandou Social and Environmental Impact Assessment (SEIA). Volume V. Social and Environmental Management Plan*. Conakry, Republic of Guinea, and London, UK: Rio Tinto Simfer S.A.

Rioja-Lang, F., Bacon, H., Connor, M. and Dwyer, C.M. (2020a). Prioritisation of animal welfare issues in the UK using expert consensus. *Veterinary Record*, **187**(12), 490. DOI: 10.1136/vr.105964.

Rioja-Lang, F.C., Connor, M., Bacon, H.J., Lawrence, A.B. and Dwyer, C.M. (2020b). Prioritization of farm animal welfare issues using expert consensus. *Frontiers in Veterinary Science*, **6**, 495. DOI: 10.3389/fvets.2019.00495.

Riva, H.G., Zordan, M.A. and Sánchez, C.R. (2020). The current state of zoological medicine in zoos and aquariums in Latin America. *International Zoo Yearbook*, **54**(1), 202–18. DOI: 10.1111/izy.12251.

Rivas, M.L., Albion, I., Bernal, B., *et al.* (2022). The plastic pandemic: COVID-19 has accelerated plastic pollution, but there is a cure. *Science of The Total Environment*, **847**, 157555. DOI: 10.1016/j.scitotenv.2022.157555.

Rivera, S.N., Knight, A. and McCulloch, S.P. (2021). Surviving the wildlife trade in Southeast Asia: reforming the "disposal" of confiscated live animals under CITES. *Animals*, **11**(2), 439. DOI: 10.3390/ani11020439.

RNZ (2020). Covid impact: Auckland Zoo receives almost $3m from government. *RNZ* [Radio New Zealand], September 27, 2020. Available at: https://www.rnz.co.nz/news/national/427019/covid-impact-auckland-zoo-receives-almost-3m-from-government.

Robbins, A.M., Manguette, M.L., Breuer, T., *et al.* (2022). Population dynamics of western gorillas at Mbeli Bai. *PLoS ONE*, **17**(10), e0275635. DOI: 10.1371/journal.pone.0275635.

Robbins, A.M., Stoinski, T., Fawcett, K. and Robbins, M.M. (2011a). Lifetime reproductive success of female mountain gorillas. *American Journal of Physical Anthropology*, **146**(4), 582–93. DOI: 10.1002/ajpa.21605.

Robbins, M.M. (2011). Gorillas: diversity in ecology and behavior. In *Primates in Perspective*, ed. C. J. Campbell, A. Fuentes, K. C. MacKinnon, S. Bearder and R. M. Stumpf. Oxford, UK: Oxford University Press, pp. 326–39.

Robbins, M.M. (2021). Assessing attitudes towards gorilla conservation via employee interviews. *American Journal of Primatology*, **83**(4), e23191. DOI: 10.1002/ajp.23191.

Robbins, M.M. and Boesch, C., ed. (2011). *Among African Apes: Stories and Photos from the Field*. Berkeley, CA: University of California Press.

Robbins, M.M. and Robbins, A.M. (2018). Variation in the social organization of gorillas: life history and socio-ecological perspectives. *Evolutionary Anthropology: Issues, News, and Reviews*, **27**, 218–33. DOI: 10.1002/evan.21721.

Robbins, M.M., Gray, M., Fawcett, K.A., *et al.* (2011b). Extreme conservation leads to recovery of the Virunga mountain gorillas. *PLoS ONE*, **6**(6), 1–10. DOI: 10.1371/journal.pone.0019788.

Robbins, M.M., Gray, M., Kagoda, E. and Robbins, A.M. (2009). Population dynamics of the Bwindi mountain gorillas. *Biological Conservation*, **142**(12), 2886–95. DOI: 10.1016/j.biocon.2009.07.010.

Robbins, M.M., Ortmann, S. and Seiler, N. (2022). Dietary variability of western gorillas (*Gorilla gorilla gorilla*). *PLoS ONE*, **17**(8), e0271576. DOI: 10.1371/journal.pone.0271576.

Roberts, L. (2019). A prescription for Madagascar's broken health system: data and a focus on details. *Science Magazine*, February 18, 2019. Available at: https://www.sciencemag.org/news/2019/02/prescription-madagascar-s-broken-health-system-data-and-focus-details.

Robertson, B.H. and Margolis, H.S. (2002). Primate hepatitis B viruses – genetic diversity, geography and evolution. *Reviews in Medical Virology*, **12**(3), 133–41. DOI: 10.1002/rmv.348.

Robins, J.G., Husson, S., Fahroni, A., *et al.* (2019). Implanted radio telemetry in orangutan reintroduction and post-release monitoring and its application in other ape species. *Frontiers in Veterinary Science*, **6**, 111. DOI: 10.3389/fvets.2019.00111.

Robson, S.L. and Wood, B. (2008). Hominin life history: reconstruction and evolution. *Journal of Anatomy*, **212**(4), 394–425. DOI: 10.1111/j.1469-7580.2008.00867.x.

Rodriguez, M., Pascual, M., Wingard, J., *et al.* (2019). *Legal Protection of Great Apes & Gibbons: Compilation of Country Profiles for 17 Range Countries*. Missoula, MT: Legal Atlas, LLC. DOI: 10.13140/RG.2.2.13189.88800.

Rodriguez-Morales, A.J. and Schlagenhauf, P. (2014). Zoonoses and travel medicine: "one world – one health". *Travel Medicine and Infectious Disease*, **12**(6, Part A), 555–6. DOI: 10.1016/j.tmaid.2014.11.003.

Roe, D. and Booker, F. (2019). Engaging local communities in tackling illegal wildlife trade: a synthesis of approaches and lessons for best practice. *Conservation Science and Practice*, **1**(5), e26. DOI: 10.1111/csp2.26.

Roe, D. and Urquhart, P. (2001). Pro-poor tourism: harnessing the world's largest industry for the world's poor. Presented at: *World Summit on Sustainable Development, Johannesburg, South Africa*. International Institute for Environment and Development (IIED) in collaboration with the Regional and International Networking Group (RING).

Roger, F., Caron, A., Morand, S., *et al.* (2016). One Health and EcoHealth: the same wine in different bottles? *Infection Ecology & Epidemiology*, **6**(1), 30978. DOI: 10.3402/iee.v6.30978.

Rohr, J.R., Barrett, C.B., Civitello, D.J., *et al.* (2019). Emerging human infectious diseases and the links to global food production. *Nature Sustainability*, **2**(6), 445–56. DOI: 10.1038/s41893-019-0293-3.

Romero, L.M., Dickens, M.J. and Cyr, N.E. (2009). The reactive scope model – a new model integrating homeostasis, allostasis, and stress. *Hormones and Behavior*, **55**(3), 375–89. DOI: 10.1016/j.yhbeh.2008.12.009.

Romero, L.M. and Wingfield, J.C. (2015). *Tempests, Poxes, Predators, and People: Stress in Wild Animals and How They Cope*. Oxford, UK: Oxford University Press. DOI: 10.1093/acprof:oso/9780195366693.001.0001.

Romero-Alvarez, D., Peterson, A.T., Salzer, J.S., *et al.* (2020). Potential distributions of *Bacillus anthracis* and *Bacillus cereus* biovar *anthracis* causing anthrax in Africa. *PLoS Neglected Tropical Diseases*, **14**(3), e0008131. DOI: 10.1371/journal.pntd.0008131.

Ronfot, D. (2016). *Animals in limbo: the importance of recognizing welfare of confiscated wild animals. An investigation in Thai governmental wildlife confiscation facilities*. MA thesis. Exeter, UK: University of Exeter.

Rose, A.L. (2011). Bonding, biophilia, biosynergy, and the future of primates in the wild. *American Journal of Primatology*, **73**(3), 245–52. DOI: 10.1002/ajp.20888.

Rosenblum, I.Y. and Coulston, F. (1983). Impaired renal function in diabetic chimpanzees (*Pan troglodytes*). *Experimental Molecular Pathology*, **38**(2), 224–9. DOI: 10.1016/0014-4800(83)90087-4.

Ross, S.R. (2020). Chimpanzee welfare in the context of science, policy, and practice. In *Chimpanzees in Context: A Comparative Perspective on Chimpanzee Behavior, Cognition, Conservation, and Welfare*, ed. L. M. Hopper and S. R. Ross. Chicago, IL: University of Chicago Press, pp. 552–84. DOI: 10.7208/chicago/9780226728032.003.0024.

Ross, S.R., Hansen, B.K., Hopper, L.M. and Fultz, A. (2019). A unique zoo-sanctuary collaboration for chimpanzees. *American Journal of Primatology*, **81**(5), e22941. DOI: 10.1002/ajp.22941.

Ross, S.R. and Leinwand, J.G. (2020). A review of research in primate sanctuaries. *Biology Letters*, **16**(4), 20200033. DOI: 10.1098/rsbl.2020.0033.

Ross, S.R., Lukas, K.E., Lonsdorf, E.V., *et al.* (2008). Inappropriate use and portrayal of chimpanzees. *Science*, **319**(5869), 1487. DOI: 10.1126/science.1154490.

Ross, S.R., Vreeman, V.M. and Lonsdorf, E.V. (2011). Specific image characteristics influence attitudes about chimpanzee conservation and use as pets. *PLoS ONE*, **6**(7), e22050. DOI: 10.1371/journal.pone.0022050.

Ross, S.R., Wagner, K.E., Schapiro, S.J. and Hau, J. (2010). Ape behavior in two alternating environments: comparing exhibit and short-term holding areas. *American Journal of Primatology*, **72**(11), 951–9. DOI: 10.1002/ajp.20857.

Roth, J.A. (2011). Veterinary vaccines and their importance to animal health and public health. *Procedia in Vaccinology*, **5**, 127–36. DOI: 10.1016/j.provac.2011.10.009.

Roth, T.S., Rianti, P., Fredriksson, G.M., Wich, S.A. and Nowak, M.G. (2020). Grouping behavior of Sumatran orangutans (*Pongo abelii*) and Tapanuli orangutans (*Pongo tapanuliensis*) living in forest with low fruit abundance. *American Journal of Primatology*, **82**(5), e23123. DOI: 10.1002/ajp.23123.

Rouquet, P., Froment, J.M., Bermejo, M., *et al.* (2005). Wild animal mortality monitoring and human Ebola outbreaks, Gabon and Republic of Congo, 2001–2003. *Emerging Infectious Diseases*, **11**(2), 283–90. DOI: 10.3201/eid1102.040533.

Rowe, M.L., Whiteley, P.L. and Carver, S. (2019). The treatment of sarcoptic mange in wildlife: a systematic review. *Parasites & Vectors*, **12**(1), 99. DOI: 10.1186/s13071-019-3340-z.

Roxana (2021). La fondation Mona: Un sanctuaire de chimpanzés en Espagne. *Sh Barcelone*, December 3, 2021. Available at: https://www.shbarcelona.fr/blog/fr/la-fondation-mona-un-sanctuaire-de-chimpanzes-en-espagne/.

Royal Commission into National Natural Disaster Arrangements (2020a). Chapter 10. Community education. In *Royal Commission into National Natural Disaster Arrangements Report 28 October 2020*, ed. Royal Commission into National Natural Disaster Arrangements. Canberra, Australia: Commonwealth of Australia, pp. 245–51. Available at: https://naturaldisaster.royalcommission.gov.au/publications/html-report/chapter-10.

Royal Commission into National Natural Disaster Arrangements (2020b). National information systems. In *Interim Observations 31 August 2020*, ed. Royal Commission into National Natural Disaster Arrangements. Canberra, Australia: Commonwealth of Australia, pp. 12–13. Available at: https://naturaldisaster.royalcommission.gov.au/publications/interim-observations-1/interim-observations-4.

RSPO (2020). *RSPO Principle Criteria for the Production of Sustainable Palm Oil 2018, Revised 01 February 2020*. Kuala Lumpar, Malyasia: Roundtable on Sustainable Palm Oil (RSPO). Available at: https://rspo.org/resources/?category=rspo-principle-criteria-for-the-production-of-sustainable-palm-oil-2018.

RSPO (n.d.). *Our Impact: Outcomes and Impacts*. Kuala Lumpar, Malyasia: Roundtable on Sustainable Palm Oil (RSPO). Available at: https://rspo.org/our-impact/outcomes-and-impacts/. Accessed: December, 2022.

Ruckert, A., Zinszer, K., Zarowsky, C., Labonté, R. and Carabin, H. (2020). What role for One Health in the COVID-19 pandemic? *Canadian Journal of Public Health*, **111**(5), 641–4. DOI: 10.17269/s41997-020-00409-z.

Rudicell, R.S., Holland Jones, J., Wroblewski, E.E., *et al.* (2010). Impact of simian immunodeficiency virus infection on chimpanzee population dynamics. *PLoS Pathogens*, **6**(9), e1001116. DOI: 10.1371/journal.ppat.1001116.

Rüegg, S.R., Häsler, B. and Zinsstag, J. (2018). *Integrated Approaches to Health: A Handbook for the Evaluation of One Health*. Wageningen, the Netherlands: Wageningen Academic Publishers. DOI: 10.3920/978-90-8686-875-9.

Runhovde, S.R. (2022). Mind the gap! Decoupling between policy and practice in the policing of illegal wildlife trade. *International Journal of Offender Therapy and Comparative Criminology*, **66**(4), 369–88. DOI: 10.1177/0306624x20967953.

Rushmore, J., Caillaud, D., Matamba, L., *et al.* (2013). Social network analysis of wild chimpanzees provides insights for predicting infectious disease risk. *Journal of Animal Ecology*, **82**(5), 976–86. DOI: 10.1111/1365-2656.12088.

Russon, A.E. (2004). Aristotle's rubicon. In *Orangutans: Wizards of the Rainforest*, ed. A. E. Russon. Toronto, Canada: Key Porter Publications.

Russon, A.E. (2009). Orangutan rehabilitation and reintroduction: successes, failures and role in conservation. In *Orangutans: Geographic Variation in Behavioral Ecology and Conservation*, ed. S. A. Wich, S. S. Utami-Atmoko, T. Mitra Setia and C. P. van Schaik. Oxford, UK: Oxford University Press, pp. 327–50.

Russon, A.E., Kuncoro, P. and Ferisa, A. (2015). Orangutan behavior in Kutai National Park after drought and fire damage: adjustments to short- and long-term natural forest regeneration. *American Journal of Primatology*, **77**(12), 1276–89. DOI: 10.1002/ajp.22480.

Russon, A.E., Smith, J.J. and Adams, L. (2016). Managing human–orangutan relationships in rehabilitation. In *Ethnoprimatology: Primate Conservation in the 21st Century*, ed. M. Waller. Cham, Switzerland: Springer, pp. 233–58. DOI: 10.1007/978-3-319-30469-4_13.

Russon, A.E. and Susilo, A. (2014). Orangutan tourism and conservation: 35 years' experience. In *Primate Tourism: A Tool for Conservation?*, ed. A. E. Russon and J. Wallis. Cambridge, UK: Cambridge University Press, pp. 76–97. DOI: 10.1017/CBO9781139087407.007.

Russon, A.E. and Wallis, J. (2014a). Primate tourism as a conservation tool: a review of the evidence, implications, and recommendations. In *Primate Tourism: A Tool for Conservation?*, ed. A. E. Russon and J. Wallis. Cambridge, UK: Cambridge University Press, pp. 313–32. DOI: 10.1017/CBO9781139087407.024.

Russon, A.E. and Wallis, J. (2014b). Reconsidering primate tourism as a conservation tool: an introduction to the issues. In *Primate Tourism: A Tool for Conservation?*, ed. A. E. Russon and J. Wallis. Cambridge, UK: Cambridge University Press, pp. 3–18. DOI: 10.1017/CBO9781139087407.002.

Russon, A.E., Wich, S.A., Ancrenaz, M., *et al.* (2009). Geographic variation in orangutan diets. In *Orangutans: Geographic Variation in Behavioral Ecology and Conservation*, ed. S. A. Wich, S. Utami-Atmoko, T. Mitra Setia and C. P. van Schaik. Oxford, UK: Oxford University Press, pp. 135–56.

Rwanda Development Board (2017). Increase of gorilla permit tariffs. *Rwanda Development Board*, May 6, 2017. Available at: https://rdb.rw/increase-of-gorilla-permit-tariffs/.

Rwego, I.B., Isabirye-Basuta, G., Gillespie, T.R. and Goldberg, T.L. (2008). Gastrointestinal bacterial transmission among humans, mountain gorillas, and livestock in Bwindi Impenetrable National Park, Uganda. *Conservation Biology*, **22**(6), 1600–7.

Ryan, S.J. and Walsh, P.D. (2011). Consequences of non-intervention for infectious disease in African great apes. *PLoS ONE*, **6**(12), e29030. DOI: 10.1371/journal.pone.0029030.

Sabuhoro, E., Wright, B., Munanura, I.E., Nyakabwa, I.N. and Nibigira, C. (2017). The potential of ecotourism opportunities to generate support for mountain gorilla conservation among local communities neighboring Volcanoes National Park in Rwanda. *Journal of Ecotourism*, **20**(1), 1–17. DOI: 10.1080/14724049.2017.1280043.

Sadler, B., Dusik, J., Fischer, T., *et al.* (2010). *Handbook of Strategic Environmental Assessment*, 1st edn. London, UK: Routledge.

Sakamoto, M., Sasaki, D., Ono, Y., Makino, Y. and Kodama, E.N. (2020). Implementation of evacuation measures during natural disasters under conditions of the novel coronavirus (COVID-19) pandemic based on a review of previous responses to complex disasters in Japan. *Progress in Disaster Science*, **8**, 100127. DOI: 10.1016/j.pdisas.2020.100127.

Sakulwira, K., Theamboonlers, A., Charoonrut, P., Ratanakorn, P. and Poovorawan, Y. (2002). Serological evidence of herpesvirus infection in gibbons. *BMC Microbiology*, **2**, 11. DOI: 10.1186/1471-2180-2-11.

Sakurai, M. and Murayama, Y. (2019). Information technologies and disaster management – benefits and issues. *Progress in Disaster Science*, **2**, 100012. DOI: 10.1016/j.pdisas.2019.100012.

Samuni, L., Preis, A., Deschner, T., Crockford, C. and Wittig, R.M. (2018). Reward of labor coordination and hunting success in wild chimpanzees. *Communications Biology*, **1**, 138. DOI: 10.1038/s42003-018-0142-3.

Samuni, L., Wegdell, F. and Surbeck, M. (2020). Behavioural diversity of bonobo prey preference as a potential cultural trait. *eLife*, **9**, e59191. DOI: 10.7554/eLife.59191.

Sanchez, C.R. and Hidalgo-Hermoso, E. (2022). *Mycobacterium tuberculosis* sensu stricto in African apes, what is its true health impact? *Pathogens*, **11**(5), 484. DOI: 10.3390/pathogens11050484.

Sanchez, K.L., Greenwood, A.D., Nielsen, A., *et al.* (2022). *Plasmodium pitheci* malaria in Bornean orang-utans at a rehabilitation centre in West Kalimantan, Indonesia. *Malaria Journal*, **21**(1), 280. DOI: 10.1186/s12936-022-04290-8.

Sánchez-Vizcaíno, F., Muniesa, A., Singleton, D.A., *et al.* (2018). Use of vaccines and factors associated with their uptake variability in dogs, cats and rabbits attending a large sentinel network of veterinary practices across Great Britain. *Epidemiology & Infection*, **146**(7), 895–903. DOI: 10.1017/S0950268818000754.

Sandbrook, C.G. (2010). Putting leakage in its place: the significance of retained tourism revenue in the local context in rural Uganda. *Journal of International Development*, **22**(1), 124–36. DOI: 10.1002/jid.1507.

Sandbrook, C. and Semple, S. (2006). The rules and the reality of mountain gorilla *Gorilla beringei beringei* tracking: how close do tourists get? *Oryx*, **40**(4), 428–33. DOI: 10.1017/S0030605306001323.

Sandosham, A.A. (1951). On two helminths from the orang utan, *Leipertrema rewelli* n.g., n.sp. and *Dirofilaria immitis* (Leidy, 1856). *Journal of Helminthology*, **25**(1–2), 19–26. DOI: 10.1017/S0022149X00018927.

Santiago-Ávila, F.J. and Lynn, W.S. (2020). Bridging compassion and justice in conservation ethics. *Biological Conservation*, **248**, 108648. DOI: 10.1016/j.biocon.2020.108648.

Santos, W.J., Guiraldi, L.M. and Lucheis, S.B. (2020). Should we be concerned about COVID-19 with nonhuman primates? *American Journal of Primatology*, **82**(8), e23158. DOI: 10.1002/ajp.23158.

Sapolsky, R., Romero, L.M. and Munck, A.U. (2000). How do glucocorticoids influence stress responses? Integrating permissive, suppressive, stimulatory, and preparative actions. *Endocrine Reviews*, **21**(1), 55–89. DOI: 10.1210/edrv.21.1.0389.

Sapolsky, R., Uno, H., Rebert, C. and Finch, C. (1990). Hippocampal damage associated with prolonged glucocorticoid exposure in primates. *The Journal of Neuroscience*, **10**(9), 2897–902. DOI: 10.1523/jneurosci.10-09-02897.1990.

Sarma, K., Krishna, M. and Kumar, A. (2015). Fragmented populations of the vulnerable eastern hoolock gibbon *Hoolock leuconedys* in the Lower Dibang Valley district, Arunachal Pradesh, India. *Oryx*, **49**(1), 133–9. DOI: 10.1017/S0030605312001299.

Sarmiento, E.E. (1985). *Functional differences in the skeleton of wild and captive orangutans and their adaptive significance.* PhD thesis. New York, NY: New York University.

Satchell, G.H. and Harrison, R.A. (1953). II. Experimental observations on the possibility of transmission of yaws by wound-feeding Diptera, in Western Samoa. *Transactions of the Royal Society of Tropical Medicine and Hygiene*, **47**(2), 148–53. DOI: 10.1016/0035-9203(53)90068-6.

Saudale, V. (2015). Ministry: Indonesia has only four decent zoos. *Jakarta Globe*, February 8, 2015. Available at: https://www.todayonline.com/world/asia/indonesia-has-only-4-decent-zoos-ministry.

Save the Chimps (n.d.). *Philosophy of Care.* Internal unpublished document. Fort Pierce, FL: Save the Chimps.

Sayektiningsih, T., Sari., U.K., Yassir, I. and Ma'ruf, A. (2020). Students and orangutan conservation: high school students' perceptions of orangutan sanctuary establishment in Balikpapan Bay, East Kalimantan, Indonesia. *Buletin Eboni*, **2**(1), 35–46. DOI: 10.20886/buleboni.5570.

Schaffner, C.M., Rebecchini, L., Ramos-Fernandez, G., Vick, L.G. and Aureli, F. (2012). Spider monkeys (*Ateles geoffroyi yucatenensis*) cope with the negative consequences of hurricanes through changes in diet, activity budget, and fission–fusion dynamics. *International Journal of Primatology*, **33**(4), 922–36. DOI: 10.1007/s10764-012-9621-4.

Schaller, G.B. (1963). *The Mountain Gorilla: Ecology and Behavior.* Chicago, IL: University of Chicago Press.

Schapiro, S.J., Bloomsmith, M.A. and Laule, G.E. (2003). Positive reinforcement training as a technique to alter nonhuman primate behavior: quantitative assessments of effectiveness. *Journal of Applied Animal Welfare Science*, **6**(3), 175–87. DOI: 10.1207/S15327604JAWS0603_03.

Schaumburg, F., Mugisha, L., Peck, B., *et al.* (2012). Drug-resistant human *Staphylococcus aureus* in sanctuary apes pose a threat to endangered wild ape populations. *American Journal of Primatology*, **74**(12), 1071–5. DOI: 10.1002/ajp.22067.

Scheffer, M., Bolhuis, J.E., Borsboom, D., *et al.* (2018). Quantifying resilience of humans and other animals. *Proceedings of the National Academy of Sciences*, **115**(47), 11883–90. DOI: 10.1073/pnas.1810630115.

Scherl, L.M., Wilson, A., Wild, R., *et al.* (2004). *Can Protected Areas Contribute to Poverty Reduction? Opportunities and Limitations.* Gland, Switzerland, and Cambridge, UK: International Union for Conservation of Nature (IUCN). Available at: https://portals.iucn.org/library/sites/library/files/documents/2004-047.pdf.

Schoenle, L.A., Downs, C.J. and Martin, L.B. (2018). An introduction to ecoimmunology. In *Advances in Comparative Immunology*, ed. E. L. Cooper. Cham, Switzerland: Springer International Publishing, pp. 901–32. DOI: 10.1007/978-3-319-76768-0_26.

Scholfield, K.A. (2013). *Transnational (dis)connections: mountain gorilla conservation in Rwanda and the DRC.* PhD thesis. Manchester, UK: University of Manchester.

Schovancová, K., Pomajbíková, K., Procházka, P., *et al.* (2013). Preliminary insights into the impact of dietary starch on the ciliate, *Neobalantidium coli*, in captive chimpanzees. *PLoS ONE*, **8**(11), e81374. DOI: 10.1371/journal.pone.0081374.

Schubert, G., Achi, V., Ahuka, S., *et al.* (2021). The African network for improved diagnostics, epidemiology and management of common infectious agents. *BMC Infectious Diseases*, **21**(1), 539. DOI: 10.1186/s12879-021-06238-w.

Schuenemann, V.J., Avanzi, C., Krause-Kyora, B., *et al.* (2018). Ancient genomes reveal a high diversity of *Mycobacterium leprae* in medieval Europe. *PLoS Pathogens*, **14**(5), e1006997. DOI: 10.1371/journal.ppat.1006997.

Schulman, F.Y., Farb, A., Virmani, R. and Montali, R.J. (1995). Fibrosing cardiomyopathy in captive western lowland gorillas (*Gorilla gorilla gorilla*) in the United States: a retrospective study. *Journal of Zoo and Wildlife Medicine*, **26**(1), 43–51.

Scorpion (2019). Document of strategy and action plan for Indonesian orangutan conservation 2019–2029 is launched. *Scorpionmonitor News*, 13 August 2019. Available at: http://scorpionmonitor.org/news/document-of-strategy-and-action-plan-for-indonesian-orangutan-conservation-2019-2029-is-launched-august-12-2019-444.html.

SCS (2017). *NSW Rural Fire Service Fire Trail Design, Construction and Maintenance Manual.* Parramatta, Australia: Soil Conservation Service (SCS). Available at: https://www.rfs.nsw.gov.au/__data/assets/pdf_file/0009/97569/Fire-Trail-Design-Construction-and-Maintenance-Manual-FINAL_reducedsize.pdf.

Scully, E.J., Basnet, S., Wrangham, R.W., *et al.* (2018). Lethal respiratory disease associated with human rhinovirus C in wild chimpanzees, Uganda, 2013. *Emerging Infectious Diseases*, **24**(2), 267–74. DOI: 10.3201/eid2402.170778.

Second Chance Chimpanzee Refuge Liberia (2020). Second Chance Chimpanzee Refuge Liberia/Save the Abandoned Chimps. *Facebook Post*, October 14, 2020. Available at: https://www.facebook.com/abandonedchimps/posts/2806313472914574.

Seifert, S.N., Fischer, R.J., Kuisma, E., *et al.* (2022). Zaire Ebola virus surveillance near the Bikoro region of the Democratic Republic of the Congo during the 2018 outbreak reveals presence of seropositive bats. *PLoS Neglected Tropical Diseases*, **16**(6), e0010504. DOI: 10.1371/journal.pntd.0010504.

Seiler, B.M., Dick Jr, E.J., Guardado-Mendoza, R., *et al.* (2009). Spontaneous heart disease in the adult chimpanzee (*Pan troglodytes*). *Journal of Medical Primatology*, **38**(1), 51–8. DOI: 10.1111/j.1600-0684.2008.00307.x.

Seiler, N., Boesch, C., Mundry, R., Stephens, C. and Robbins, M.M. (2017). Space partitioning in wild, non-territorial mountain gorillas: the impact of food and neighbours. *Royal Society Open Science*, **4**(11), 170720. DOI: 10.1098/rsos.170720.

Seiler, N., Boesch, C., Stephens, C., *et al.* (2018). Social and ecological correlates of space use patterns in Bwindi mountain gorillas. *American Journal of Primatology*, **80**(4), e22754. DOI: 10.1002/ajp.22754.

Seiler, N. and Robbins, M.M. (2016). Factors influencing ranging on community land and crop raiding by mountain gorillas. *Animal Conservation*, **19**(2), 176–88. DOI: 10.1111/acv.12232.

Seiler, N. and Robbins, M.M. (2020). Ecological correlates of space use patterns in wild western lowland gorillas. *American Journal of Primatology*, **82**(9), e23168. DOI: 10.1002/ajp.23168.

Seimon, T.A., Olson, S.H., Lee, K.J., *et al.* (2015). Correction: Adenovirus and herpesvirus diversity in free-ranging great apes in the Sangha region of the Republic of Congo. *PLoS ONE*, **10**(11), e0142766. DOI: 10.1371/journal.pone.0142766.

Sekerka, L.E. and Bagozzi, R.P. (2007). Moral courage in the workplace: moving to and from the desire and decision to act. *Business Ethics: A European Review*, **16**(2), 132–49. DOI: 10.1111/j.1467-8608.2007.00484.x.

Seneviratne, S.I., Nicholls, N., Easterling, D., *et al.* (2012). Changes in climate extremes and their impacts on the natural physical environment. In *Managing the Risks of Extreme Events and Disasters to Advance Climate*

Change Adaptation A Special Report of Working Groups I and II of the Intergovernmental Panel on Climate Change (IPCC), ed. IPCC. Cambridge, UK, and New York, NY: Cambridge University Press, pp. 109–230.

Serckx, A., Huynen, M.-C., Bastin, J.-F., *et al.* (2014). Nest grouping patterns of bonobos (*Pan paniscus*) in relation to fruit availability in a forest–savannah mosaic. *PLoS ONE*, **9**(4), e93742. DOI: 10.1371/journal.pone.0093742.

Sergio, F., Blas, J. and Hiraldo, F. (2018). Animal responses to natural disturbance and climate extremes: a review. *Global and Planetary Change*, **161**, 28–40. DOI: 10.1016/j.gloplacha.2017.10.009.

Shaffer, C.A., Yukuma, C., Marawanaru, E. and Suse, P. (2018). Assessing the sustainability of Waiwai subsistence hunting in Guyana by comparison of static indices and spatially explicit, biodemographic models. *Animal Conservation*, **21**(2), 148–58. DOI: 10.1111/acv.12366.

Sharp, P.M. and Hahn, B.H. (2011). Origins of HIV and the AIDS pandemic. *Cold Spring Harbor Perspectives in Medicine*, **1**(1), a006841. DOI: 10.1101/cshperspect.a006841.

Shave, R., Oxborough, D., Somauroo, J., *et al.* (2014). Echocardiographic assessment of cardiac structure and function in great apes: a practical guide. *International Zoo Yearbook*, **48**(1), 218–33. DOI: 10.1111/izy.12026.

Shchelkunov, S.N., Totmenin, A.V., Babkin, I.V., *et al.* (2001). Human monkeypox and smallpox viruses: genomic comparison. *FEBS Letters*, **509**(1), 66–70. DOI: 10.1016/S0014-5793(01)03144-1.

Sherman, J., Ancrenaz, M. and Meijaard, E. (2020). Shifting apes: conservation and welfare outcomes of Bornean orangutan rescue and release in Kalimantan, Indonesia. *Journal for Nature Conservation*, **55**, 125807. DOI: 10.1016/j.jnc.2020.125807.

Sherman, J., Ancrenaz, M., Voigt, M., *et al.* (2020). Envisioning a future for Bornean orangutans: conservation impacts of action plan implementation and recommendations for improved population outcomes. *Biodiversitas*, **21**(2), 456–77.

Sherman, J., Brent, L. and Farmer, K. (2016). Poster: A picture is worth a thousand words: an analysis of animal images posted on the internet by African ape sanctuaries. Presented at: *International Primatological Society, 26th Congress, August 23, 2016, Chicago, IL*. International Primatological Society.

Sherman, J. and Greer, D. (2018). The status of captive apes. I. Beyond capacity: sanctuaries and the status of captive apes in shrinking natural habitats. In *State of the Apes: Infrastructure Development and Ape Conservation*, ed. Arcus Foundation. Cambridge, UK: Cambridge University Press, pp. 227–55. Available at: https://www.stateoftheapes.com/themes/ch-8-the-status-of-captive-apes/.

Sherman, J., Unwin, S., Travis, D.A., *et al.* (2021). Disease risk and conservation implications of orangutan translocations. *Frontiers in Veterinary Science*, **8**, 749547. DOI: 10.3389/fvets.2021.749547.

Sherwen, S.L., Hemsworth, L.M., Beausoleil, N.J., Embury, A. and Mellor, D.J. (2018). An animal welfare risk assessment process for zoos. *Animals*, **8**(8), 130. DOI: 10.3390/ani8080130.

Shin, N.S., Kwon, S.W., Han, D.H., *et al.* (1995). *Mycobacterium tuberculosis* infection in an orangutan (*Pongo pygmaeus*). *Journal of Veterinary Medical Science*, **57**(5), 951–3. DOI: 10.1292/jvms.57.951.

Shue, H. (1996). *Basic Rights: Subsistence, Affluence, and U.S. Foreign Policy*. Princeton, NJ: Princeton University Press.

Shutt, K.A. (2014). *Wildlife tourism and conservation: an interdisciplinary evaluation of gorilla ecotourism in Dzanga-Sangha, Central African Republic*. PhD thesis. Durham, UK: Durham University.

Shutt, K., Heistermann, M., Kasim, A., *et al.* (2014). Effects of habituation, research and ecotourism on faecal glucocorticoid metabolites in wild western lowland gorillas: implications for conservation management. *Biological Conservation*, **172**, 72–9. DOI: 10.1016/j.biocon.2014.02.014.

Sierra Maestra (2020). *Zoológico de Santiago de Cuba: un atractivo citadino*. Havana, Cuba: Sierra Maestra. Available at: http://www.sierramaestra.cu/index.php/especiales/32854-zoologico-de-santiago-de-cuba-un-atractivo-citadino.

Silva, C. (2018). Animal welfare law of Costa Rica. *The Costa Rica News*, September 18, 2018. Available at: https://thecostaricanews.com/animal-welfare-law-of-costa-rica/#:~:text=The%20law%20establishes%20punishments%20of,with%20them%20or%20practices%20vivisection.

Sim, S.H., Ong, C.E.L., Gan, Y.H., *et al.* (2018). Melioidosis in Singapore: clinical, veterinary, and environmental perspectives. *Tropical Medicine and Infectious Disease*, **3**(1), 31. DOI: 10.3390/tropicalmed3010031.

Sinclair, M. and Phillips, C.J.C. (2018a). International Animal Protection Society leadership: the right people for the right issues. *Animals*, **8**(6), 89. DOI: 10.3390/ani8060089.

Sinclair, M. and Phillips, C.J.C. (2018b). Key tenets of operational success in international animal welfare initiatives. *Animals*, **8**(6), 92. DOI: 10.3390/ani8060092.

Singer, P. (2011). *Practical Ethics.* Cambridge, UK: Cambridge University Press.

Singer, T. and Klimecki, O.M. (2014). Empathy and compassion. *Current Biology*, **24**(18), R875–8. DOI: 10.1016/j.cub.2014.06.054.

Singleton, I., Knott, C.D., Morrogh-Bernard, H.C., Wich, S.A. and van Schaik, C.P. (2009). Ranging behavior of orangutan females and social organization. In *Orangutans: Geographic Variation in Behavioral Ecology and Conservation*, ed. S. A. Wich, S. Utami-Atmoko, T. Mitra Setia and C. P. van Schaik. Oxford, UK: Oxford University Press, pp. 205–13.

Singleton, I., Wich, S.A., Nowak, M., Usher, G. and Utami-Atmoko, S.S. (2017). Pongo abelii *(errata version published in 2018)*. *The IUCN Red List of Threatened Species 2017: e.T121097935A123797627.* Gland, Switzerland: International Union for Conservation of Nature (IUCN). DOI: 10.2305/IUCN.UK.2017-3.RLTS.T121097935A115575085.en.

Siregar, J.E., Faust, C.L., Murdiyarso, L.S., *et al.* (2015). Non-invasive surveillance for *Plasmodium* in reservoir macaque species. *Malaria Journal*, **14**(1), 404. DOI: 10.1186/s12936-015-0857-2.

Skinner, M.F. (1986). Enamel hypoplasia in sympatric chimpanzee and gorilla. *Human Evolution*, **1**(4), 289–312. DOI: 10.1007/BF02436704.

Sklenovská, N. and Van Ranst, M. (2018). Emergence of monkeypox as the most important orthopoxvirus infection in humans. *Frontiers in Public Health*, **6**, 241. DOI: 10.3389/fpubh.2018.00241.

Slater, O.M., Terio, K.A., Zhang, Y., *et al.* (2014). Human metapneumovirus infection in chimpanzees, United States. *Emerging Infectious Diseases*, **20**(12), 2115–18. DOI: 10.3201/eid2012.140408.

Sloan, S., Alamgir, M., Campbell, M.J., Setyawati, T. and Laurance, W.F. (2019). Development corridors and remnant-forest conservation in Sumatra, Indonesia. *Tropical Conservation Science*, **12**, 1940082919889509. DOI: 10.1177/1940082919889509.

Sloan, S., Supriatna, J., Campbell, M.J., Alamgir, M. and Laurance, W.F. (2018). Newly discovered orangutan species requires urgent habitat protection. *Current Biology*, **28**(11), R650–1. DOI: 10.1016/j.cub.2018.04.082.

Smiley Evans, T., Barry, P.A., Gilardi, K.V., *et al.* (2015). Optimization of a novel non-invasive oral sampling technique for zoonotic pathogen surveillance in nonhuman primates. *PLoS Neglected Tropical Diseases*, **9**(6), e0003813. DOI: 10.1371/journal.pntd.0003813.

Smiley Evans, T., Gilardi, K.V.K., Barry, P.A., *et al.* (2016). Detection of viruses using discarded plants from wild mountain gorillas and golden monkeys. *American Journal of Primatology*, **78**(11), 1222–34. DOI: 10.1002/ajp.22576.

Smiley Evans, T., Lowenstine, L.J., Gilardi, K.V., *et al.* (2017). Mountain gorilla lymphocryptovirus has Epstein-Barr virus-like epidemiology and pathology in infants. *Scientific Reports*, **7**, 5352. DOI: 10.1038/s41598-017-04877-1.

Smith, P.C., Yuill, T.M., Buchanan, R.D., Stanton, J.S. and Chaicumpa, V. (1969). The gibbon (*Hylobates lar*); a new primate host for *Herpesvirus hominia*. I. A natural epizootic in a laboratory colony. *Journal of Infectious Diseases*, **120**(3), 292–7. DOI: 10.1093/infdis/120.3.292.

Smithsonian Institute (2022). *What Does it Mean to be Human?* Washington DC: Smithsonian Institution. Available at: http://humanorigins.si.edu/evidence/genetics.

Smits, W.T.M., Heriyanto and Ramono, W.S. (1995). A new method for rehabilitation of orangutans in Indonesia. In *The Neglected Ape*, ed. R. D. Nadler, B. F. M. Galdikas, L. K. Sheeran and N. Rosen. Boston, MA: Springer, pp. 69–77. DOI: 10.1007/978-1-4899-1091-2_8.

SOC (n.d.). *Meet Orangutan.* East Kalimantan, Indonesia: Singtan Orangutan Center (SOC). Available at: https://www.soc.or.id/sintang-orangutan-center/meet-orangutan/. Accessed: October, 2020.

SOCP (n.d.-a). *Creating New Wild Populations.* Medan, Indonesia: Sumatran Orangutan Conservation Program (SOCP). Available at: https://www.sumatranorangutan.org/our-work/creating-new-wild-populations/. Accessed: August, 2021.

SOCP (n.d.-b). *Jantho Orangutan Reintroduction Centre.* Medan, Indonesia: Sumatran Orangutan Conservation Program (SOCP). Available at: https://www.sumatranorangutan.org/our-work/creating-new-wild-populations/reintroduction/jantho/. Accessed: September, 2022.

SOCP (n.d.-c). *Our Work.* Medan, Indonesia: Sumatran Orangutan Conservation Program (SOCP). Available at: https://www.sumatranorangutan.org/our-work/creating-new-wild-populations/. Accessed: September, 2022.

SOCP (n.d.-d). *Rehabilitation*. Medan, Indonesia: Sumatran Orangutan Conservation Program (SOCP). Available at: https://www.sumatranorangutan.org/our-work/creating-new-wild-populations/rehabilitation/. Accessed: September, 2022.

Solleveld, H.A., van Zwieten, M.J., Heidt, P.J. and van Eerd, P.M. (1984). Clinicopathologic study of six cases of meningitis and meningoencephalitis in chimpanzees (*Pan troglodytes*). *Laboratory Animal Science*, **34**(1), 86–90.

Sollund, R. (2022). Wildlife trade and law enforcement: a proposal for a remodeling of CITES incorporating species justice, ecojustice, and environmental justice. *International Journal of Offender Therapy and Comparative Criminology*, **66**(9), 1017–35. DOI: 10.1177/0306624x221099492.

Somerville, K. (2020). Focus: Wildlife & pandemics: COVID-19, bushmeat and poaching in Africa. *Global Geneva*, September 14, 2020. Available at: https://www.global-geneva.com/focus-wildlife-pandemics-covid-19-bushmeat-and-poaching-in-africa/.

Soorae, P.S., Al Hemeri, A., Al Shamsi, A. and Al Suwaidi, K. (2008). A survey of the trade in wildlife as pets in the United Arab Emirates. *TRAFFIC Bulletin*, **22**(1), 41–6.

Southern, L.M., Deschner, T. and Pika, S. (2021). Lethal coalitionary attacks of chimpanzees (*Pan troglodytes troglodytes*) on gorillas (*Gorilla gorilla gorilla*) in the wild. *Scientific Reports*, **11**, 14673. DOI: 10.1038/s41598-021-93829-x.

Species360 (n.d.). *ZIMS by Species360*. Minneapolis, MN: Species360. Available at: https://zims.species360.org. Accessed: August, 2020.

Spehar, S.N., Sheil, D., Harrison, T., *et al.* (2018). Orangutans venture out of the rainforest and into the Anthropocene. *Science Advances*, **4**(6), 1–13. DOI: 10.1126/sciadv.1701422.

Spelman, L.H., Gilardi, K.V.K., Lukasik-Braum, M., *et al.* (2013). Respiratory disease in mountain gorillas (*Gorilla beringei beringei*) in Rwanda, 1990–2010: outbreaks, clinical course, and medical management. *Journal of Zoo and Wildlife Medicine*, **44**(4), 1027–35. DOI: 10.1638/2013-0014R.1.

Spenceley, A., Habyalimana, S., Tusabe, R. and Mariza, D. (2010). Benefits to the poor from gorilla tourism in Rwanda. *Development Southern Africa*, **27**(5), 647–62. DOI: 10.1080/0376835X.2010.522828.

Spencer, J., Amony, I. and Dube, C. (2020). The impacts of mountain gorilla tourism in Uganda: can participating stakeholders benefit? In *3rd International Conference on Tourism Research, Universidad Europea de Valencia, Spain, 27–28 March 2020*, ed. J. Martí-Parreño, R. Gómez-Calvet and J. Muñoz de Prat. Sonning Common, UK: Academic Conferences and Publishing International Ltd, pp. 355–62.

Spessa, A. and Field, R. (2015). Indonesia at risk from huge fires because of El Niño. *The Conversation*, June 16, 2015. Available at: https://theconversation.com/indonesia-at-risk-from-huge-fires-because-of-el-nino-43072.

Spessa, A.C., Field, R.D., Pappenberger, F., *et al.* (2015). Seasonal forecasting of fire over Kalimantan, Indonesia. *Natural Hazards Earth Systems Sciences*, **15**(3), 429–42. DOI: 10.5194/nhess-15-429-2015.

Spillmann, B., van Noordwijk, M.A., Willems, E.P., *et al.* (2015). Validation of an acoustic location system to monitor Bornean orangutan (*Pongo pygmaeus wurmbii*) long calls. *American Journal of Primatology*, **77**(7), 767–76. DOI: 10.1002/ajp.22398.

Spillmann, B., Willems, E.P., van Noordwijk, M.A., Setia, T.M. and van Schaik, C.P. (2017). Confrontational assessment in the roving male promiscuity mating system of the Bornean orangutan. *Behavioral Ecology and Sociobiology*, **71**(1), 20. DOI: 10.1007/s00265-016-2252-6.

Špinka, M. and Wemelsfelder, F. (2018). Environmental challenge and animal agency. In *Animal Welfare*, ed. M. C. Appleby, I. A. S. Olsson and F. Galindo. Wallingford, UK: CABI International, pp. 39–55. DOI: 10.1079/9781786390202.0039.

SPOTT (n.d.). *Timber and Pulp: ESG Policy Transparency*. SPOTT. Available at: https://www.spott.org/timber-pulp/. Accessed: December, 2022.

Sprague, L.D. and Neubauer, H. (2004). Melioidosis in animals: a review on epizootiology, diagnosis and clinical presentation. *Journal of Veterinary Medicine, Series B*, **51**(7), 305–20. DOI: 10.1111/j.1439-0450.2004.00797.x.

Spruijt, B.M., van den Bos, R. and Pijlman, F.T.A. (2001). A concept of welfare based on reward evaluating mechanisms in the brain: anticipatory behaviour as an indicator for the state of reward systems. *Applied Animal Behaviour Science*, **72**(2), 145–71. DOI: 10.1016/S0168-1591(00)00204-5.

Srivathsan, A., Lee, L., Katoh, K., *et al.* (2021). ONTbarcoder and MinION barcodes aid biodiversity discovery and identification by everyone, for everyone. *BMC Biology*, **19**(1), 217. DOI: 10.1186/s12915-021-01141-x.

Staupe-Delgado, R. (2019). Analysing changes in disaster terminology over the last decade. *International Journal of Disaster Risk Reduction*, **40**, 101161. DOI: 10.1016/j.ijdrr.2019.101161.

Steinmetz, H.W. and Zimmermann, N.E. (2012). Computed tomography for the diagnosis of sinusitis and air sacculitis in orangutans. In *Fowler's Zoo and Wild Animal Medicine, Current Therapy, Volume 7*, ed. R. E. Miller and M. Fowler. St Louis, MO: Elsevier Saunders, pp. 422–30. DOI: 10.1016/B978-1-4377-1986-4.00055-X.

Steinmetz, R., Srirattanaporn, S., Mor-Tip, J. and Seuaturien, N. (2014). Can community outreach alleviate poaching pressure and recover wildlife in south-east Asian protected areas? *Journal of Applied Ecology*, **51**(6), 1469–78. DOI: 10.1111/1365-2664.12239.

Stephen, C. and Karesh, W.B. (2014). Is One Health delivering results? Introduction. *Revue Scientifique et Technique de l'Office International des Épizooties*, **33**(2), 375–92. DOI: 10.20506/rst.33.2.2301.

Stephens, N., Vogelnest, L., Lowbridge, C., *et al.* (2013). Transmission of *Mycobacterium tuberculosis* from an Asian elephant (*Elephas maximus*) to a chimpanzee (*Pan troglodytes*) and humans in an Australian zoo. *Epidemiology & Infection*, **141**(7), 1488–97. DOI: 10.1017/s095026881300068x.

Stevens, J. (2020). *EAZA Best Practice Guidelines – Bonobo (*Pan panicus*).* Amsterdam, the Netherlands: European Association of Zoos and Aquaria (EAZA) Great Ape Taxon Advisory Group (TAG). Available at: https://www.eaza.net/assets/Uploads/CCC/BPG-2020/Bonobo-BPG-final-version-2020.pdf.

Stevens, J.M., Alonso, A.S., Aerts, T. and Vervaecke, H. (2008). The behaviour of a group of chimpanzees: influence of spatial crowding and visitor numbers. Presented at: *Proceedings of the Tenth Annual Symposium on Zoo Research, Hull, UK, 15–16 July 2008*. London, UK: British and Irish Association of Zoos and Aquariums (BIAZA).

Stewart, K. (1988). Suckling and lactational anoestrus in wild gorillas (*Gorilla gorilla*). *Journal of Reproduction and Fertility*, **83**(2), 627–34.

Stewart, M.C. and Wilson, G.B. (2016). The dynamic role of social media during Hurricane #Sandy: An introduction of the STREMII model to weather the storm of the crisis lifecycle. *Computers in Human Behavior*, **54**, 639–46. DOI: 10.1016/j.chb.2015.07.009.

Stibbe, A. (2001). Language, power and the social construction of animals. *Society & Animals*, **9**(2), 145–61. DOI: 10.1163/156853001753639251.

Stoinski, T.S., Perdue, B.M., Breuer, T. and Hoff, M.P. (2013). Variability in the developmental life history of the genus *Gorilla. American Journal of Physical Anthropology*, **152**(2), 165–72.

Stokes, E.J. and Byrne, R.W. (2006). Effect of snare injuries on the fig-feeding behavior of chimpanzees of the Budongo Forest, Uganda. In *Primates of Western Uganda*, ed. N. E. Newton-Fisher, H. Notman, J. D. Paterson and V. Reynolds. New York, NY: Springer, pp. 281–97. DOI: 10.1007/978-0-387-33505-6_16.

Stop Animal Selfies (n.d.). *Home*. Costa Rica: Stop Animal Selfies. Available at: https://stopanimalselfies.org/en/home/. Accessed: March, 2021.

Strindberg, S., Maisels, F., Williamson, E.A., *et al.* (2018). Guns, germs, and trees determine density and distribution of gorillas and chimpanzees in western Equatorial Africa. *Science Advances*, **4**(4), eaar2964. DOI: 10.1126/sciadv.aar2964.

Strong, V.J., Grindlay, D., Redrobe, S., Cobb, M. and White, K. (2016). A systematic review of the literature relating to captive great ape morbidity and mortality. *Journal of Zoo and Wildlife Medicine*, **47**(3), 697–710. DOI: 10.1638/2015-0240.1.

Strong, V.J., Martin, M., Redrobe, S., White, K. and Baiker, K. (2018). A retrospective review of great ape cardiovascular disease epidemiology and pathology. *International Zoo Yearbook*, **52**(1), 113–25. DOI: 10.1111/izy.12193.

Strong, V., Moittié, S., Sheppard, M.N., *et al.* (2020). Idiopathic myocardial fibrosis in captive chimpanzees (*Pan troglodytes*). *Veterinary Pathology*, **57**(1), 183–91. DOI: 10.1177/0300985819879442.

Strum, S.C. (2005). Measuring success in primate translocation: a baboon case study. *American Journal of Primatology*, **65**(2), 117–40. DOI: 10.1002/ajp.20103.

Stuart, P., Yalcindag, E., Ali, I.K.M., *et al.* (2020). *Entamoeba histolytica* infections in wild and semi-wild orangutans in Sumatra and Kalimantan. *American Journal of Primatology*, **82**(5), e23124. DOI: 10.1002/ajp.23124.

Subudhi, S., Rapin, N. and Misra, V. (2019). Immune system modulation and viral persistence in bats: understanding viral spillover. *Viruses*, **11**(2), 192. DOI: 10.3390/v11020192.

Sumarga, E. (2017). Spatial indicators for human activities may explain the 2015 fire hotspot distribution in central Kalimantan Indonesia. *Tropical Conservation Science*, **10**. DOI: 10.1177/1940082917706168.

Susman, R.L. (1984). The locomotor behavior of *Pan paniscus* in the Lomako Forest. In *The Pygmy Chimpanzee: Evolutionary Biology and Behavior*, ed. R. L. Susman. Boston, MA: Springer, pp. 369–93. DOI: 10.1007/978-1-4757-0082-4_15.

Suzuki, K., Tanigawa, K., Kawashima, A., Miyamura, T. and Ishii, N. (2011). Chimpanzees used for medical research shed light on the pathoetiology of leprosy. *Future Microbiology*, **6**(10), 1151–7. DOI: 10.2217/fmb.11.97.

Swaisgood, R.R. (2010). The conservation–welfare nexus in reintroduction programmes: a role for sensory ecology. *Animal Welfare*, **19**(2), 125–37. DOI: 10.1017/S096272860000138X.

Szentiks, C.A., Köndgen, S., Silinski, S., Speck, S. and Leendertz, F.H. (2009). Lethal pneumonia in a captive juvenile chimpanzee (*Pan troglodytes*) due to human-transmitted human respiratory syncytial virus (HRSV) and infection with *Streptococcus pneumoniae*. *Journal of Medical Primatology*, **38**(4), 236–40. DOI: 10.1111/j.1600-0684.2009.00346.x.

Tabor, P.D. (2011). Vicarious traumatization: concept analysis. *Journal of Forensic Nursing*, **7**(4), 203–8. DOI: 10.1111/j.1939-3938.2011.01115.x.

Tacugama Chimpanzee Sanctuary (n.d.). *About Us*. Freetown, Sierra Leone: Tacaguma Chimpanzee Sanctuary. Available at: https://www.tacugama.com/about-us/. Accessed: October, 2020.

Tangtrongsup, S., Sripakdee, D., Malaivijitnond, S., Angkuratipakorn, R. and Lappin, M.R. (2019). Intestinal parasites and the occurrence of zoonotic *Giardia duodenalis* genotype in captive gibbons at Krabokkoo Wildlife Breeding Center, Thailand. *Frontiers in Veterinary Science*, **6**, 110. DOI: 10.3389/fvets.2019.00110.

Tapanes, E., Detwiler, K.M. and Cords, M. (2016). Bat predation by cercopithecus monkeys: implications for zoonotic disease transmission. *EcoHealth*, **13**(2), 405–9. DOI: 10.1007/s10393-016-1121-0.

Tapper, R. (2006). *Wildlife Watching and Tourism: A Study on the Benefits and Risks of a Fast Growing Tourism Activity and its Impacts on Species*. Bonn, Germany: United Nations Environment Programme (UNEP)/ Convention on Migratory Species (CMS) Secretariat.

TAWIRI (2018). *Tanzania Chimpanzee Conservation Action Plan 2018–2023*. Arusha, Tanzania: Tanzania Wildlife Research Institute (TAWIRI). Available at: https://www.researchgate.net/publication/332865978_Tanzania-Chimpanzee-Conservation-Action-Plan-2018.

Tayleur, C., Balmford, A., Buchanan, G.M., *et al.* (2017). Global coverage of agricultural sustainability standards, and their role in conserving biodiversity. *Conservation Letters*, **10**(5), 610–18. DOI: 10.1111/conl.12314.

Tchakoudeu Kehou, S., Daïnou, K. and Lagoute, P. (2021). The reasons great ape populations are still abundant in logged concessions: environmental drivers and the influence of management plans. *Forest Ecology and Management*, **483**, 118911. DOI: 10.1016/j.foreco.2020.118911.

Teare, J.A. and Loomis, M.R. (1982). Epizootic of balantidiasis in lowland gorillas. *Journal of the American Veterinary Medical Association*, **181**(11), 1345–7.

Teixeira, C.P., de Azevedo, C.S., Mendl, M., Cipreste, C.F. and Young, R.J. (2007). Revisiting translocation and reintroduction programmes: the importance of considering stress. *Animal Behaviour*, **73**(1), 1–13. DOI: 10.1016/j.anbehav.2006.06.002.

Teo, H.C., Lechner, A.M., Sagala, S. and Campos-Arceiz, A. (2020). Environmental impacts of planned capitals and lessons for Indonesia's new capital. *Land*, **9**(11), 438. DOI: 10.3390/land9110438.

Teo, S.Z., Tuen, A.A., Madinah, A., Aban, S. and Chong, Y.L. (2019). Occurrence of gastrointestinal nematodes in captive nonhuman primates at Matang Wildlife Centre, Sarawak. *Tropical biomedicine*, **36**(3), 594–603.

Testamenti, V.A., Surya, M., Saepuloh, U., *et al.* (2020). Characterization of *Burkholderia pseudomallei* from spontaneous melioidosis in a Bornean orangutan. *Veterinary World*, **13**, 2459–68. DOI: 10.14202/vetworld.2020.2459-2468.

Thangavelu, K., Jamir, I., Ellappan, K., *et al.* (2021). Comparison of MGIT 960 with Lowenstein Jensen media for recovery of mycobacteria from extrapulmonary specimens in southern India. *Journal of Clinical and Diagnostic Research*, **15**(3), DC01–4. DOI: 10.7860/jcdr/2021/47238.14603.

The Chimpanzee Sequencing and Analysis Consortium (2005). Initial sequence of the chimpanzee genome and comparison with the human genome. *Nature*, **437**(7055), 69–87. DOI: 10.1038/nature04072.

The Ebola Gbalo Research Group (2019). Responding to the Ebola virus disease outbreak in DR Congo: when will we learn from Sierra Leone? *The Lancet*, **393**(10191), 2647–50. DOI: 10.1016/S0140-6736(19)31211-5.

The Republic of Rwanda (2018). *Ebola Virus Disease (EVD) Contingency Plan*. The Republic of Rwanda. Available at: https://www.preventionweb.net/files/63524_rwandaebolavirusdiseaseevdcontingen.pdf.

Thinh, V.N., Mootnick, A.R., Thanh, V.N., Nadler, T. and Roos, C. (2010). A new species of crested gibbon, from the central Annamite mountain range. *Vietnamese Journal of Primatology*, **4**, 1–12.

Thornton, S.M., Walker, S. and Zuckerman, J.N. (2001). Management of hepatitis B virus infections in two gibbons and a western lowland gorilla in a zoological collection. *Veterinary Record*, **149**(4), 113–15. DOI: 10.1136/vr.149.4.113.

Thorpe, W.H. (1963). *Learning and Instinct in Animals*. London, UK: Methuen.

Tindana, P., Molyneux, C.S., Bull, S. and Parker, M. (2014). Ethical issues in the export, storage and reuse of human biological samples in biomedical research: perspectives of key stakeholders in Ghana and Kenya. *BMC Medical Ethics*, **15**(1), 76. DOI: 10.1186/1472-6939-15-76.

Toft, J.D., II (1982). The pathoparasitology of the alimentary tract and pancreas of nonhuman primates: a review. *Veterinary Pathology*, **19**(S7), 44–92.

Toft, J.D., II (1986). The pathoparasitology of nonhuman primates: a review. In *Primates: The Road to Self-Sustaining Populations*, ed. K. Benirschke. New York, NY: Springer, pp. 571–679. DOI: 10.1007/978-1-4612-4918-4_45.

Tolbert, S., Makambo, W., Asuma, S., Musema, A. and Mugabukomeye, B. (2019). The perceived benefits of protected areas in the Virunga-Bwindi Massif. *Environmental Conservation*, **46**(1), 76–83. DOI: 10.1017/S0376892918000309.

Tong, L.J., Flach, E.J., Sheppard, M.N., *et al.* (2014). Fatal arrhythmogenic right ventricular cardiomyopathy in 2 related subadult chimpanzees (*Pan troglodytes*). *Veterinary Pathology*, **51**(4), 858–67. DOI: 10.1177/0300985813501333.

Toppenberg-Pejcic, D., Noyes, J., Allen, T., *et al.* (2019). Emergency risk communication: lessons learned from a rapid review of recent gray literature on Ebola, Zika, and yellow fever. *Health Communication*, **34**(4), 437–55. DOI: 10.1080/10410236.2017.1405488.

TRAFFIC (2022). Financial flows toolkit to tackle illegal wildlife trade. *TRAFFIC NEWS*, March 3, 2022. Available at: https://www.traffic.org/news/uk-iwt-financial-flows-toolkit-launch/.

Travis, D.A., Lonsdorf, E.V. and Gillespie, T.R. (2018). The grand challenge of great ape health and conservation in the anthropocene. *American Journal of Primatology*, **80**(1), e22717. DOI: 10.1002/ajp.22717.

Trayford, H.R. and Farmer, K.H. (2012). An assessment of the use of telemetry for primate reintroductions. *Journal for Nature Conservation*, **20**, 311–25. DOI: 10.1016/j.jnc.2012.07.004.

Tremaroli, V. and Bäckhed, F. (2012). Functional interactions between the gut microbiota and host metabolism. *Nature*, **489**(7415), 242–9. DOI: 10.1038/nature11552.

Trivedy, C. (2020). Is 2020 the year when primatologists should cancel fieldwork? A reply. *American Journal of Primatology*, **82**(8), e23173. DOI: 10.1002/ajp.23173.

Trogisch, L. and Fletcher, R. (2022). Fortress tourism: exploring dynamics of tourism, security and peace around the Virunga transboundary conservation area. *Journal of Sustainable Tourism*, **30**(2–3), 352–71. DOI: 10.1080/09669582.2020.1857767.

Truelove, M.A., Martin, J.E., Langford, F.M. and Leach, M.C. (2020). The identification of effective welfare indicators for laboratory-housed macaques using a Delphi consultation process. *Scientific Reports*, **10**, 20402. DOI: 10.1038/s41598-020-77437-9.

Truman, R. (2005). Leprosy in wild armadillos. *Leprosy Review*, **76**(3), 198–208.

Tshibangu, G.M. (2018). An analysis of strategic environmental assessment legislation and regulations in African countries. *Journal of Environmental Assessment Policy and Management*, **20**(1), 1–26. DOI: https://www.jstor.org/stable/90020684.

Tsujino, R., Yumoto, T., Kitamura, S., Djamaluddin, I. and Darnaedi, D. (2016). History of forest loss and degradation in Indonesia. *Land Use Policy*, **57**, 335–47. DOI: 10.1016/j.landusepol.2016.05.034.

Tumusiime, D. and Vedeld, P. (2012). False promise or false premise? Using tourism revenue sharing to promote conservation and poverty reduction in Uganda. *Conservation and Society*, **10**(1), 15–28. DOI: 10.4103/0972-4923.92189.

Turnbaugh, P.J., Bäckhed, F., Fulton, L. and Gordon, J.I. (2008). Diet-induced obesity is linked to marked but reversible alterations in the mouse distal gut microbiome. *Cell Host & Microbe*, **3**(4), 213–23. DOI: 10.1016/j.chom.2008.02.015.

Turner, W.C., Kausrud, K.L., Krishnappa, Y.S., *et al.* (2014). Fatal attraction: vegetation responses to nutrient inputs attract herbivores to infectious anthrax carcass sites. *Proceedings of the Royal Society B: Biological Sciences*, **281**(1795), 20141785. DOI: 10.1098/rspb.2014.1785.

Tutin, C.E.G., Ancrenaz, M., Paredes, J., *et al.* (2001). Conservation biology framework for the release of wild-born orphaned chimpanzees into the Conkouati Reserve, Congo. *Conservation Biology*, **15**(5), 1247–57. DOI: 10.1111/j.1523-1739.2001.00046.x.

Tutin, C.E.G. and Fernandez, M. (1991). Responses of wild chimpanzees and gorillas to the arrival of primatologists: behaviour observed during habituation. In *Primate Responses to Environmental Change*, ed. H. O. Box. Dordrecht, the Netherlands: Springer, pp. 187–97. DOI: 10.1007/978-94-011-3110-0_10.

Tweh, C.G., Lormie, M.M., Kouakou, C.Y., *et al.* (2015). Conservation status of chimpanzees *Pan troglodytes verus* and other large mammals in Liberia: a nationwide survey. *Oryx*, **49**(4), 710–18. DOI: 10.1017/S0030605313001191.

Twycross Zoo (n.d.). *Ape Heart Project*. Twycross, UK: Twycross Zoo. Available at: https://twycrosszoo.org/conservation/research-at-twycross-zoo/ape-heart-project/. Accessed: January, 2022.

Tyler, A.D., Mataseje, L., Urfano, C.J., *et al.* (2018). Evaluation of Oxford Nanopore's MinION sequencing device for microbial whole genome sequencing applications. *Scientific Reports*, **8**, 10931. DOI: 10.1038/s41598-018-29334-5.

UBOS and ICF (2018). *Uganda Demographic and Health Survey 2016*. Kampala, Uganda, and Rockville, MD: Uganda Bureau of Statistics (UBOS) and ICF. Available at: https://dhsprogram.com/pubs/pdf/FR333/FR333.pdf.

UN [United Nations] (2019). Increased community-based engagement seen as critical to build climate action and achieve the Sustainable Development Goals. *Sustainable Development Goals*, July 19, 2019. Available at: https://www.un.org/sustainabledevelopment/blog/2019/07/increased-community-based-engagement-seen-as-critical-to-build-climate-action-and-achieve-the-sustainable-development-goals/.

UN DESA (n.d.). *The 17 Goals*. New York, NY: United Nations (UN) Department of Economic and Social Affairs (DESA). Available at: https://sdgs.un.org/goals. Accessed: March, 2021.

UNCST (2020). *National Guidelines for Conduct of Research During Coronavirus Disease 2019 (COVID-19) Pandemic*. Kampala, Uganda: Uganda National Council for Science and Technology (UNCST).

UNDP (n.d.). *Human Development Index (HDI)*. United Nations Development Programme (UNDP). Available at: http://hdr.undp.org/en/content/human-development-index-hdi. Accessed: January, 2021.

UNDRR (2015). *Sendai Framework for Disaster Risk Reduction 2015–2030*. Geneva, Switzerland: United Nations Office for Disaster Risk Reduction (UNDRR). Available at: https://www.preventionweb.net/files/43291_sendaiframeworkfordrren.pdf.

UNDRR (n.d.-a). *Sendai Framework Terminology on Disaster Risk Reduction: Contingency Planning*. Geneva, Switzerland: United Nations Office for Disaster Risk Reduction (UNDRR). Available at: https://www.undrr.org/terminology/contingency-planning. Accessed: July, 2022.

UNDRR (n.d.-b). *Sendai Framework Terminology on Disaster Risk Reduction: Disaster*. Geneva, Switzerland: United Nations Office for Disaster Risk Reduction (UNDRR). Available at: https://www.undrr.org/terminology/disaster. Accessed: July, 2022.

UNDRR (n.d.-c). *What is the Sendai Framework for Disaster Risk Reduction?* Geneva, Switzerland: United Nations Office for Disaster Risk Reduction (UNDRR). Available at: https://www.undrr.org/implementing-sendai-framework/what-sendai-framework. Accessed: July, 2022.

UNDRR DesInventar Sendai (n.d.). *Sendai Framework for Disaster Risk Reduction. DesInventar as a Disaster Information Management System*. Geneva, Switzerland: United Nations Office for Disaster Risk Reduction (UNDRR). Available at: https://www.desinventar.net/what_is.html. Accessed: July, 2022.

UNEP (2020). Virus which causes COVID-19 threatens great ape conservation. Interview with Johannes Refisch, United Nations Great Apes Survival Partnership Programme, Manager and Coordinator *UN Environment Programme News and Stories*, March 25, 2020. Available at: https://www.unep.org/news-and-stories/story/virus-which-causes-covid-19-threatens-great-ape-conservation. Accessed: July, 2022.

UNEP (2022). *UNEP/EA.5/Res.1. Resolution Adopted by the United Nations Environment Assembly on 2 March 2022 5/1*. Nairobi, Kenya: United Nations Environment Assembly of the United Nations Environment Programme (UNEP). Available at: https://wedocs.unep.org/bitstream/handle/20.500.11822/39795/ANIMAL%20WELFARE%e2%80%93ENVIRONMENT%e2%80%93SUSTAINABLE%20DEVELOPMENT%20NEXUS.%20English.pdf?sequence=1&isAllowed=y. Accessed: July, 2022.

UNEP and ILRI (2020). *Preventing the Next Pandemic: Zoonotic Diseases and How to Break the Chain of Transmission*. Nairobi, Kenya: United Nations Environment Programme (UNEP) and International Livestock Research Institute (ILRI). Available at: https://www.unep.org/resources/report/preventing-future-zoonotic-disease-outbreaks-protecting-environment-animals-and.

UNEP-WCMC (2021a). *Protected Area Profile for Cameroon from the World Database of Protected Areas*, October 2021. Cambridge, UK: UN Environment Programme World Conservation Monitoring Centre (UNEP-WCMC) and International Union for Conservation of Nature (IUCN). Available at: https://www.protectedplanet.net.

UNEP-WCMC (2021b). *Protected Area Profile for Central African Republic from the World Database of Protected Areas*, October 2021. Cambridge, UK: United Nations Environment Programme World Conservation Monitoring Centre (UNEP-WCMC) and International Union for Conservation of Nature (IUCN). Available at: https://www.protectedplanet.net.

UNEP-WCMC (2021c). *Protected Area Profile for Democratic Republic of Congo from the World Database of Protected Areas*, October 2021. Cambridge, UK: United Nations Environment Programme World Conservation Monitoring Centre (UNEP-WCMC) and International Union for Conservation of Nature (IUCN). Available at: https://www.protectedplanet.net.

UNEP-WCMC (2021d). *Protected Area Profile for Indonesia from the World Database of Protected Areas*, October 2021. Cambridge, UK: United Nations Environment Programme World Conservation Monitoring Centre (UNEP-WCMC) and International Union for Conservation of Nature (IUCN). Available at: https://www.protectedplanet.net.

UNEP-WCMC (2021e). *Protected Area Profile for Republic of Congo from the World Database of Protected Areas*, October 2021. Cambridge, UK: United Nations Environment Programme World Conservation Monitoring Centre (UNEP-WCMC) and International Union for Conservation of Nature (IUCN). Available at: https://www.protectedplanet.net.

UNEP-WCMC (2021f). *Protected Area Profile for Rwanda from the World Database of Protected Areas*, October 2021. Cambridge, UK: United Nations Environment Programme World Conservation Monitoring Centre (UNEP-WCMC) and International Union for Conservation of Nature (IUCN). Available at: https://www.protectedplanet.net.

UNEP-WCMC (2021g). *Protected Area Profile for Senegal from the World Database of Protected Areas*, October 2021. Cambridge, UK: United Nations Environment Programme World Conservation Monitoring Centre (UNEP-WCMC) and International Union for Conservation of Nature (IUCN). Available at: https://www.protectedplanet.net.

UNEP-WCMC (2021h). *Protected Area Profile for Tanzania from the World Database of Protected Areas*, October 2021. Cambridge, UK: United Nations Environment Programme World Conservation Monitoring Centre (UNEP-WCMC) and International Union for Conservation of Nature (IUCN). Available at: https://www.protectedplanet.net.

UNEP-WCMC (2021i). *Protected Area Profile for Uganda from the World Database of Protected Areas*, October 2021. Cambridge, UK: United Nations Environment Programme World Conservation Monitoring Centre (UNEP-WCMC) and International Union for Conservation of Nature (IUCN). Available at: https://www.protectedplanet.net.

UNESCO (2020). UNESCO supports the development of a regional contingency plan for protecting mountain gorillas, conservation personnel, tourists and park adjacent communities from SARS CoV-19. *UNESCO [United Nations Educational, Scientific and Cultural Organization] Press Release*, May 4, 2020. Available at: https://en.unesco.org/news/unesco-supports-development-regional-contingency-plan-protecting-mountain-gorillas-0.

UNESCO World Heritage Convention (2020). Safeguarding the endangered mountain gorilla during COVID-19 crisis. *UNESCO World Heritage Convention News*, June 23, 2020. Available at: http://whc.unesco.org/en/news/2125/.

UNGA (2019). *A/74/199. Third Industrial Development Decade for Africa (2016–2025). Note / by the Secretary General.* New York, NY: United Nations General Assembly (UNGA). Available at: https://digitallibrary.un.org/record/3824252?ln=en.

UNHCR (2015). *Emergency Response Planning (ERP). Draft for Field Testing. Emergency Handbook.* Geneva, Switzerland: Inter-Agency Standing Committee (IASC) United Nations High Commissioner for Refugees

(UNHCR). Available at: https://cms.emergency.unhcr.org/documents/11982/54224/Emergency+Response+Preparedness+July+2015/cc602e5b-7084-483d-becb-ea72286cc00e#_ga=2.170040233.841944529.1658964322-1951150208.1658964322.

UNISDR (2010). *Early Warning Practices can Save Many Lives: Good Practices and Lessons Learned.* Bonn, Germany: United Nations Secretariat of the International Strategy for Disaster Reduction (UNISDR) Platform for the Promotion of Early Warning. Available at: https://www.unisdr.org/files/15254_EWSBBLLfinalweb.pdf.

University of Birmingham (n.d.). *Enclosure Design Tool.* Birmingham, UK: University of Birmingham. Available at: https://www.birmingham.ac.uk/schools/biosciences/research/showcase/enclosure-design-tool/index.aspx. Accessed: May, 2022.

University of Minnesota (n.d.-a). *Canvas Login.* St Paul, MN: University of Minnesota, College of Veterinary Medicine. Available at: https://umnadvet.instructure.com/login/canvas. Accessed: October, 2022.

University of Minnesota (n.d.-b). *Non Human Primate COVID-19 Information Hub.* St Paul, MN: University of Minnesota, College of Veterinary Medicine. Available at: https://umnadvet.instructure.com/courses/324. Accessed: September, 2022.

UNOOSA (n.d.). *Information Management for Disaster-Risk Reduction.* Vienna, Austria: United Nations Office for Outer Space Affairs (UNOOSA) UN-SPIDER Knowledge Portal. Available at: http://www.un-spider.org/risks-and-disasters/disaster-risk-management/information-management. Accessed: July, 2022.

Unwin, S., Chatterton, J. and Chantrey, J. (2013). Management of severe respiratory tract disease caused by human respiratory syncytial virus and *Streptococcus pneumoniae* in captive chimpanzees (*Pan troglodytes*). *Journal of Zoo and Wildlife Medicine*, **44**(1), 105–15.

Unwin, S., Commitante, R., Moss, A., *et al.* (2022). Evaluating the contribution of a wildlife health capacity building program on orangutan conservation. *American Journal of Primatology*, **84**(4–5), e23273. DOI: 10.1002/ajp.23273.

USAID (n.d.). *Emerging Pandemic Threats.* Washington DC: United States Agency for International Development (USAID). Available at: https://www.usaid.gov/news-information/fact-sheets/emerging-pandemic-threats-program. Accessed: January, 2021.

USDA (2020). *USDA to Launch Updated Animal Welfare Act Compliance Database and Public Search Tool.* Washington DC: US Department of Agriculture (USDA). Available at: https://www.aphis.usda.gov/aphis/newsroom/stakeholder-info/sa_by_date/sa-2020/sa-08/updated-awa-database.

Utami-Atmoko, S.S., Singleton, I., van Noordwijk, M.A., van Schaik, C.P. and Mitra Setia, T. (2009). Male–male relationships in orangutans. In *Orangutans: Geographic Variation in Behavioral Ecology and Conservation*, ed. S. A. Wich, S. S. Utami-Atmoko, T. Mitra Setia, C. P. van Schaik and M. A. van Noordwijk. Oxford, UK: Oxford University Press, pp. 225–33.

Utami-Atmoko, S.S., Traylor-Holzer, K., Rifqi, M.A., *et al.*, ed. (2017). *Orangutan Population and Habitat Viability Assessment: Final Report.* Apple Valley, MN: International Union for Conservation of Nature (IUCN) Species Survival Commission (SSC) Conservation Breeding Specialist Group. Available at: https://www.cbsg.org/sites/cbsg.org/files/documents/2016%20Orangutan%20PHVA.pdf.

Utermohlen, M. and Baine, P. (2018). *In Plane Sight: Wildlife Trafficking in the Air Transport Sector.* Washington DC: Center for Advanced Defense (C4ADS). Available at: https://www.traffic.org/publications/reports/in-plane-sight.

UWA (2020a). *Standard Operating Procedures for Tourism Services and Research Activities in UWA Estates and the Reopening of the Protected Areas to the General Public during Covid-19 Pandemic.* Kampala, Uganda: Uganda Wildlife Authority (UWA).

UWA (2020b). *Uganda Wildlife Authority Ranger Based Monitoring Data for 2020, Bwindi Impenetrable National Park, Buhoma, Kanungu.* Internal data seen by author. Kampala, Uganda: Uganda Wildlife Authority (UWA).

UWA (2022). *Conservation Tariff July 2022 to June 2024.* Kampala, Uganda: Uganda Wildlife Authority (UWA). Available at: https://ugandawildlife.org/uwa-rates/.

Vale (n.d.). *Biodiversity.* Rio de Janeiro, Brazil: Vale. Available at: https://www.vale.com/web/esg/biodiversity. Accessed: October, 2022.

Van Hamme, G., Svensson, M.S., Morcatty, T.Q., Nekaris, K.A.-I. and Nijman, V. (2021). Keep your distance: using Instagram posts to evaluate the risk of anthroponotic disease transmission in gorilla ecotourism. *People and Nature*, **3**(2), 325–34. DOI: 10.1002/pan3.10187.

Van Herck, K., Castelli, F., Zuckerman, J., *et al.* (2004). Knowledge, attitudes and practices in travel-related infectious diseases: the European Airport Survey. *Journal of Travel Medicine*, **11**(1), 3–8. DOI: 10.2310/7060.2004.13609.

Van Heuverswyn, F., Li, Y., Bailes, E., *et al.* (2007). Genetic diversity and phylogeographic clustering of SIVcpzPtt in wild chimpanzees in Cameroon. *Virology*, **368**(1), 155–71.

van Noordwijk, M.A., Arora, N., Willems, E.P., *et al.* (2012). Female philopatry and its social benefits among Bornean orangutans. *Behavioral Ecology and Sociobiology*, **66**, 823–34.

van Noordwijk, M.A., Sauren, S.E.B., Nuzuar, *et al.* (2009). Development of independence: Sumatran and Bornean orangutans compared. In *Orangutans: Geographic Variation in Behavioral Ecology and Conservation*, ed. S. A. Wich, S. S. Utami-Atmoko, T. Mitra Setia and C. P. van Schaik. Oxford, UK: Oxford University Press, pp. 189–203.

van Noordwijk, M.A., Utami-Atmoko, S.S., Knott, C.D., *et al.* (2018). The slow ape: high infant survival and long interbirth intervals in wild orangutans. *Journal of Human Evolution*, **125**, 38–49. DOI: 10.1016/j.jhevol.2018.09.004.

van Noordwijk, M.A., Willems, E.P., Utami-Atmoko, S.S., Kuzawa, C.W. and van Schaik, C.P. (2013). Multi-year lactation and its consequences in Bornean orangutans (*Pongo pygmaeus wurmbii*). *Behavioral Ecology and Sociobiology*, **67**(5), 805–14. DOI: 10.1007/s00265-013-1504-y.

van Schaik, C.P. (1999). The socioecology of fission–fusion sociality in orangutans. *Primates*, **40**(1), 69–86. DOI: 10.1007/BF02557703.

Varkey, B. (2021). Principles of clinical ethics and their application to practice. *Medical Principles and Practice*, **30**(1), 17–28. DOI: 10.1159/000509119.

Varner, G. (1998). *In Nature's Interests? Interests, Animal Rights, and Environmental Ethics*. New York, NY: Oxford University Press.

Vaz, M., Sridhar, T.S. and Pai, S.A. (2016). The ethics of research on stored biological samples: outcomes of a workshop. *Indian Journal of Medical Ethics*, **1**(2), 118–22. DOI: 10.20529/ijme.2016.032.

Veasey, J.S. (2020a). Assessing the psychological priorities for optimising captive Asian elephant (*Elephas maximus*) welfare. *Animals*, **10**(1), 39. DOI: 10.3390/ani10010039.

Veasey, J.S. (2020b). Can zoos ever be big enough for large wild animals? A review using an expert panel assessment of the psychological priorities of the Amur tiger (*Panthera tigris altaica*) as a model species. *Animals*, **10**(9), 1536. DOI: 10.3390/ani10091536.

Venter, O., Sanderson, E.W., Magrach, A., *et al.* (2016). Sixteen years of change in the global terrestrial human footprint and implications for biodiversity conservation. *Nature Communications*, **7**, 12558. DOI: 10.1038/ncomms12558.

Verburg-van Kemenade, B.M.L., Cohen, N. and Chadzinska, M. (2017). Neuroendocrine–immune interaction: evolutionarily conserved mechanisms that maintain allostasis in an ever-changing environment. *Developmental & Comparative Immunology*, **66**, 2–23. DOI: 10.1016/j.dci.2016.05.015.

Verweij, M. and Bovenkerk, B. (2016). Ethical promises and pitfalls of One Health. *Public Health Ethics*, **9**(1), 1–4. DOI: 10.1093/phe/phw003.

Viciunaite, V. and Alfnes, F. (2020). Informing sustainable business models with a consumer preference perspective. *Journal of Cleaner Production*, **242**, 118417. DOI: 10.1016/j.jclepro.2019.118417.

Vidal, J. (2015). Indonesia's forest fires threaten a third of world's wild orangutans. *The Guardian*, October 26, 2015. Available at: https://www.theguardian.com/environment/2015/oct/26/indonesias-forest-fires-threaten-a-third-of-worlds-wild-orangutans.

Videan, E.N., Fritz, J. and Murphy, J. (2008). Effects of aging on hematology and serum clinical chemistry in chimpanzees (*Pan troglodytes*). *American Journal of Primatology*, **70**(4), 327–38. DOI: 10.1002/ajp.20494.

Videan, E.N., Heward, C.B., Fritz, J., *et al.* (2007). Relationship between sunlight exposure, housing condition, and serum vitamin D and related physiologic biomarker levels in captive chimpanzees (*Pan troglodytes*). *Comparative Medicine*, **57**(4), 402–6.

Virunga National Park (n.d.-a). *The History of Virunga National Park*. Virunga, Democratic Republic of Congo: Virunga National Park. Available at: https://virunga.org/about/. Accessed: April, 2022.

Virunga National Park (n.d.-b). *Virunga National Park Temporarily Closes Mountain Gorilla Tourism Due to COVID-19*. Kinshasa, DRC: Virunga National Park. Available at: https://virunga.org/news/virunga-national-park-temporarily-closes-mountain-gorilla-tourism-due-to-covid-19/. Accessed: April, 2022.

Virunga National Park Congo (n.d.). *Virunga National Park Reopens for Tourism.* Virunga National Park Congo. Available at: https://www.virungaparkcongo.com/information/virunga-national-park-reopens-for-tourism/. Accessed: April, 2022.

Visit Rwanda (n.d.). *Gorilla Tracking.* Kigali, Rwanda: Rwanda Development Board. Available at: https://www.visitrwanda.com/interests/gorilla-tracking/. Accessed: August, 2022.

Vitone, N.D., Altizer, S. and Nunn, C.L. (2004). Body size, diet and sociality influence the species richness of parasitic worms in anthropoid primates. *Evolutionary Ecology Research*, **6**(2), 183–99.

Vogel, E.R. (2018). Wildfire smoke could have lasting effects for endangered orangutans. *Laboratory for Primate Dietary Ecology and Physiology*, May 15, 2018. Available at: https://erinvogelphd.wordpress.com/2018/05/15/wildfire-smoke-could-have-lasting-effects-for-endangered-orangutans/.

Voigt, M., Wich, S.A., Ancrenaz, M., *et al.* (2018). Global demand for natural resources eliminated more than 100,000 Bornean orangutans. *Current Biology*, **28**(5), 761–9. DOI: 10.1016/j.cub.2018.01.053.

von Magnus, P., Andersen, E.K., Petersen, K.B. and Birch-Andersen, A. (1959). A pox-like disease in cynomolgus monkeys. *Acta Pathologica Microbiologica Scandinavica*, **46**(2), 156–76. DOI: 10.1111/j.1699-0463.1959.tb00328.x.

Vucetich, J.A., Burnham, D., Macdonald, E.A., *et al.* (2018). Just conservation: what is it and should we pursue it? *Biological Conservation*, **221**, 23–33. DOI: 10.1016/j.biocon.2018.02.022.

Walaga, P. and Mashoo, E. (2009). Uganda earns Shs488 billion from gorilla tourism. *The Daily Monitor*, April 20, 2009.

Walhisumut, O. (2021). WALHI North Sumatra files lawsuit against PT. Nuansa Alam Nusantara for illegally keeping animals in a zoo without permits. *WALHI North Sumatra Press Release*, April 13, 2021. Available at: http://walhisumut.org/2021/04/13/walhi-north-sumatra-files-lawsuit-against-pt-nuansa-alam-nusantara-for-illegally-keeping-animals-in-a-zoo-without-permits/.

Walker, K.K., Walker, C.S., Goodall, J. and Pusey, A.E. (2018). Maturation is prolonged and variable in female chimpanzees. *Journal of Human Evolution*, **114**, 131–40. DOI: 10.1016/j.jhevol.2017.10.010.

Walker, S.L., Withington, S.G. and Lockwood, D.N.J. (2014). Leprosy. In *Manson's Tropical Infectious Diseases*, 23rd edn, ed. J. Farrar, P. J. Hotez, T. Junghanss, *et al.* London, UK: W.B. Saunders, pp. 506–18.e1. DOI: 10.1016/B978-0-7020-5101-2.00042-X.

Wallace, R.G., Bergmann, L., Kock, R., *et al.* (2015). The dawn of Structural One Health: a new science tracking disease emergence along circuits of capital. *Social Science & Medicine*, **129**, 68–77. DOI: 10.1016/j.socscimed.2014.09.047.

Wallach, A.D., Batavia, C., Bekoff, M., *et al.* (2020). Recognizing animal personhood in compassionate conservation. *Conservation Biology*, **34**(5), 1097–106. DOI: 10.1111/cobi.13494.

Wallach, A.D., Bekoff, M., Batavia, C., Nelson, M.P. and Ramp, D. (2018). Summoning compassion to address the challenges of conservation. *Conservation Biology*, **32**(6), 1255–65. DOI: 10.1111/cobi.13126.

Wallis, J. and Lee, D.R. (1999). Primate conservation: the prevention of disease transmission. *International Journal of Primatology*, **20**, 803–26. DOI: 10.1023/A:1020879700286.

Walraven, E. and Duffy, S. (2017). Embedding animal welfare in staff culture: the Taronga Conservation Society Australia experience. *International Zoo Yearbook*, **51**(1), 203–14. DOI: 10.1111/izy.12149.

Walsh, P.D., Abernethy, K.A., Bermejo, M., *et al.* (2003). Catastrophic ape decline in western equatorial Africa. *Nature*, **422**(6932), 611–14. DOI: 10.1038/nature01566.

Walsh, P.D., Kurup, D., Hasselschwert, D.L., *et al.* (2017). The final (oral Ebola) vaccine trial on captive chimpanzees? *Scientific Reports*, **7**, 43339. DOI: 10.1038/srep43339.

Waltner-Toews, D., Kay, J.J. and Lister, N.M.E. (2008). *The Ecosystem Approach: Complexity, Uncertainty, and Managing for Sustainability.* New York, NY: Columbia University Press.

Wang, C.-B., Zhao, L.-X., Jin, C.-Z., *et al.* (2014). New discovery of Early Pleistocene orangutan fossils from Sanhe Cave in Chongzuo, Guangxi, southern China. *Quaternary International*, **354**, 68–74. DOI: 10.1016/j.quaint.2014.06.020.

WAP (2017). *A Close Up on Cruelty: The Harmful Impact of Wildlife Selfies in the Amazon.* London, UK: World Animal Protection (WAP).

WAP (2019). *The Show Can't Go On: End the Suffering of Wild Animals at Cruel Visitor Attractions in Zoos and Aquariums.* London, UK: World Animal Protection (WAP). Available at: https://www.changeforanimals.org/help-end-animal-abuse-in-top-zoos.

WAP (n.d.-a). *Animal Protection Index.* London, UK: World Animal Protection (WAP). Available at: https://api.worldanimalprotection.org/. Accessed: October, 2020.

WAP (n.d.-b). *Animal Protection Index: China.* London, UK: World Animal Protection (WAP). Available at: https://api.worldanimalprotection.org/country/china. Accessed: October, 2020.

WAP (n.d.-c). *Methodology.* London, UK: World Animal Protection (WAP). Available at: https://api.worldanimalprotection.org/methodology. Accessed: December, 2020.

Ward, S.J., Williams, E., Groves, G., Marsh, S. and Morgan, D. (2020). Using zoo welfare assessments to identify common issues in developing country zoos. *Animals*, **10**(11), 2101. DOI: 10.3390/ani10112101.

Warfield, K.L., Goetzmann, J.E., Biggins, J.E., *et al.* (2014). Vaccinating captive chimpanzees to save wild chimpanzees. *Proceedings of the National Academy of Sciences*, **111**(24), 8873–6. DOI: 10.1073/pnas.1316902111.

Wark, J.D., Cronin, K.A., Niemann, T., *et al.* (2019). Monitoring the behavior and habitat use of animals to enhance welfare using the ZooMonitor App. *Animal Behavior and Cognition*, **6**(3), 158–67. DOI: 10.26451/abc.06.03.01.2019.

Warren, C.E., Bellows, B., Marcus, R., *et al.* (2021). Strength in diversity: integrating community in primary health care to advance universal health coverage. *Global Health: Science and Practice*, **9**(S1), S1–5. DOI: 10.9745/ghsp-d-21-00125.

Warren, K.S. (2001). *Orang-utan conservation: epidemiological aspects of health management and population genetics.* PhD thesis. Mudoch, Australia: Murdoch University.

Warren, K.S., Heeney, J.L., Swan, R.A., Heriyanto and Verschoor, E.J. (1999). A new group of hepadnaviruses naturally infecting orangutans (*Pongo pygmaeus*). *Journal of Virology*, **73**(9), 7860–5. DOI: 10.1128/JVI.73.9.7860-7865.1999.

Wasser, S.K., Sewall, G. and Soules, M.R. (1993). Psychosocial stress as a cause of infertility. *Fertility and Sterility*, **59**(3), 685–9. DOI: 10.1016/S0015-0282(16)55824-5.

Waters, S., Hansen, M.F., Setchell, J.M., *et al.* (2023). *Responsible Primate-Watching for Tourists.* Gland, Switzerland: International Union for Conservation of Nature (IUCN) Species Survival Commission (SSC) Primate Specialist Group (PSG) Section for Human–Primate Interactions (SHPI). Available at: https://humanprimateinteractions.files.wordpress.com/2023/09/responsible-primate-watching-for-tourists.pdf.

Waters, S., Setchell, J.M., Maréchal, L., *et al.* (2021). *Best Practice Guidelines for Responsible Images of Non-Human Primates.* International Union for Conservation of Nature (IUCN) Species Survival Commission (SSC) Primate Specialist Group (PSG) Section for Human–Primate Interactions (SHPI). Available at: https://www.arcusfoundation.org/wp-content/uploads/2021/09/Best-Practices-for-Responsible-Images-of-Nonhuman-Primates.pdf.

Watson, J.E.M., Evans, T., Venter, O., *et al.* (2018). The exceptional value of intact forest ecosystems. *Nature Ecology & Evolution*, **2**(4), 599–610. DOI: 10.1038/s41559-018-0490-x.

Watters, J.V., Margulis, S.W. and Atsalis, S. (2009). Behavioral monitoring in zoos and aquariums: a tool for guiding husbandry and directing research. *Zoo Biology*, **28**(1), 35–48. DOI: 10.1002/zoo.20207.

Watts, D.P. (1989). Infanticide in mountain gorillas: new cases and a reconsideration of the evidence. *Ethology*, **81**(1), 1–18. DOI: 10.1111/j.1439-0310.1989.tb00754.x.

Watts, D.P., Muller, M., Amsler, S.J., Mbabazi, G. and Mitani, J.C. (2006). Lethal intergroup aggression by chimpanzees in Kibale National Park, Uganda. *American Journal of Primatology*, **68**(2), 161–80. DOI: 10.1002/ajp.20214.

Watts, J. (2019). Chinese dam project in Guinea could kill up to 1,500 chimpanzees. *The Guardian*, February 28, 2019. Available at: https://www.theguardian.com/world/2019/feb/28/chinese-dam-project-in-guinea-could-kill-up-to-1500-chimpanzees.

Waugh, W.L. and Liu, C.Y. (2014). Disasters, the whole community, and development as capacity building. In *Disaster and Development: Examining Global Issues and Cases*, ed. N. Kapucu and K. T. Liou. Cham, Switzerland: Springer International Publishing, pp. 167–79. DOI: 10.1007/978-3-319-04468-2_10.

WAZA (2019). WAZA works. 2nd WAZA animal welfare evaluation summit. *WAZA [World Association of Zoos and Aquariums] News*, **2019**(2), 29. DOI: https://www.waza.org/wp-content/uploads/2019/11/WAZA-magazine-02-final2.pdf.

WAZA (n.d.). *How to Become a WAZA Member.* Barcelona, Spain: World Association of Zoos and Aquariums (WAZA). Available at: https://www.waza.org/members/how-to-become-a-waza-member/. Accessed: May, 2022.

WCS (n.d.-a). *The 2019 Berlin Principles on One Health.* New York, NY: Wildlife Conservation Society (WCS). Available at: https://oneworldonehealth.wcs.org/About-Us/Mission/The-2019-Berlin-Principles-on-One-Health. aspx#:~:text=Fifteen%20years%20ago%2C%20the%20Wildlife,%2C%20animal%2C%20and%20ecosystem%20 health. Accessed: September, 2022.

WCS (n.d.-b). *The Conservation, Mitigation and Biodiversity Offset (COMBO) Program.* New York, NY: Wildlife Conservation Society (WCS). Available at: https://comboprogram.org/. Accessed: December, 2022.

WCS (n.d.-c). *Home.* Conakry, Republic of Guinea: Winning Consortium Simandou (WSC). Available at: https:// wcsglobal.com/en/. Accessed: December, 2022.

WCS (n.d.-d). *Project Description.* Conakry, Republic of Guinea: Winning Consortium Simandou (WCS). Available at: https://wcsglobal.com/en/csr_part/project-description. Accessed: June, 2022.

Webber, B. and Vedder, A. (2001). *In the Kingdom of Gorillas.* New York, NY: Simon and Shuster.

Weber, A., Kalema-Zikusoka, G. and Stevens, N.J. (2020). Lack of rule-adherence during mountain gorilla tourism encounters in Bwindi Impenetrable National Park, Uganda, places gorillas at risk from human disease. *Frontiers in Public Health*, **8**, February 13, 2020. DOI: 10.3389/fpubh.2020.00001.

Wedana, M., Masnur, I., Ibrahim, S., *et al.* (2021). Reinforcement of an isolated Javan silvery gibbon population on Mt. Tilu, West Java. In *Global Conservation Translocation Perspectives: 2021. Case Studies From Around the Globe*, ed. P. S. Soorae. Gland, Switzerland: International Union for Conservation of Nature (IUCN) Species Survival Commission (SSC) Conservation Translocation Specialist Group, Environment Agency, Abu Dhabi, and Calgary Zoo, Canada, pp. 235–40. Available at: https://portals.iucn.org/library/sites/library/files/ documents/2021-007-En.pdf.

Wendler, D. (2014). Should protections for research with humans who cannot consent apply to research with non-human primates? *Theoretical Medicine and Bioethics*, **35**(2), 157–73. DOI: 10.1007/s11017-014-9285-5.

Wenker, C., Hoby, S., Wyss, F., *et al.* (2019). Alveolar echinococcosis in western lowland gorillas (*Gorilla gorilla gorilla*): albendazole was not able to stop progression of the disease. *Journal of Zoo and Wildlife Medicine*, **50**(1), 243–53. DOI: 10.1638/2018-0064.

Werdenich, D., Dupain, J., Arnheim, E., *et al.* (2003). Reactions of chimpanzees and gorillas to human observers in a non-protected area in south-eastern Cameroon. *Folia Primatologica*, **74**(2), 97–100. DOI: 10.1159/000070005.

Wertheim, J.O., Smith, M.D., Smith, D.M., Scheffler, K. and Kosakovsky Pond, S.L. (2014). Evolutionary origins of human herpes simplex viruses 1 and 2. *Molecular Biology and Evolution*, **31**(9), 2356–64. DOI: 10.1093/molbev/msu185.

Wessling, E.G., Kühl, H.S., Mundry, R., Deschner, T. and Pruetz, J.D. (2018). The costs of living at the edge: seasonal stress in wild savanna-dwelling chimpanzees. *Journal of Human Evolution*, **121**, 1–11. DOI: 10.1016/j.jhevol. 2018.03.001.

Westlund, K. (2015). Training laboratory primates: benefits and techniques. *Primate Biology*, **2**(1), 119–32. DOI: 10.5194/ pb-2-119-2015.

Weston-Murphy, H. (2015). Great apes. In *Fowler's Zoo and Wild Animal Medicine, Volume 8*, ed. R. E. Miller and M. E. Fowler. St Louis, MO: W.B. Saunders, pp. 336–54. DOI: 10.1016/B978-1-4557-7397-8.00038-4.

WFA (2022). *Historic UN Resolution Recognizes Animal Welfare's Role in Sustainability.* World Federation for Animals (WFA). Available at: https://wfa.org/historic-un-resolution-recognizes-animal-welfares-role-in-sustainability/.

WFA (n.d.). *Achieving Global Impact For Animals – Together.* World Federation for Animals (WFA). Available at: https://wfa.org/. Accessed: May, 2022.

WFEN (n.d.). *Wildlife Friendly TM Tourism.* Wildlife Friendly Enterprise Network (WFEN). Available at: https:// wildlifefriendly.org/wildlife-friendly-tourism/. Accessed: April, 2022.

WHA (2018). *Guidelines for Management of an Emergency Wildlife Disease Response. Working Draft November 2018.* Mosman, Australia: Wildlife Health Australia (WHA). Available at: https://wildlifehealthaustralia.com. au/WHADocuments.aspx.

Whitfort, A. (2019). Wildlife crime and animal victims: improving access to environmental justice in Hong Kong. *Journal of International Wildlife Law & Policy*, **22**(3), 203–30. DOI: 10.1080/13880292.2019.1677055.

Whitham, J. and Wielebnowski, N. (2015). WelfareTRAK. A tool for capturing zookeepers' assessment of animal welfare. *CONNECT (AZA News)*, January, 16–17.

Whitham, J.C. and Miller, L.J. (2016). Using technology to monitor and improve zoo animal welfare. *Animal Welfare*, **25**(4), 395–409. DOI: 10.7120/09627286.25.4.395.

Whitham, J.C. and Wielebnowski, N. (2009). Animal-based welfare monitoring: using keeper ratings as an assessment tool. *Zoo Biology*, **28**(6), 545–60. DOI: 10.1002/zoo.20281.

Whittaker, D. and Knight, R.L. (1998). Understanding wildlife responses to humans. *Wildlife Society Bulletin*, **26**, 312–17.

Whittaker, M. and Laule, G. (2012). Training techniques to enhance the care and welfare of nonhuman primates. *Veterinary Clinics of North America: Exotic Animal Practice*, **15**(3), 445–54. DOI: 10.1016/j.cvex.2012.06.004.

Whittier, C.A., Nutter, F.B., Johnson, P.L.F., *et al.* (2022). Population structure, intergroup interaction, and human contact govern infectious disease impacts in mountain gorilla populations. *American Journal of Primatology*, **84**(4–5), e23350. DOI: 10.1002/ajp.23350.

WHO (2012). *Social and Environmental Determinants*. Geneva, Switzerland: World Health Organization (WHO) Available at: https://www.euro.who.int/__data/assets/pdf_file/0006/185217/Social-and-environmental-determinants-Fact-Sheet.pdf.

WHO (2014). *Early Detection, Assessment and Response to Acute Public Health Events: Implementation of Early Warning and Response With a Focus on Event-Based Surveillance. Interim Version*. Geneva, Switzerland: World Health Organization (WHO). Available at: https://www.who.int/publications/i/item/WHO-HSE-GCR-LYO-2014.4.

WHO (2017a). *A Strategic Framework for Emergency Preparedness*. Geneva, Switzerland: World Health Organization (WHO). Available at: https://www.who.int/publications/i/item/a-strategic-framework-for-emergency-preparedness.

WHO (2017b). *WHO Simulation Exercise Manual*. Geneva, Switzerland: World Health Organization (WHO. Available at: https://www.who.int/publications/i/item/WHO-WHE-CPI-2017.10.

WHO (2018). *WHO Guidance for Contingency Planning*. Geneva, Switzerland: World Health Organization (WHO). Available at: https://apps.who.int/iris/bitstream/handle/10665/260554/WHO-WHE-CPI-2018.13-eng.pdf?ua=1.

WHO (2019). *Burn-Out an "Occupational Phenomenon": International Classification of Diseases*. Geneva, Switzerland: World Health Organization (WHO). Available at: https://www.who.int/mental_health/evidence/burn-out/en/.

WHO (2020a). *Basic Documents. Forty-ninth Edition. Including Amendments Adopted up to 31 May 2019*. Geneva, Switzerland: World Health Organization (WHO). Available at: https://apps.who.int/gb/bd/pdf_files/BD_49th-en.pdf.

WHO (2020b). *COVID-19 Public Health Emergency of International Concern (PHEIC). Global Research and Innovation Forum: Towards a Research Roadmap*. Geneva, Switzerland: World Health Organization (WHO). Available at: https://www.who.int/publications/m/item/covid-19-public-health-emergency-of-international-concern-(pheic)-global-research-and-innovation-forum.

WHO (2020c). *Global Tuberculosis Report*. Geneva, Switzerland: World Heath Organization (WHO). Available at: https://apps.who.int/iris/bitstream/handle/10665/336069/9789240013131-eng.pdf.

WHO (2020d). *Glossary of Health Emergency and Disaster Risk Management Terminology*. Geneva, Switzerland: World Health Organization (WHO). Available at: https://www.who.int/publications/i/item/9789240003699.

WHO (n.d.). *Tuberculosis*. Geneva, Switzerland: World Health Organization (WHO). Available at: https://www.who.int/health-topics/tuberculosis#tab=tab_1. Accessed: September, 2022.

WHO, FAO and OIE (2019). *Taking a Multisectoral, One Health Approach: A Tripartite Guide to Addressing Zoonotic Diseases in Countries*. Geneva, Switzerland: World Health Organization (WHO), Food and Agriculture Organization of the United Nations (FAO) and World Organisation for Animal Health (OIE). Available at: https://www.who.int/publications/i/item/9789241514934.

WHO/EHA (2002). *Disasters and Emergencies Definitions Training Package*. Addis Ababa, Ethiopia: World Health Organization (WHO)/Emergency and Humanitarian Action (EHA). Available at: https://apps.who.int/disasters/repo/7656.pdf.

Wich, S.A., de Vries, H., Ancrenaz, M., *et al.* (2009a). Orangutan life history variation. In *Orangutans: Geographic Variation in Behavioral Ecology and Conservation*, ed. S. A. Wich, S. S. Utami-Atmoko, T. Mitra Setia and C. P. van Schaik. Oxford, UK: Oxford Academic, pp. 65–75. DOI: 10.1093/acprof:oso/9780199213276.003.0005.

Wich, S.A., Fredriksson, G., Usher, G., Kühl, H.S. and Nowak, M.G. (2019). The Tapanuli orangutan: status, threats, and steps for improved conservation. *Conservation Science and Practice*, **1**(6), e33. DOI: 10.1111/csp2.33.

Wich, S.A., Fredriksson, G.M., Usher, G., *et al.* (2012a). Hunting of Sumatran orang-utans and its importance in determining distribution and density. *Biological Conservation*, **146**(1), 163–9. DOI: 10.1016/j.biocon.2011.12.006.

Wich, S.A., Garcia-Ulloa, J., Kühl, Hjalmar S., *et al.* (2014a). Will oil palm's homecoming spell doom for Africa's great apes? *Current Biology*, **24**(14), 1659–63. DOI: 10.1016/j.cub.2014.05.077.

Wich, S.A., Gaveau, D., Abram, N., *et al.* (2012b). Understanding the impacts of land-use policies on a threatened species: is there a future for the Bornean orang-utan? *PLoS ONE*, **7**(11), e49142. DOI: 10.1371/journal.pone.0049142.

Wich, S.A., Geurts, M.L., Mitra Setia, T. and Utami-Atmoko, S.S. (2006). Influence of fruit availability on Sumatran orangutan sociality and reproduction. In *Feeding Ecology in Apes and Other Primates: Ecological, Physiological and Behavioural Aspects*. Cambridge Studies in Biological and Evolutionary Anthropology Volume 48, ed. G. Hohmann, M. M. Robbins and C. Boesch. Cambridge, UK: Cambridge University Press, pp. 337–58.

Wich, S.A. and Piel, A.K., ed. (2021). *Conservation Technology*. Oxford, UK: Oxford University Press. DOI: 10.1093/oso/9780198850243.001.0001.

Wich, S.A., Singleton, I., Nowak, M.G., *et al.* (2016). Land-cover changes predict steep declines for the Sumatran orangutan (*Pongo abelii*). *Science Advances*, **2**(3), e1500789. DOI: 10.1126/sciadv.1500789.

Wich, S.A., Usher, G., Peters, H.H., *et al.* (2014b). Preliminary data on the highland Sumatran orangutans (*Pongo abelii*) of Batang Toru. In *High Altitude Primates*, ed. N. B. Grow, S. Gursky-Doyen and A. Krzton. Cambridge, UK: Springer, pp. 265–83.

Wich, S.A., Utami-Atmoko, S., Mitra Setia, T. and van Schaik, C.P., ed. (2009b). *Orangutans: Geographic Variation in Behavioral Ecology and Conservation*. Oxford, UK: Oxford University Press.

Wiederholt, R. and Post, E. (2010). Tropical warming and the dynamics of endangered primates. *Biology Letters*, **6**(2), 257–60. DOI: 10.1098/rsbl.2009.0710.

Wiedmann, T., Lenzen, M., Keyßer, L.T. and Steinberger, J.K. (2020). Scientists' warning on affluence. *Nature Communications*, **11**(1), 3107. DOI: 10.1038/s41467-020-16941-y.

Wikelski, M. and Cooke, S.J. (2006). Conservation physiology. *Trends in Ecology & Evolution*, **21**(1), 38–46. DOI: 10.1016/j.tree.2005.10.018.

Wilcox, B.A., Aguirre, A.A., De Paula, N., Siriaroonrat, B. and Echaubard, P. (2019). Operationalizing One Health employing social-ecological systems theory: lessons from the Greater Mekong Sub-region. *Frontiers in Public Health*, **7**, 85. DOI: 10.3389/fpubh.2019.00085.

Wildlife Rescue Center Jogja (n.d.). *Giving Day For Apes*. Wildlife Rescue Center Jogja. Available at: https://wrcjogja.org/giving-day-for-apes-2020/. Accessed: October, 2020.

Wilkinson, D.A., Marshall, J.C., French, N.P. and Hayman, D.T.S. (2018). Habitat fragmentation, biodiversity loss and the risk of novel infectious disease emergence. *Journal of The Royal Society Interface*, **15**(149), 20180403. DOI: 10.1098/rsif.2018.0403.

Williams, D.R., Clark, M., Buchanan, G.M., *et al.* (2021). Proactive conservation to prevent habitat losses to agricultural expansion. *Nature Sustainability*, **4**(4), 314–22. DOI: 10.1038/s41893-020-00656-5.

Williams, J.L. and Behie, A.M. (2020). Northern yellow-cheeked crested gibbons (*Nomascus annamensis*) travel and scan more at the cost of rest when in the presence of tourists. *Animal Biology*, **70**(4), 427–43. DOI: 10.1163/15707563-bja10040.

Williams, J.M., Lonsdorf, E.V., Wilson, M.L., *et al.* (2008). Causes of death in the Kasekela chimpanzees of Gombe National Park, Tanzania. *American Journal of Primatology*, **70**(8), 766–77. DOI: 10.1002/ajp.20573.

Williamson, E.A. and Butynski, T.M. (2013a). *Gorilla beringei* eastern gorilla. In *Mammals of Africa. Volume II: Primates*, ed. T. M. Butynski, J. Kingdon and J. Kalina. London, UK: Bloomsbury Publishing, pp. 45–53.

Williamson, E.A. and Butynski, T.M. (2013b). *Gorilla gorilla* western gorilla. In *Mammals of Africa. Volume II: Primates*, ed. T. M. Butynski, J. Kingdon and J. Kalina. London, UK: Bloomsbury Publishing, pp. 39–45.

Williamson, E.A. and Feistner, A.T.C. (2011). Habituating primates: processes, techniques, variables and ethics. In *Field and Laboratory Methods in Primatology: A Practical Guide*, ed. D. J. Curtis and J. M. Setchell. Cambridge, UK: Cambridge University Press, pp. 33–50. DOI: 10.1017/CBO9780511921643.004.

Williamson, E.A., Maisels, F.G., Groves, C.P., *et al.* (2013). Hominidae. In *Handbook of the Mammals of the World. Volume 3: Primates*, ed. R. A. Mittermeier, A. B. Rylands and D. E. Wilson. Barcelona, Spain: Lynx Edicions, pp. 792–854.

Williamson, E.A., Strindberg, S. and Maisels, F. (2018). New population estimate for western lowland gorillas. *Gorilla Journal*, **56**, 18–19. DOI: https://www.berggorilla.org/en/home/news-archive/article-view/new-population-estimate-for-western-lowland-gorillas/.

Williamson, E.A., Tutin, C.E.G., Rogers, M.E. and Fernandez, M. (1990). Composition of the diet of lowland gorillas at Lopé in Gabon. *American Journal of Primatology*, **21**(4), 265–77. DOI: 10.1002/ajp.1350210403.

Williamson, L. (2001). Mountain gorilla tourism: some costs and benefits. *Gorilla Journal*, **22**, 35–7.

Wilson, D. and Reeder, D. (2005). *Mammal Species of the World: A Taxonomic and Geographic Reference*, 3rd edn. Baltimore, MD: Johns Hopkins University Press.

Wilson, H.B., Meijaard, E., Venter, O., Ancrenaz, M. and Possingham, H.P. (2014a). Conservation strategies for orangutans: reintroduction versus habitat preservation and the benefits of sustainably logged forest. *PLoS ONE*, **9**(7), e102174. DOI: 10.1371/journal.pone.0102174.

Wilson, M.L., Boesch, C., Fruth, B., *et al.* (2014b). Lethal aggression in *Pan* is better explained by adaptive strategies than human impacts. *Nature*, **513**, 414–17. DOI: 10.1038/nature13727.

Wilson, P., Weavers, E., West, B., *et al.* (1984). *Mycobacterium bovis* infection in primates in Dublin Zoo: epidemiological aspects and implications for management. *Laboratory Animals*, **18**(4), 383–7. DOI: 10.1258/002367784780865351.

Wilson, R.P. and McMahon, C.R. (2006). Measuring devices on wild animals: what constitutes acceptable practice? *Frontiers in Ecology and the Environment*, **4**(3), 147–54. DOI: 10.1890/1540-9295(2006)004[0147:MDOWAW]2.0.CO;2.

Winders, D.J. (2017). Captive wildlife at a crossroads – sanctuaries, accreditation, and humane-washing. *Animal Studies Journal*, **6**(2), 161–78. DOI: https://ro.uow.edu.au/asj/vol6/iss2/9.

Winter, G., Hart, R.A., Charlesworth, R.P.G. and Sharpley, C.F. (2018). Gut microbiome and depression: what we know and what we need to know. *Reviews in the Neurosciences*, **29**(6), 629–43. DOI: 10.1515/revneuro-2017-0072.

Wise, S.M. (2010). Legal personhood and the nonhuman rights project. *Animal Law*, **17**(1), 1–11. DOI: https://www.animallaw.info/article/legal-personhood-and-nonhuman-rights-project.

Wise, S., Durham, D. and Banes, G.L. (2020). The campaign for non-human rights and the status of captive apes. In *State of the Apes: Killing, Capture, Trade and Conservation*, ed. Arcus Foundation. Cambridge, UK: Cambridge University Press, pp. 231–62. Available at: https://www.stateoftheapes.com/volume-4-killing-capture-trade/.

Wittig, R.M., Crockford, C., Weltring, A., *et al.* (2016). Social support reduces stress hormone levels in wild chimpanzees across stressful events and everyday affiliations. *Nature Communications*, **7**, 13361. DOI: 10.1038/ncomms13361.

Wiysonge, C.S. (2019). Vaccine hesitancy, an escalating danger in Africa. *Think Global Health*, December 17, 2019. Available at: https://www.thinkglobalhealth.org/article/vaccine-hesitancy-escalating-danger-africa.

WOAH (2021). *OIE Wildlife Health Framework: Protecting Wildlife Health to Achieve One Health*. Paris, France: World Organisation for Animal Health (WOAH/OIE). Available at: https://www.woah.org/fileadmin/Home/eng/Internationa_Standard_Setting/docs/pdf/WGWildlife/A_Wildlifehealth_conceptnote.pdf.

Wolf, T.M., Sreevatsan, S., Singer, R.S., *et al.* (2016). Noninvasive tuberculosis screening in free-living primate populations in Gombe National Park, Tanzania. *EcoHealth*, **13**(1), 139–44. DOI: 10.1007/s10393-015-1063-y.

Wolf, T.M., Sreevatsan, S., Travis, D., Mugisha, L. and Singer, R.S. (2014). The risk of tuberculosis transmission to free-ranging great apes. *American Journal of Primatology*, **76**(1), 2–13. DOI: 10.1002/ajp.22197.

Wolfensohn, S., Shotton, J., Bowley, H., *et al.* (2018). Assessment of welfare in zoo animals: towards optimum quality of life. *Animals*, **8**(7), 110. DOI: 10.3390/ani8070110.

Wong, S.L. (2020). When Covid resets ecotourism. *Earth Journalism Network*, September 8, 2020. Available at: https://earthjournalism.net/stories/when-covid-resets-ecotourism.

Wood, M.E. (2002). *Ecotourism: Principles, Practices and Policies for Sustainability*. Paris, France: United Nations Environment Programme (UNEP).

Wood, W. (1998). Interactions among environmental enrichment, viewing crowds, and zoo chimpanzees (*Pan troglodytes*). *Zoo Biology*, **17**(3), 211–30. DOI: 10.1002/(SICI)1098-2361(1998)17:3<211::AID-ZOO5>3.0.CO;2-C.

Woodford, M.H., Butynski, T.M. and Karesh, W.B. (2002). Habituating the great apes: the disease risks. *Oryx*, **36**(2), 153–60. DOI: 10.1017/S0030605302000224.

World Bank (2016). *The Cost of Fire: An Economic Analysis of Indonesia's 2015 Fire Crisis. Indonesia Sustainable Landscapes Knowledge Note 1*. Jakarta, Indonesia: The World Bank. Available at: http://documents.worldbank. org/curated/en/776101467990969768/The-cost-of-fire-an-economic-analysis-of-Indonesia-s-2015-fire-crisis.

World Bank (2017). *The Growing Role of Minerals and Metals for a Low Carbon Future*. Washington DC: World Bank Group. Available at: https://documents.worldbank.org/en/publication/documents-reports/documentdetail/207371500386458722/The-Growing-Role-of-Minerals-and-Metals-for-a-Low-Carbon-Future.

World Bank (2018). *Closing the Potential–Performance Divide in Ugandan Agriculture*. Washington DC: World Bank Group. Available at: http://documents.worldbank.org/curated/en/996921529090717586/Closing-the-potential-performance-divide-in-Ugandan-agriculture.

World Bank (2021). Safeguarding animal, human and ecosystem health: One Health at the World Bank. *World Bank*, June 3, 2021. Available at: https://www.worldbank.org/en/topic/agriculture/brief/safeguarding-animal-human-and-ecosystem-health-one-health-at-the-world-bank.

World Bank Group (2018). *One Health: Operational Framework for Strengthening Human, Animal, and Environmental Public Health Systems at their Interface*. Washington DC: World Bank. Available at: http://documents1. worldbank.org/curated/en/703711517234402168/pdf/123023-REVISED-PUBLIC-World-Bank-One-Health-Framework-2018.pdf.

Wrangham, R.W. (1974). Artificial feeding of chimpanzees and baboons in their natural habitat. *Animal Behaviour*, **22**(1), 83–93. DOI: 10.1016/S0003-3472(74)80056-4.

Wrangham, R.W. (1986). Ecology and social relationships in two species of chimpanzee. In *Ecological Aspects of Social Evolution: Birds and Mammals*, ed. D. I. Rubenstein and R. W. Wrangham. Princeton, NJ: Princeton University Press, pp. 352–78.

Wright, E., Grueter, C.C., Seiler, N., *et al.* (2015). Energetic responses to variation in food availability in the two mountain gorilla populations (*Gorilla beringei beringei*). *American Journal of Physical Anthropology*, **158**(3), 487–500. DOI: 10.1002/ajpa.22808.

WTTC (2020). *Travel and Tourism: Economic Impact*. London, UK: World Travel & Tourism Council (WTTC).

WWF (2018). New Barclays policy to protect World Heritage sites: a welcome first step. *WWF [World Wide Fund for Nature] News*, May 1, 2018. Available at: https://wwf.panda.org/?327030/New-Barclays-policy-to-protect-World-Heritage-sites-a-welcome-first-step.

Wyatt, T., Maher, J., Allen, D., Clarke, N. and Rook, D. (2022). The welfare of wildlife: an interdisciplinary analysis of harm in the legal and illegal wildlife trades and possible ways forward. *Crime, Law and Social Change*, **77**(1), 69–89. DOI: 10.1007/s10611-021-09984-9.

Xie, L. (2021). *Valuing Inclusion and Diversity, Embracing Uncertainty: Ways Forward for Nature-based Solutions*. London, UK: The British Academy. DOI: 10.5871/bacop26/9780856726712.001.

Xie, T., Liu, W., Anderson, B.D., Liu, X. and Gray, G.C. (2017). A system dynamics approach to understanding the One Health concept. *PLoS ONE*, **12**(9), e0184430. DOI: 10.1371/journal.pone.0184430.

Xie, X., Li, Y., Chwang, A.T.Y., Ho, P.L. and Seto, W.H. (2007). How far droplets can move in indoor environments – revisiting the Wells evaporation-falling curve. *Indoor Air*, **17**(3), 211–25. DOI: 10.1111/j.1600-0668.2007.00469.x.

Yaguchi, Y., Okabayashi, S., Abe, N., *et al.* (2014). Genetic analysis of *Enterobius vermicularis* isolated from a chimpanzee with lethal hemorrhagic colitis and pathology of the associated lesions. *Parasitology Research*, **113**(11), 4105–9. DOI: 10.1007/s00436-014-4080-9.

Yamagiwa, J. and Basabose, A.K. (2009). Fallback foods and dietary partitioning among *Pan* and *Gorilla*. *American Journal of Physical Anthropology*, **140**(4), 739–50. DOI: 10.1002/ajpa.21102.

Yang, X.L., Zhang, Y.Z., Jiang, R.D., *et al.* (2017). Genetically diverse filoviruses in *Rousettus* and *Eonycteris* spp. bats, China, 2009 and 2015. *Emerging Infectious Diseases*, **23**(3), 482–6. DOI: 10.3201/eid2303.161119.

Yang, Y. and Jobin, C. (2014). Microbial imbalance and intestinal pathologies: connections and contributions. *Disease Models & Mechanisms*, **7**(10), 1131–42. DOI: 10.1242/dmm.016428.

Yeager, C.P. (1997). Orangutan rehabilitation in Tanjung Puting National Park, Indonesia. *Conservation Biology*, **11**(3), 802–5.

YEL (n.d.-a). *Environmental Education.* Medan, Indonesia: Yayasan Ekosistem Lestari (YEL). Available at: https://www.yel.or.id/environmental-education/. Accessed: September, 2020.

YEL (n.d.-b). *Yayasan Ekosistem Lestari: The Foundation for a Sustainable Ecosystem.* Medan, Indonesia: Yayasan Ekosistem Lestari (YEL). Available at: https://www.yel.or.id/en/who-we-are. Accessed: September, 2020.

Yersin, H., Asiimwe, C., Voordouw, M.J. and Zuberbühler, K. (2017). Impact of snare injuries on parasite prevalence in wild chimpanzees (*Pan troglodytes*). *International Journal of Primatology,* **38**(1), 21–30. DOI: 10.1007/s10764-016-9941-x.

Yin, J., Lampert, A., Cameron, M., Robinson, B. and Power, P. (2012). Using social media to enhance emergency situation awareness. *IEEE Intelligent Systems,* November/December 2012, 52–9. DOI: https://delvalle.bphc.org/pluginfile.php/847/mod_resource/content/2/YIN-IS2012.pdf.

Yon, L., Williams, E., Harvey, N.D. and Asher, L. (2019). Development of a behavioural welfare assessment tool for routine use with captive elephants. *PLoS ONE,* **14**(2), e0210783. DOI: 10.1371/journal.pone.0210783.

Yu, E. and Fan, R. (2007). A Confucian view of personhood and bioethics. *Journal of Bioethical Inquiry,* **4**(3), 171–9. DOI: 10.1007/s11673-007-9072-3.

Yu, X. and Jia, W. (2015). *Moving Targets: Tracking Online Sales of Illegal Wildlife Products in China.* Cambridge, UK: TRAFFIC. Available at: https://www.traffic.org/publications/reports/moving-targets-tracking-online-sales-of-illegal-wildlife-products-in-china/.

ZAHN (2011). *Lessons Learned Annex.* Silver Spring, MD: Zoo Animal Health Network (ZAHN). Available at: https://zahp.org/wp-content/uploads/2020/11/Lessons_Learned_Chart.pdf.

ZAHP (2017). *Contingency Planning for the Exotic Animal Industry Workbook.* Silver Spring, MD: Zoo and Aquarium All Hazards Partnership (ZAHP). Available at: https://zahp.org/all-hazards/.

ZAHP (n.d.). *Is Your Facility Prepared?* Silver Spring, MD: Zoo and Aquarium All Hazards Partnership (ZAHP). Available at: https://zahp.org/. Accessed: November, 2021.

Zander, K.K., Pang, S.T., Jinam, C., Tuen, A.A. and Garnett, S.T. (2014). Wild and valuable? Tourist values for orang-utan conservation in Sarawak. *Conservation and Society,* **12**(1), 27–42.

ZBPWG (2011). *Zoological Best Practices Working Group Planning Roadmap – A Basic Guide for Emergency Planners for Managed Wildlife Facilities.* Zoo Best Practices Working Group for Disaster Preparedness and Contingency Planning (ZBPWG). Zoo Animal Health Network. Available at: https://www.yumpu.com/en/document/read/51212560/zoological-best-practices-working-group-planning-roadmap.

Zenda, C. (2020). COVID-19 sees increase in wildlife poaching in Southern Africa. *Fair Planet,* September 28, 2020. Available at: https://www.fairplanet.org/story/covid-19-sees-increase-in-wildlife-poaching-in-southern-africa/.

Zhang, D., Fei, H.-L., Yuan, S.-D., *et al.* (2014). Ranging behavior of eastern hoolock gibbon (*Hoolock leuconedys*) in a northern montane forest in Gaoligongshan, Yunnan, China. *Primates,* **55**(2), 239–47. DOI: 10.1007/s10329-013-0394-y.

Zhang, F. and Zhu, L. (2019). Enhancing corporate sustainable development: stakeholder pressures, organizational learning, and green innovation. *Business Strategy and the Environment,* **28**(6), 1012–26. DOI: 10.1002/bse.2298.

Zhang, L., Ameca, E.I., Cowlishaw, G., *et al.* (2019). Global assessment of primate vulnerability to extreme climatic events. *Nature Climate Change,* **9**(7), 554–61. DOI: 10.1038/s41558-019-0508-7.

Zheng, L., Shen, C., Tang, L., *et al.* (2013). Data mining meets the needs of disaster information management. *IEEE Transactions on Human–Machine Systems,* **43**(5), 451–64. DOI: 10.1109/THMS.2013.2281762.

Zhou, J., Wei, F., Li, M., Pui Lok, C.B. and Wang, D. (2008). Reproductive characters and mating behaviour of wild *Nomascus hainanus. International Journal of Primatology,* **29**(4), 1037–46. DOI: 10.1007/s10764-008-9272-7.

Zhu, P., Garber, P.A., Wang, L., *et al.* (2020). Comprehensive knowledge of reservoir hosts is key to mitigating future pandemics. *The Innovation,* **1**(3), 100065. DOI: 10.1016/j.xinn.2020.100065.

Zimmerman, D.M., Mitchell, S.L., Wolf, T.M., *et al.* (2022). Great ape health watch: enhancing surveillance for emerging infectious diseases in great apes. *American Journal of Primatology,* **84**(4–5), e23379. DOI: 10.1002/ajp.23379.

Zimmermann, F., Köhler, S.M., Nowak, K., *et al.* (2017). Low antibody prevalence against *Bacillus cereus* biovar *anthracis* in Taï National Park, Côte d'Ivoire, indicates high rate of lethal infections in wildlife. *PLoS Neglected Tropical Diseases,* **11**(9), e0005960. DOI: 10.1371/journal.pntd.0005960.

Zimmermann, N., Pirovino, M., Zingg, R., *et al.* (2011). Upper respiratory tract disease in captive orangutans (*Pongo* sp.): prevalence in 20 European zoos and predisposing factors. *Journal of Medical Primatology*, **40**(6), 365–75. DOI: 10.1111/j.1600-0684.2011.00490.x.

Zinsstag, J., Schelling, E., Waltner-Toews, D. and Tanner, M. (2011). From "one medicine" to "one health" and systemic approaches to health and well-being. *Preventive Veterinary Medicine*, **101**(3), 148–56. DOI: 10.1016/j.prevetmed.2010.07.003.

Zommers, Z., Macdonald, D.W., Johnson, P.J. and Gillespie, T.R. (2013). Impact of human activities on chimpanzee ground use and parasitism (*Pan troglodytes*). *Conservation Letters*, **6**(4), 264–73. DOI: 10.1111/j.1755-263X.2012.00288.x.

ZooLeón (n.d.). *Mapa Zoologico de Léon.* Léon, Mexico: ZooLeón. Available at: http://www.zooleon.org.mx/mapa-zoologico-de-leon/. Accessed: December, 2020.

Zoological Society of Milwaukie (n.d.). *Bonobo Species Survival Plan.* Milwaukee, WI: Zoological Society of Milwaukie. Available at: https://www.zoosociety.org/Conservation/BonoboSSP.php. Accessed: October, 2020.

Zoológico de Culiacán (2020). Zoológico de Culiacán Zoo. *Facebook Post*, November 7, 2020. Available at: https://www.facebook.com/zoologicoculiacan/posts/pfbid0s9SV9rjNq33FMmQ37hjNK89qYHraKx-pU6xrNKLdN727TNGhauB6FSpvKxiy470M5l.

ZSL (2016). *Boîte à outils pour la prise en compte de la faune dans les forêts de production du bassin du Congo.* London, UK: Zoological Society of London (ZSL). Available at: https://www.zsl.org/sites/default/files/media/2016-10/Toolkit%20Report-v6-2-screen-LR_0.pdf.

ZSL (n.d.). *Wildlife Wood Project.* London, UK: Zoological Society of London (ZSL). Available at: https://www.zsl.org/conservation/regions/africa/wildlife-wood-project. Accessed: October, 2019.

Zulfikri, M., Ridwan, Y. and Cahyaningsih, U. (2018). Prevalence of intestinal helminth parasites in wild and soft-release Bornean orangutan (*Pongo pygmaeus*) in Lamandau Wildlife reserve, Central Kalimantan. *IOP Conference Series: Materials Science and Engineering*, **434**(1), 012135. DOI: 10.1088/1757-899X/434/1/012135.

Zumla, A., Valdoleiros, S.R., Haider, N., *et al.* (2022). Monkeypox outbreaks outside endemic regions: scientific and social priorities. *The Lancet Infectious Diseases*, **22**(7), 929–31. DOI: 10.1016/s1473-3099(22)00354-1.

INDEX

For EU product safety concerns, contact us at Calle de José Abascal, 56–1°,
28003 Madrid, Spain or eugpsr@cambridge.org.

www.ingramcontent.com/pod-product-compliance
Ingram Content Group UK Ltd.
Pitfield, Milton Keynes, MK11 3LW, UK
UKHW050901071225
465726UK00006B/259